AF881132

Organoselenium Chemistry Between Synthesis and Biochemistry

Editor

Claudio Santi

Department of Pharmaceutical Sciences
University of Perugia
Via del Liceo 1 – 06100 Perugia
Italy

DEDICATION

On the memory of Prof. Marcello Tiecco and to Prof. Lorenzo Testaferri in the occasion of his retirement

CONTENTS

FOREWORD

Selenium as a versatile and highly interesting element has primarily concerned chemists but, since the discovery of selenium being an essential trace element, also biologists and biochemists. This book highlights all aspects of current organoselenium chemistry and the editor, Prof. Claudio Santi, has attracted many outstanding contributions to emphasise selenium research at the borderline to biology and medicine. In addition, catalytic reactions are emphasized in many chapters demonstrating the need for sustainable chemistry while still exploiting the unique and advantageous properties of selenium.

This book will be a very valuable source of information for every chemist – in education, in research, and in industry.

Thomas Wirth

Cardiff University

Cardiff, Wales

UK

PREFACE

In 1817, Jakob Berzelius discovered selenium in the sludge of the lead chambers of the sulphuric acid chamber process of a plant at Gripsholm. Because of its characteristic smelling of tellurium, (*tellus* = hearth), he named the new element from the Greek word "σελήυη" (Selȇnȇ) meaning goddess of the Moon.

For a long time selenium was considered mainly as a toxic element, and it is currently debated if Marco Polo was the first to write about selenium poisoning. He went to China in 1271 *via* the ancient Silk Road and in his travelogue in 1295, he wrote that a toxic wild plant collected in the today's Gansu Province killed many cattle.

In 1957 it was found to be an essential trace element as component of the 21th amino acid, selenocysteine and in 1973 the first selenoenzyme, glutathione peroxidase, was isolated. And characterized the development of the organoselenium chemistry has been strongly increased during the last fifty years and nowadays these compounds are efficiently used as intermediates and/or catalysts in a number of organic synthesis: At the same time organoselenium derivatives are currently object of wide investigations for their biological activity, especially, as antiviral, antimicrobial, anticancer, anti-inflammatory and GPx-like mimetics.

This book cover the most recent developments in the field of the classical synthetic application of organoselenium reagents such as electrophilic, nucleophilic and radical reagents focusing the discussion also on the synthesis and the synthetic applications of some emerging classes of compounds such as the hypervalent selenium species and the selenoamides. The use of organoselenium reagents as catalysts is a common thread that runs through the chapters of the book introducing new important aspects of the modern organoselenium chemistry: organocatalysis, green chemistry, bioinspiration, antioxidant activity.

The book also addresses more biological issues such as the antimicrobial activity of organoselenium derivatives and the biochemistry of selenium from the aliments to the selenoproteins.

I'd like to thanks and underline the precious contribute of the 27 distinguished scientists that decided to share their expertise on producing this volume that is the first multidisciplinary approach to the constantly in progress world of organoselenium compounds.

I hope that this book will be read with interest not only by experts but also by students and any other researcher and chemist.

Claudio Santi

Department of Pharmaceutical Sciences
Lab of Catalysis and Green Chemistry
University of Perugia *Via* del Liceo 1 06100 Perugia
Italy

List of Contributors

Adrian Zając

Centre of Molecular and Macromolecular Studies, Polish Academy of Sciences Łódź, Poland

Antonio L. Braga

Federal University of Santa Catarina, Florianópolis, SC, Brazil

Barahman Movassag

Department of Chemistry, K.N. Toosi University of Technology, Tehran, Iran

Bartolini Desirèe

Department of Internal Medicine, Nutrition and Clinical Biochemistry Lab University of Perugia, Perugia, Italy

Caterina Tidei

Department of Pharmaceutical Sciences, University of Perugia, Perugia, Italy

Ciffolilli Silvia

Department of Internal Medicine, Nutrition and Clinical Biochemistry Lab University of Perugia, Perugia, Italy

Claudio Santi

Department of Pharmaceutical Sciences, University of Perugia, Perugia, Italy

Donatella Pietrella

Department of Pharmaceutical Sciences, University of Perugia, Perugia, Italy

Francesca Marini

Department of Pharmaceutical Sciences, University of Perugia, Perugia, Italy

Francesco Galli

Department of Internal Medicine, Nutrition and Clinical Biochemistry Lab University of Perugia, Perugia, Italy

Jacek Ścianowski

Department of Organic Chemistry, Faculty of Chemistry, Nicolaus Copernicus University -Torun, Poland

Jarosław Lewkowski

University of Łódź, Faculty of Chemistry, Department of Organic Chemistry Łódź, Poland

Józef Drabowicz

Centre of Molecular and Macromolecular Studies, Polish Academy of Sciences, Łódź, Poland and Department of Chemistry, Environment Protection and Biotechnology, Jan Dlugosz University in Czestochowa, 42-200- Czestochowa Poland

Luana Bagnoli

Department of Pharmaceutical Sciences, University of Perugia, Perugia, Italy

Michio Iwaoka

Department of Chemistry, School of Science, Tokai University-Tokay, Japan

Mozhgan Navidi

Department of Chemistry, K.N. Toosi University of Technology, Tehran, Iran

Murdolo Giuseppe

Department of Internal Medicine, Nutrition and Clinical Biochemistry Lab University of Perugia, Perugia, Italy

Oscar E.D. Rodrigues

Federal University of Santa Maria, Santa Maria, RS, Brazil

Patrycja Pokora-Sobczak

Centre of Molecular and Macromolecular Studies, Polish Academy of Sciences Łódź, Poland

Piotr Kiełbasiński

Centre of Molecular and Macromolecular Studies, Polish Academy of Sciences, Łódź, Poland

Piroddi Marta

Department of Internal Medicine, Nutrition and Clinical Biochemistry Lab University of Perugia, Perugia, Italy

Ricardo S. Schwab

Federal University of São Carlos, São Carlos, SP, Brazil

Satoko Hayashi

Department of Material Science and Chemistry, Faculty of Systems Engineering, Wakayama University, Japan

Silvia Sternativo

Department of Pharmaceutical Sciences, University of Perugia, Perugia, Italy

Tortoioli Cristina

Department of Internal Medicine, Nutrition and Clinical Biochemistry Lab University of Perugia, Perugia, Italy

Toshiaki Murai

Department of Chemistry, Faculty of Engineering, Gifu University, Yanagido Gifu, Japan

Waro Nakanishi

Department of Material Science and Chemistry, Faculty of Systems Engineering Wakayama University, Japan

Zbigniew Rafiński

Department of Organic Chemistry, Faculty of Chemistry, Nicolaus Copernicus University -Torun, Poland

CHAPTER 1

Introduction

Claudio Santi*

Department of Pharmaceutical Sciences, Lab of Catalysis and Green Chemistry, University of Perugia, Via del Liceo 1 06100 Perugia, Italy

Abstract: Organoselenium compounds attracted in the last twenty years a growing scientific interest. From a chemical point of view, organoselenium reagent can be used in mild conditions to effect chemio, regio and stereoselective reactions and, in a number of examples as efficient and selective catalysts.

More recently also the biological aspect has been deeply explored and some organoselenium derivatives showed interesting and evaluable pharmacological properties.

Keywords: Selenium, Selenoxides, Selenones, Selenurane, Seleniranium, Seleneic Acid, Perselenenic Acid, Selenurea, Selenocysteine, NMR.

1.1. OVERVIEW ON ORGANOSELENIUM COMPOUNDS

Gunther and Klayman edited the first extensive review on organoselenium reagents in 1973 [1] and this could be considered the beginning of the scientific interest on this field of chemistry. The first organoselenium compounds (alkyl selenols, selenides and diselenides) were known since nineteenth century but their unpleasant odor and instability represented for a long time an obstacle to the development of the research and the synthetic application of this class of compounds.

Probably the *syn*-elimination of selenoxides studied starting from 1968 [2] and proposed by Sharpless in 1970 [3] as a straightforward stereospecific method for the olefin generation (Scheme **1.1**), is nowadays one of the most commonly used application of organoselenium chemistry also for the synthesis of complex structures.

***Address correspondence to Claudio Santi:** Department of Pharmaceutical Sciences, Lab of Catalysis and Green Chemistry, University of Perugia *Via* del Liceo 1 06100 Perugia, Italy; Tel: +39 0755855102; Fax: +39 075 5855116; E-mail: claudio.santi@unipg.it

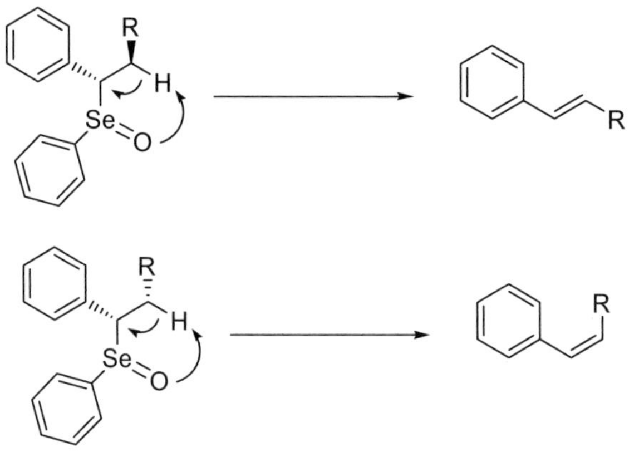

Scheme 1.1

All the results produced in the wake of these pioneering discoveries have been the object of numerous reviews and books appeared around the eighties. During the same period, there has been a rapid progress in this field of chemical research with interesting applications in the total synthesis of natural compounds [4].

More recently a practical guide on organoselenium chemistry has been reported by Back [5] and two books edited by Thomas Wirth in 2000 and 2012 demonstrate a continuous interest of the scientific community [6].

Despite the close position in the periodic table of the elements, selenium presents some chemical difference with the related sulfur. Selenium, compared to sulfur, establish a weaker and longer bond with carbon atoms (234 kJ/mol *vs* 272 KJ/mol) and generally, organoselenium derivatives are more nucleophilic and more acidic than the corresponding sulfur analogues.

These chemical differences are fundamental from a biological point of view and for these reasons selenocysteine, the 21th amino acid, plays a crucial role in the place of cysteine in some redox enzyme deputed to the cell protection against the oxidative stress [7].

Schwartz and Foltz demonstrated in 1957 the role of selenium as micronutrient necessary to prevent necrotic liver degeneration in animals but also a number of other human diseases such as Keshan and Kashin- Beck disease, neurodegenerations and cancers [8-13].

The second naturally occurring selenium containing amino acid is the selenomethionine but, contrary to selenocysteine, it is the result of a post-transcriptional modification and not specifically encoded in the DNA (Scheme **1.2**).

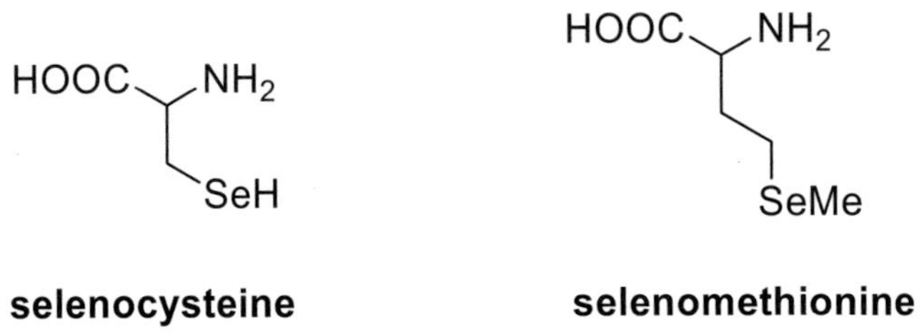

Scheme 1.2

Organoselenium compound can be classified, based on the reactivity of the chalcogen, as stable compound, nucleophilic reagents, electrophilic reagent and some oxidized species of selenium were isolated and characterized.

Selenides (R-Se-R) and diselenides (R-Se-Se-R) are generally the most stable and easily handling compounds. The selenium-selenium bond on diselenides can be cleaved heterolytically by oxidation or reduction affording electrophilic or nucleophilic organoselenium species, respectively. Under radical activating conditions, the same cleavage can be affected in homolytic way affording selenium-centered radicals.

The nucleophilic reagents are represented by selenols (R-SeH) and their corresponding anions, selenolates (R-Se-M). Structurally equivalents of alcohols and thiols the selenols and the selenolates are highly unstable and particularly smelling compounds.

Aryl and alkyl selenenyl halides (R-Se-Cl, R-Se-Br) are reactive electrophiles broadly applied in organic synthesis. They can be easily preparer by halogenation of diselenides and the phenylselenenyl derivatives are nowadays commercially available.

Concerning the oxidized species selenides can be converted in to the corresponding sulfoxide (R-Se(O)-R) and sulfones (R-Se(O)$_2$-R), reactive intermediates that were variously employed in synthesis.

The oxidation of selenols or selenolates afford sequentially selenenic acid (R-Se-OH), seleninic acid (R-Se(O)-OH) and perseleninic acid (R-Se(O)-OOH).

In addition are known some hypervalent species of selenium, such as selenuranes, cationic three-membered ring called seleniranes in analogy with the corresponding thiiranes an some species with a selenium carbon (Sp2) double bond such as selones and selenurea (Scheme **1.3**).

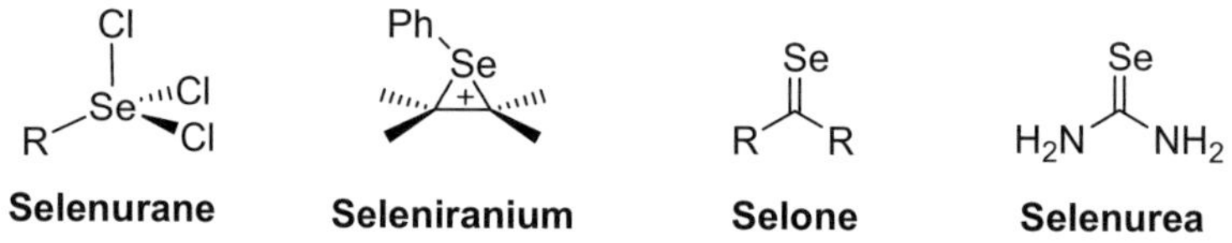

Scheme 1.3

Among the analytical methods commonly used to study and characterize different organoselenium derivatives, the 77Se-NMR spectroscopy offers extremely powerful tools.

Selenium 77 is approximately 3 times more sensitive than 13C and the resonance are spread in a broad range of ppm (up to 1000 ppm) allowing the easily characterization of selenium atoms in different functional groups.

Some representative examples are reported in Scheme **1.4.**

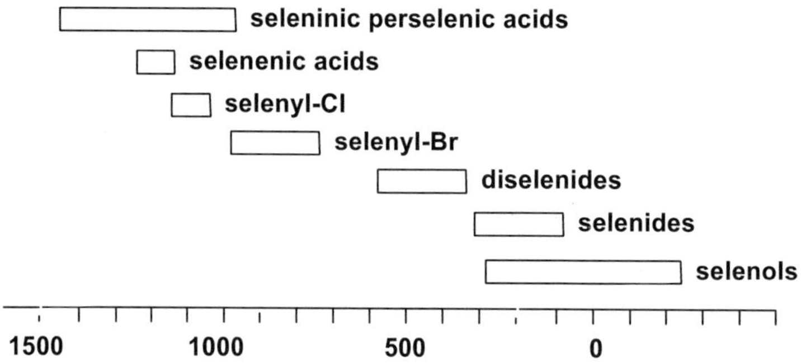

Scheme 1.4

ACKNOWLEDGEMENT

Declared none.

CONFLICT OF INTEREST

The authors confirm that this chapter contents have no conflict of interest.

REFERENCES

[1] *Organoselenum compounds: Their Chemistry and Biology*; Kayman, D. L.; Gunther, W. H. H. (eds); Wiley, New York, **1973**.

[2] Huguet, J. L. Oxidation of olefines catalyzed by Selenium. *Adv. Chem.,* **1968**, *76*, 345-351.

[3] (a) Sharples, K.B.; Lauer, R. F. Mild procedure for the conversion of epoxides to allylic alcohols. First organoselenium reagent. *J. Am. Chem. Soc.*, **1973**, *95*, 2697-2699. (b) Sharples, K.B.; Lauer, R. F.; Teranisky, A. Y. Electrophilic and nucleophilic organoselenium reagents. New routes to.alpha.,.beta.-unsaturated carbonyl compounds. *J. Am. Chem. Soc.*, **1973**, *95*, 6137-6139.

[4] (a) *Organoselenum Chemistry*; Liotta, D. (ed); Wiley, New York, **1983**. (b) *The Chemistry of Organic Selenium and Tellurium Compounds*; Patai, S.; Rappoport, Z. (eds); Wiley, Chichester **1986** (Vol.1), and **1987** (Vol 2). (c) *Organoselenum Chemistry*; Krief, A.; Havesi, L. (eds); Springer, Berlin, **1988**.

[5] *Organoselenium Chemistry: A Practical Approach*, Back, T.G. Ed.; Oxford University Press: Oxford, **1999**.

[6] (a) Organoselenium Chemistry; Wirth, T. Ed.; *Top. Curr. Chem.*, **2000**, Vol. *208*, p.p. 143–176

[7] Santi, C.; Tidei, C.; Scalera, C., M. Piroddi, F. Galli, Selenium Containing Compounds from Poison to Drug Candidates: A Review on the GPx-like Activity, *Curr. Chem. Biol,* **2013***, 7*, 25-36.

[8] Schwarz K., Foltz CM. Selenium as an integral part of factor 3 against dietary necrotic liver degeneration. *J Am Chem Soc* **1957**; *79*: 3292-93.

[9] Yang GQ, Ge KY, Chen JS, Chen XS. Selenium-related endemic diseases and the daily selenium requirement of humans. *World Rev Nutr Diet* **1988**; *55*: 98-152.

[10] Peng A, Yang C, Rui H, Li H. Study on the pathogenic factors of Kashin-Beck disease. *J Toxicol Environ Health* **1992**; *35*: 79-90.

[11] Hoffmann PR, Berry MJ. The influence of selenium on immune responses. *Mol Nutr Food Res* **2008**; *52*: 1273-80.

[12] Salonen JT. Selenium in ischaemic heart disease. *Int J Epidemiol* **1987**; *16*: 323-28.

[13] Gromadzinska J, Reszka E, Bruzelius K, Wasowicz W, Akesson B. Selenium and cancer: biomarkers of selenium status and molecular action of selenium supplements. *Eur J Nutr* **2008**; *47* (Supp.2): 29-50.

Send Orders for Reprints to reprints@benthamscience.net

CHAPTER 2

Electrophilic Selenium Reagents: Addition Reactions to Double Bonds and Selenocyclizations

Jacek Ścianowski* **and Zbigniew Rafiński**

Department of Organic Chemistry, Faculty of Chemistry, Nicolaus Copernicus University, 7 Gagarin Street, 87-100 Torun, Poland

Abstract: Addition reactions of electrophilic selenium reagents to the double bonds, selenocyclizations, and their stereochemical aspects are described. The synthesis of electrophilic organoselenium reagents, including optically active examples, the formation of new carbon-oxygen, carbon-nitrogen, carbon-carbon bonds, and applications in the synthesis of heterocycles are presented.

Keywords: Addition, Electrophiles, Selenides, Oxyselenenylation, Azaselenenylation, Carboselenenylation, Asymmetric Synthesis, Cyclofunctionalization, Heterocycles, Stereoselectivity.

2.1. GENERAL ASPECTS OF SELENENYLATION AND SELENOCYCLIZATION REACTIONS

Selenofunctionalization of organic molecules plays an important role in modern organic synthesis. The interest in organoselenium compounds is due to their specific reactivity, enabling the introduction and transformations of functional groups with good chemo-, regio-, and stereoselectivity, under mild reaction conditions. An additional advantage is the possibility of functionalization by nucleophilic, electrophilic or radical organoselenium reagents [1].

Organoselenium compounds are used as catalysts and can be apply to new synthetic protocols [2]. They have also found applications in asymmetric reactions [3]. New methodologies of synthesis of organoselenium compounds are very important, due to their biological and pharmacological functions [4].

***Address correspondence to Jacek Ścianowski:** Department of Organic Chemistry, Faculty of Chemistry, Nicolaus Copernicus University, 7 Gagarin Street, 87-100 Torun, Poland; Tel: +48 56 611 4532; Fax: +48 56 654 2477; E-mail: jsch@chem.umk.pl

This chapter highlights the developments of the addition reactions of electrophilic selenium reagents to double bounds and selenocyclizations, including stereoselective reactions.

Selenenylation reaction involve the introduction of a selenium function with simultaneous formation of new carbon-halogen, carbon-oxygen, carbon-nitrogen, or carbon-carbon bonds (Scheme **2.1**, path a). Intramolecular reaction (selenocyclization) gives cyclic and heterocyclic compounds (path b), and optically active electrophilic selenium reagents lead to the formation of new stereogenic centers [5].

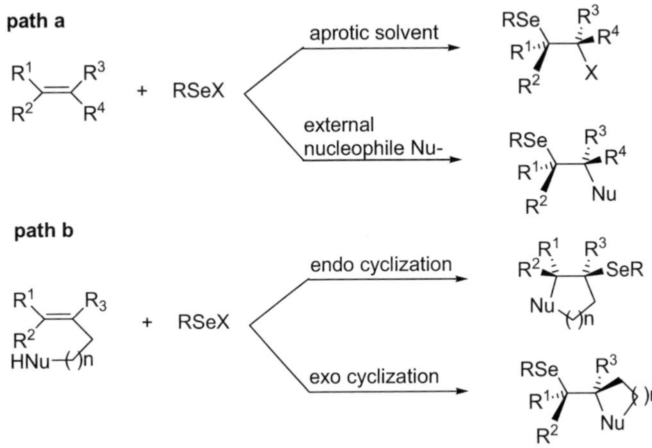

Scheme 2.1

An important advantage of the introduction of the selenium function to organic molecules is the possibility of its further transformations with the removal of the organoselenium moiety through substitution or elimination pathways (Scheme **2.2**). For example, reactions with tin hydrides in the presence of AIBN, lead to the homolityc cleavage of carbon selenium bonds and to radical reactions [6]. Probably the most studied deselenenylation is the oxidation of aryl selenyl groups to selenoxides, followed by spontaneous *syn*-elimination and formation of a C-C double bond [7]. Less common is the oxidation of aryl selenyl moiety *via* selenoxides to selenones, which were recognized as excellent leaving groups in the nucleophilic substitution [8]. The selenonyl group reacts with various nucleophiles, such as NaN_3, KCN, $NaSCH_3$, NaI producing the corresponding azido, cyano, methylthio, and iodo derivatives [9, 10]. Oxidation of the phenylselenyl group with MCPBA in the presence of a base, *via* an intramolecular nucleophilic substitution, gives aziridines [11], azetidines [12], and 1,3-oxazolidin-2-ones [13, 14] respectively.

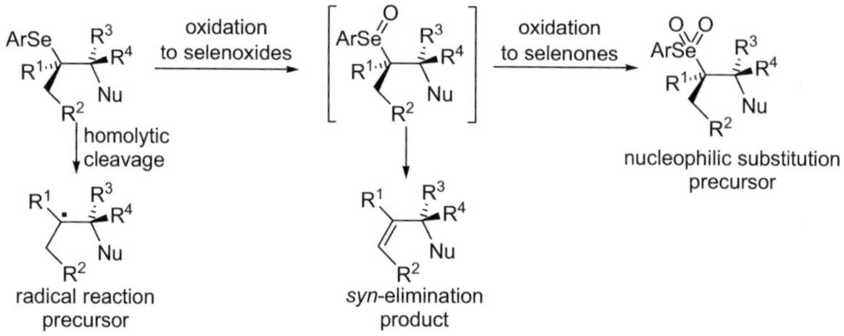

Scheme 2.2

Organoselenium chemistry has found many interesting applications in organic synthesis, including the preparation of some natural compounds.

An intramolecular diastereoselective aminoselenenylation, utilizing the electrophilic selenium reagent **1**, has been used as a key step in the synthesis of (*S*)-Salsolidine **3**, a naturally occurring tetrahydroisoquinoline alkaloid (Scheme **2.3**) [15]. The arylselenyl group was removed from **2** using Ph_3SnH and AIBN and the subsequent deprotection of the amino function afforded the product (-)-**3** in good yield.

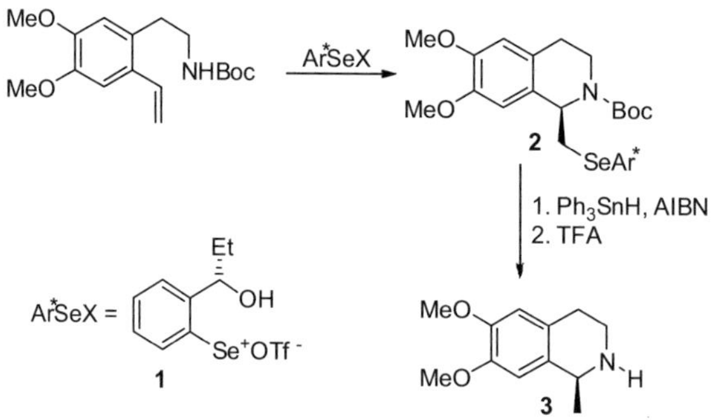

Scheme 2.3

Interesting selenium-assisted aminocyclization of the methyl ester **4** to azabicyclo compounds **5** was applied to the synthesis of **7**, an analogue of anatoxin-a (a highly toxic alkaloid produced by cyanobacteria) [16]. The phenylselenyl group was quantitatively reduced by Raney Nickel in MeOH to the intermediate **6** (Scheme **2.4**).

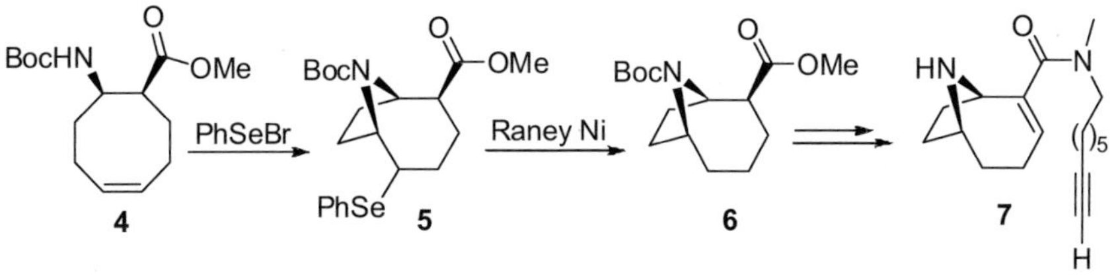

Scheme 2.4

The electrophilic reagent **1** was successfully used for the synthesis of (+)-Samin **12**, the precursor of furofuran lignans (Scheme **2.5**) [17]. Oxyselenenylation of alkene **8** with allenylic alcohol **9** gave adduct **10**, which was converted to product **11** *via* 5-*exo-trig* radical cyclization using Ph_3SnH/AIBN. Further transformations gave (+)-Samin **12**.

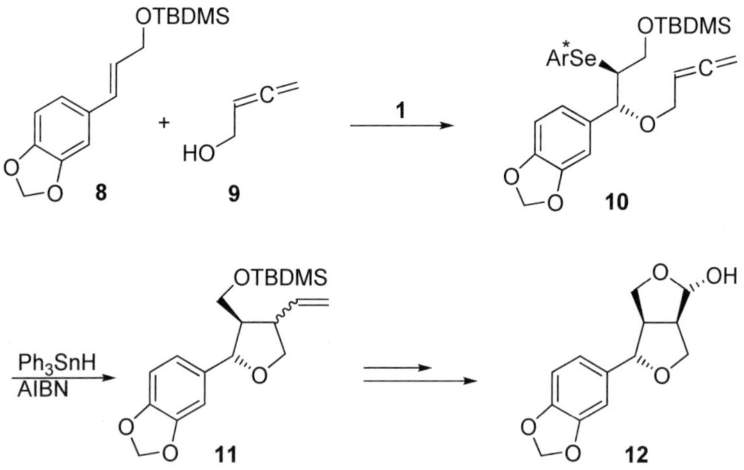

Scheme 2.5

Organoselenium cyclofunctionalization of compounds **13**, induced by PhSeBr/AgOTf and followed by an oxidative-elimination of the selenium moiety with H_2O_2, were used for the synthesis of enantiomerically pure original series of 1,4-diazepine-2,5-diones **14** that act as antagonists of the HDM2-p53 protein-protein interaction (Scheme **2.6**) [18].

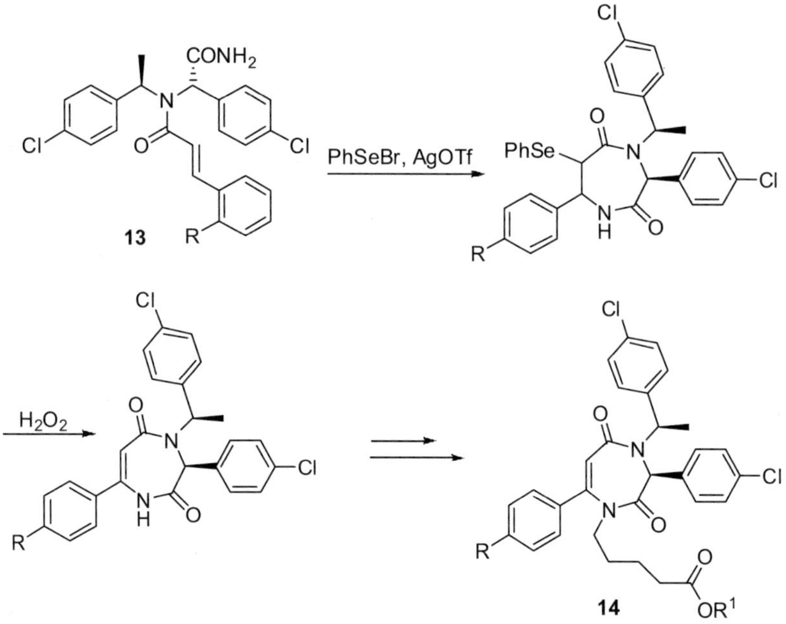

Scheme 2.6

A series of interesting enantiopure 2,5-disubstituted pyrrolidines, containing azido **18**, methylthio **19**, cyano **20** and iodo **21** substituents, were synthesized from *N*-Boc protected δ-alkenyl amine **15** *via* the substrate-controlled asymmetric 5-*exo*-*trig* cyclization promoted by *N*-(phenylseleno)phtalimide (*N*-PSP), then oxidation of the phenylselenyl group with Oxone to selenones **17**, and substitution by the corresponding nucleophiles (Scheme **2.7**) [9].

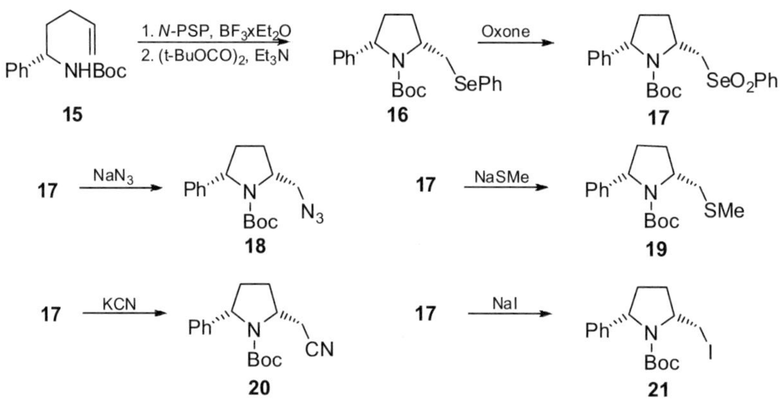

Scheme 2.7

2.2 SYNTHESIS OF ELECTROPHILIC SELENIUM REAGENTS

The most commonly used electrophilic selenium reagents (RSeX), were prepared from diaryl or diakyl diselenides. The methodology of their synthesis mainly depends on the structure of the organic scaffold (R = alkyl or aryl group) and the relative counteranion (X^-). Starting from diselenides, three alternative routes were developed (Scheme **2.8**):

a) Selenium-selenium bond of diselenides (**22**) can be oxidatively clived with bromine, chlorine or $SOCl_2$. The commercially available phenylselenyl bromide **23**, and phenylselenyl chloride **24** can be also prepared starting from diphenyl diselenide. To avoid the presence of nucleophilic halide anion during the selenenylation or selenocyclization reaction and due to the possible reversibility of these reactions [19], it is possible to exchange the halides *in situ* with less nuclephilic anions by the addition of silver salts (path a). Generally clean addition reactions at low temperature were observed with a very efficient triflate anion [20], although the stoichiometric amount of trifluoromethanesulfonic acid is formed during the reaction. Other silver salts such as $AgPF_6$ [21], $AgSbF_6$ [21] and AgOTs [22] were also used. Recently, an efficient reagent, *N*-PSSac **25,** with saccharine counteranion was employed for these reactions [23]. By reacting phenylselenyl chloride with potassium phtalimide, *N*-(phenylseleno)phtalimide (*N*-PSP) **26** was obtained [24] and similarly the less stable *N*-(phenylseleno)succinimide **27** [25]. Chiral silver salts derived from BINOL **28-29**, camphor **30** and proline **31** were used for the generation of electrophilic phenylselenyl reagents with chiral counteranions [26].

b) Diselenides can be also oxidized by ammonium persulfate producing phenylselenenyl sulfate (PSS) **32** [27] (path b), as well as by other inorganic and organic oxidizing agents: KNO_3 [28], $CuSO_4$ [29], $Ce(NH_4)_2(NO_2)_6$ [30], $Mn(OAc)_3$ [31], [*bis*(trifluoroacetoxy)-iodo]benzene [32], (diacetoxyiodo)benzene [33], 2,3-dichloro-5,6-dicyano-1,4-benzoquinone [34].

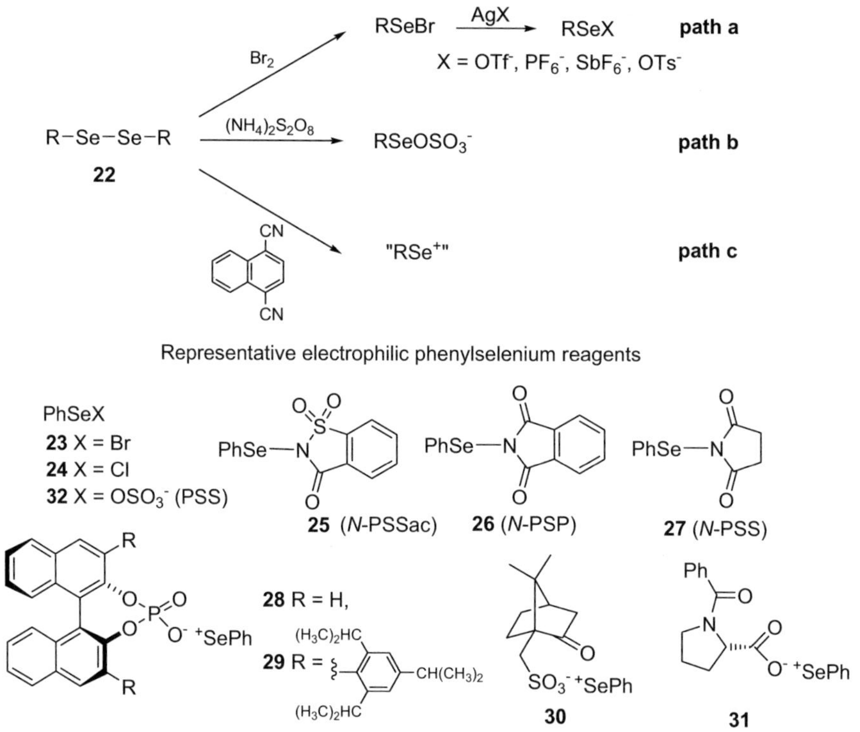

Scheme 2.8

c) Alternatives to above mentioned methods are: the electrochemical procedure that generates phenylselenyl bromide from diphenyl diselenide [35] or the photoinduced electron transfer (PET) from diselenide to 1,4-dicyanonaphthalene (path c) [36].

During the last two decades a number of optically active diselenides was prepared and used for the asymmetric selenenylation of double bonds and selenocyclization reactions.

Optically active binaphtyl diselenides **33** were first reported by Tomoda and co-workers. They were obtained from chiral binaphtylamines by a two-step procedure [37]. This work signed the beginning of a rapid growth in the methodologies for the synthesis of optically active diselenides. Selected diselenides reported by Deziel **34**, **35** [38], Uemura **36** [39], Tomoda **37** [40], Wirth **38-46**, [41], Tiecco **47-49** [42], Back **50**, **51** [43], and Scianowski **52-55** [44], are presented in Fig. **2.1**.

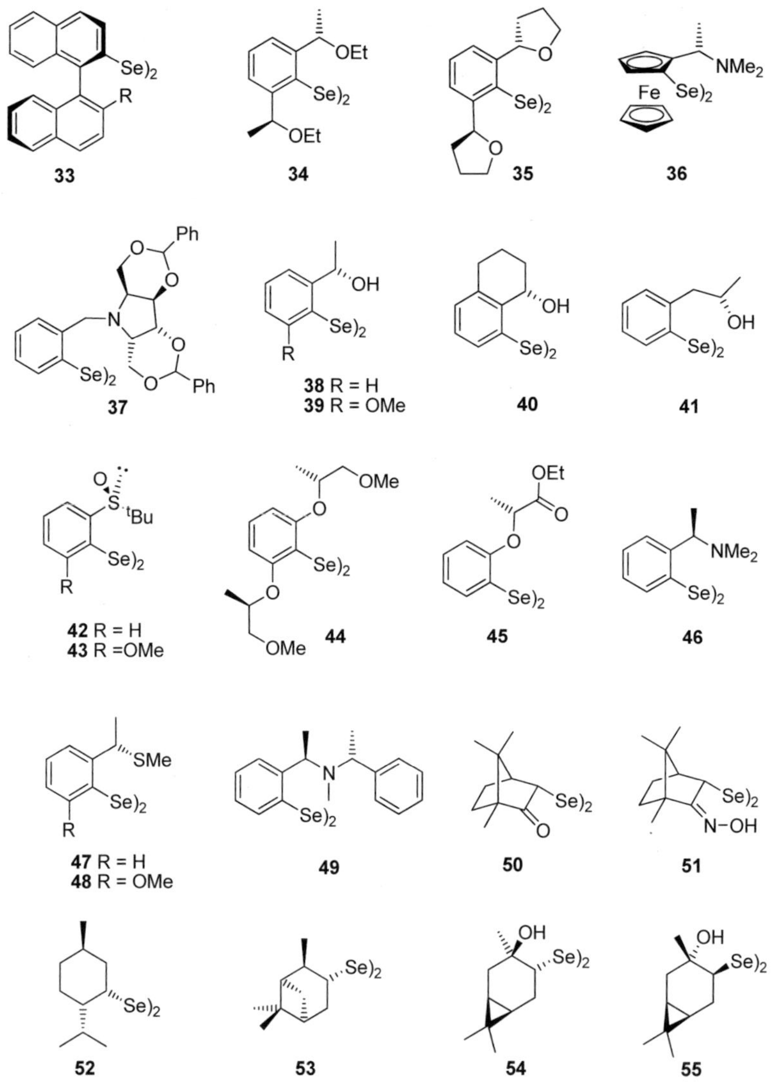

Figure 2.1: Structures of chiral diselenides.

Most diaryl diselenides were prepared from the Grignard reagents or from *ortho*-lithiated intermediates by the addition of selenium powder and air oxidation [38-41]. Dialkyl diselenides, derived from camphor, were prepared by the reaction with LDA, followed by the addition of selenium and oxidation [43]. Other dialkyl diselenides were obtained by the reaction of alkyl tosylates, chlorides and epoxides with sodium diselenide [44].

An interesting polymer-supported electrophilic selenium reagent **56** attached to polystyrene, TentaGel, or mesoporous silica as a solid support, was prepared by Wirth and coworkers and was used for the enantioselective electrophilic addition reactions and selenocyclizations Fig. **2.2** [45].

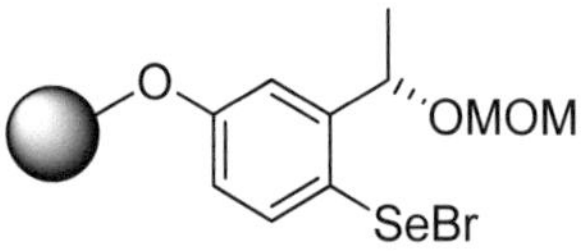

56

Figure 2.2: Polymer - supported electophilic reagent.

2.3. ADDITION REACTIONS TO DOUBLE BONDS

2.3.1. Mechanism of Selenenylation of Alkenes

The selenenylation of alkenes proceeds *via* two steps, generation of seleniranium ion **57**, and *anti*-addition of nucleophile (Scheme **2.9**). The regiochemistry of the addition reaction favores the termodynamicaly stable Markovnikov product **58**.

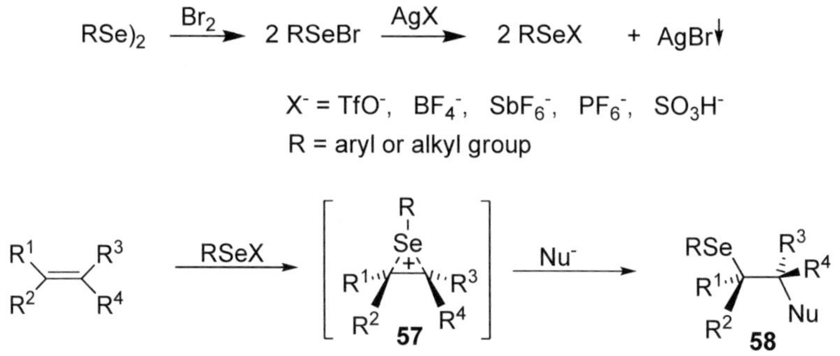

Scheme 2.9

When chiral electrophiles are used, a facial selectivity of the addition reaction to the unsymmetrically substituted alkenes is observed (path a, Scheme **2.10**). At the first step, the formation of two diastereomeric seleniranium ions is possible. Although this step is reversible, it is under kinetic control at low temperatures. The reversibility and mechanism of this reaction was confirmed by experimental and theoretical studies [46]. The differentiation between the *Re* and *Si* – faces depends on the reaction conditions, steric and electronic interactions. The type of counteranion, structure of electrophilic selenium reagents, nonbonding selenium – heteroatom interactions, and the nature of external nucleophiles (oxygen, nitrogen, carbon-centerd nucleophiles) are the key factors influencing the overall stereoselectivity.

In the case of (*Z*)-alkenes, the addition of selenium electrophile to the double bond leads to the formation of one seleniranium ion and nucleophilic attack, in the second step of the reaction, determines the stereochemistry of the products (path b, Scheme **2.10**). Generally, lower selectivities were observed for these substrates.

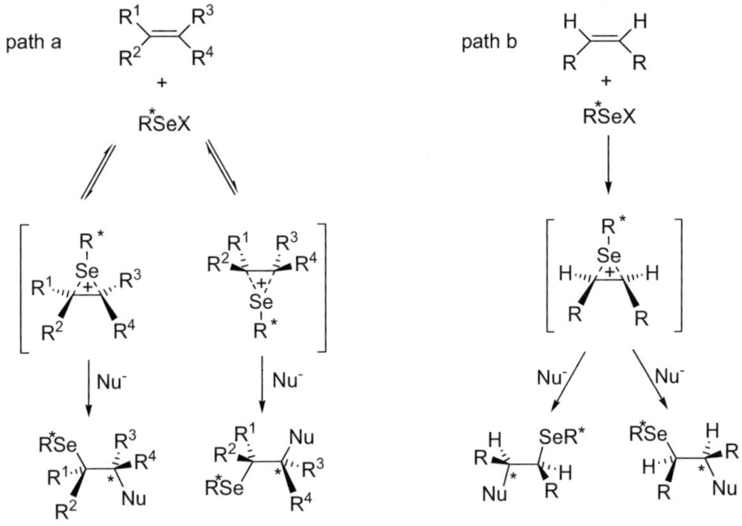

Scheme 2.10

2.3.2. Oxyselenenylation

Oxyselenenylation of alkenes with electrophilic selenium reagents proceeds with the formation of new carbon-oxygen bonds. When alcohol, water or acetic acid were used as external nucleophiles, the corresponding alkoxy, hydroxy and acetoxy groups [47] were introduced to organic molecules. Typically procedures of hydroxy- and methoxyselenenylations of alkenes with phenylselenium electrophilic reagents, like PhSeCl, PhSeOTf and $PhSeOSO_3^-$, were collected by Beaulieu and Deziel [48].

Oxyselenenylations are used as standard procedures in synthetic organic transformations. For example, the methoxyselenenylation reaction has been used in one step of the synthesis of (+)-Obtusenyne [49]. The reaction of *exo*-cyclic enol ether **59** with PhSeCl gave a mixture of selenoacetals **60** in 55% yield, which react with titanium tetrachloride and triethylsilane, followed by quenching with methanol to give 2-oxabicyclo[4.3.1]decane **61** (Scheme **2.11**).

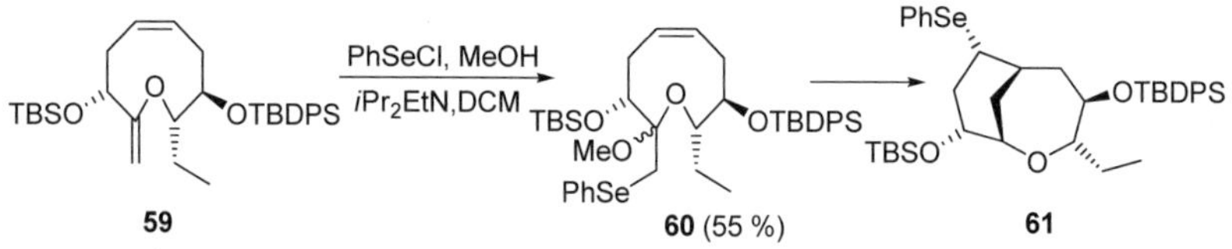

Scheme 2.11

During the transformation of the tricyclic skeleton of the natural Celastraceae sesquiterpenoids, alkene **62** reacted with phenylselenenyl chloride and silver trifluoroacetate to give the corresponding trifluoroacetate **63**. Hydrolysis and oxidation with H_2O_2 produced dienon **64** (Scheme **2.12**) [50].

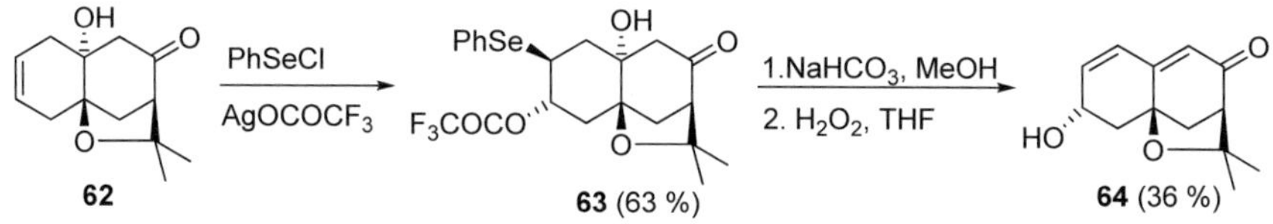

Scheme 2.12

When chiral alkenes were used for the selenenylation reaction, the corresponding chiral products were obtained. Optically active terpenes such as *p*-menthane **65** and 3-carane **66** gave the corresponding hydroxyphenylselenides **67** and **68** in good yields, by the reaction with *N*-(phenylseleno)succinimide **27** in the presence of water and a catalytic amount of camphorsulfonic acid (Scheme **2.13**) [51].

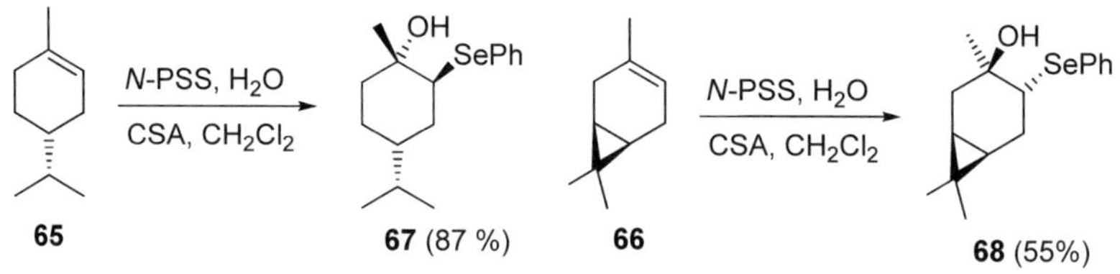

Scheme 2.13

The reaction of 2-vinylperhydro-1,3-benzoxazines **69** and cinnamylamines attached to a chiral perhydrobenzoxazines **70** with benzeneselenenyl chloride and methanol afforded the methoxy selenides **71-74** in high yields, and very good regio- and stereoselectivities (Scheme **2.14**) [52, 53]. Coordination of selenium atom to oxygen of the heterocycle, in the first case, or to nitrogen, in the second case, during the methoxyselenylation reaction was essential for stereoselection (diastereomeric ratio > 96%). The presence and relevance of nonbonding interaction between a divalent selenium atom and an electron donor heteroatom have been recently reviewed [54].

Oxyselenenylation reactions using enantiomerically pure nucleophiles, *e.g.*, an optically active diol **75**, are also known and were employed for the synthesis of dioxanes **77a,b** (Scheme **2.15**) [55]. In the first step, two enantiomerically pure diastereomeric alkoxyselenides **76a,b** were obtained by the addition of **75** to alkene, promoted by *N*-(phenylseleno)phthalimide. These were separated by chromatography and final 1,4-dioxanes were accessed by oxidative elimination of phenylselenyl group, followed by the reaction with NaH.

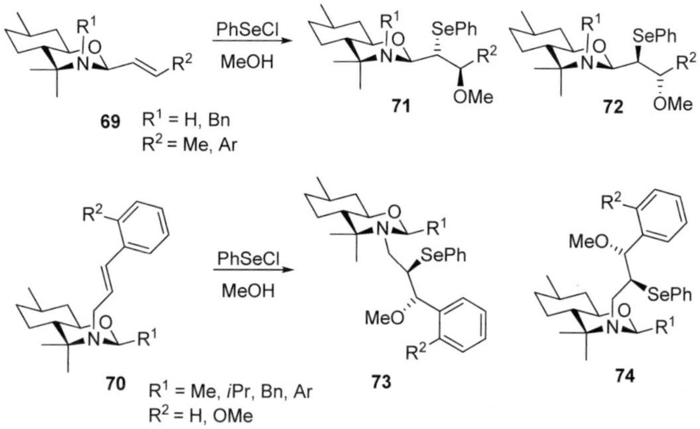

Scheme 2.14

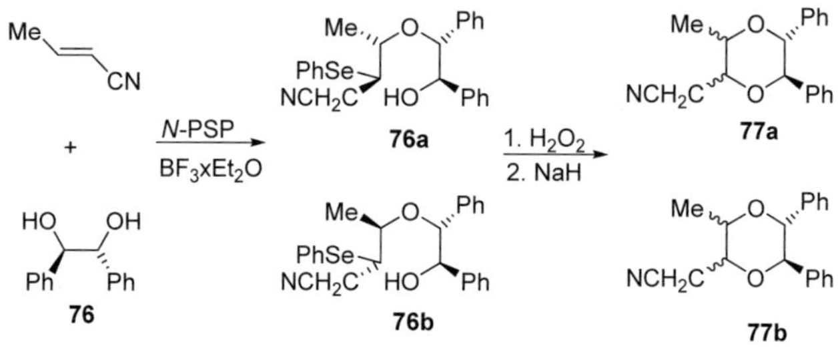

Scheme 2.15

Other chiral nucleophiles were used for the synthesis of cyclitols [56], carbohydrate derivatives [57], and morpholines [58].

In the last two decades, the most extensive research was conducted on the use of chiral electrophiles in the asymmetric methoxy- and hydroxyselenenylation reactions.

The first examples of methoxyselenenylation reactions of simple alkenes with 1,1'-binaphtalene-2-selanyl bromide **33a** was reported by Tomoda and coworkers in 1988 [37a]. Diastereoselectivities for these reactions were relatively low, as an example the methoxyselenenylation of styrene (*de* 49%). Later, these authors used *N*-substituted amido-2-binaphthyl selenium electrophiles **33b-d** for methoxyselenenylation of (*E*)-*β*-methylstyrene, and the corresponding chiral products **78a-d** were obtained in good yields and *de* (Scheme **2.16**) [40b].

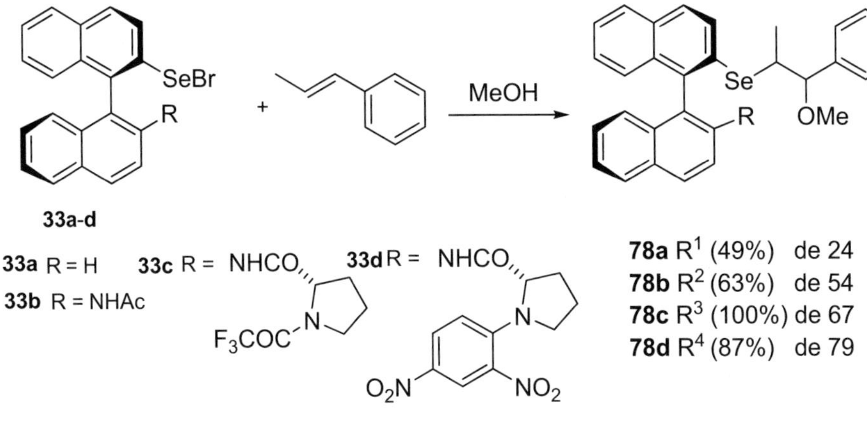

Scheme 2.16

In further investigations the methoxyselenenylation reaction with diaryl diselenide, derived from *D*-Mannitol **37**, Tomoda and coworkers showed that the nucleophilicity of the counterion can affect the selectivity of selenenylation reactions (Scheme **2.17**) [40d, 40f]. The results of methoxyselenenylation of *trans*-*β*-methylstyrene with various counteranions are collected in Table **2.1**. Based on these results, they suggested that lower nucleophilicity of the counteranion may be effective to enhance the diastereomeric excess and these results were confirmed by other research groups [59].

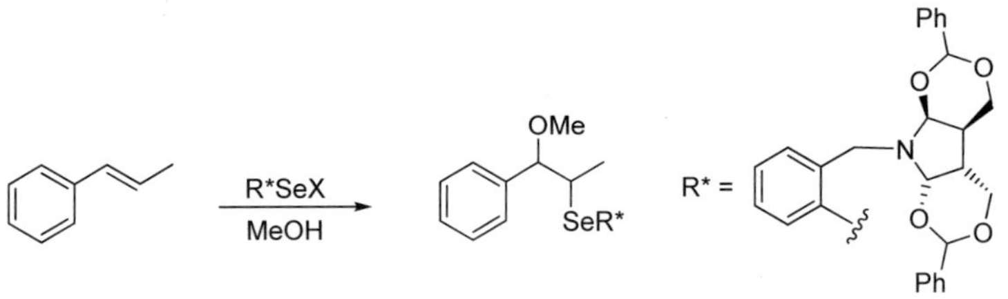

Scheme 2.17

Table 2.1: Effect of counteranions on the methoxyselenenylatin reaction of *trans*-β-methylstyrene

X^-	*de* [%]	Yield [%]
Br^-	52	85
ClO_4^-	80	47
TfO^-	89	68
BF_4^-	90	67
SbF_6^-	94	64
PF_6^-	95	58

The selenenylating reagents with chiral counteranions, **28-31** were applied to the methoxyselenenylation of styrene and selenocyclization reactions, leading to the racemic products [26].

Subsequently, many attempts to improve the stereoselectivity were reported. As a test reaction to compare the diastereoselectivity induced by these reagents, the methoxyselenenylation of styrene can be taken in consideration (Scheme **2.18**). The results of these reactions with the use of selected chiral diselenides **34-56** are summarized in Table **2.2**.

Scheme 2.18

Based on the results presented in Table **2.2**, the diastereoselectivity of the methoxyselenenylation reaction can be affected by a number of factors.

Table 2.2: Asymmetric methoxyselenenylation of styrene

Entry	Diselenide	X^-	T [°C]	*de* [%]	Yield [%]	Ref.
1	**34**	OTf^-	-78	77	88	38a
2	**35**	OTf^-	-78	94	73	38d
3	**36**	OTf^-	-78	35	97	39c
4	**38**	OTf^-	-100	83	67	41c
5	**39**	OTf^-	-100	96	55	41e
6	**40**	OTf^-	-100	93	28	41h
7	**41**	OTf^-	-78	60	63	41c
8	**42**	OTf^-	-78	70	52	41j
9	**44**	OTf^-	-78	89	68	41k
10	**45**	OTf^-	-78	86	62	41l
11	**46**	OTf^- $HOSO_3^-$	25 25	35 61	65 83	41f
12	**47**	OTf^-	-78	92	80	42a
13	**48**	$HOSO_3^-$	-30	96	72	42c
14	**49**	OTf^- $HOSO_3^-$	-78 25	94 90	40 70	42b 42b
15	**50**	OTf^-	-78	48	77	43e
16	**51**	OTf^-	-78	88	52	43g
17	**52**	OTf^-	-78	40	93	44b
18	**53**	OTf^-	-78	64	57	44b
19	**54**	OTf^-	-78	80	54	44h
20	**55**	OTf^-	-78	48	58	44h
21	**56**	Br^-	-78	51	60	45

The reduced conformational flexibility of electrophilic selenium reagents increases the selectivity of the reaction, *e.g.*, cyclic tetrahydrofuran substituents **(35a)** gave better selectivities then acyclic groups **(34a)** (Table **2.2**, entries 1-2, Fig. **2.3**).

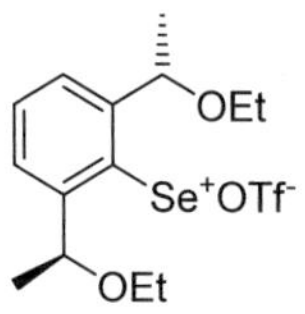

34a (de = 77%)

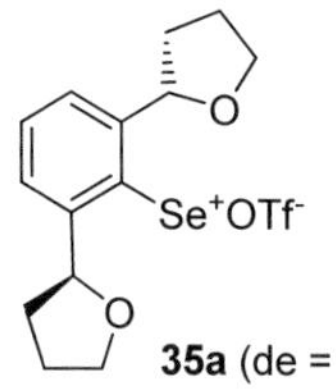

35a (de = 94%)

Figure 2.3: Comparison of selectivity of electrophilic selenenyl reagents **34a** and **35a**.

The same effect was observed by Wirth and coworkers for the cyclic compound **40** and its chain analog **38** [41c]. Higher diastereoselectivities were observed for bicyclic terpenes **50, 53** than for more flexible monocyclic terpene **52** (entries 15,17-18). Increase in the selectivity for *p*-menthane derivatives was correlated with the DFT calculated stability of the chair conformers of their selenenyl bromides [44d].

The introduction of methoxy groups at the *orto* position respect to selenium atom in some cases increased the reaction selectivity (entries 4, 5, 12, 13). Interestingly, a significant decrease of selectivity was observed in the case of diselenide **43** [41j]. Comparing the selectivity of electrophilic reagents **46a** and **49a**, the introduction of the second chiral center in some cases can improve the enantioselectivity of the reaction (entries 11, 14).

The most important factor influencing the stereoselectivity of the methoxyselenenylation reaction has been proven to be the possibility to form a nonbonding interactions between heteroatoms and the electrophilic selenium atom. The occurring of these interactions in crystal structures were confirmed by X-ray crystallography, DFT calculations, and in solution by NMR experiments [54, 60]. Strong interactions result in greater conformational rigidity which is necessary for a transfer of chirality in asymmetric methoxyselenenylation reactions. The lack of heteroatom may results in the lack of stereoselectivity as observed for the selenium electrophile **82** [60] (Fig. **2.5**). An important role plays the type of heteroatoms and their location in the side chain. The presented aromatic selenium electrophiles contain heteroatoms separated by four (**79**), five (**80**) or six (**81**) bonds from the selenium atom (Fig. **2.4**).

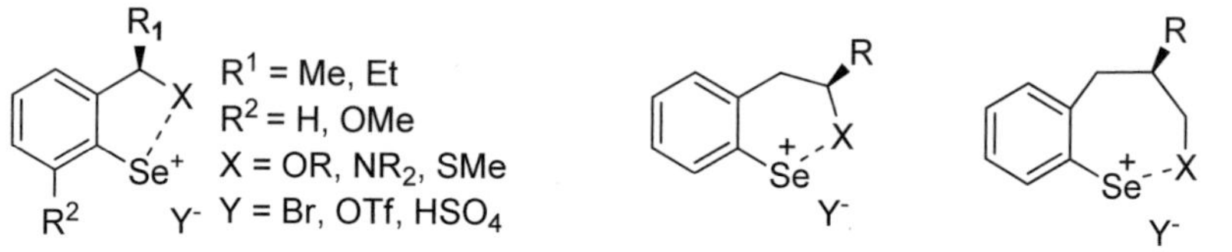

79 **80** **81**

Figure 2.4: Location of heteroatom in a side chain.

Comparing the selenium electrophiles possesing an oxygen **38a**, nitrogen **46a** or sulfur **47a** heteroatom, it was demonstrated that the sulfur-selenium interaction is the most effective on inducing diastereoselectivity (entries 4, 11, 12). Decrease in selectivity was observed for the selenium electrophile **41a** (entry 7). Recently, Wirth and coworkers presented the syntheses of new diselenides like **44** and **45,** and tested them in the methoxyselenenylation of styrene (entries 9, 10) [41k,l]. Again, an increase in stereoselectivity of the reaction was observed, and a new type of selenium-oxygen interaction was proposed (Fig. **2.5**).

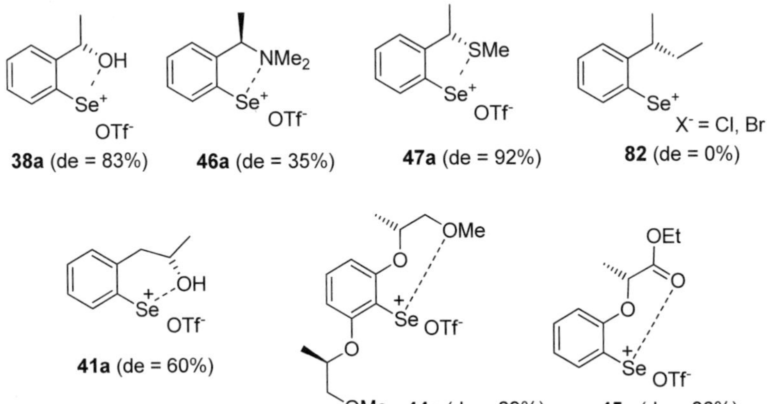

38a (de = 83%) **46a** (de = 35%) **47a** (de = 92%) **82** (de = 0%)

41a (de = 60%) **44a** (de = 89%) **45a** (de = 86%)

Figure 2.5: Selenium electrophiles containing a heteroatom in the side chain.

Alkyl selenium electrophiles also showed higher selectivities when they had heteroatoms which can interract with the electrophilic selenium atom (entries 15, 16 and 19, 20). Replacement of the 2-keto group in **50a** with the oxime substituent **51a** makes possible an interaction with the oxime hydroxyl group [43f,g] (Fig. **2.6**). Increased diastereoselectivity on methoxyselenenylation of styrene produced by the electrophile **54a** is the result of an equatorial-equatorial arrangement of hydroxyl and selenide groups in the carane system for which the effective interaction is possible. This arrangement was confirmed by the X-ray crystallography and DFT calculations. In **55a** this interaction is not possible [44g].

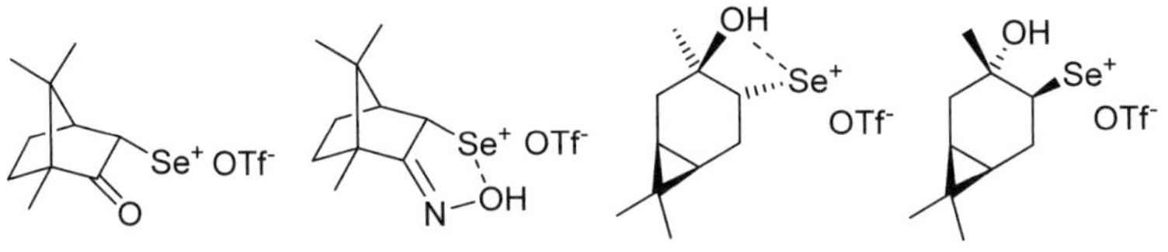

50a (de = 48%) **51a** (de = 88%) **54a** (de = 80%) **55a** (de = 48%)

Figure 2.6: Selectivity of alkyl selenium electrophiles.

Recently, Santi and coworkers presented the asymmetric selenenylation of β-aryl α,β-unsaturated aldehydes with the electrophilic selenium reagent **83**, obtained from diselenide **47** by the reaction with SO_2Cl_2 (Scheme **2.19**) [61].

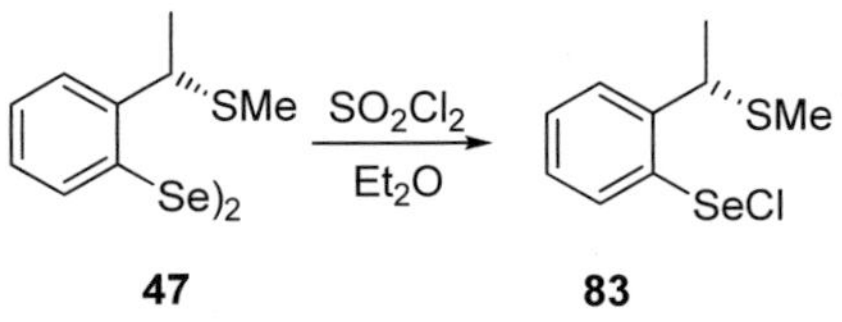

Scheme 2.19

These reactions led to the formation of four enantiomerically pure diastereoisomers, *e.g.* the reaction of cinnamaldehyde **84** in the presence of $MgSO_4$ gave products **85a-d** in 91% yield and in ratio 81:9:5:5. Higher yield of product **85a** was explained by the formation of a strong interaction between seleniranium ion and the hydroxyl group (intermediate **87**) deriving from the reaction of hemiacetal **86** and selenium reagent **83** (Scheme **2.20**).

Chiral electrophilic selenium reagents were used for asymmetric hydroxyselenenylation of alkenes [42a-c, 62]. Typically these reactions were carried out in THF or acetonitrile in the presence of water, using electrophilic selenium triflates or sulfates as chiral reagents (Scheme **2.21**). The choice of a counteranion depended on the reaction conditions. Low temperatures below -30° C required the triflate counteranion. The results of hydroxyselenenylation of styrene with chiral diselenides **47-50**, **88**, are summarized in brackets in Fig. **2.7**.

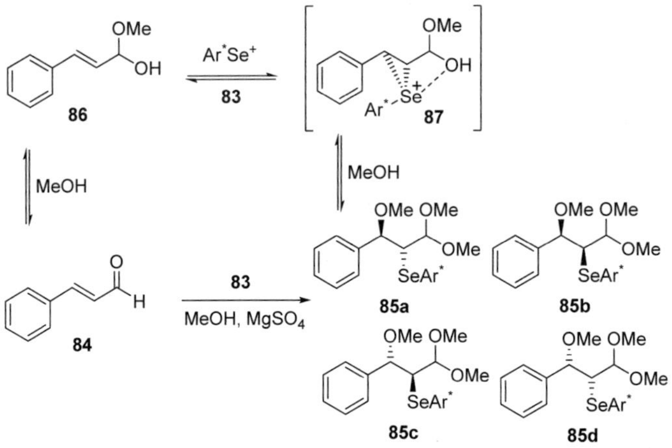

Scheme 2.20

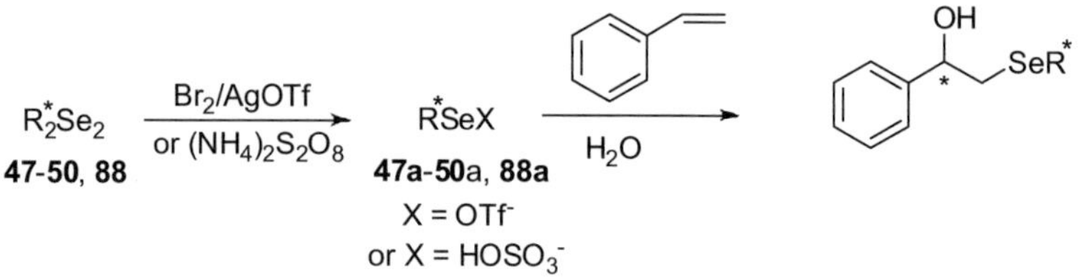

Scheme 2.21

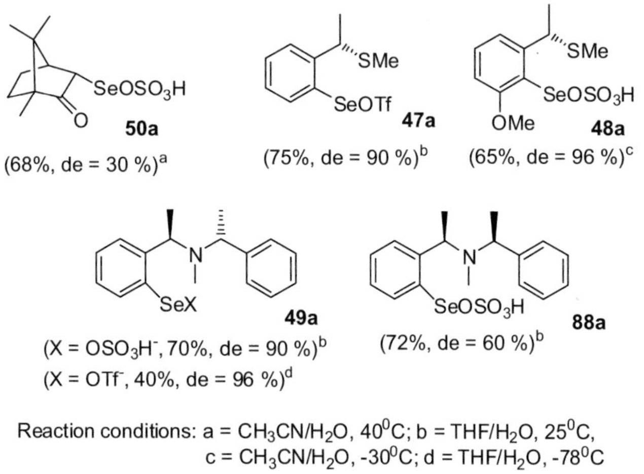

Figure 2.7: Results of asymmetric hydroxyselenenylation of styrene.

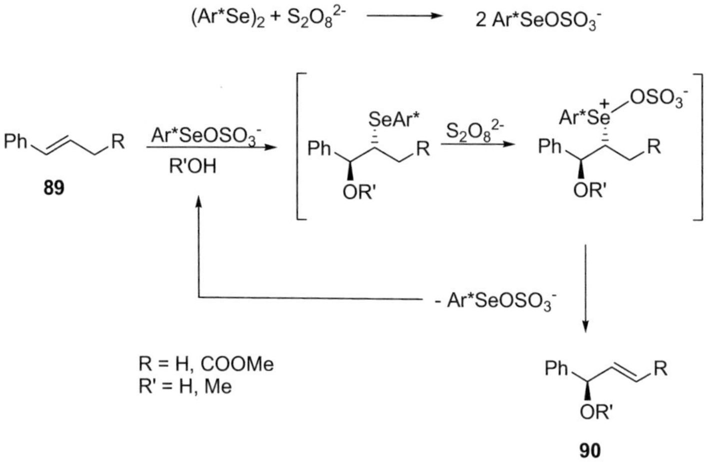

Scheme 2.22

Chiral electrophilic selenium reagents were also used as catalysts for asymmetric oxyselenenylation alkenes with simultaneous formation of double bonds (oxyselenenylation-elimination reactions). A catalytic amount of chiral selenium reagents and the stoichiometric amount of peroxodisulfate, activated a sequence of the oxyselenenylation reaction and oxidative elimination of alkenes **89** producing the corresponding allylic alcohols or allylic ethers **90** (Scheme **2.22**). Wirth and coworkers developed this reaction using *β*-methylstyrene as a substrate, using 10% of diselenide **46** or **91** (Table **2.3**, entries 1, 2) [63]. The same reaction sequence was used by Tiecco and his research group for oxyselenenylation-elimination of *β*,*γ*-unsaturated methyl esters, in the presence of 10% of diselenide **49** or 5% of diselenide **48** (entries 3, 4) [42b, 42c, 64].

Table 2.3: Catalytic oxyselenenylation-elimination reaction of alkenes

Entry	Diselenide	Cat. [%]	R	R'	Yield 90 [%]	Ee [%]
1	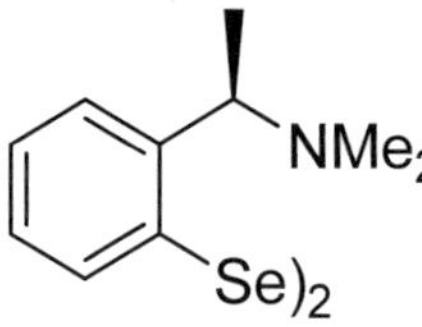 **46**	10	H	Me	30	58
2	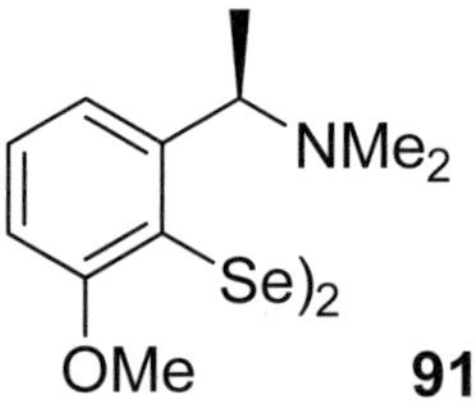 **91**	10	H	Me	23	75
3	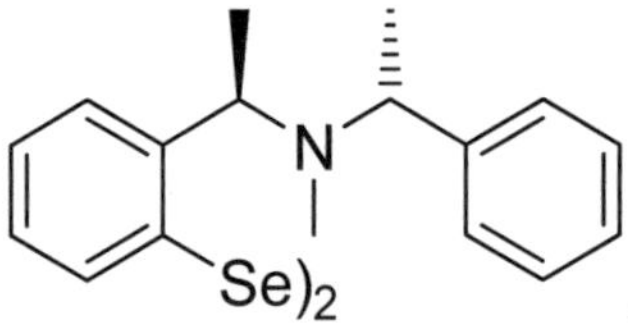 **49**	10	COOMe	Me	12	94
4	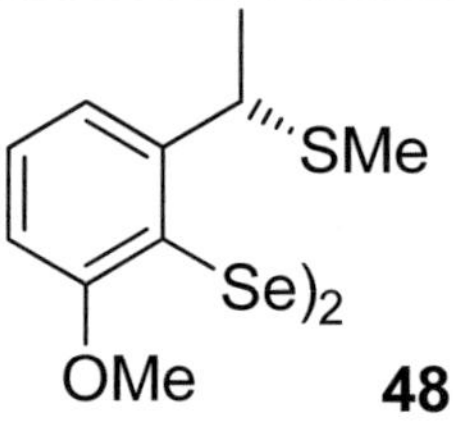 **48**	5	COOMe	Me H	98 98	78 82

Other applications of organoselenium compounds as catalysts in organic transformations are presented in Chapter 8.

2.3.3. Azaselenenylation

Addition reactions of electrophilic selenium reagents and nitrogen nucleophiles to the double bonds play an important role in organic synthesis due to the further transformations of the selenium and nitrogen functions. Two types of nitrogen nucleophiles, nitriles (amidoselenenylation reaction) or the azide anion (azidoselenenylation reaction) were mostly developed.

The first *trans-β*-amidoalkyl phenyl selenides **92** were reported by Toshimitsu and coworkers [65]. Cyclic alkenes, *e.g.*, cyclohexene, reacted with PhSeCl in various nitriles in the presence of water and trifluoromethane sulfonic acid giving the products in good yields as illustrated in Scheme **2.23**.

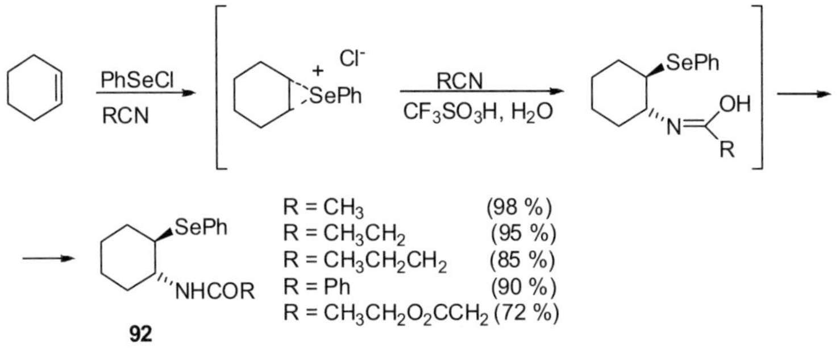

Scheme 2.23

Terminal olefins **93** (1-hexene, 1-octene, 1-decene), reacted with nitriles affording predominantly the regioisomers **94a** and ~ 15% of the regioisomers **94b** (Scheme **2.24**). Styrene gave only one product **94a** (R = Ph) in 36 % yield.

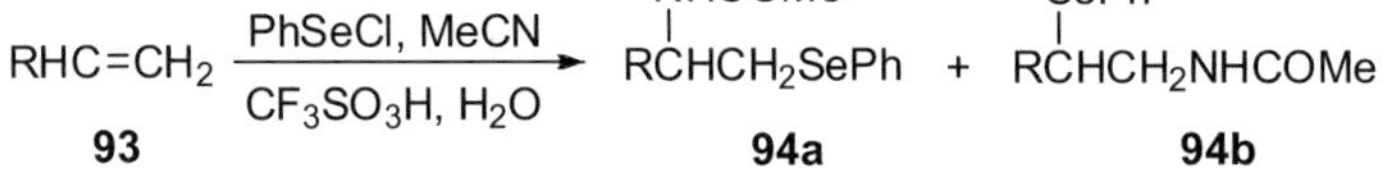

Scheme 2.24

Similar results were obtained by Tiecco and coworkers using diphenyl diselenide and ammonium peroxydisulfate for the reaction with olefins in acetonitrile in the presence of trifluoromethane sulfonic acid and water [66]. Results of amidoselenenylation of cyclohexene, *E*-3-hexene and 1-octene are summarized in Fig. **2.8**.

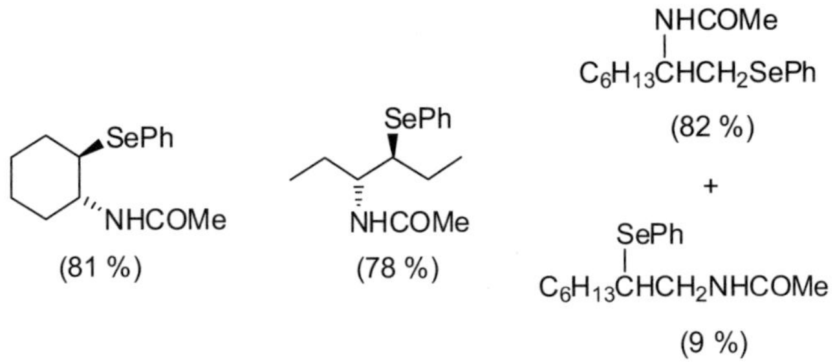

Figure 2.8: Results of aminoselenenylation of olefins with PSS 31.

Trans-β-amidoalkyl phenyl selenides were oxidized with 30% H_2O_2 in THF giving the corresponding allylic amides in very good yields [65].

Tiecco and coworkers have developed asymmetric amidoselenenylation reaction of simple alkenes **95** with acetonitrile, butyronitrile or benzonitrile with camphorselenenyl sulfate **50a** [67], and a mixture of two diastereomers **96** was obtained. The diastereoisomers of the acetamido selenides **96a**, **96b** could be readily separated by medium pressure column chromatography (Scheme **2.25**). Examples of acetamidoselenenylation reaction are summarized in Table **2.4**. Selenides **96a**, **96b** were transformed into the corresponding oxazolines **97**, and **98** by the reaction with PhSeOTf or SO_2Cl_2. Treatment of the diastereomeric mixtures of selenides with the Lawesson's reagent gave mixtures of thioamido selenides **99** [68]. Thioamido selenides **99a**, **b** were separated by flash chromatography, and were efficiently converted into the corresponding enantiomerically pure thiazolines **100**, **101** by the reaction with PhSeCl.

The reaction of PhSeCl and sodium azide in DMSO with alkenes, reported by Hassner and coworkers, was the first example of the azidoselenenylation reaction [69]. Cyclohexene produced stereospecifically the *trans* adduct **103** in 91% yield. Terminal olefins, such as 1-hexene or styrene, gave a mixture of regioisomers **104a,b** and **105a,b**. Giuliano and Duarte modified the reaction conditions using *N*-(phenylseleno)phtalimide and aziditrimethylsilane for the azidoselenenylation of exocyclic alkenes [70]. In the case of methylenecyclohexane, an increase of the regioselectivity of products **106a,b** was observed (Fig. **2.9**).

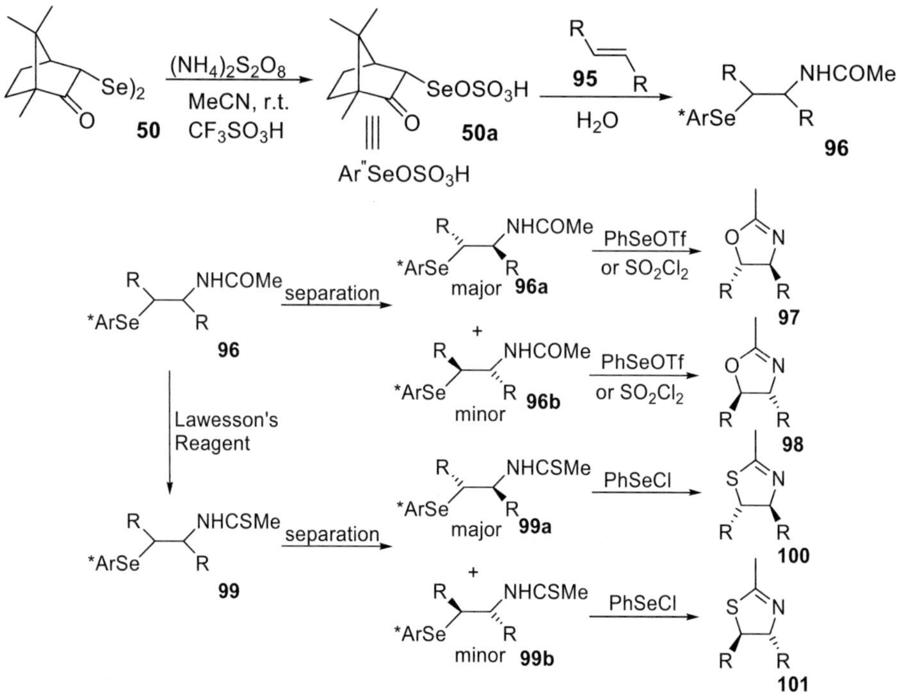

Scheme 2.25

Table 2.4: Synthesis of acetamido selenides

Entry	Alkene	Product 96a + 96b	Yield [%]	Dr
1	(*E*)-3-Hexene	**96a** + **96b** R = C_2H_5	82	73:27
2	(*E*)-4-Octene	**96a** + **96b** R = C_3H_7	78	75:25
3	(*E*)-5-Decene	**96a** + **96b** R = C_4H_9	80	80:20
4	Styrene	**96** R = C_6H_5	50	53:47
5	Cyclohexene	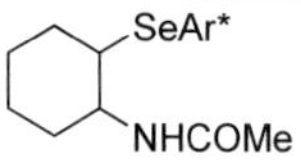 **102**	61	65:35

Tiecco and coworkers reported that azidoselenenylation of phenyl substituted alkenes can be carried out with phenylselenenyl triflate and sodium azide in acetonitrile with complete regio- and stereoselectivity [71], *e.g.*, styrene gave only one product **105a** in 70% yield. The azidoselenenylation of *p*-menthane **65** and 3-carane **66**, under the same reaction conditions, gave the β-azidophenylselenides **107** and **108** (Scheme **2.26**) [72].

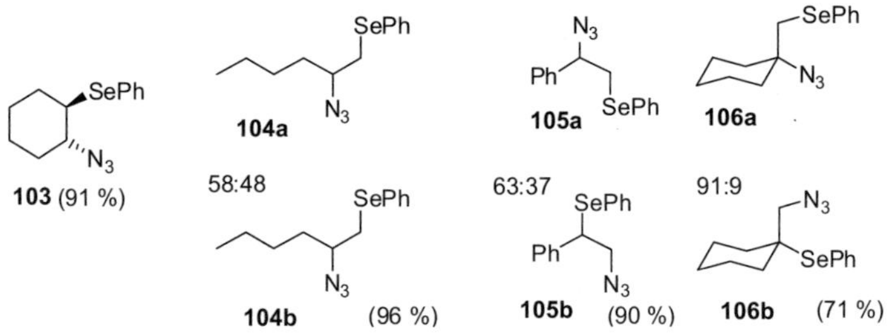

Figure 2.9: First examples of azidoselenenylation of alkenes.

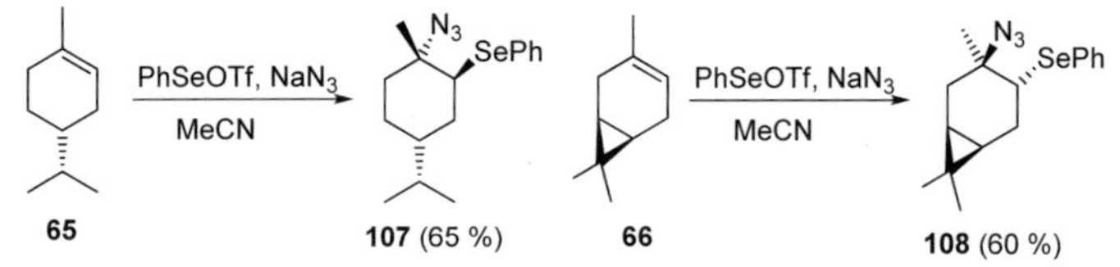

Scheme 2.26

The first example of asymmetric azidoselenenylation of olefins proceeding with a very high facial selectivity was reported by Tiecco group [73]. Five chiral diselenides **47-50**, **109** were tested in the asymmetric azidoselenenylation of styrene (Scheme **2.27**).

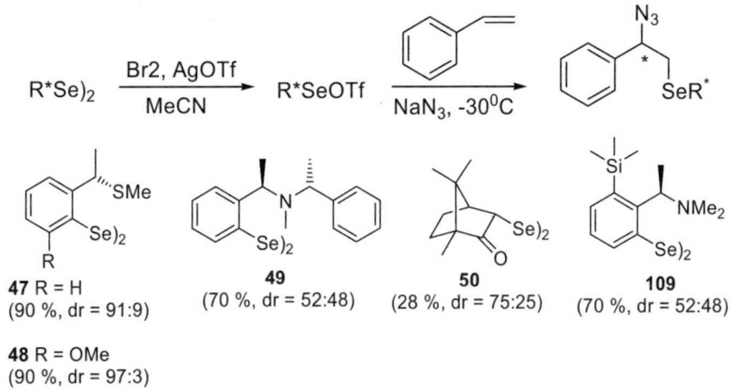

Scheme 2.27

Diselenide **47** was selected for the azidoselenenylation of *β*-methylstyrene, *α*-methylstyrene, (*E*)-4-octene, (*E*)-5-decene and 1-methyl-1-cyclohexene, proceeding with excellent selectivities (de up to 98 %) and good yields.

The azidoselenenylated products **102** were used for the synthesis of chiral aziridines **110**, oxazolines **111**, and triazoles **112** (Scheme **2.28**).

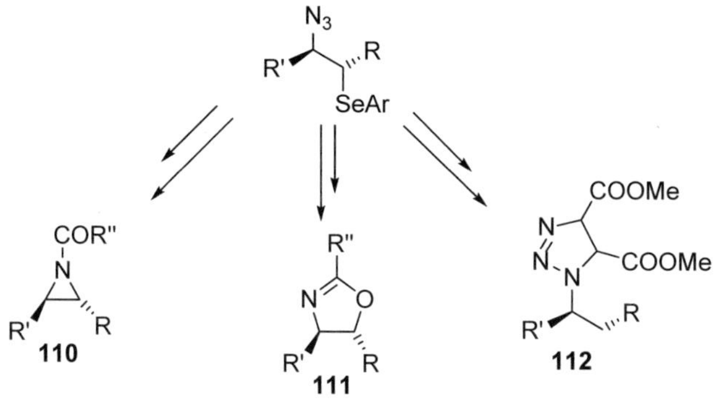

Scheme 2.28

2.3.4. Carboselenenylation

Nishibayashi and coworkers developed a new carbon-carbon bond formation, based on the reaction of simple alkenes with aromatic compounds using C_2-asymmetric electrophilic selenenyl triflate **113** [74]. The asymmetric carboselenenylation reaction of (*E*)-*β*-methylstyrene with 2-methylfuran **114**, and **113**, proceded with high diastereoselectivity and good yield (Scheme **2.29** – product **115**). The reaction can be considered as a new type of the asymmetric Friedel-Crafts alkylation of an aromatic compound with an alkene. The reaction products **116-121**, obtained from (*E*)-*β*-methylstyrene and aromatic compounds such as 2,3-dimethylfuran, 2-methylthiophene, *N*-methylpyrrole, 1,3,5-trimethoxybenzene, *N*,*N*-dimethylaniline, and azulene, are presented in Fig. **2.10**.

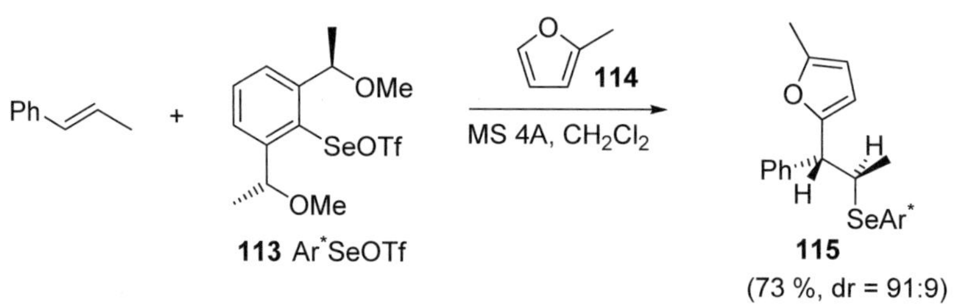

Scheme 2.29

Recently, Wirth and coworkers presented selenenylation of alkenes with styrene nucleophiles [75]. The selenenylation of styrene with styrene as the nucleophile was performed using diphenyl diselenide in the presence of 20% of iodine. The product **122** was obtained in 55 % yield (Scheme **2.30**).

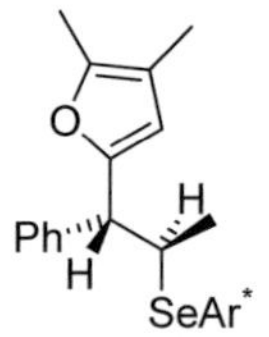

116
(63 %, dr = 92:8)

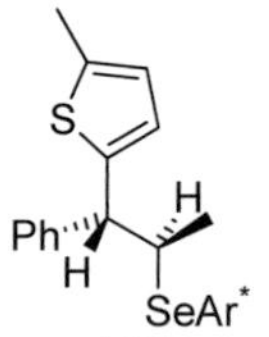

117
(52 %, dr = 91:9)

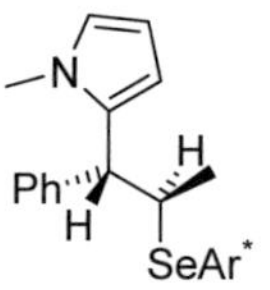

118
(48 %, dr = 89:11)

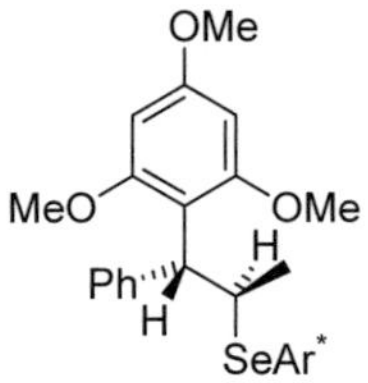

119
(77 %, dr = 95:5)

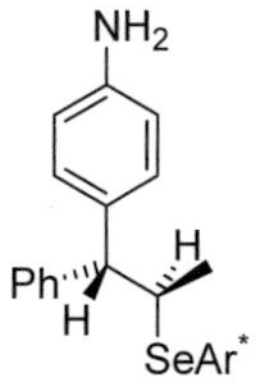

120
(51 %, dr = 90:10)

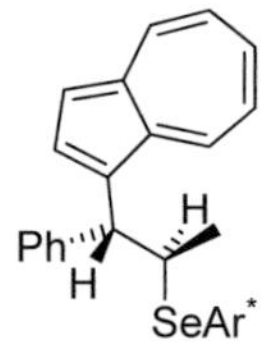

121
(34 %, dr = 91:9)

Figure 2.10: Carboselenenylation (E)-β-methylstyrene with various aromatic compounds.

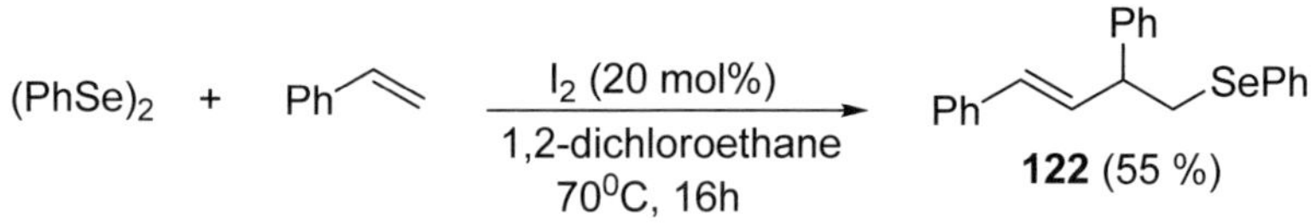

122 (55 %)

Scheme 2.30

Analogous products were obtained from 2-, 3-, and 4-methylstyrenes in good yields (55-62 %). In the case of 4-vinyl-1,1-binaphtyl and 2-vinylnaphtalene deselenylated side products were observed.

2.4. SELENOCYCLIZATIONS

A nucleophilic functional group, contained in the unsaturated hydrocarbon molecule, makes possible the intramolecular reaction under the selenenylation reaction conditions. The mechanism of selenocyclization reactions is similar to the previously presented mechanism of the selenenylation reactions with an external nucleophile. The electrophilic addition of selenonium groups to the carbon-carbon double bond provides the seleniranium intermediates which are opened by the intramolecular attack of the nucleophile. Depending on the position of a nucleophilic group and the double bond, the seleniranium intermediates can undergo either the *endo*-cyclization or *exo*-cyclization to form heterocyclic derivatives with different ring size (Scheme **2.31**).

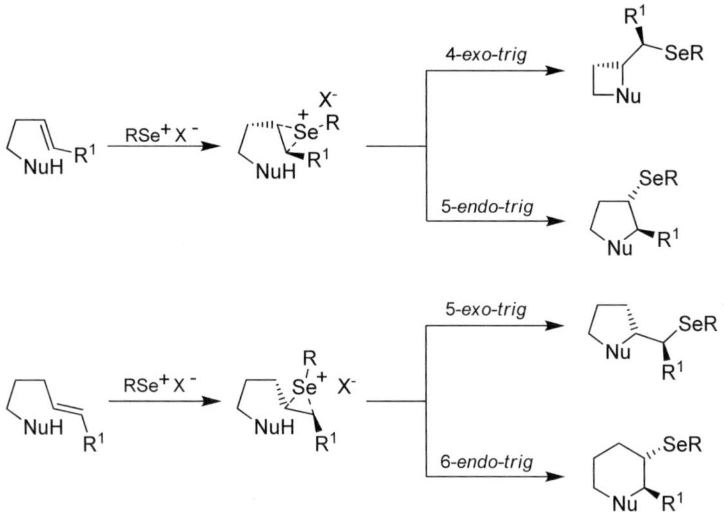

Scheme 2.31

Similarly to other electrophilic additions of selenium reagents to the carbon-carbon double bonds, selenocyclization reactions proceed in the stereospecific *anti* manner. Factors such as, the nature of the electrophilic selenium reagent, the counteranion, solvents, and external additives coordinating to the selenium atom are influencing the course of cyclizations [76]. Recently, Denmark and coworkers carried detailed mechanistic studies on the selenocyclization of β,γ-unsaturated acids and derivatives with PhSeCl or PhSeBr. They showed that in the first step a haloselenenylation intermediate is formed as a kinetic product which is converted into thermodynamically favoured cyclization product [77]. Further studies led to the development of the first catalytic selenofunctionalization of the unactivated olefins, catalysed by chiral and achiral Lewis bases [78]. Among many examples described in the literature, the majority of products has five- or six-membered rings, formed by *5-endo- trig*, *5-exo-trig*, or *6-endo-trig* pathways.

As already mentioned, selenocyclization reactions allow the construction of various heterocycles. The presence of a selenium function in the cyclization product allows its further functionalization. For example, it can be removed or substituted in various ways. The most commonly used methods are the radical deselenylation, the oxidation followed by substitution or by *syn*-oxidative elimination *via* the selenoxide.

Due to the nature of the internal nucleophile, selenocyclization reactions can be generally divided into three types. In the most popular, oxygen and nitrogen nucleophiles, such as hydroxyl, carboxy, amino, oxime groups are used and make possible the formation of cyclic ethers, lactones, amines and cyclic 1,2-oxazines or nitrones, respectively. In the last type, carbon nucleophiles enable the formation of a new carbon-carbon bond.

The ring closure reactions, described below, involve the formation of carbon-oxygen, carbon-nitrogen and carbon-carbon bonds.

2.4.1. Oxyselenocyclization

The selenocyclization reactions of olefins bearing the hydroxyl or carboxy group are the most known and explored selenium-mediated cyclofunctionalizations. The products are cyclic ethers and lactones, respectively. General examples described below show the potential of the oxyselenocyclizations. *Cis*-fused tetrahydrofurans **124** are produced when 2-cycloalkenyl-substituted ethanol **123** derivatives are cyclized *via* the 5-*exo* ring closure. The related cycloalkenyl acids **125** systems also yield *cis*-fused tetrahydrofuranones **126** upon the reaction with phenylselenenyl reagents, similar to their alkenol analogs. The cyclization of the acid **125** leads to γ-lactones **126** arising from the *anti*-addition across the double bond [79]. Fused ring γ-lactones may also be formed by cyclizations of the acid **127** in organic media under very mild conditions and low temperature [80]. Nicolaou showed that in the phenylselenolactonization of 1-cyclohexeneacetic acid **127** an unstable β-lactone is formed rearranging readily to the more stable γ-lactone **128** [80]. Cyclofunctionalization of cyclo-2,4-dieneacetic acids **129** results in the 1,4-addition to form *cis*-fused γ-lactones **130** as shown in Scheme **2.32**, entry d [81]. In this reaction, the phenylselenyl group is in *trans* relation to the lactone ring. Interesting products **132**, **134**, and **136** were also obtained by *O*-selenocyclizations of cyclic allylic amine derivatives **131**, **133**, **136** (Scheme **2.32**, entries e-g) [82].

Furthermore, alkenyl nitrones **137** reacted with the selenium electrophile providing the cyclic products through the intramolecular nucleophilic attack of the oxygen atom. The obtained iminium salts **138** can undergo reduction by $NaBH_4$, or can be directly treated with nucleophilic reagents, affording *N*-alkyl **139** or *N*-unsubstituted **140** six-membered 1,2-oxazines, respectively (Scheme **2.33**) [83].

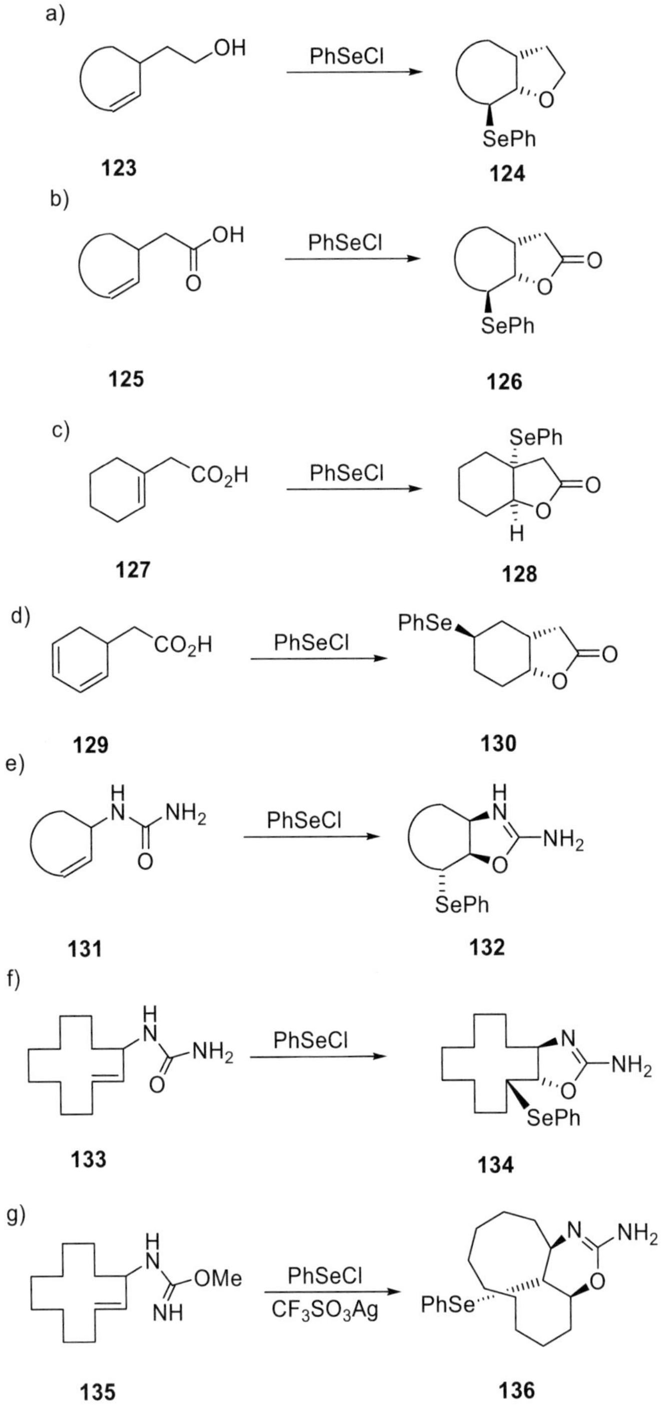
a)
OH
PhSeCl
O
SePh
123
124
b)
OH
O
PhSeCl
O
O
SePh
125
126
c)
CO2H
PhSeCl
SePh
O
O
H
127
128
d)
CO2H
PhSeCl
PhSe
O
O
129
130
e)
H
N
NH2
O
PhSeCl
H
N
NH2
O
SePh
131
132
f)
H
N
NH2
O
PhSeCl
N
NH2
O
SePh
133
134
g)
H
N
OMe
NH
PhSeCl
CF3SO3Ag
N
NH2
O
PhSe
135
136

Scheme 2.32

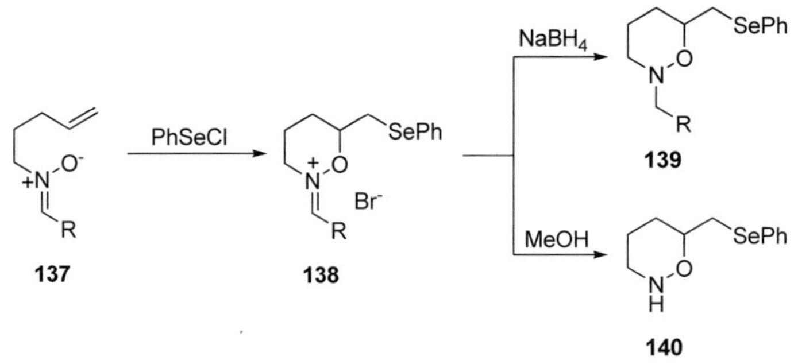

Scheme 2.33

The cyclization of terminal alkenyl oximes **141** leads to six-membered dihydro-1,2-oxazines **142**, in competition to tetrahydro derivatives, and/or to five/six-membered cyclic nitrones **143**, **144**, depending on the substituents and the geometry of the starting oximes (Scheme **2.34**) [84].

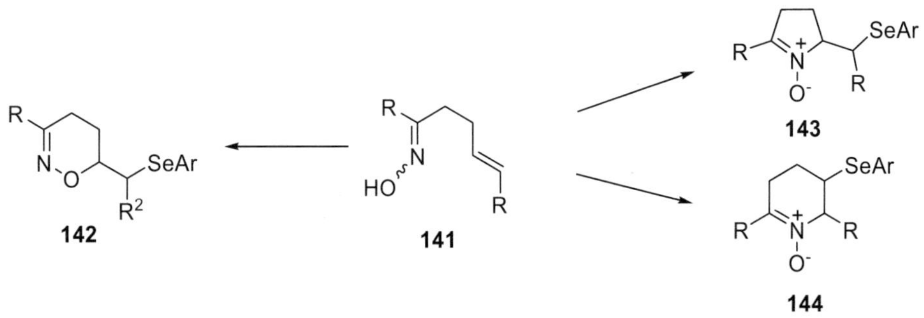

Scheme 2.34

In recent years, Wirth has developed a convenient synthesis of lactones **145-147**, based on the *one-pot* catalytic addition-elimination reaction using PhSeSePh and hypervalent iodine as an oxidant. The procedure has been used for the synthesis of five and six-membered butenolines **145** [85], 2*H*-pyran-2-ones **146** [86], and isocoumarines **147** [87] (Scheme **2.35**), which are biologically active compounds, and their structural motifs occur in a number of natural products. A similar sequence was described by Tiecco and coworkers, which employed diphenyl diselenide and peroxydisulfate as an oxidant, to obtain the corresponding lactones in good yields [88]. Tiecco also found that (diacetoxyiodo)benzene reacts with diphenyl diselenide in acetonitrile to give cleanly the corresponding lactone [89]. The same transformation has been investigated by Denmark in great detail [77].

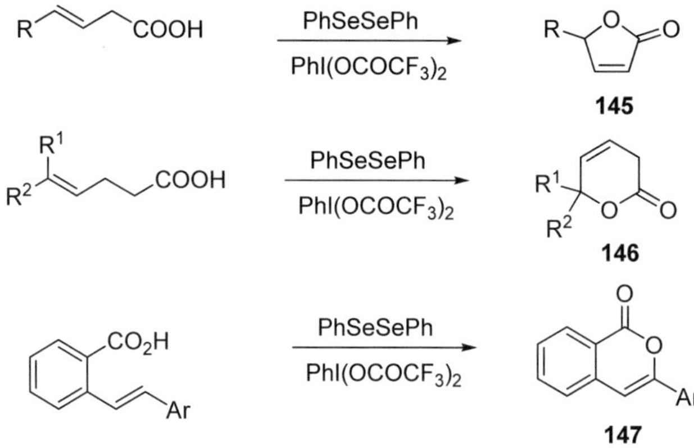

Scheme 2.35

More recently, the concept of selenocyclofunctionalization by the "Lewis base activation of Lewis acids" has been introduced [90]. In 2007 Denmark reported the first example of a Lewis base catalyzed selenolactonization reaction [78c]. The reaction of unsaturated acids **148** with *N*-phenylselenenylsuccinimide (NPSS) **27** in the presence of a sub-stoichiometric amount of a Lewis base afforded seleno lactones **149** in excellent yields (Scheme **2.36**). A variety of Lewis bases, such as DMPU, DMPU(S), $PPh_3(S)$, HMPT, HMPA, and HMPA(chalcogen), served as efficient catalysts for this transformation. The selenolactonization reactions were investigated with a wide variety of enantioenriched Lewis bases. The products, obtained in good to excellent yields, were racemic [78b].

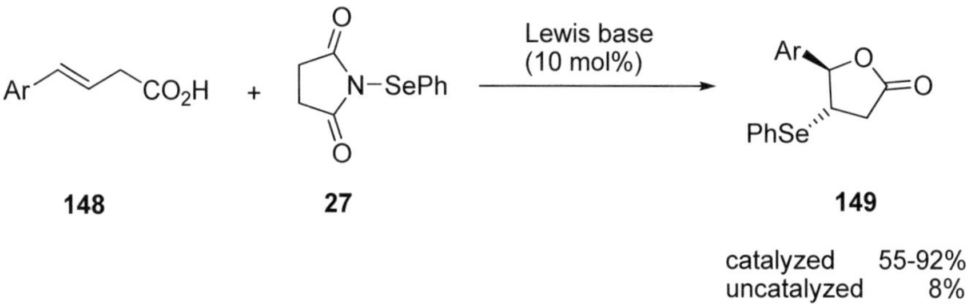

Scheme 2.36

Selenocyclization of the alkenols prompted by Lewis base catalysts was also investigated [78b,c]. Detailed studies on the mechanism revealed that the addition of a Brønsted acid is essential for this reaction. The screening of the catalyst structures revealed that the BINAM-derived thiophosphoramides **150** catalyzes the enantioselective selenoetherification of a variety of olefins in modest to good enantioselectivities. Among the catalysts examined, the thiophosphoryl(V) Lewis bases are the most efficient in catalyzing the selenoetherifications (Scheme **2.37**) [78b].

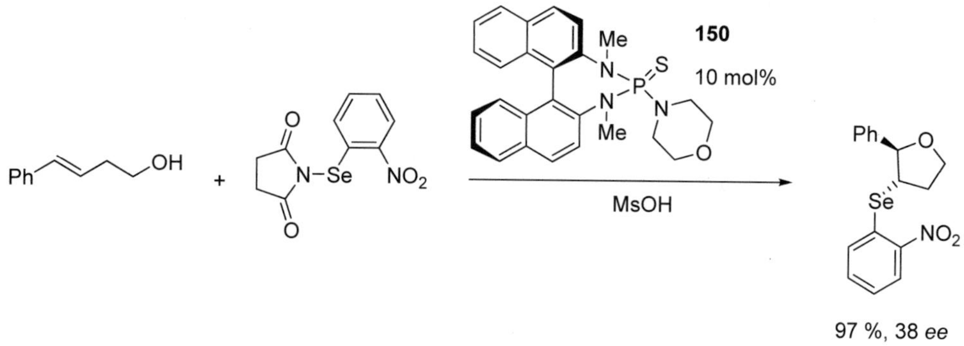

Scheme 2.37

The selenocyclofunctionalization reaction is also an important tool for the construction of building blocks and is widely used in total syntheses. Nicolaou first reported a synthetic route to a highly functionalized core structure of garsubellin A employing a selenocyclization approach [91]. Initially, the precursor **152** was synthesized from commercially available 1,3-cyclohexanedione **151** in eight steps (Scheme **2.38**). The selenium-mediated cyclization in the presence of *N*-(phenylseleno)phthalimide (*N*-PSP) and $SnCl_4$ furnished the selenide **153**. Selective reduction of the bridged ketone **153** produced a single alcohol which, upon alkylation with *trans*-1,2-*bis*(phenylsulfonyl)ethylene, yielded vinylogous sulfone **154**. The construction of tetracycle **155** was achieved by the use of *n*-Bu_3SnH and AIBN, and further transformations led to garsubellin A **156**.

Another example of the utility of selenocyclization is the synthesis of leucascandrolide A **159** (Scheme **2.39**) [92]. The use of selenium electrophiles was crucial in the selenoetherification leading to 2,6-*trans*-tetrahydropyran **158**. Unexpectedly, the iodine-based electrophiles show little diastereoselectivity, however, it is dramatically improved by the use of TIPPSeBr – (2,4,6-triisopropylphenyl)selenenyl bromide.

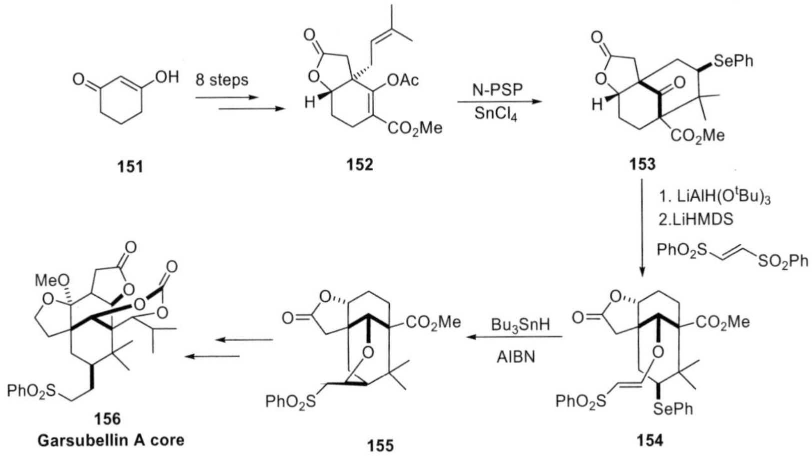

Scheme 2.38

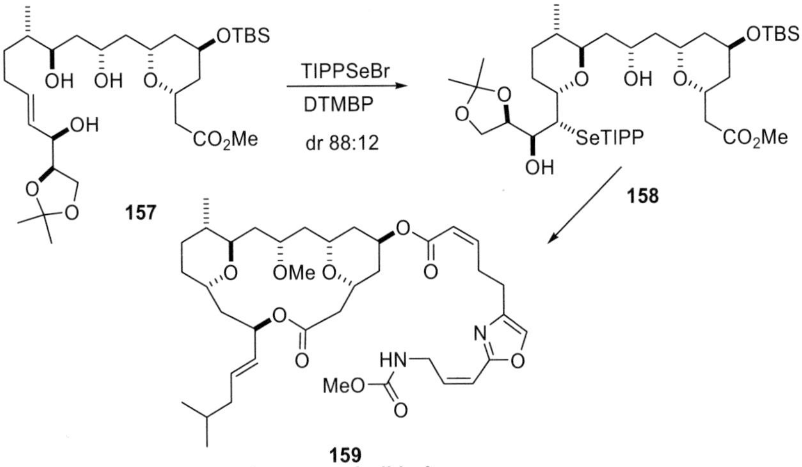

Scheme 2.39

Selenocyclization reactions have been known since 60's, and proceed stereospecifically with good regiochemistry, however, to the mid 90's, asymmetric variant of this reaction has not been elucidated.

The first report on asymmetric selenocyclizations appeared in 1995 [93]. Uemura conducted selenoetherification and selenolactonization using chiral selenyl electrophile, derived from the ferrocene system to give the optically active cyclic ethers and lactones. Products were obtained with moderate diastereoselectivity, except the derivatives of pentenoic acid and *o*-allylphenol, obtained with an excellent diastereomeric excess (Scheme **2.40**, entries a and d). Uemura and coworkers proposed a plausible reaction scheme of the cyclization where the chiral selenylating agent approached the double bond of the substrate from the less sterically congested direction, to give a chiral episelenonium ion. The intramolecular back-side attack of the nucleophile completed the reaction [94].

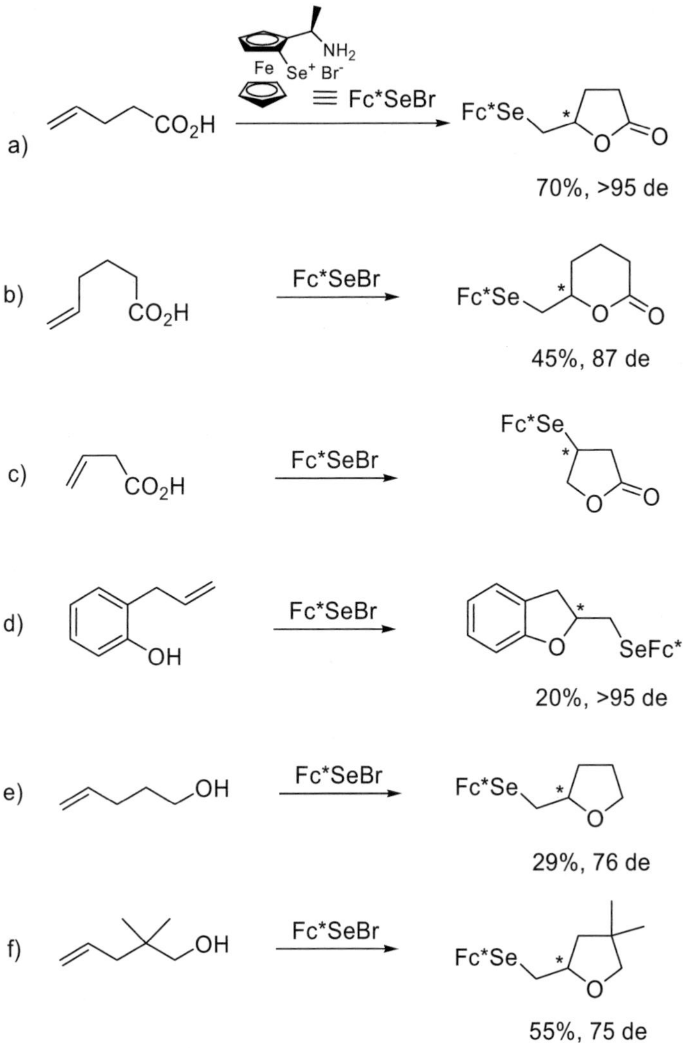

Scheme 2.40

In subsequent years, particular attention was given to chiral aromatic selenenyl electrophiles. Tomoda, Wirth and Tiecco developed a series of aryl selenenyl reagents with a chiral *orto* substituent **37a, 47a, 160** [95-97]. Various unsymmetrical alkenols and alkenoic acids were employed, but generally selenocyclization reaction proceeded efficiently with high selectivities. Selected examples are presented in Table **2.5**.

Table 2.5: Selenocyclization with the use of chiral aryl selenenyl reagents

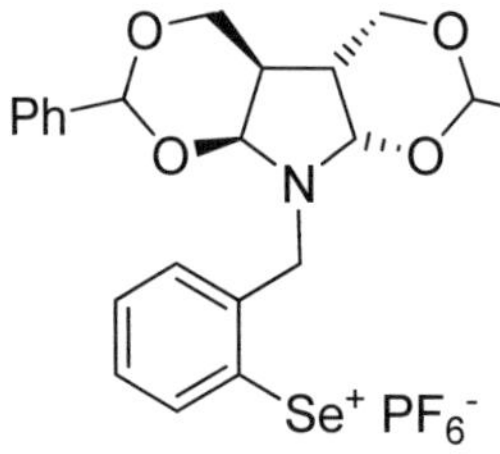 **37a**	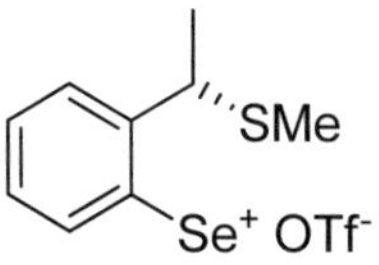 **47a**	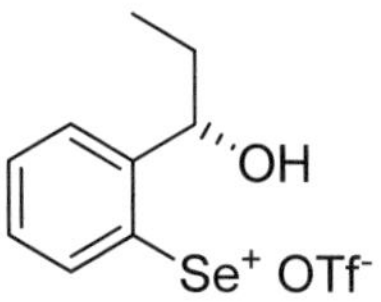 **160**		
Alkene	**Product**	**Yield, % Diastereomeric ratio, *dr***		
		Electrophile 37a	**Electrophile 47a**	**Electrophile 160**
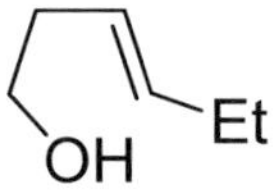	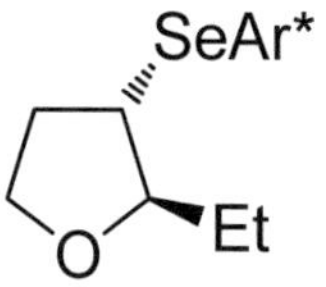	56 80:20	86 >99:1	60 0
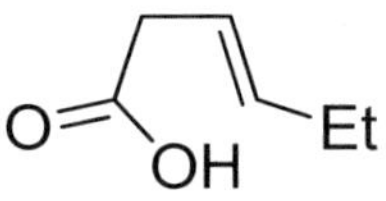	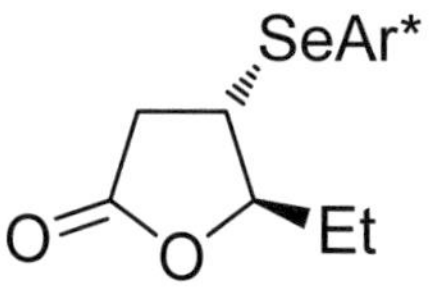	75 89:11	90 >99:1	----- -----
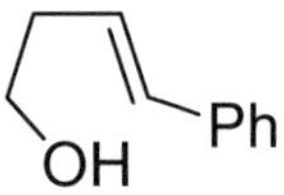	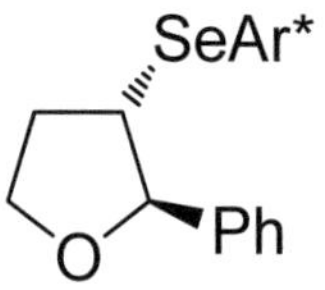	89 93:7	88 97:3	87 92:8
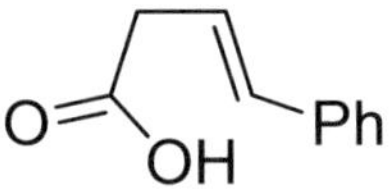	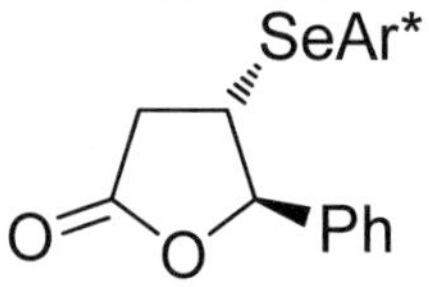	75 95:5	87 96:4	41 86:14

Another approach to the asymmetric selenocyclofunctionalizations was the use of optically active selenenyl reagents **51a**, **52a**, **161**, **162**, based on chiral natural precursors. Back and Scianowski reported an effective cyclization of homoallylic alcohols and unsaturated acids using monoterpenes, such as camphor, *p*-menthane, carane and pinane derivatives as the chirality sources [43e, 98, 99]. Various counterions and additives were studied. In the case of the camphor skeleton, the best results were obtained with the chloride counterion. However, triflate gave better stereoselection then chloride in the menthane system [99]. Selected examples are presented in Table **2.6**.

Table 2.6: Selenocyclization promoted by terpenylselenenyl reagents

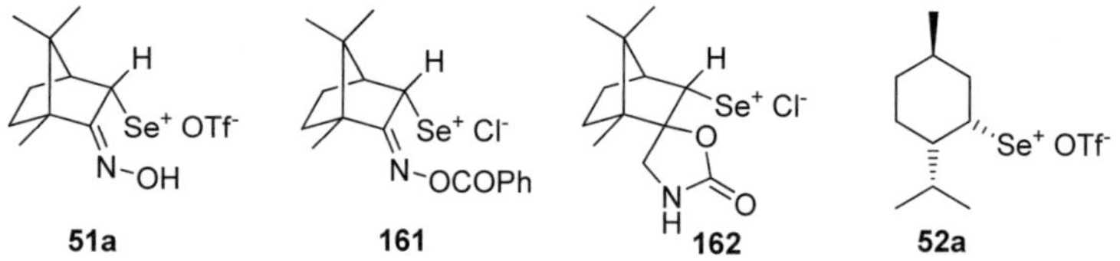

51a **161** **162** **52a**

Alkene	**Product**	**Yield, % Diastereomeric ratio, *dr***			
		Electrophile 51a	**Electrophile 161**	**Electrophile 162**	**Electrophile 52a**
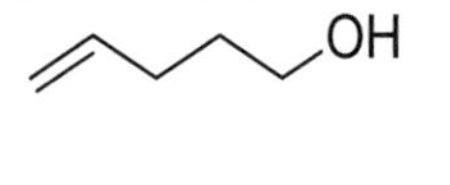	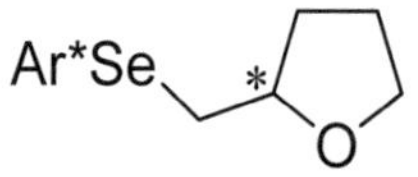	55 60:40	63 67:33	87 84:16	------
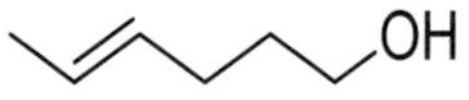	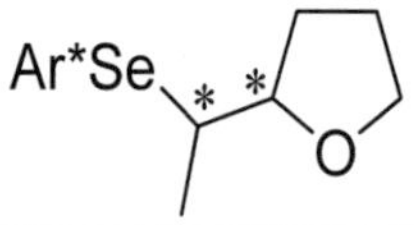	-------	99 71:29	92 85:29	50 56:44
	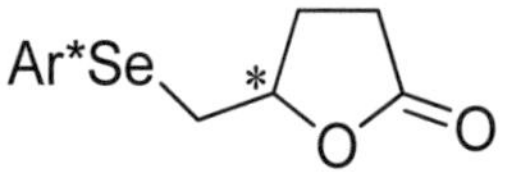	84 58:42	64 56:44	81 >95:5	87 65:35
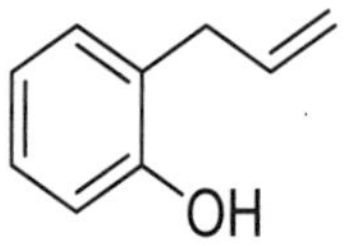	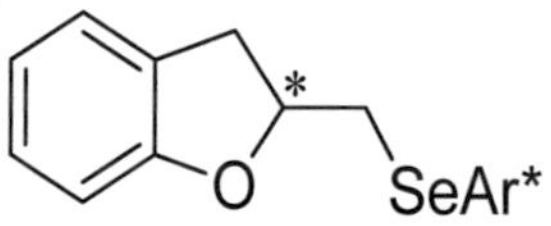	74 53:47	72 66:34	93 92:8	57 70:30

Several research groups have shown that factors, such as the structure of the selenium electrofile, its counterion, solvents, and additives coordinating to selenium species, influence the course of selenocyclization reaction [100]. Wirth and coworkers presented selective cyclization of **163**, bearing a hydroxy and carboxyl groups as a internal nucleophiles [101]. The reaction with phenylselenenyl triflate and 10 equivalents of acetic acid leads exclusively to tetrahydrofurans **164/165**, whereas with phenylselenenyl hexafluorophosphate and 10 equivalents of methanol, as external additive, only lactones **166/167** are formed (Scheme **2.41**).

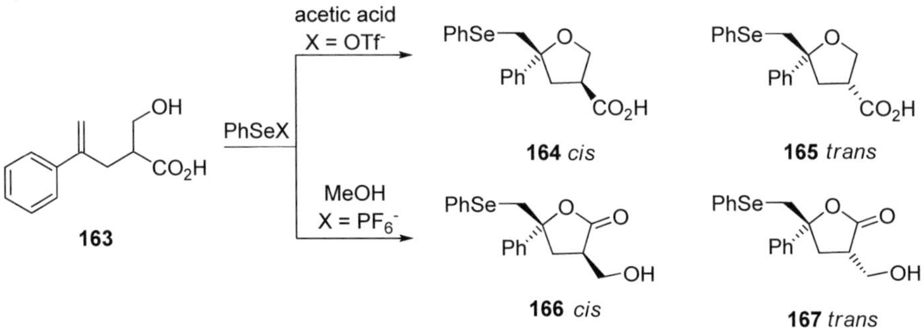

Scheme 2.41

2.4.2. Azaselenocyclization

Selenocyclofunctionalization of olefins using nitrogen nucleophiles is a powerful tool in the construction of nitrogen-containing heterocycles. Azaselenocyclizations have often been used in the total synthesis of natural products. First examples of the selenium promoted cyclization of alkenes containing internal nitrogen nucleophiles were reported by Clive [102] and Danishefsky [103], who showed that the ring closure reaction with a primary amino group does not take place. However, the selenocyclization proceeds easily when an electron-withdrawing group is linked to the nitrogen atom. Various nucleophiles, such as amines, carbamates, oximes and imines have been used as a internal nucleophiles in these reactions [104], and selected examples are shown on Scheme **2.42**.

Tiecco and coworkers showed that *O*-allyl oximes **168** are efficient reagents for the cyclization reactions. The reaction with a phenylselenenyl reagent generats the intermediate iminium ion **170** which can be reduced with sodium borohydride to *N*-alkyl isoxazolidines **171** or hydrolyzed to give the isoxazolidine **172** and starting ketones (Scheme **2.43**) [105].

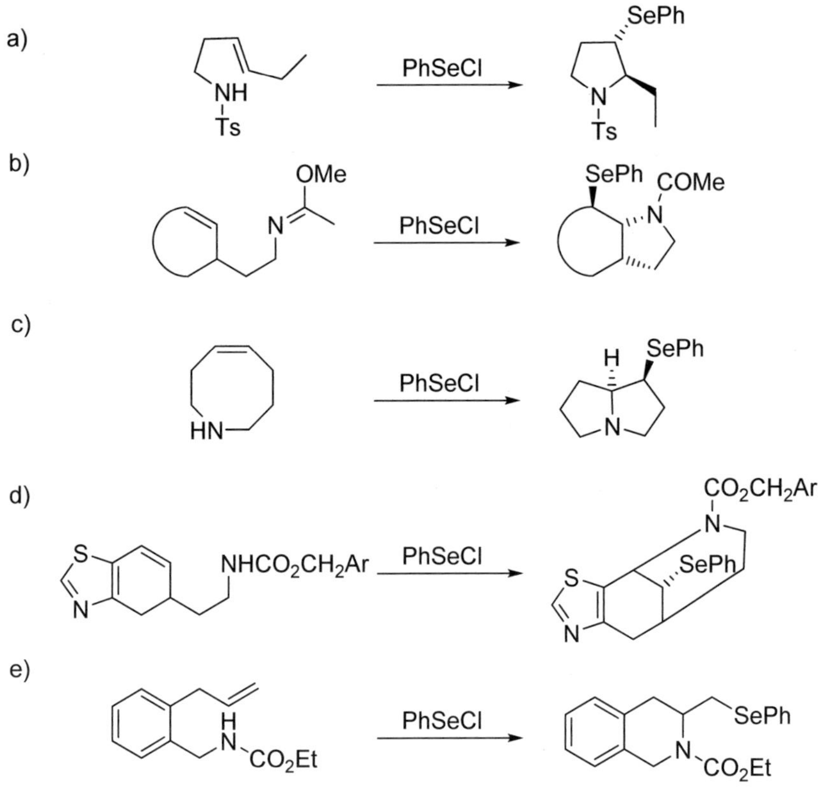

Scheme 2.42

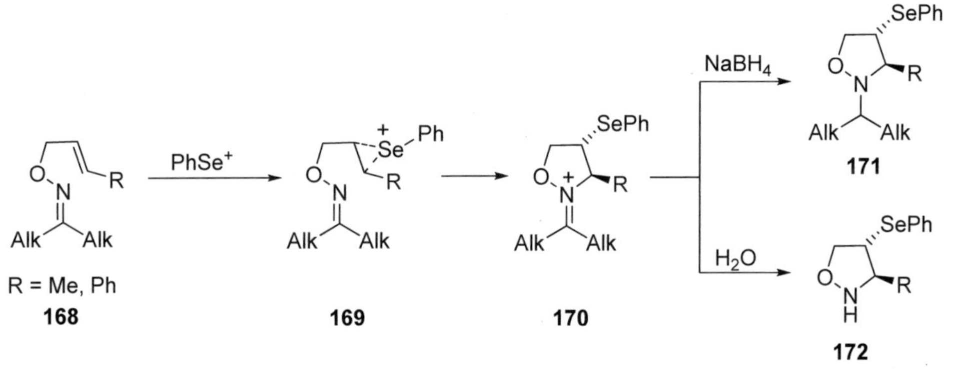

Scheme 2.43

Furthermore, the same group conducted the asymmetric version of these reactions using the chiral selenenylation agent **47a**. Isoxazolidines **173** were obtained in good to excellent yields and diastereoselectivities up to 97:3 (Scheme **2.44**) [106].

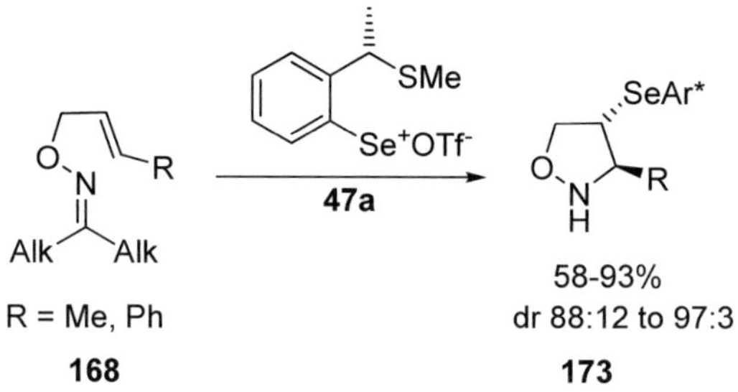

Scheme 2.44

Similarly to oximes, the secondary ketimine groups **174** can also be effective substrates for the aza-selenocyclization reactions. The mechanism is similar to the one previously presented for the isoxazolidines and leads to the formation of the cyclic iminium ion **176** which can be reduced to the corresponding amine **177** by treatment with sodium borohydride (Scheme **2.45**) [107].

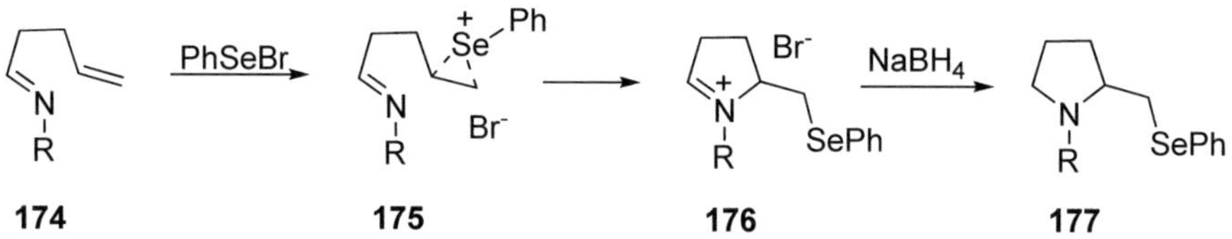

Scheme 2.45

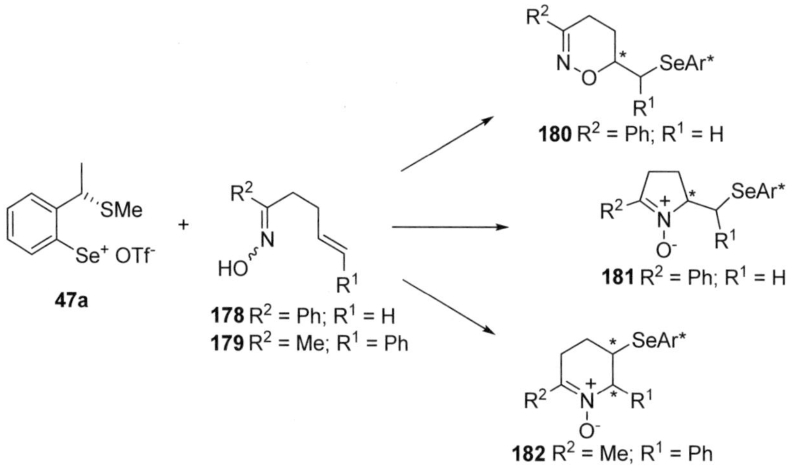

Scheme 2.46

Selenocyclization with the use of *O*-homoallyl oximes was also investigated (Scheme **2.46**) [84]. It was observed that the nature of the reaction products depends on the geometry of the starting oximes **178**, **179**. The *Z* isomers **178** gave 1,2-oxazines **180**, while the more stable *E* isomers gave cyclic nitrones **181**. In the case of oxime **179**, the phenyl substituent controls the regioselectivity of cyclization. Using this reagent, only six-membered nitrones **182** were obtained. These products are observed only in the presence of the phenyl group. The configuration of **179** does not affect the reaction product, and the *anti*-addition occurs always in the benzylic position with high diastereoselectivity, 95:5 *dr*.

More recently, an interesting application of selenium reagents was used in the total synthesis of NW-G01 **185**, its epimer 34-epi-NW-G01 **186**, and other alkaloids, based on the 34-epi-NW-G01 scaffold, such as chloptosin **187** and himastatin **188**, antitumor dimeric cyclohexapeptides. NW-G01 **185** exhibits a potent antibacterial activity against gram-positive bacteria, including methicillin-resistant Staphylococcus aureus (MRSA) representing a cyclic hexapeptide antibiotic (Scheme **2.47**) [108].

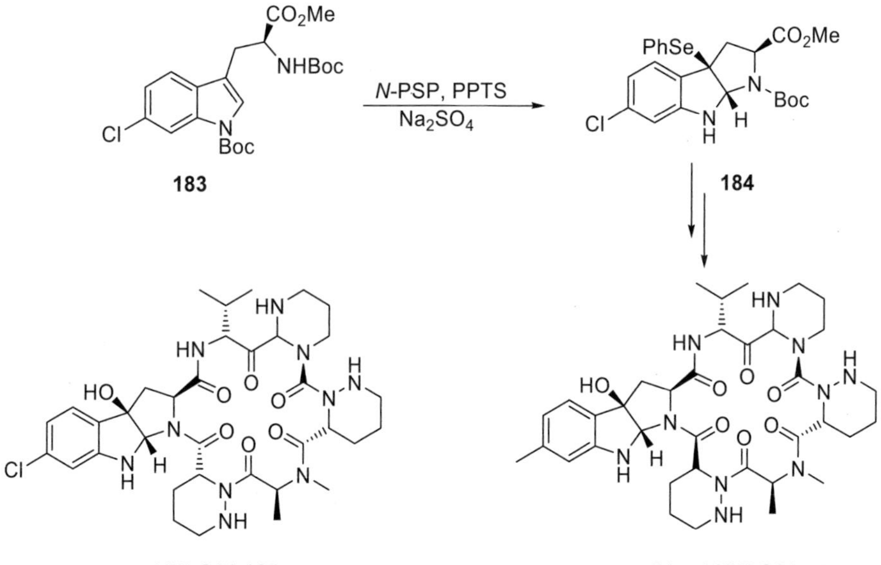

183 **184**

NW-G01 **185** 34-epi-NW-G01 **186**

contd...

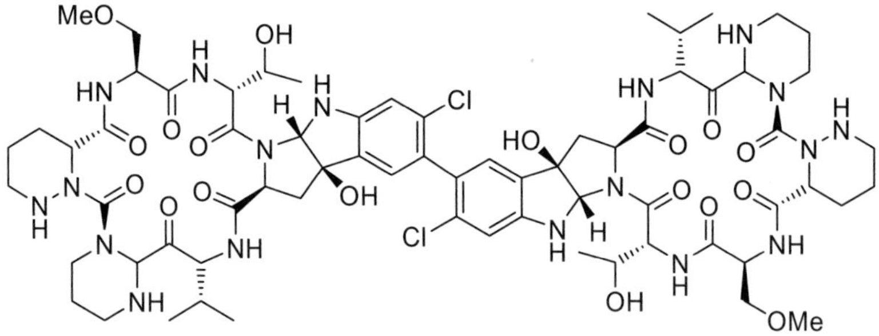

Chloptosin **187**

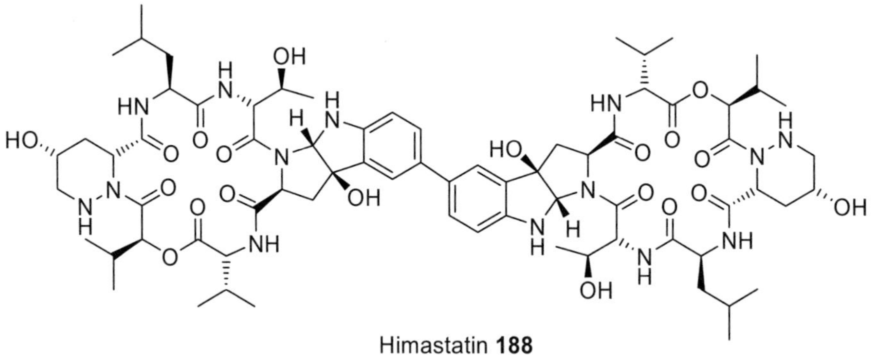

Himastatin **188**

Scheme 2.47

2.4.3. Carboselenocyclization

Selenocyclizations with the use of carbon nucleophile have not been extensively studied. The first report describing this type of reaction appeared in the eighties, when the *β*-dicarbonyl compounds were cyclized to various functionalized products by the use of electrophilic selenenylating reagents [109]. In 1998, Deziel and coworker described the asymmetric addition chiral selenenyl triflate **35a** to **189** in dichloromethane and methanol (Scheme **2.48**). The reaction gave an equimolar mixture of the methoxyselenenylation product **190** and carbocyclization product **191** with 98% *de*. Treatment of the methoxyselenenylation product **190** with triflic acid leads to complete conversion to **191** *via* the formation of seleniranium intermediate in 70% yield and without the loss of *de* [110].

More recently, Wirth developed the synthesis of the biaryl scaffold **193** by the cyclization of **192** with selenium electrophile **35a** in the presence of Lewis acid, and subsequent 1,2-rearrangement of the aryl group (Scheme **2.49**) [111].

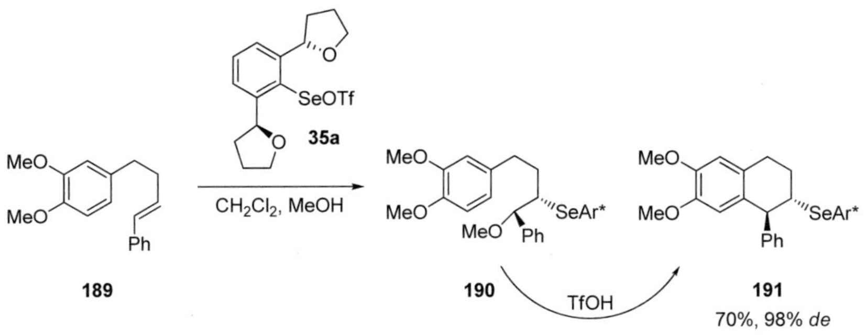

Scheme 2.48

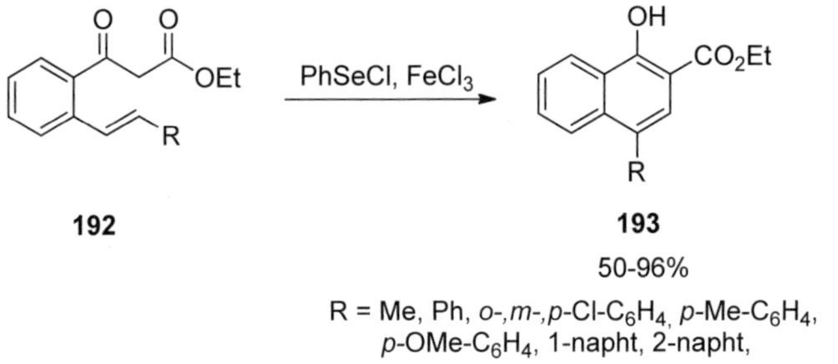

Scheme 2.49

ACKNOWLEDGEMENT

Declared none.

CONFLICT OF INTEREST

The authors confirm that this chapter contents have no conflict of interest.

REFERENCES

[1] (a) *Selenium Reagents and Intermediates in Organic Synthesis*; Paulmier, C. Ed.; Pergamon Press: Oxford, **1986**; (b) *The Chemistry of Organic Selenium and Tellurium Compounds*; Patai, S.; Rappoport Z. Eds.; John Wiley & Sons: New York, Vol. *1*, **1986**; (c) *Organoselenium Chemistry*; Liotta, D. Ed.; John Wiley & Sons: New York, **1987**; (d) *The Chemistry of Organic Selenium and Tellurium Compounds*; Patai, S.; Rappoport, Z. Eds.; John Wiley & Sons: New York, Vol. *2*, **1987**; (e) *Organoselenium Chemistry: A Practical Approach*, Back, T.G. Ed.; Oxford University Press: Oxford, **1999**; (f) Drabowicz, J.; Mikołajczyk, M. In: Organoselenium Chemistry; Wirth, T. Ed.; *Top. Curr. Chem.*, **2000**, Vol. *208*, p.p. 143–176. (g) Zhu, C.; Huang, Y. Asymmetric Synthesis of Chiral Organoselenium Compounds. *Curr. Org. Chem.*, **2006**, *10*, 1905–1920; (h) Braga, A.L.; Lüdtke, D.S.; Vargas, F. Enantioselective Synthesis Mediated by Catalytic Chiral

Organoselenium Compounds. *Curr. Org. Chem.*, **2006**, *10*, 1921–1938. (i) Wirth, T. In: *Comprehensive Organometallic Chemistry III*; Crabtree, R.H.; Mingos, D.M.P. Eds.; Elsevier: Oxford, **2007**, Vol. *9*, p.p. 457–500;

[2] (a) Freundendahl, D. M.; Santoro, S.; Shahzad, S. A.; Santi, C.; Wirth, T. Green Chemistry with Selenium Reagents: Development of Efficient Catalytic Reactions. *Angew. Chem. Int. Ed.*, **2009**, *48*, 8409-8411; (b) Santi, C.; Santoro, S.; Battistelli, B. Organoselenium Compounds as Catalysts in Nature and Laboratory. *Curr. Org. Chem.*, **2010**, *14*, 2442–2462. (c) Santi, C.; Tiecco, M.; Testaferri, L.; Tomassini, C.; Santoro, S.; Bizzoca, G. Diastereo and Enantioselective Synthesis of 1,2-Diols Promoted by Electrophilic Selenium Reagents. *Phosphorus Sulfur.*, **2008**, 183, 956-960. (d) Santoro, S.; Santi, C.; Sabatini, M.; Testaferri, L.; Tiecco, M. Eco-Friendly Olefin Dihydroxylation Catalyzed by Diphenyl Diselenide. *Adv. Synth. Catal.*, **2008**, 350, 2881–2884. (e) Santi, C.; Di Lorenzo, R.; Tidei, C.; Bagnoli, L.; Wirth, T. Stereoselective selenium catalyzed dihydroxylation and hydroxymethoxylation of alkenes. *Tetrahedron*, **2012**, *68*, 10530–10535.

[3] (a) Wirth, T. Chiral Selenium Compounds in Organic Synthesis. *Tetrahedron*, **1999**, *55*, 1–28; (b) Wirth, T. *Angew. Chem. Int. Ed.*, **2000**, *39*, 3740–3749; (c) Braga, A.L.; Ludtke, D.S.; Vargas, F.; Braga, R.C. Catalytic Applications of Chiral Organoselenium Compounds in Asymmetric Synthesis. *Synlett*, **2006**, 1453–1466; (d) Browne, D.M.; Wirth, T. New Developments with Chiral Electrophilic Selenium Reagents. *Curr. Org. Chem.*, **2006**, *10*, 1893–1903; (e) Freudendahl, D.M.; Shahzad, S.A.; Wirth, T. Recent Advances in Organoselenium Chemistry. *Eur. J. Org. Chem.*, **2009**, *11*, 1649–1664.

[4] a) Mugesh, G.; Singh, H.B. Synthetic organoselenium compounds as antioxidants: glutathione peroxidase activity. *Chem. Soc. Rev.*, **2000**, *29*, 347–357; b) Mugesh, G.; du Mont, W.-W.; Sies, H. Chemistry of Biologically Important Synthetic Organoselenium Compounds. *Chem. Rev.*, **2001**, *101*, 2125–2179; c) Wójtowicz, H.; Chojnacka, M.; Młochowski, J.; Palus, J.; Syper, L.; Hudecowa, D.; Uher, M.; Piasecki, E.; Rybka, M. Functionalized alkyl and aryl diselenides as antimicrobial and antiviral agents: synthesis and properties. *Il Farmaco* **2003**, *58*, 1235–1242; d) Meotti, F.C.; Stangherlin, G.Z.; Nogueira, C. W.; Rocha, J.B.T. Protective role of aryl and alkyl diselenides on lipid peroxidation. *Environ. Res.*, **2004**, *94*, 276–282; e) Nogueira, C.W.; Zeni, G.; Rocha, J.B.T. Organoselenium and Organotellurium Compounds: Toxicology and Pharmacology *Chem. Rev.*, **2004**, *104*, 6255-6285; f) Młochowski, J.; Kloc, K.; Lisiak, R.; Potaczek, P.; Wójtowicz, H. Developments in the chemistry of selenaheterocyclic compounds of practical importance in synthesis and medicinal biology. *ARKIVOC*, **2007**, *vi*, 14-46; g) Narajji, C.; Karvekar, M.D.; Das, A.K. Biological importance of organoselenium compounds. *Indian J. Pharm. Sci.*, **2007**, *69*, 344-351; (h) Bhuyan, B.J.; Mugesh G. In: *Organoselenium Chemistry: Synthesis and Reactions*; Wirth, T. Ed.; Wiley-VCH: Weinheim, **2012**; p.p. 361–396.

[5] (a) Singh, F.V.; Wirth, T. *In: PATAI'S Chemistry of Functional Groups: The Chemistry of Organic Selenium and Tellurium Compound;* Rappoport, Z. Ed.; John Wiley & Sons: Chichester, **2012**; Vol. *3,* p.p. 303–355. (b) Santi, C.; Santoro, S. In: *Organoselenium Chemistry: Synthesis and Reactions*; Wirth, T. Ed.; Wiley-VCH, Weinheim, **2012**; p.p. 1–51.

[6] Bowman, W.R. In: *Organoselenium Chemistry: Synthesis and Reactions*; Wirth, T. Ed.; Wiley-VCH: Weinheim, **2012**; p.p. 111–146.

[7] (a) Huguet, J. L. In: *Oxidation of Organic Compounds;* Mayo, F.R. Ed.; *Adv. Chem. Ser.*, **1968**, Vol. *76*, p.p 345–351; (b) Jones, D.N.; Mundy, D.; Whitehouse, R.D. Steroidal Selenoxides Diastereomeric at Selenium; *syn*-Elimination, Absolute Configuration, and Optical Rotatory Dispersion Characteristic. *J. Chem. Soc., Chem. Commun.*, **1970**, 86–87; (c) Walter, R.; Roy, J. Selenomethionine, a Potential Catalytic Antioxidant in Biological Systems. *J. Org. Chem.*, **1971**, *36*, 2561–2563.

[8] Drabowicz, J.; Lewkowski, J.; Ścianowski J. In: *Organoselenium Chemistry: Synthesis and Reactions*; Wirth, T. Ed.; Wiley-VCH: Weinheim, **2012**; p.p. 191–256.

[9] Tiecco, M.; Testaferri, L.; Bagnoli, L.; Scarponi, G.; Temperini, A.; Marini, F.; Santi, C. Selenium promoted synthesis of enantiopure pyrrolidines starting from chiral aminoalcohols. *Tetrahedron: Asymmetry*, **2007**, *18*, 2758–2767.

[10] Marini, F.; Sternativo, S.; Del Verme, F.; Testaferri, L.; Tiecco, M. A New Stereoselective Synthesis of Cyclopropanes Containing Quaternary Stereocenters *via* Organocatalytic Michael Addition to Vinyl Selenones. *Adv. Synth. Catal.*, **2009**, *351*, 103–106.

[11] Ward, V.R.; Cooper, M.A.; Ward, A.D. Synthesis of *N*-acylaziridines from β-amido selenides. *J. Chem. Soc., Perkin Trans 1*, **2001**, 944–945.

[12] Tiecco, M.; Testaferri, L.; Temperini, A.; Terlizzi, R.; Bagnoli, L.; Marini, F.; Santi, C. Stereocontrolled synthesis of substituted *N*-arenesulfonyl azetidines from γ-(phenylseleno)alkyl arylsulfonamides. *Org. Biomol. Chem.*, **2007**, *5*, 3510–3519.

[13] Tiecco, M.; Testaferri, L.; Temperini, A.; Bagnoli, L.; Marini, F.; Santi, C. Ring Closure Reactions through Intramolecular Displacement of the Phenylselenonyl Group by Nitrogen Nucleophiles: A New Stereospecific Synthesis of *N*-Tosyl and *N*-Benzoyl-1,3-oxazolidin-2-ones from β-Hydroxyalkyl Phenyl Selenides. *Chem. Eur. J.*, **2004**, *10*, 1752–1764.

[14] Sheng, S.-R.; Luo, H.-R.; Huang, Z.-Z.; Sun, W.-K.; Liu, X-L. Facile One-Pot Synthesis of Oxazolidin-2-ones from Phenyl 2-Hydroxyalkyl Selenides. *Synth. Commun.*, **2007**, *37*, 2693–2699.

[15] Wirth, T.; Fragale, G. Stereoselective Isoquinoline Alkaloid Synthesis with New diselenides. *Synthesis*, **1998**, 162–166.

[16] Marc, M.; Outurquin, F.; Renard, P-Y.; Créminon, C.; Franck, X. Synthesis of a (+)-anatoxin-a analogue for monoclonal antibodies production. *Tetrahedron Lett.*, **2009**, *50*, 4554–4557.

[17] Wirth, T.; Kulicke, K.J.; Fragale, G. Chiral Diselenides in the Total Synthesis of (+)-Samin. *J. Org. Chem.*, **1996**, *61*, 2686–2689.

[18] Raboisson, P,; Maruga´n, J.J.; Schubert, C.; Koblish, H.K.; Lu, T.; Zhao, S.; Player, M.R.; A. Maroney, C.; Reed, R.L.; Huebert, N.D.; Lattanze, J.; Parks, D.J.; Cummings, M.D. Structure-based design, synthesis, and biological evaluation of novel 1,4-diazepines as HDM2 antagonists. *Bioorg. Med. Chem. Lett.*, **2005**, *15*, 1857–1861.

[19] Raucher, S. The regioselective synthesis of vinyl phenylselenides. *J. Org. Chem.*, **1977**, *42*, 2950–2951.

[20] (a) Murata, S.; Suzuki, T. Stereoselective Selenolactonization by Superelectrophilic Benzeneselenenyl Triflate. *Chem. Lett.*, **1987**, 849–852; (b) Murata, S.; Suzuki, T. Cyclization of Olefinic Alcohol by Benzeneselenenyl triflate. *Tetrahedron Lett.*, **1987**, *28*, 4297–4298; (c) Murata, S.; Suzuki, T. Stereoselective formation of Tetrahydrofuran and –pyran by Benzeneselenenyl triflate. *Tetrahedron Lett.*, **1987**, *28*, 4415–4416.

[21] Jackson, W.P.; Ley, S.V.; Whittle, A.J. Selenium-mediated Cyclization Reactions of Alkenyl-substituted *β*-ketoesters. *J. Chem. Soc., Chem. Commun.*, **1980**, 1173–1174.

[22] Back, T.G.; Muralidharan, K.R. Formation and electrophilic reactions of benzeneselenenyl p-toluenesulfonate. Preparation and properties of addition products with acetylenes. *J. Org. Chem.*, **1991**, *56*, 2781–2787.

[23] Tingoli, M.; Diana, M.; Panunzi, B. *N*-Phenylselenosaccharin (NPSSac): a new electrophilic selenium-containing reagent. *Tetrahedron Lett.*, **2006**, *47*, 7529–7531.

[24] Nicolaou, K.C.; Petatis, N.A.; Claremon, D.A. N-Phenylselenophtalimide (NPSP) A Valuable Selenenylating Agent. *Tetrahedron*, **1985**, *41*, 4835–4841;

[25] Nicolaou, K.C.; Claremon, D.A.; Barnette, W.E.; Seitz, S.P. N-Phenylselenophthalimide (N-PSP) and N-Phenylselenosuccinimide (N-PSS). Two Versatile Carriers of the Phenylseleno Croup. Oxyselenation of Olefins and a Selenium-Based Macrolide Synthesis. *J. Am. Chem. Soc.*, **1979**, *101*, 3704–3706.

[26] Freudendahl, D.M.; Wirth, T. In: *Selenium and Tellurium Chemistry*; Woollins, J.D.; Laitinen R.S. Eds.; Springer-Verlag: Berlin Heidelberg; **2011**; p.p. 41–55.

[27] Tiecco, M.; Testaferri, L.; Tingoli, M.; Bartoli, D. The reaction of diphenyl diselenide with peroxydisulphate ions in methanol a convenient procedure to effect the methoxyselenenylation of alkenes. *Tetrahedron Lett.*, **1989**, *30*, 1417–1420.

[28] Tiecco, M.; Testaferri, L.; Tingoli, M.; Chianelli, D.; Bartoli, D. Selenium catalyzed conversion of vinyl halides into α-alkoxy acetals. *Tetrahedron*, **1988**, *44*, 2273–2282.

[29] Tiecco, M.; Testaferri, L.; Tingoli, M.; Bagnoli, L.; Marini, F. Ring-closure reactions of alkenyl oximes induced by persulfate anion oxidation of diphenyl diselenide. Formation of 1,2-oxazines and cyclic nitrones. *J. Chem. Soc., Perkin Trans. 1* **1993**, 1989–1993.

[30] Bosman, C.; D'Annibale, A.; Resta, S.; Trogolo, C. Oxidation of diphenyl diselenide with ceric ammonium nitrate: a Novel route for functionalization of olefins. *Tetrahedron Lett.*, **1994**, *35*, 6525–6528.

[31] Lee, D.H.; Kim, Y.H. Regioselective Phenylselenenylation at the 5-Position of Pirymidine Nucleosides Mediated by Manganese(III) Acetate. *Synlett*, **1995**, 349–350.

[32] Roh, K.R.; Chang, H.K.; Kim, Y.H. Efficient Phenylsulfenylation and Phenylselenenylation at the 5-Position of Uracil Nucleosides with Disulfide and Diselenide Mediated by [Bis(trifluoroacetoxy)-iodo]benzene. *Heterocycles*, **1998**, *48*, 437–441.

[33] Tingoli, M.; Tiecco, M.; Testaferri, L.; Temperini, A. Iodosobenzene Diacetate and Diphenyl Diselenide: An Electrophilic Selenenylating Agent of Double Bonds. *Synth. Commun.*, **1998**, *28*, 1769–1778.

[34] Tiecco, M.; Testaferri, L.; Temperini, A.; Bagnoli, L.; Marini, F.; Santi, C. Oxidation of Diphenyl Diselenide with 2,3-Dichloro-5,6-dicyanobenzoquinone (DDQ). A New Method for the Electrophilic Phenylselenenylation of Alkenes under Mild Conditions. *Synlett*, **2001**, 1767–1771.

[35] Niyomura, O.; Cox, M.; Wirth, T. Electrochemical Generation and Catalytic Use of Selenium Electrophiles. *Synlett*, **2006**, 251–254.

[36] a) Pandey, G.; Rao, V.J.; Bhalerao, U.T. Formation of Electrophilic Selenium Species ($PhSe^+$) by Photo-oxidative (Single-electron Transfer) Cleavage of Diphenyl Diselenide. *J. Chem. Soc., Chem.Commun.*, **1989**, 416–417; b) Pandey, G.; Sekhar, B.B.V.S. Enyne Cyclization *via* Photoinduced Electron Transfer (PET) Generated Electrophilic Selenium Species: a New Carbon-Carbon Bond Formation Strategy. *J. Chem. Soc.,Chem. Commun.*, **1993**, 780–782.

[37] (a) Tomoda, S.; Iwaoka, M. Asymmetric *Trans*-addition Reactions of a Chiral Selenobinaphthyl with Prochiral Olefins. The Case of Methoxyselenenylation. *Chem. Lett.*,

1988, 1895–1898; (b) Tomoda, S.; Iwaoka, M.; Yakushi, K.; Kawamoto, A.; Tanaka, J. Synthesis of optically active 2,2′-diselenocyanato-1,1′-binaphthyl and its molecular structure. Hypervalent nature of divalent selenium atom in crystal state. *J. Phys. Org. Chem.*, **1988**, *1*, 179–184.

[38] (a) Déziel, R.; Goulet, S.; Grenier, L.; Bordeleau, J.; Bernier, J. Asymmetric Selenomethoxylation of Olefins Involving a Chiral C_2 Symmetrical Electrophilic Organoselenium Reagent. *J. Org. Chem.*, **1993**, *58*, 3619–3621; (b) Déziel, R.; Malenfant, E. Asymmetric Ring Closure Reactions Mediated by a Chiral C_2 Symmetrical Organoselenium Reagent. *J. Org. Chem.*, **1995**, *60*, 4660–4662; (c) Déziel, R.; Malenfant, E.; Bélanger, G. Practical Synthesis of (*R,R*)- and (*S,S*)-Bis[2,6-bis(1-ethoxyethyl)phenyl]Diselenide. *J. Org. Chem.*, **1996**, *61*, 1875–1876; (d) Déziel, R.; Malenfant, E.; Thibault, C.; Fréchette, S.; Gravel, M. 2,6-Bis[(2S)-tetrahydrofuran-2-yl]phenyl Diselenide: An Effective Reagent for Asymmetric Electrophilic Addition Reactions to Olefins. *Tetrahedron Lett.*, **1997**, *38*, 4753–4756.

[39] (a) Nishibayashi, Y.; Singh, J.D.; Uemura, S.; Fukuzawa, S. Synthesis of Chiral Diferrocenyl Diselenides and Their Application to Asymmetric Reactions. *Tetrahedron Lett.*, **1994**, *35*, 3115–3118; (b) Nishibayashi, Y.; Singh, J.D.; Fukuzawa, S.; Uemura, S. Synthesis of [R,S;R,S]- and [S,R;S,R]-Bis[2-[1-(dimethylamino)ethyl] ferrocenyl] Diselenides and Their Application to Asymmetric Selenoxide Elimination and [2,3]Sigmatropic Rearrangement. *J. Org. Chem.*, **1995**, *60*, 4114–4120; (c) Fukuzawa, S.; Takahashi, K.; Kato, H.; Yamazaki, H. Asymmetric Methoxyselenenylation of Alkenes with Chiral Ferrocenylselenium Reagents. *J. Org. Chem.*, **1997**, *62*, 7711–7716.

[40] (a) Tomoda, S.; Iwaoka, M. Synthesis of Selenium-containing Binaphtyls and their Aplication to the Asymmetric Ring-opening of Cyclohexene Oxide. *J. Chem. Soc., Chem. Commun*,. **1988**, 1283–1284; (b) Tomoda, S.; Fujita, K.; Iwaoka, M. Double Differentiation in Asymmetric Methoxyselenenylation of *trans-β*-Methylstyrene. *J. Chem. Soc., Chem. Commun.*, **1990**, 129–131; (c) Tomoda, S.; Fujita, K.; Iwaoka, M. Asymmetric Oxyselenenylation of Olefins Using Optically Active Selenobinaphthyls and *d*-Menthol as a Nucleophile. *Chem. Lett.*, **1992**, 1123–1124; (d) Tomoda, S.; Fujita, K.; Iwaoka, M. Asymmetric *Trans*-Addition Reactions Using Chiral Selenobinaphthyls. *Phosphorus Sulfur*, **1992**, *67*, 247–252; (e) Fujita, K.; Iwaoka, M.; Tomoda, S. Synthesis of Diaryl Diselenides Having Chiral Pyrrolidine Rings with C_2 Symmetry. Their Application to the Asymmetric Methoxyselenenylation of *trans-β*-Methylstyrenes. *Chem. Lett.*, **1994**, 923–926; (f) Fujita, K.; Murata, K.; Iwaoka M.; Tomoda, S. Asymmetric Methoxyselenenylation of Olefins Usin an Optically Active Diaryl Diselenide Derived from D-Mannitol. *Tetrahedron:Lett.*, **1995**, *36*, 5219–5222. (g) Fujita, K.; Murata, K.; Iwaoka M.; Tomoda, S. Design of Optically Active Selenium Reagents Having a Chiral Tetriary Amino Group and Their Applications to Asymmetric Inter- and Intramolecular Oxyselenenylations. *Tetrahedron*, **1997**, *53*, 2029–2048.

[41] (a) Wirth, T. Asymmetric Reaction of Arylalkenes with Diselenides. *Angew. Chem. Int. Ed.*, **1995**, *34*, 1726–1728; (b) Wirth, T. Enantioselective Alkylation of Aldehydes Catalyzed by New Chiral Diselenides. *Tetrahedron Lett.*, **1995**, *36*, 7849–7852; (c) Wirth, T.; Fragale, G. Asymmetric Addition Reactions with Optimized Selenium Electrophiles *Chem. Eur. J.*, **1997**, *3*, 1894–1902; (d) Wirth, T. Chiral Selenium Compounds: Versatile Reagents in Organic Synthesis. *Liebigs Ann./Recueil*, **1997**, *11*, 2189–2196 ; (e) Fragale, G.; Neuburger, M.; Wirth, T. New and efficient selenium reagents for stereoselective

selenenylation reactions. *Chem. Commun.*, **1998**, 1867–1868; (f) Santi, C.; Fragale, G.; Wirth, T. Synthesis of new chiral nitrogen containing diselenide as a precursor for selenium electrophiles. *Tetrahedron: Asymmetry*, **1998**, *9*, 3625–3628. (g) Santi, C.; Wirth, T. Synthesis of Non Racemic Nitrogen Containing Diselenides as Efficient Precursor Catalysts in the Diethylzinc Addition to Benzaldehyde. *Tetrahedron: Asymmetry* **1999**, *10*, 1019–1023; (h) Uehlin, L.; Fragale, G.; Wirth, T. New and Efficient Chiral Selenium Electrophiles. *Chem. Eur. J.*, **2002**, *8*, 1125–1133; (i) Cox, M.; Wirth, T. Synthesis of a Selenium-Substituted Diselenides. *Phosphorus Sulfur*, **2005**, *180*, 659–666. (j) Freudendahl, D.M.; Iwaoka, M.; Wirth, T. Synthesis of New Sulfoxide-Containing Diselenides and Unexpected Cyclization Reactions to 2,3-Dihydro-1,4-benzoselenothiine 1-Oxides. *Eur. J. Org. Chem.*, **2010**, *20*, 3934–3944 (k) Zhao, L.; Li, Z.; Wirth, T. Preparation of Novel Chiral Non-Racemic Diselenides and Applications in Asymmetric Synthesis. *Eur. J. Org. Chem.*, **2011**, 176–182; (l) Zhao, L.; Li Z.; Wirth, T. Asymmetric Methoxyselenenylations with Chiral Selenium Electrophiles. *Eur. J. Org. Chem.* **2011**, 7080–7082.

[42] (a) Tiecco, M.; Testaferri, L.; Bagnoli, L.; Marini, F.; Temperini, A.; Tomassini, C.; Santi, C. Efficient asymmetric selenomethoxylation and selenohydroxylation of alkenes with a new sulfur containing chiral diselenide. *Tetrahedron Lett.*, **2000**, *41*, 3241–3245; (b) Tiecco, M.; Testaferri, L.; Santi, C.; Tomassini, C.; Bagnoli, L.; Marini, F.; Temperini, A. New nitrogen containing chiral diselenides: synthesis and asymmetric addition reactions to olefins. *Tetrahedron: Asymmetry*, **2000**, *11*, 4645–4650; (c) Tiecco, M.; Testaferri, L.; Santi, C.; Tomassini, C.; Marini, F.; Bagnoli, L.; Temperini, A. Preparation of a New Chiral Non-Racemic Sulfur-Containing Diselenide and Applications in Asymmetric Synthesis. *Chem. Eur. J.*, **2002**, *8*, 1118–1124.

[43] (a) Back, T.G.; Dyck, B.P.; Parvez, M. Unexpected Formation of 1,3-Diselenetanes from the Reaction of Camphor Enolate with Selenium. *J. Chem. Soc., Chem. Commun.*, **1994**, 515–516; (b) Back, T.G.; Dyck, B.P.; Parvez, M. 1,3-Diselenetanes and 1,3-Dithietanes Derived from Camphor. Formation, Structure, Stereochemistry, and Oxidation to Selenoxide and Sulfoxide Products. *J. Org. Chem.*, **1995**, *60*, 703–710; (c) Back, T.G.; Dyck, B.P. Asymmetric cyclization of unsaturated alcohols and carboxylic acids with camphor-based selenium electrophiles. *J. Chem. Soc., Chem. Commun.*, **1996**, 2567–2568; (d) Back, T.G.; Nan, S. Asymmetric methoxyselenenylations with camphor-based selenium electrophiles. *J. Chem. Soc., Perkin Trans 1,* **1998**, 3123–3138; (e) Back, T.G.; Dyck, B.P.; Nan, S. Asymmetric Electrophllic Methoxyselenenylations and Cyclizations with 3-Camphorseleno Derivatives. *Tetrahedron*, **1999**, *55*, 3191–3194; (f) Back, T.G.; Moussa, Z. New Chiral Auxiliaries for Highly Stereoselective Asymmetric Methoxyselenenylations. *Organic Lett.*, **2000**, *19*, 3007–3009. (g) Back, T.G.; Moussa, Z.; Parvez, M. Asymmetric Methoxyselenenylations and Cyclizations with 3-Camphorseleno Electrophiles Containing Oxime Substituents at C-2. Formation of an Unusual Oxaselenazole from an Oxime–Substituted Selenenyl Bromide. *J. Org. Chem.*, **2002**, *67,* 499–509.

[44] (a) Ścianowski, J. Convenient route to dialkyl diselenides from alkyl tosylates. Synthesis of di(*cis*-myrtanyl) diselenide. *Tetrahedron Lett.*, **2005**, *46*, 3331–3334; (b) Ścianowski, J.; Rafiński, Z.; Wojtczak, A. Syntheses and Reactions of New Optically Active Terpene Dialkyl Diselenides. *Eur. J. Org. Chem.*, **2006**, 3216–3225; (c) Rafiński, Z.; Ścianowski, J.; Wojtczak, A. Asymmetric selenocyclization with the use of dialkyl monoterpene diselenides. *Tetrahedron: Asymmetry*, **2008**, *19*, 223–230; (d) Rafiński, Z.; Ścianowski, J.

Synthesis and reactions of enantiomerically pure dialkyl diselenides from *p*-menthane group. *Tetrahedron:Asymmetry*, **2008**, *19*, 1237–1244; (e) Rafiński, Z.; Ścianowski, J.; Wojtczak, A. Synthesis and Reactions of the Optically Active Dialkyl Diselenides from the Pinane Group. *Lett. Org. Chem.*, **2009**, *6*, 321–328; (f) Ścianowski, J.; Rafiński, Z.; Szuniewicz, A.; Wojtczak, A. New chiral selenium electrophiles derived from functionalized terpenes. *Tetrahedron*, **2009**, *65*, 10162–10174; (g) Ścianowski, J.; Rafiński, Z.; Wojtczak, A.; Burczyński, K. Syntheses and reactions of terpene β-hydroxyselenides and β-hydroxydiselenides. *Tetrahedron:Asymmetry*, **2009**, *20*, 2871–2879.

[45] Uehlin, L.; Wirth, T. Novel Polymer-Bound Chiral Selenium Electrophiles. *Org. Lett.*, **2001**, *3*, 2931–2933.

[46] (a) Wirth, T.; Fragale G.; Spichty, M. Mechanistic Course of the Asymmetric Methoxyselenenylation Reaction. *J. Am. Chem. Soc.*, **1998**, *120*, 3376–3381; (b) Wang, X.; Houk, K.N.; Spichty, M.; Wirth, T. Origin of Stereoselectivities in Asymmetric Alkoxyselenenylations. *J. Am. Chem. Soc.*, **1999**, *121*, 8567–8576; (c) Fragale, M.; Spichty, G.; Wirth, T. Theoretical Investigations on the Stereoselective Selenenylation Reaction of Alkenes. *J. Am. Chem. Soc.*, **2000**, *122*, 10914–10916. (d) Denmark, S.E.; Collins, W.R.; Cullen, M.D. Observation of Direct Sulfenium and Selenenium Group Transfer from Thiiranium and Seleniranium Ions to Alkenes. *J. Am. Chem. Soc.*, **2009**, *131*, 3490–3492.

[47] Engman, L. Acetoxyselenenylation of Olefins for the Preparation of Vinylic and Allylic Acetates. *J. Org. Chem.*, **1989**, *54*, 884–889.

[48] Beaulieu, P.L.; Deziel, R. In: *Organoselenium Chemistry: A Practical Approach*, Back, T.G., Ed.; Oxford University Press: Oxford, **1999**, p.p. 35–66.

[49] Mak, S.Y.F.; Curtis, N.R.; Payne, A.N.; Congreve, M.S.; Wildsmith, A.J.; Francis, C.L.; Davies, J.E.; Pascu, S.I.; Burton, J.W.; Holmes A.B. Synthesis of (+)-Obtusenyne. *Chem. Eur. J.*, **2008**, *14*, 2867–2885.

[50] Siwicka, A.; Cuperly, D.; Tedeschi, L.; Le Vezouet, R.; Whitea A. J.P.; Barretta, A.G.M. Total synthesis of the tricyclic skeleton of the natural Celastraceae sesquiterpenoids and related synthetic analogs. *Tetrahedron*, **2007**, *63*, 5903–5917.

[51] Ścianowski, J.; Wełniak, M. Syntheses of the Optically Active Terpene Hydroxyphenylselenides. *Phosphorus, Sulfur,* **2009**, *184*, 1440-1447.

[52] Pedrosa, R.; Andres, C.; Arias, R.; Mendiguchia, P.; Nieto, J. Regio- and Stereoselective Methoxyselenenylation of Chiral 2-Vinyl Perhydro-1,3-benzoxazines Promoted by Selenium-Heteroatom Nonbonded Interactions. *J. Org. Chem.*, **2006**, *71*, 2424-2428.

[53] Pedrosa, R.; Andres, C.; Arias, R.; Mendiguchia, P.; Nieto, J. 1,4-Asymmetric Induction in Methoxyselenenylation of Double Bonds at the Nitrogen Side of Chiral Perhydro-1,3-benzoxazines Promoted by Nonbonded Se-N Interactions. *J. Org. Chem.*, **2006**, *71*, 5388-5391.

[54] (a) Mukherjee, A.J.; Zade, S.S.; Singh, H.B.; Sunoj, R.B. Organoselenium Chemistry: Role of Intramolecular Interactions. *Chem. Rev.*, **2010**, *110*, 4357–4416. (b) Coles, M.P. Utilization of Nonbonded Interactions Involving Organoselenium Compounds. *Curr. Org. Chem.*, **2006**, *10*, 1993–2005;

[55] Tiecco, M.; Testaferri, L.; Marini, F.; Sternativo, S.; Santi, C.; Bagnoli, L.; Temperini, A. Synthesis of enantiomerically pure 1,4-dioxanes from alkenes promoted by organoselenium reagents. *Tetrahedron:Asymmetry*, **2003**, *14*, 1095–1102.

[56] (a) Kim, K.S.; Park, J.I.; Moon, H.K.; Yi, H. Transformation of cyclohexene to enantiopure cyclitols mediated by sequential oxyselenenylation with (*S*,*S*)-hydrobenzoin: synthesis of D-*chiro*-inositol and *muco*-quercitol. *Chem. Commun.*, **1998**, 1945–1946; (b) Kim, K.S.; Park, J.I.; Ding, P. Synthesis of enantiopure cyclopentitols and aminocyclopentitols mediated by oxyselenenylation of cyclopentene with (*R*,*R*)-hydrobenzoin. *Tetrahedron Lett.*, **1998**, *39*, 6471–6474.

[57] Kim, S.K.; Moon, C.W.; Park, J.I.; Han, S.H. Enantiopure synthesis of carbohydrates mediated by oxyselenenylation of 3,4-dihydro-2*H*-pyran. *J. Chem. Soc., Perkin Trans. 1*, **2000**, 1341–1343.

[58] Tiecco, M.; Testaferri, L.; Marini, F.; Sternativo, S.; Santi, C.; Bagnoli, L.; Temperini, A. Selenium-promoted synthesis of enantiomerically pure substituted morpholines starting from alkenes and chiral aminoalcohols. *Tetrahedron:Asymmetry*, **2003**, *14*, 2651–2657.

[59] Tiecco, M.; Testaferri, L.; Santi, C.; Marini, F.; Bagnoli, L.; Temperini, A. Asymmetric Selenometoxylation of Alkenes with Camphorselenenyl Sulfate. *Tetrahedron Lett.*, **1998**, *39*, 2809–2812; (b) Khokhar, S.S.; Wirth, T. Selenocyclizations: Control by Coordination and by the Counterion. *Angew. Chem. Int. Ed.*, **2004**, *43*, 631–633.

[60] Tiecco, M.; Testaferri, L.; Santi, C.; Tomassini, C.; Santoro, S.; Marini, F.; Bagnoli, L.; Temperini, A.; Costantino F. Intramolecular Nonbonding Interactions between Selenium and Sulfur – Spectroscopic Evidence and Importance in Asymmetric Synthesis. *Eur. J. Org. Chem.*, **2006**, 4867–4873.

[61] Santi, C.; Santoro, S.; Tomassini, C.; Pascolini, F.; Testaferri, L.; Tiecco, M. Enantioselective Methoxyselenenylation of α,β–Unsatureted Aldehydes. *Synlett*, **2009**, 743–746.

[62] Tiecco, M.; Testaferri, L.; Santi, C.; Marini, F.; Bagnoli, L.; Temperini, A.; Tomassini C. Asymmetric Selenohydroxylation of Alkenes with Camphorselenenyl Sulfate. *Eur. J. Org. Chem.*, **1998**, 2275–2277.

[63] Wirth, T.; Häuptli, S.; Leuenberger, M. Catalytic asymmetric oxyselenenylation–elimination reactions using chiral selenium compounds. *Tetrahedron: Asymmetry*, **1998**, *9*, 547–550.

[64] Tiecco, M.; Testaferri, L.; Marini, F.; Santi, C.; Bagnoli L.; Temperini, A. Asymmetric oxyselenenylation–deselenenylation reactions of alkenes induced by camphor diselenide and ammonium persulfate. A convenient one-pot synthesis of enantiomerically enriched allylic alcohols and ethers. *Tetrahedron:*Asymmetry, **1999**, *10*, 747–757.

[65] Toshimitsu, A.; Aoai, T.; Uemura, S.; Okano, M. Simple Procedure for the Aminoselenation of Olefins. *J. Chem. Soc., Chem.Commun.*, **1980**, 2567–2568.

[66] Tiecco, M.; Testaferri, L.; Tingoli, M. Bartoli, D. Phenylseleno-Lactonization of Olefinic Nitriles Promoted by Peroxydisulphate Ion Oxidation of Diphenyl Diselenide. *Tetrahedron*, **1989**, *65*, 6819–6832.

[67] Tiecco, M.; Testaferri, L. Santi, C.; Tomassini, C.; Marini, F.; Bagnoli, L. Temperini, A. Asymmetric Amidoselenenylation of Alkenes Promoted by Camphorselenenyl Sulfate: A Useful Synthetic Route to Enantiopure Oxazolines. *Eur. J. Org. Chem.*, **2000**, 3451–3457.

[68] Tiecco, M.; Testaferri, L.; Santi, C.; Tomassini, C.; Marini, F.; Bagnoli, L.; Temperini, A. Asymmetric synthesis of thioamido selenides. A simple synthetic route to enantiopure thiazolines. *Tetrahedron: Asymmetry*, **2002**, *13*, 429–435.

[69] Hassner, A.; Amarasekara, A.S. Phenylselenium Azide Addition to Alkenes. A New and Stereospecific Introduction of Se and N Into Organic Molecules. *Tetrahedron Lett.*, **1987**, *28*, 5185–5188.

[70] Giuliano, R.M.; Duarte, F. Addition of Nitrogen Functionality to Exocyclic Alkenes: A New Procedure for Azidoselenenylation. *Synlett*, **1992**, 419–421.

[71] Tiecco, M.; Testaferri, L.; Temperini, A.; Bagnoli, L.; Marini, F.; Santi, C. Electrophilic Azido Selenenylation of Alkenes. A Simple Synthetic Route to Racemic Taxol Side Chain. *Synth. Commun.*, **1998**, *28*, 2167–2179.

[72] Tiecco, M.; Testaferri, L.; Santi, C.; Tomassini, C.; Santoro, S.; Marini, F.; Bagnoli, L.; Temperini, A. Synthesis of enantiomerically pure *β*-azidoselenides starting from natural terpenes. *Tetrahedron*, **2007**, *63*, 12373–12378.

[73] Tiecco, M.; Testaferri, L.; Santi, C.; Tomassini, C.; Marini, F.; Bagnoli, L.; Temperini, A. Asymmetric Azidoselenenylation of Alkenes: A Key Step for the Synthesis of Enantiomerically Enriched Nitrogen-Containing Compounds. *Angew. Chem. Int. Ed.*, **2003**, *42*, 3131–3133.

[74] Okamoto, K.; Nishibayashi, Y.; Uemura, S. Toshimitsu, A. Asymmetric Carboselenenylation Reaction of Alkenes with Aromatic Compounds. *Angew. Chem. Int. Ed.* **2005**, *44*, 3588–3591.

[75] Gabriele, E.; Singh, F.V.; Freudendahl, D.M.; Wirth T. Selenenylations of alkenes with styrene nucleophiles. *Tetrahedron*, **2012**, *68*, 10573–10576.

[76] Tiecco, M.; Testaferri, L.; Temperini, A.; Bagnoli, L.; Marini, F.; Santi, C. A New Synthesis of α-Phenylseleno γ- and δ-Lactones from Terminal Alkynes. *Synlett*, **2003**, 655–658; (b) Aprile, C.; Gruttadauria, M.; Amato, M.E.; Anna; F.D.; Meo, P.L.; Riela, S.; Noto, R. Studies on the stereoselective selenolactonization, hydroxy and methoxy selenenylation of α- and β-hydroxy acids and esters. Synthesis of δ- and γ-lactones. *Tetrahedron*, **2003**, *59*, 2241–2251; (c) Petragnani, N.; Stefani, H.A.; Valduga, C.J. Recent advances in selenocyclofunctionalization reactions. *Tetrahedron*, **2001**, *57*, 1411–1448.

[77] Denmark, S.E.; Edwards M.G. On the Mechanism of the Selenolactonization Reaction with Selenenyl Halides. *J. Org. Chem.*, **2006**, *7*, 7293–7306.

[78] (a) Denmark, S.E.; Burk, M.T. Enantioselective Bromocycloetherification by Lewis Base/Chiral Brønsted Acid Cooperative Catalysis. *Org. Lett.*, **2012**, *14*, 256–259; (b) Denmark, S.E.; Kalyani, D.; Collins, W.R. Preparative and Mechanistic Studies toward the Rational Development of Catalytic, Enantioselective Selenoetherification Reactions. *J. Am. Chem. Soc.*, **2010**, *132*, 15752–15765; (c) Denmark, S.E.; Collins, W.R. Lewis Base Activation of Lewis Acids: Development of a Lewis Base Catalyzed Selenolactonization. *Org. Lett.*, **2007**, *9*, 380–384.

[79] Clive, D.L.J.; Chittattu, G.; Wong, C.K. A new route to ring-fused tetrahydrofurans: reaction of Δ^4-unsaturated alcohols with phenylselenenyl chloride. *Can. J. Chem.*, **1987**, *55*, 3894–3897.

[80] (a) Nicolaou, K.C.; Lysenko, Z. Phenylselenolactonization. An extremely mild and synthetically useful cyclization process. *J. Am. Chem. Soc.*, **1977**, *99*, 3185–3187; (b) Nicolaou, K.C.; Seitz, S.P.; Sipio, W.J.; Blount, J.F. Phenylseleno- and Phenylsulfenolactonizations, Two Highly Efficient and Synthetically Useful Cyclization Procedures. *J. Am. Chem. Soc.*, **1979**, *101*, 3884–3893.

[81] Pearson, A.J.; Ray, T.; Richards, I.C.; Clardy, J.; Silveira, L. Conjugate Phenylselenolactonization Coupled with Allylic Selenoxide Rearrangement for Functionalizing Dienylacetic Acids. *Tetrahedron Letters.*, **1983**, *24*, 5827–5830.

[82] Betancor, C.; Leon, E.I.; Prange, T.; Salazar, J.A.; Suarez, E. 2-Oxazolines (4,5-Dihydro-oxazoles) by Organoselenium-induced Cyclisation of Allylic Ureas. *J. Chem. Soc., Chem. Commun.*, **1989**, 450–452.

[83] Tiecco, M.; Testaferri, L.; Bagnoli, L. Alkenyl nitrones cyclizations induced by phenylselenenyl bromide. A convenient synthetic route to 1,2-oxazines. *Tetrahedron*, **1996**, *52*, 6811–6822.

[84] Tiecco, M.; Testaferri, L.; Bagnoli, L.; Purgatorio, V.; Temperini, A.; Marini F.; Santi, C. Efficient asymmetric selenocyclizations of alkenyl oximes into cyclic nitrones and 1,2-oxazines promoted by sulfur containing diselenides. *Tetrahedron:Asymmetry*, **2001**, *12*, 3297–3304.

[85] Browne, D.M.; Niyomura, O.; Wirth, T. Catalytic Use of Selenium Electrophiles in Cyclizations. *Org. Lett.*, **2007**, *9*, 3169–3171.

[86] Singh F.V.; Wirth. T. Selenium-Catalyzed Regioselective Cyclization of Unsaturated Carboxylic Acids Using Hypervalent Iodine Oxidants. *Org. Lett.*, **2011**, *13*, 6504–6507.

[87] Shahzad, S.A.; Venin, C.; Wirth, T. Diselenide- and Disulfide-Mediated Synthesis of Isocoumarins. *Eur. J. Org. Chem.*, **2010**, 3465–3472.

[88] Tiecco, M.; Testaferri, L.; Tingoli, M.; Bagnoli, L.; Santi, C. Selenium Catalysed Conversion of β,γ-Unsaturated Acids into Butenolides. *Synlett*, **1993**, 798–800.

[89] Tingoli, M.; Tiecco, M.; Testaferri, L.; Temperini, Iodosobenzene Diacetate and Diphenyl Diselenide: An Electrophilic Selenenylating Agent of Double Bonds. A. *Synth. Commun.* **1998**, *28*, 1769–1778.

[90] Denmark, S.E.; Beutner, G.L.; Wynn, T.; Eastgate, M.D. Lewis Base Activation of Lewis Acids: Catalytic, Enantioselective Addition of Silyl Ketene Acetals to Aldehydes. *J. Am. Chem. Soc.*, **2005**, *127*, 3774–3789.

[91] Nicolaou, K.C.; Pfefferkorn, J.A.; Kim, S.; Wei, H.X. Synthesis of the Fully Functionalized Bicyclic Core of Garsubellin A. *J. Am. Chem. Soc.*, **1999**, *121*, 4724–4725.

[92] Fettes, A.; Carreira, E.M. Leucascandrolide A: Synthesis and Related Studies. *J. Org. Chem.*, **2003**, *68*, 9274–9283.

[93] Nishibayashi, Y.; Srivastava, S.K.; Takada, H.; Fukuzawa, S. Uemura, S. Highly selective asymmetric intramolecular selenocyclisation. *J. Chem. Soc., Chem. Commun.*, **1995**, 2321–2322.

[94] Takada, H.; Nishibayashi Y.; Uemura, S. The use of chiral diferrocenyl diselenides for highly selective asymmetric intramolecular selenocyclisation. *J. Chem. Soc., Perkin Trans. 1*, **1999**, 1511–1516.

[95] Fujita, K.; Murata, K.; lwaoka M.; Tomoda, S. Asymmetric Intramolecular Selenoetherification and Selenolactonization using an Optically Active Diaryl Diselenide derived from D-Mannitol. *J. Chem. Soc., Chem. Commun.*, **1995**, 1641–1642.

[96] Tiecco, M.; Testaferri, L.; Marini, F.; Sternativo, S.; Bagnoli, L.; Santi, C.; Temperini, A. A sulfur-containing diselenide as an efficient chiral reagent in asymmetric selenocyclization reactions. *Tetrahedron: Asymmetry*, **2001**, *12*, 1493–1502.

[97] Fragale, G.; Wirth, T. Chiral Diselenides in Asymmetric Cyclization Reactions. *Eur. J. Org. Chem.* **1998**, 1361–1369.

[98] Back, T.G.; Dyck, B.P. Asymmetric cyclization of unsaturated alcohols and carboxylic acids with camphor-based selenium electrophiles. *Chem. Commun.*, **1996**, 2567–2568.

[99] Rafiński, Z.; Ścianowski, J.; Wojtczak, A. Asymmetric selenocyclization with the use of dialkyl monoterpene diselenides. *Tetrahedron: Asymmetry*, **2008**, *19*, 223–230.

[100] (a) Tiecco, M.; Testaferri, L.; Temperini, A.; Bagnoli, L.; Marini F.; Santi, C. A New Synthesis of α-Phenylseleno γ- and δ-Lactones from Terminal Alkynes. *Synlett*, **2003**, 655–658. (b) Tiecco, M.; Testaferri, L.; Marini, F.; Sternativo, S.; Santi, C.; Bagnoli L.;

Temperini, A. Conjugated Additions of Selenium Containing Enolates to Enones – Enantioselective Synthesis of δ-Oxo-α-Seleno Esters and Their Facile Transformations. *Eur. J. Org. Chem.*, **2005**, 543–551.

[101] (a) Khokhar, S.S.; Wirth, T. Selenocyclizations: Control by Coordination and by the Counterion. *Angew. Chem. Int. Ed.*, **2004**, *43*, 631–633. (b) Khokhar, S.S.; Wirth, T. Nucleophile-Selective Selenocyclizations. *Eur. J. Org. Chem.*, **2004**, *22*, 4567–4581.

[102] Clive, D.L.J.; Wong, C.K.; Kiel, W.A.; Menchen, S.M. Cyclofunctionalisation of Olefinic Urethanes with Benzeneselenenyl Reagents: a New General Synthesis of Nitrogen Heterocycles. *J. Chem. Soc., Chem. Commun.*, **1978**, 379–380. (b) Clive, D.J.; Farina, V.; Singh, A.; Wong, C.K.; Kiel, W.A.; Menchen, S.M. Cyclofunctionalization of Olefinic Urethanes Using Benzeneselenenyl Chloride in the Presence of Silica Gel: A General Route to Nitrogen Heterocycles. *J. Org. Chem.*, **1980**, *45*, 2120–2126.

[103] Webb, R.R., II; Danishefsky, S. Ureidoallylation of double bonds. *Tetrahedron Lett.*, **1983**, *24*, 1357–1360.

[104] (a) Toshimitsu, A.; Aoai, T.; Owada, H.; Uemura, S.; Okano, M. Simple Procedure for the Aminoselenation of Olefins. *J. Chem. Soc., Chem. Commun.*, **1980**, 1041–1042; (b) Toshimitsu, A.; Aoai, T.; Owada, H.; Uemura, S.; Okano, M. Amidoselenation of Olefins and Its Utilization for Synthesis of Allylic Amides. *J. Org. Chem.*, **1981**, ***46***, 4727–4733; (c) Toshimitsu, A.; Hayashi, G.; Terao, K.; Uemura, S. Amidoselenation of Olefins *via* Hydroxyselenation: Reactions using Nitriles in Reagent Quantity and the Synthesis of β-(Acrylamido)alkyl Phenyl Selenides *J. Chem. Soc., Perkin Trans. 1*, **1986**, 343–347; (d) Francisco, C.G.; Hernandez, R.; Leon, E.I.; Salazar J.A.; Suarez, E. Aminoselenenylation of alkenes: syntheses of β-phenylseleno carbamates and β-phenylseleno cyanamides. *J. Chem. Soc., Perkin Trans. 1*, **1990**, 2417–2424. (e) Toshimitsu, A.; Kusumoto, T.; Oida, T.; Tanimoto, S. Amidoselenation of Olefins Using p-Toluenesulfonamide as a Nitrogen Nucleophile. *Bull. Chem. Soc. Jpn.*, **1991**, *64*, 2148–2152. (f) Katsuura, K.; Mitsuhashi, K., Thiazole Analogs of Benzomorphans. III. Syntheses of 4,5,6,7,8,9-Hexahydro-4,8-methano-5-methylthiazolo<4,5-c>azocine and 4,5,6,7,8,9-Hexahydro-4,8-methano-5-methylthiazolo <5,4-c>azocine. *Chem. Pharm. Bull.*, **1983**, *31*, 2094–2102. (g) Wilson, S.R.; Sawicki, R.A. Transannular cyclizations of 1-aza-4-cyclooctene. *J. Org. Chem.*, **1979**, *44*, 287–291.

[105] (a) Tiecco, M.; Testaferri L.; Tingoli, M.; Bagnoli, L.; Santi, C. Selenium-induced cyclization of O-allyl oximes as a synthetic route to N-alkyl isoxazolidines. *Tetrahedron*, **1995**, *51,* 1277–1284. (b) Tiecco, M.; Testaferri, L.; Tingoli, M.; Bagnoli, L. Organoselenium-induced stereoselective cyclisation of *O*-allyl oximes: a new synthetic route to isoxazolidines. *J. Chem. Soc., Chem. Commun.*, **1995**, 235–236.

[106] Tiecco, M.; Testaferri, L.; Marini, F.; Sternativo, S.; Santi, C.; Bagnoli, L.; Temperini, A. Optically active isoxazolidines and 1,3-amino alcohols by asymmetric selenocyclization reactions of *O*-allyl oximes. *Tetrahedron: Asymmetry*, **2001**, *12*, 3053–3059.

[107] De Kimpe, N.; Boelens, M. Organoselenium-induced cyclization of γ,δ-alkenimine to nitrogen heterocycles. *J. Chem. Soc., Chem. Commun.*, **1993**, 916–918. (b) De Smaele, D.; De Kimpe, N. Synthesis of functionalized pyrrolidines from *N*-(benzylidene)- and *N*(alkylidene)-homoallylamines. *J. Chem. Soc., Chem. Commun.*, **1995**, 2029–2030.

[108] (a) Wen-Xu, H.; Ling-Jun, C.; Chun-Long, Z.; Zhu-Jun, Y. Bidirectional Synthesis of the Central Amino Acid of Chloptosin. *Org. Lett.*, **2006**, 8, 4919–4922; (b) Shun-Ming, Y.; Wen-Xu, H.; Yuan, W.; Chun-Long, Zhong, and Zhu-Jun, Y. Total Synthesis of

Chloptosin, a Potent Apoptosis-Inducing Cyclopeptide; *Org. Lett.*, **2009**, *12*, 1124–1127; (c) Lopez, C.S.; Perez-Balado, C.; Rodrıguez-Grana, P.; de Lera, Angel R. Mechanistic Insights into the Stereocontrolled Synthesis of Hexahydropyrrolo[2,3-b]indoles by Electrophilic Activation of Tryptophan Derivatives. *Org. Lett.*, **2008**, *10*, 77–80.; (d) Shibahara, S.; Matsubara, T.; Takahashi, K.; Ishihara, J.; Hatakeyama, S. Total Synthesis of NW-G01, a Cyclic Hexapeptide Antibiotic, and 34-epi-NW-G01. *Org. Lett.*, **2011**, *13*, 4700–4703.

[109] (a) Jackson, W.P.; Ley, S.V.; Morton, J.A. New cyclization procedure for alkenyl-substituted β-dicarbonyl compounds using *N*-phenylselenophthalimide. *J. Chem. Soc., Chem. Commun.*, **1980**, 1028–1029; (b) Jackson, W.P.; Ley S.V.; Whittle, A.J. Selenium-mediated cyclization reactions of alkenyl-substituted β-ketoesters. *J. Chem. Soc., Chem. Commun.*, **1980**, 1173–1174. (c) Jackson, W.P.; Ley S.V.; Morton, J.A. Cyclisation reactions of alkenyl β-ketoesters involving a novel phenylseleno group migration. *Tetrahedron Lett.*, **1981**, *22*, 2601–2604. (d) Ley, S.V.; Lygo, B.; Molines, H.; Morton, J.A. Synthesis of (*cis*-6-methyltetrahydropyran-2-yl)acetic acid involving the use of an organoselenium-mediated cyclization reaction. *J. Chem. Soc., Chem. Commun.*, **1982**, 1251–1252.

[110] Deziel, R.; Malenfant E.; Thibault, C. Asymmetric arene-alkene cyclizations mediated by a chiral organoselenium reagent. *Tetrahedron Lett.*, **1998**, *39*, 5493–5496.

[111] Shahzad, S.A.; Vivant, C.; Wirth, T. Selenium-Mediated Synthesis of Biaryls through Rearrangement. *Org. Lett.*, **2010**, *12*, 1364–1367.

Organoselenium Chemistry Between Synthesis and Biochemistry, 2014, 61-93

CHAPTER 3

Nucleophilic Selenium: Nucleophilic Substitution

Barahman Movassagh* **and Mozhgan Navidi**

Department of Chemistry, K.N. Toosi University of Technology, P.O. Box 16315-1618, Tehran, Iran

Abstract: Nucleophilic selenium compounds are useful for a number of functional group transformations. The aim of this chapter is to present the various aspects of the preparation and the synthetic utility of the most recently used nucleophilic selenium reagents now available to academic and industrial chemists.

Keywords: Nucleophilic substitution, selenolates, selenols, selenides, stereoselectivity, vinyl selenides, selenoesters, ring opening reactions, ring closing reactions, solvent-free.

3.1. INTRODUCTION

There is no doubt that selenium-containing organic molecules have played and continue to play an important role in organic synthesis, biology and medicine. Over the last three decades, many investigators have described important chemical transformations that were efficiently achieved using organoselenium reagents. Once selenium is incorporated into a substrate, it can be removed easily either *via* selenoxide *syn*-elimination or [2,3]-sigmatropic rearrangement. In addition, the carbon-selenium bond can be replaced by carbon-hydrogen, carbon-halogen, carbon-lithium, or carbon-carbon bonds. Thus, in general, organoselenium species can be introduced, manipulated, and removed in a variety of ways under mild reaction conditions [1].

Among the methods for introducing a selenium moiety into organic molecules, the use of selenide anions (selenolates) is especially convenient and common. Although selenolates are weak bases, they are nevertheless powerful, soft nucleophiles [2] because of the high polarizability of the selenium atom.

***Address correspondence to Barahman Movassagh:** Department of Chemistry, K.N. Toosi University of Technology, P.O.Box 16315-1618, Tehran, Iran; Tel: +98-21-23064323; Fax: +98-21-22853650;
E-mails: movassagh@kntu.ac.ir; bmovass1178@yahoo.com

Certain features make selenium compounds particularly valuable. For instance, carbon-selenium bonds are weaker than carbon-sulfur bonds, and the selenoxide functionality is more strongly polarized than the analogous sulfoxide linkage. Therefore, selenoxide eliminations typically occur easily and faster than those of sulfoxides.

Additionally, selenols and selenolate ions (RSe^-) are more nucleophilic than their sulfur counterparts; in general, nucleophilicity decreases in the order $PhSe^- > PhS^- > PhO^-$. Based on the principles of hard-soft acid-base (HSAB) theory, selenols and selenide anions are predicted to be powerful soft nucleophiles because of their inherently low ionization potentials and high polarizabilities; therefore, the formation of C-Se bonds is often a simple process. Also, selenolate ions are less basic than thiolate ions (RS^-) and better leaving groups [2,3]. The literature on nucleophilic selenium has been thoroughly reviewed up to the early 2000s [1-8].

3.2. PREPARATION OF NUCLEOPHILIC SELENIUM SPECIES

In general, selenols (RSeH) are relatively strong acids (*e.g.* $pK_a = 5.9$ for PhSeH [9]) and air-sensitive compounds having an intolerable odour. The areneselenols are more stable than the alkaneselenols, but are nevertheless easily oxidized to diaryl diselenides. Selenols can be synthesized from either elemental selenium or other selenium compounds (Table **3.1**). The methods for the preparation of selenide anions (selenolates, RSe^-) include reaction of elemental selenium, selenocyanates, selenols and diselenides in various ways (Table **3.1**). Diselenide anions are mainly prepared from reduction of elemental selenium (Table **3.1**).

Table 3.1: Nucleophilic selenium reagents

Starting Material	**Reagent System**	**Nucleophilic Species**	**Refs.**
Se	RC≡CH/*n*-BuLi	RC≡CSeLi	[10]
Se	$RBr/Mg/ZnCl_2$	RSeZnCl	[11]
Se	$RMgX(ArMgX)/H_3O^+$	RSeH (ArSeH)	[12]
Se	1) Na/NH_3; 2) RX	RSeNa	[13]
Se	RLi	RSeLi	[14]
Se	RLi/Me_3SiCl	$RSeSiMe_3$	[14a]
Se	$ArBr/Mg/Me_3SiCl$	$ArSeSiMe_3$	[15]

Table 3.1: contd....

Se	R_3Al/Benzene	R_2AlSeR	[16]
Se	$LiEt_3BH$/THF	$LiHSe(BEt_3)$	[17]
Se	N_2H_4/NaOH	Se_2Na_2	[18]
Se	SmI_2/THF	“Se_2^{2-}”	[19]
H_2Se	RX	RSeH	[12]
H_2Se	$RCH=CH_2$	RCH_2CH_2SeH	[12]
H_2Se		$HOCH_2CH_2SeH$	[12]
RSeH	MeMgI	RSeMgI	[20]
RSeH	BBr_3	$(RSe)_3B$	[21]
ArSeH	Na/THF	ArSeNa	[22]
ArSeH	NaH/THF	ArSeNa	[23]
ArSeH	$Me_3SiCl/(Ph_3P)_3RhCl$	$ArSeSiMe_3$	[24]
ArSeH	1) MeLi; 2) Me_3SiCl	ArSeSiMe	[24]
ArSeH	*n*-BuLi/Benzene	ArSeLi	[25]
RSeSeR	OH^-/EtOH	RSeH	[26]
RSeSeR	$LiAlH_4/Me_3SiCl$	$RSeSiMe_3$	[27]
RSeSeR	$N_2H_4.H_2O$/NaOH/Bu_4NCl	RSeNa	[18]
RSeSeR	In/THF-H_2O	“RSe^- ”	[28]
RSeSeR	Zn/HCl-Et_2O	RSeH	[29]
RSeSeR	Zn/$RuCl_3$	“RSe^- ”	[30]
RSeSeR	Zn/$AlCl_3$	$(RSe)_2Zn$	[31]
RSeSeR	Zn/[BMIM]BF_4	“RSe^- ”	[32]
RSeSeR	La/I_2	“RSe^- ”	[33]
RSeSeR	Hg	$(RSe)_2Hg$	[34]
RSeSeR	Zn/$InBr_3$	“RSe^- ”	[35]
RSeSeR	InI	$(RSe)_2InI$	[36]
RSeSeR	Sm/Me_3SiCl/H_2O	$RSeSiMe_3$	[37]
RSeSeR	H_3PO_2/H_2O	RSeH	[38]
RSeSeR	Rongalite/CsF	RSeNa	[39]
RSeSeR	SmI_2/THF	$RSeSmI_2$	[40]
RSeSeR	NaH/THF	RSeNa	[3]
RSeSeR	$LiEt_3BH$/THF	$RSe(BEt_3)Li$	[17]
RSeSeR	$NaB(MeO)_3H$/DMF	$RSe(BOMe)_3Na$	[41]
RSeSeR	Bu_3P/THF/NaOH	RSeNa	[42]
RSeSeR	Na/Me_3SiCl/THF	$RSeSiMe_3$	[43]

Table 3.1: contd....

RSeSeR	*i*-Bu_2AlH(DIBAL)/Hexane	*i*-Bu_2AlSeR	[44]
RSeSeR	$NaBH_4$/EtOH	RSeNa	[45]
RSeSeR	$2e^-$/MeOH	"RSe^-"	[46]
RSeSeR	Pt electrode/Me_3SiCl/Et_3NOTs/ MeOH	$RSeSiMe_3$	[47]
RSeSeR	*N*-acetylcysteine/ NaOH/ MeOH	RSeNa	[48]
RSeCN	MH (M= Li, Na, K)	RSeM	[49]
RSeCN	$NaBH_4$/EtOH	RSeNa	[50]
RSeCN	Zn/H^+	RSeH	[12]
RSeCN	H_3PO_2	RSeH	[12]
ROH	$Ph_3P(SeCN)_2$/THF-CH_2Cl_2	RSeH	[51]
$RSeSiMe_3$	KF/18-Crown-6	RSeK	[43]
$RSeSiMe_3$	$MeOH/H^+$	RSeH	[24b]
RSe(O)OH	$NaBH_4$	RSeH	[12]
$RSeO_2OH$	$H_2S/SO_2/Zn/HCl$	RSeH	[12]

3.3. REACTIONS OF NUCLEOPHILIC SELENIUM SPECIES

3.3.1. Reactions with Alkyl Halides, Tosylates and Acetates

Symmetrical dialkyl selenides and diselenides are synthesized from the reaction of Se^{2-} and Se_2^{2-} with alkyl halides (eq. 1-8, Scheme **3.1**); these species can be readily prepared by reduction of elemental selenium using sodium hydroxymethane sulfinate or "Rongalite" **1** [52], $NaBH_4$ [53], alkali metal-ammonia [13,54], metallic magnesium [55] and hydrazine [56].

$$Se \xrightarrow[\text{aq. NaOH}]{HOCH_2SO_2Na\ \mathbf{1}} Na_2Se \xrightarrow{2\ RX} R_2Se \qquad (1)$$

$$Se \xrightarrow[\text{aq. NaOH}]{HOCH_2SO_2Na\ \mathbf{1}} Na_2Se_2 \xrightarrow{2\ RX} R_2Se_2 \qquad (2)$$

$$Se \xrightarrow[H_2O\text{ or EtOH}]{NaBH_4} NaHSe\text{ and }Na_2Se \xrightarrow{2\ RX} R_2Se \qquad (3)$$

$$Se \xrightarrow[H_2O\text{ or EtOH}]{NaBH_4} Na_2Se_2 \xrightarrow{2\ RX} R_2Se_2 \qquad (4)$$

$$Se \xrightarrow{Na/NH_3} Na_2Se \xrightarrow{2\ RX} R_2Se \qquad (5)$$

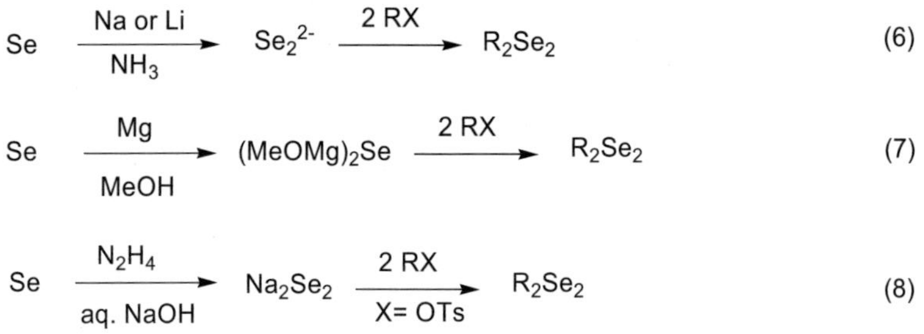

Scheme 3.1

Alkyl or aryl selenolates (RSeM, **5**) have been prepared from elemental selenium and organometallics [12, 57, 58], by reaction of selenols **2** with bases [3, 12],by reduction of selenocyanates **3** [50b] and diselenides **4** with metals [59] or metal hydrides [3, 60]. They react readily with primary and secondary alkyl halides, tosylates, and acetates to afford selenides (Scheme **3.2**).

In an interesting investigation, Krief and coworkers [61] reported the synthesis of *tert*-alkyl selenides from *tert*-alkyl halides, diorganic diselenides or selenols in the presence of zinc.

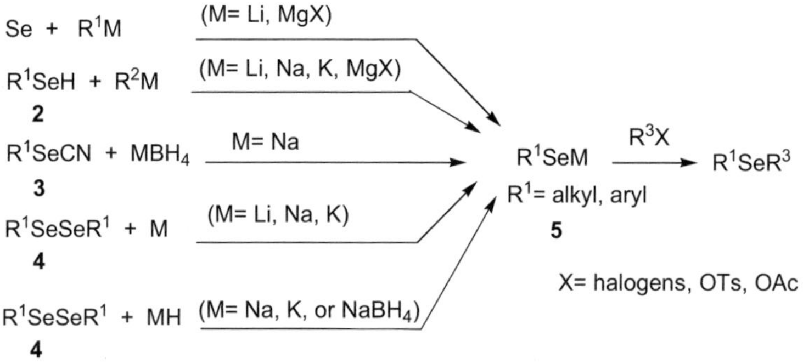

Scheme 3.2

The selenium-metal bond can also be formed by cleavage of the -Se-Se- bond in diorganodiselenides using Cd(0) [62], La(0) [33], In(0) [28a], In(I) [36], Sm(0) [63], Sm(II) [64], Sn(0) [65], Yb(II) [66], and Cu(II)/Sn(II) [67]. Several recent reports [32, 35, 68] revealed that zinc powder promotes cleavage of the Se-Se bond to form zinc selenolate **6**, which can then react with various alkyl halides to afford unsymmetrical diorganyl selenides (Scheme **3.3**). This metal is a good alternative to the previously reported processes which use Yb, La, Sm, or In salts.

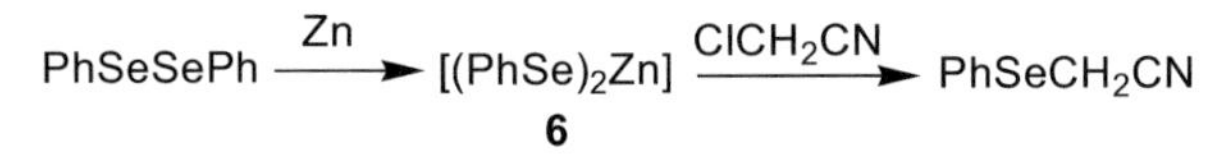

Scheme 3.3

One interesting application of this type of reaction is presented in Scheme **3.4**, in which the synthesis of the biologically important selenocysteine derivative **8** was accomplished from the corresponding bromoamino ester **7** in ionic liquids [32].

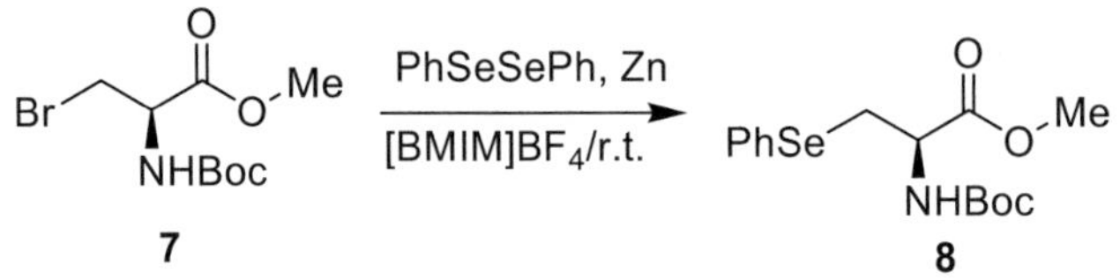

Scheme 3.4

3.3.2. Reactions with Aryl Halides, Aryl Boronic Acids and Diaryl Iodonium Salts

In 1985, Cristau reported the synthesis of diaryl selenides by the reaction of sodium benzeneselenolate **9** with various aryl iodides and bromides catalyzed by *bis*-(bipyridyl) nickel(II) bromide, $(bpy)_2NiBr_2$ (Scheme **3.5**) [69]. The synthesis of diaryl selenides from aryl halides and aryl tributylstannyl selenide ($ArSeSnBu_3$) catalyzed by copper(I) and palladium complexes has also been reported [70].

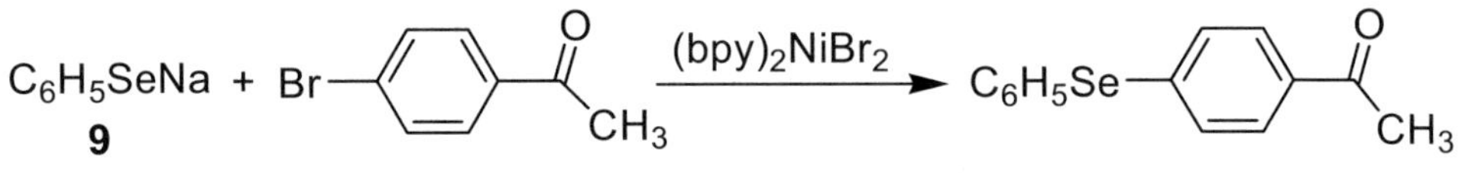

Scheme 3.5

As an alternative to aryl halides, the palladium- and zinc-catalyzed cross-coupling reactions of diaryliodonium salts **10** with areneselenyl or alkaneselenyl magnesium bromide **11** (eq. 9, Scheme **3.6**) and diaryl diselenide (eq. 10, Scheme **3.6**) were introduced [71]. These hypervalent iodonium salts are good electrophilic arylating agents and are more susceptible to nucleophilic displacement, since their positive charges should facilitate polar fission of the bonds to the aromatic systems.

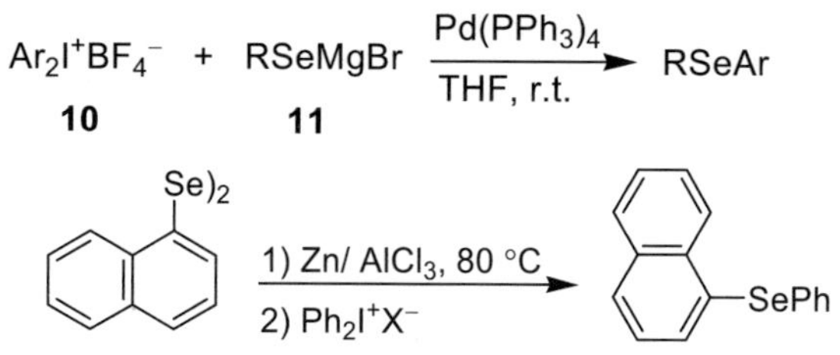

(9)

(10)

Scheme 3.6

In the last decade, several transition metals such as palladium [72], cobalt [73], copper [74], and lanthanum salts [33] in combination with various ligands have been employed in bringing out coupling reactions of aryl donors such as arylboronic acids and aryl halides with diaryl diselenides (Scheme **3.7**). Alves and coworkers [75] introduced a ligand-free cross-coupling reaction of organic diselenides with aryl boronic acids using CuO nanoparticles (Scheme **3.7**). In a separate report, Li and coworkers [76] described that the addition of Fe powder to CuS accelerated the coupling reaction of aryl halides with diaryl diselenides (Scheme **3.7**).

$$Ar^1X + Ar^2SeSeAr^2 \xrightarrow{\text{Catalyst}} Ar^1SeAr^2$$

X= I, Br, Cl, $B(OH)_2$

Catalysts: [Pd],[Co], [Cu], [La], CuONP, CuS-Fe

Scheme 3.7

In 2011, cross-coupling of aryl halides with diphenyl diselenide was developed under ligand-free conditions in the presence of magnetically separable and recyclable copper ferrite nanoparticles (Scheme **3.8**) [77].

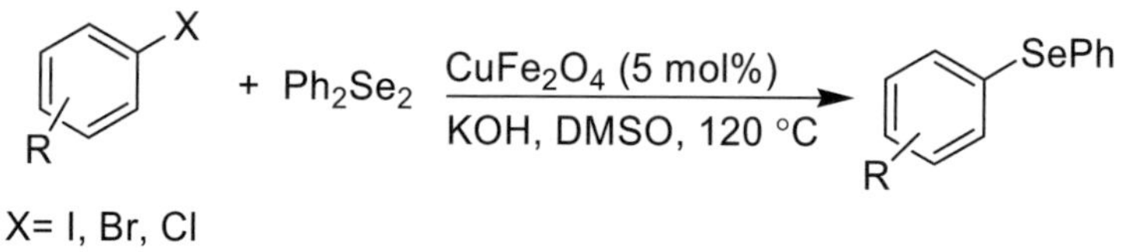

Scheme 3.8

For the synthesis of the target compounds **14** with glutathione peroxidase (GPx)-like activity, Singh and coworkers [78] used aromatic nucleophilic substitution reaction of *N*-(2-bromo-3-nitrobenzyl) anilines **12** with the *in situ* prepared *n*-

BuSeNa which afforded the corresponding *N*-[2-(butylselanyl)-3-nitrobenzyl] anilines **13** (Scheme **3.9**).

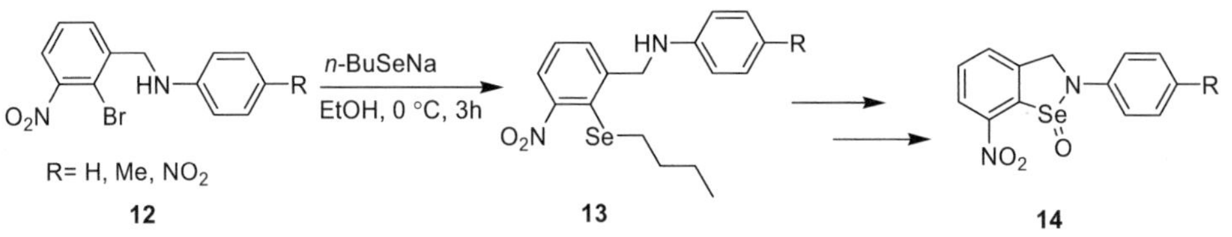

Scheme 3.9

Millois and Diaz [79] reported the preparation of a series of selenium-containing diaryl retionides **15** by a nickel(II)-catalyzed coupling of a diselenide with aryl iodide in the presence of polymer-supported borohydride (Scheme **3.10**).

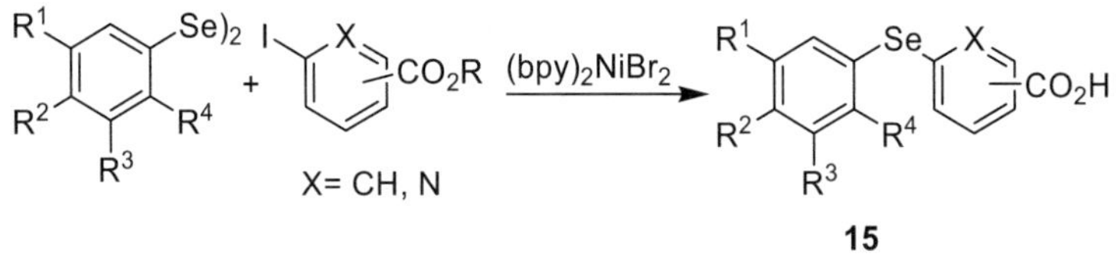

Scheme 3.10

3.3.3. Reactions with Vinyl-Halides, -Boronic Acids, -Iodonium and -Trifluroborate Salts

The most recent review article extensively discussing the preparation of vinyl selenides was presented by Perin and Lenardao [80] and was not until 1984, when the formation of vinyl selenides by vinylic substitution of unactivated vinyl halides was first described by Tiecco and coworkers [81]. In their pioneering work, they used selenide anions (RSe^-) as a nucleophilic selenium species to selectively obtain (*E*)- or (*Z*)-vinyl selenides. Since then, a wide variety of nucleophilic selenium has been used in this reaction [82].

Ranu and coworkers [72b] employed phenylselenolate anion generated by the reaction of indium(I) iodide with diphenyl diselenide for the selective synthesis of several vinylselenides **17** starting from vinyl bromides **16** (Scheme **3.11**). The conversion of (*E*)-vinyl bromides was remarkably stereoselective giving (*E*)-vinyl selenides whereas the stereoselectivity in reaction of (*Z*)-vinyl bromides was not very good.

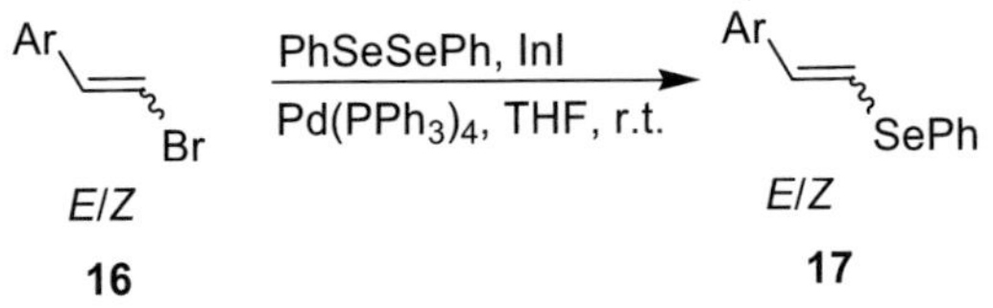

Scheme 3.11

In a recent report, Chang and Bao [83] described the zinc-induced CuI/*L*-Proline-catalyzed coupling reaction of diaryl diselenides with vinyl bromides at 110 °C in 1-Butyl-3-methylimidazolium tetrafluoroborate, [BMIM] BF_4 (Scheme **3.12**). They reused the metal catalyst immobilized in ionic liquid [IL] up to four times with no significant effect on the rate or yield of the reaction during each cycle.

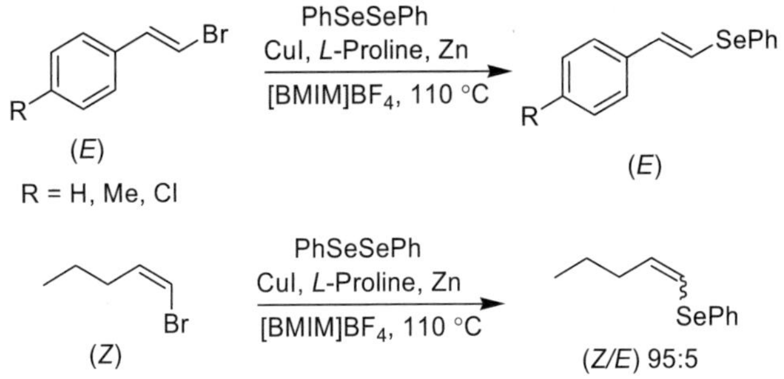

Scheme 3.12

Yan and Chen [84] developed a process for the stereoselective synthesis of vinyl selenides by the reaction of sodium selenolates, generated *in situ* from the reaction of diselenides with $NaBH_4/C_2H_5OH$ at 0 °C, with vinyl (phenyl) iodonium salts **18**. The reaction proceeds with retention or inversion of configuration (Scheme **3.13**).

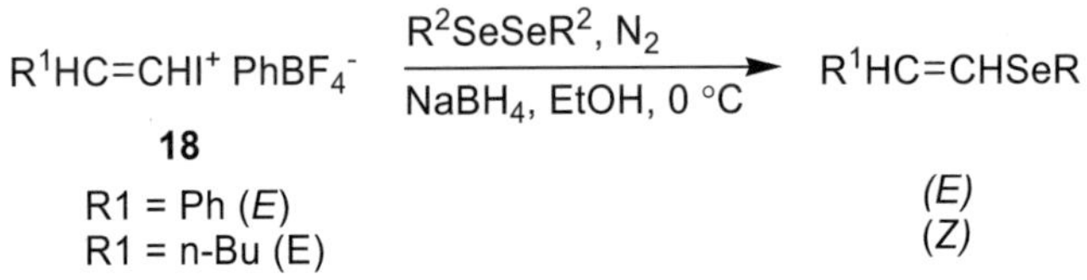

Scheme 3.13

The authors suggested that the retention of configuration occurs *via* an addition-elimination or a ligand coupling mechanism, while the inversion of the configuration could be related to an S_N2 transition state.

Potassium vinyltrifluoroborate salts **19** were recently used as new precursors in the selective synthesis of (*E*)-vinyl selenides (Scheme **3.14**) [85]. This CuI-catalyzed coupling reaction was performed with diselenides in DMSO at 100 °C giving the corresponding (*E*)-vinyl selenides in 44-98 % yields.

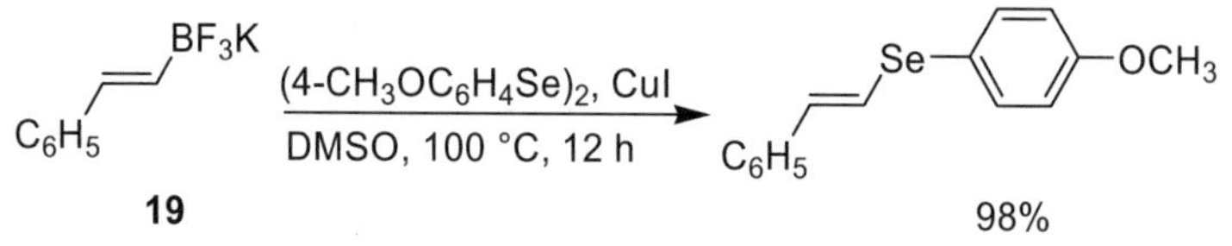

Scheme 3.14

In addition to these methods, Stefani and coworkers [86] reported the substitution of iodine and bromine atoms in 1-chalcogene-1-haloalkenes **20** by phenylselenolate anion to afford selenoketeneacetals **21** (Scheme **3.15**).

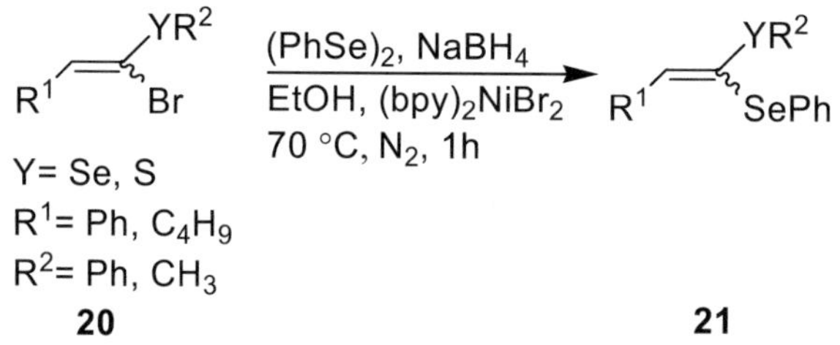

Scheme 3.15

In 2008, Santi and coworkers introduced a new solid and air-stable zinc selenolate **22** from treatment of commercially available PhSeCl with a stoichiometric amount of zinc powder in refluxing THF [87]. This reagent was then utilized for stereospecific vinylic substitution with retention of the alkene geometry; the only exception was ketone (*E*) **23** which mainly led to the formation of vinyl selenide (*Z*) **24** (Scheme **3.16**). In this report, they also showed the strong influence of zinc in the outcome of the reaction using density functional theory (DFT) calculations [88].

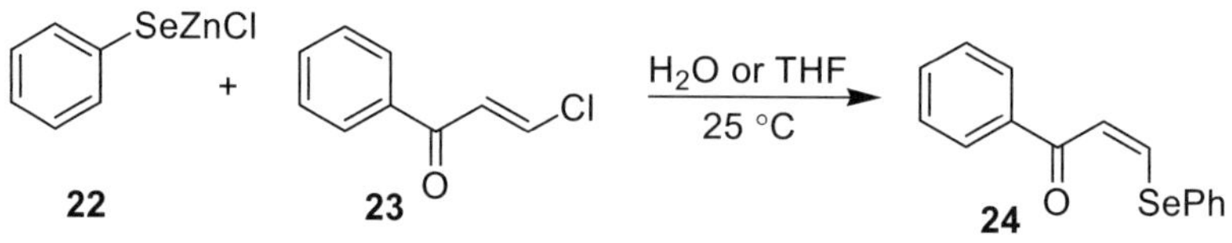

Scheme 3.16

3.3.4. Attack at an Acyl Carbon

3.3.4.1. Reactions with Acid Chlorides and Anhydrides

Due to the usefulness of selenoesters as valuable acyl transfer agents, for example, in synthesis of macrocyclic lactones and lactams [89] and of steroids and sex hormones [90], many researchers have devoted efforts toward the development of various strategies for their preparation.

Prior to the development of the new methodologies, selenoesters were produced almost solely by the reaction of selenols or their metal salts (Na, Mg, Pb, Cd, *etc.*) with acid chlorides (Scheme **3.17**) [12, 82a, 91].

$$R^1COCl + R^2SeX \xrightarrow{\text{Pyridine}} R^1COSeR^2 + XCl$$

X= H or Metal

Scheme 3.17

Other approaches to the preparation of selenoesters are based on reactions between acyl chlorides and anhydrides or esters with nucleophilic organoselenating reagents, as organoselenols in the presence of bases, tributylstannyl selenides, mercury bisarylselenolates, samarium and aluminium organoselenolates (Scheme **3.18**) [34, 92].

$$R^1COX + R^3SeY \longrightarrow R^1COSeR^3$$

X= Cl, $OCOR^1$, OR^2

Y= $SiMe_3$, $AlMe_2$, $HgSeR^3$, $SnBu_3$

Scheme 3.18

In recent years, metal catalysts have been used to activate and cleave diselenides. Samarium metal in the presence of KI [93], TMSCl [94], $CrCl_3$ [95], $CoCl_2$ [96], promoted diselenides to react with acid chlorides and acid anhydrides in one-pot to give selenoesters. Indium metal has also been proven to be efficient to cleave diselenides and reacts with acid chlorides to yield selenoesters [97]; it was suggested that indium species, $In(SePh)_3$, may coordinate with the substrate and produce a carbocation-like intermediate.

Selenoesters were also prepared from acid anhydrides or chlorides and diselenides by reductive cleavage of Se-Se bond with a $Zn/AlCl_3$ system [31b, 98].

In 2010, the application of Rongalite ($NaHSO_2.CH_2O$)/CsF in acylation of diphenyl diselenide with anhydrides to afford selenoesters was published by Dan and coworkers [39].

3.3.4.2. Reactions with Chloroformates and Carbamyl Chlorides

Selenocarbamates, $R^1SeCONR_2^2$, and selenoformates (selenocarbonates, $R^1SeCO_2R^2$) have been employed by a number of researchers as efficient sources of acyl radicals [99]. Among other applications, they have been used in ring-closing reactions to form lactones [99b], lactams [99d] and substituted tetrahydrofurans [99c], for the radical deoxygenation of alcohols [100] and for the conversion of carboxylic acids to isonitriles [99a].

In 2000, Lucas and co-workers [101] reported the preparation of 2,3,4-tri-*O*-benzyl-1,5-dideoxy-5-seleno-*D*-pentopyranose sugars (**26**) by thermolysis of selenoformate derivatives of pentose carbohydrates (**25**) starting from *D*-arabinose (Scheme **3.19**).

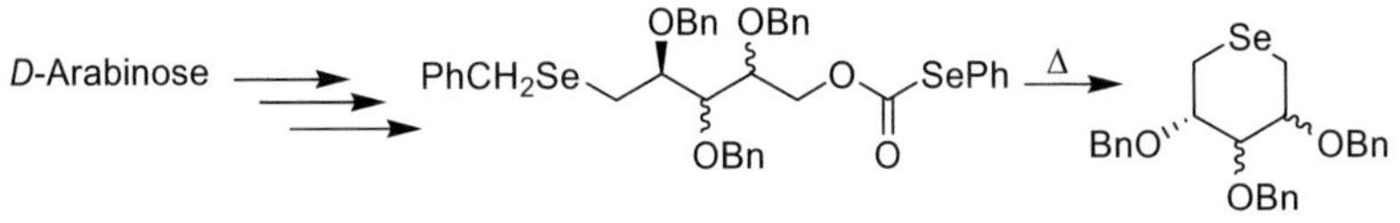

Scheme 3.19

Reinerth and Tour [102] prepared a series of selenocarbonates and selenoformates by reductive cleavage of diselenides using lithium triethylborohydride, $LiHBEt_3$, followed by reaction with diethylcarbamyl chloride, $ClCONEt_2$, or ethyl chloroformate.

There are several reported methods for the preparation of selenoformates such as the treatment of phenylselenotris(trimethylsilyl) silane [103], phenylselenol [99c] and sodium phenylselenolate [104] with chloroformates. Similar improved methods were described for the synthesis of selenoformates by reductive cleavage of Se-Se bond promoted by zinc metal [105].

Treatment of an amine with triphosgene followed by addition of phenylselenol [106], and the reaction of isocyanates with LiAlHSeH and haloalkanes [107] are other procedures for the preparation of selenocarbamates.

3.3.5. Ring Openings by Nucleophilic Selenium

3.3.5.1. Epoxides and Cyclic Ethers

The S_N2 ring-opening of epoxides, using selenium nucleophiles, is a common method for preparing *β*-hydroxy selenides. This method provides an excellent preparative route to allylic alcohols *via* selenoxide elimination [108]. *β*-Hydroxy selenides are valuable selenium intermediates in organic synthesis, since they can be converted to olefins, bromohydrines, vinyl selenides and epoxides [108,109]. Other effective reagents for epoxide-opening include phenyl-trimethylsilyl selenide [43], selenoboranes [110], aluminum selenolates [111], benzeneselenol in the presence of alumina [112] or *β*-cyclodextrin [113], tributylstannyl phenyl selenide [114] and selenolate ion generated by treatment of diphenyl diselenide with tributylphosphine in NaOH/THF [115] or $NaBH_4$/NaOH [116].

The regioselectivity of the ring-opening by nucleophilic selenium species is dependent on the steric and electronic features of the epoxide system. In the case of alkyl substituted derivatives, the nucleophilic attack occurs preferentially on the less hindered carbon atom leading to the stereospecific formation of regioisomer **27** as the sole or in some cases the major product (Scheme **3.20**); on the other hand, in the aryl substituted epoxides, *e.g.* styrene oxide, the electronic effects overshadow the steric ones and a mixture of two regioisomers **27** and **28** is obtained (Scheme **3.20**). This is presumably due to partial stabilization of developing positive charge at the benzylic carbon atom. In cyclic systems, a strong preference for approach from the axial direction may outweigh other factors.

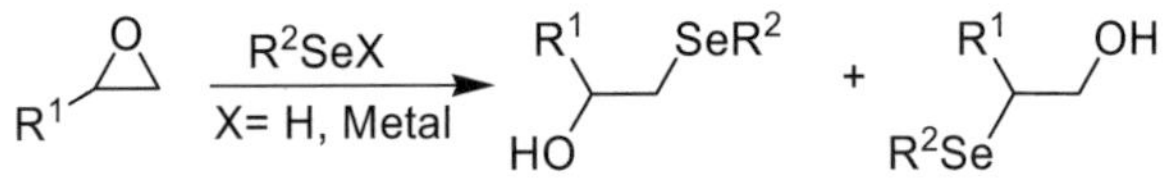

27 **28**

Scheme 3.20

Recently, several regio- and stereoselective ring opening of epoxides with diorganic diselenides mediated by active metallic indium, generated *in situ* from $Sm/InCl_3 \cdot 4\ H_2O$ [117], InI [118], and Zn [29, 31c] have been introduced.

In 2008, Santi and coworkers published the first "bench-stable" phenyl selenolate as an "on water" nucleophilic selenium reagent [87]; treatment of commercially available PhSeCl and PhSeBr with a stochiometric amount of zinc powder in refluxing THF led to the corresponding zinc selenolates **29** through an oxidative insertion of zinc into the selenium-halide bond (Scheme **3.21**).

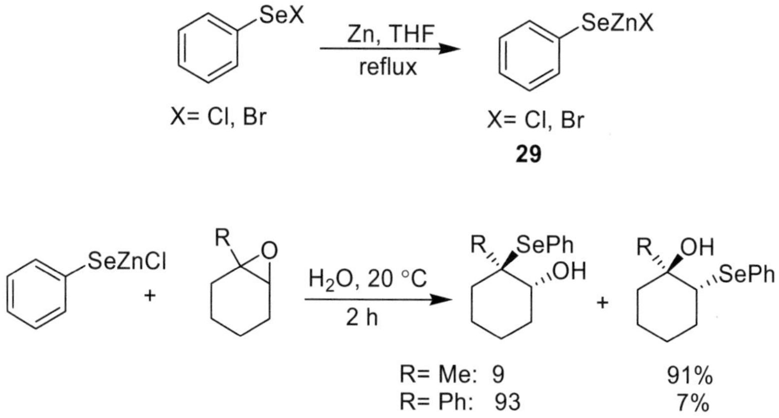

Scheme 3.21

The first example of enantioselective ring-opening reaction of *meso*-epoxides with aryl selenols to give optically active *β*-arylseleno alcohols, using a chiral Ti-Ga-Salen heterometallic catalyst was reported by Yang and coworkers (Scheme **3.22**) [119].

In 2008, Tiecco and coworkers [120] described another enantioselective ring-opening of *meso*-epoxides by (phenylseleno) silanes using salen (Cr) complexes as catalyst.

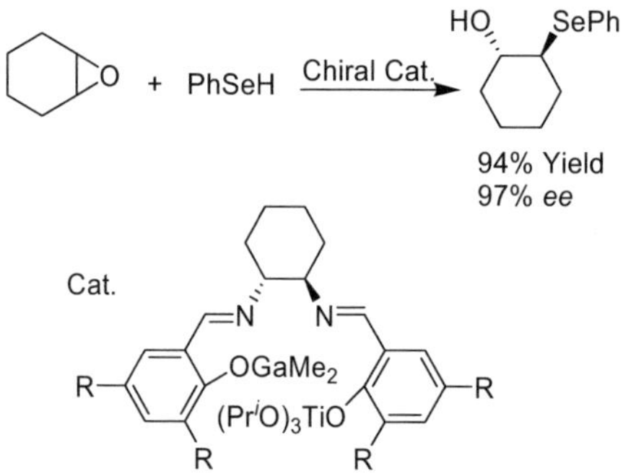

Scheme 3.22

Ferrocenyl selenide anion ($Fc^{*}Se^{-}$) produced by reduction of the chiral diferrocenyl diselenide $(Fc^{*}Se)_2$ with $LiAlH_4$ in THF was also found to act as a useful stereoselective nucleophile for ring opening of *meso*-epoxides to give the corresponding ferrocenyl *β*-hydroxy alkyl selenides with high diastereoisomeric excess (Scheme **3.23**) [121].

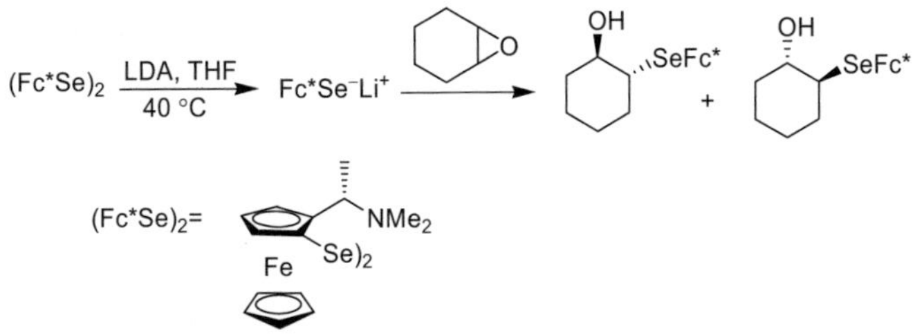

Scheme 3.23

Site-selective epoxide opening with phenyl selenide anion had been widely employed in natural products synthesis. Knapp and coworkers [122] used this reaction in the preparation of **31**, a key intermediate in 3-*C*-benzylsiastatin **32** synthesis, from the unsaturated piperidine ester **30** (Scheme **3.24**).

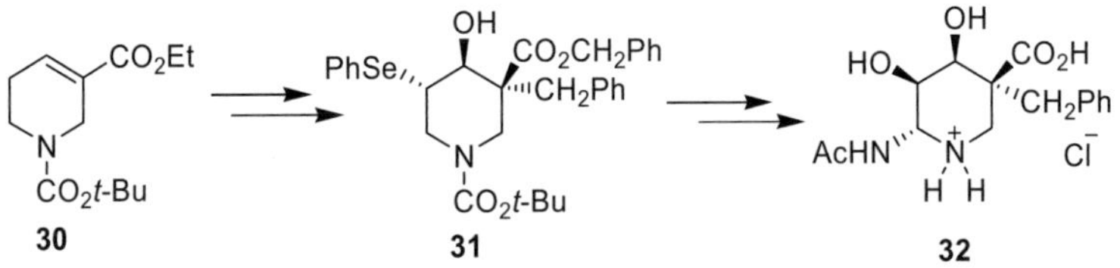

Scheme 3.24

The epoxide opening reaction has also been utilized in the synthesis of narciclasine [123], hexahydroxanthene moieties of the natural product Schweinfurthin [124], eucannabinolide [125], senepoxide [126] and in a biomimetic approach to the amphilectane diterpenes [127].

Phenyl selenide anion, prepared from diphenyl diselenide and $LiAlH_4$ in dioxane, has been employed to ring-open various oxetanes and oxolanes (tetrahydrofurans) to afford *γ*- and *δ*-phenylselenenyl alcohols (Scheme **3.25**) [128]. C-O bond cleavage of large cyclic ethers such as tetrahydrofurane and tetrahydropyran was also accomplished using tris(phenylseleno)borane in the presence of zinc iodide to produce *δ*- and *ε*- phenylseleno alcohols [129]. Abe and coworkers [130] reported the reaction of tetrahydrofurane with methaneselenolate anion in the presence of $AlCl_3$.

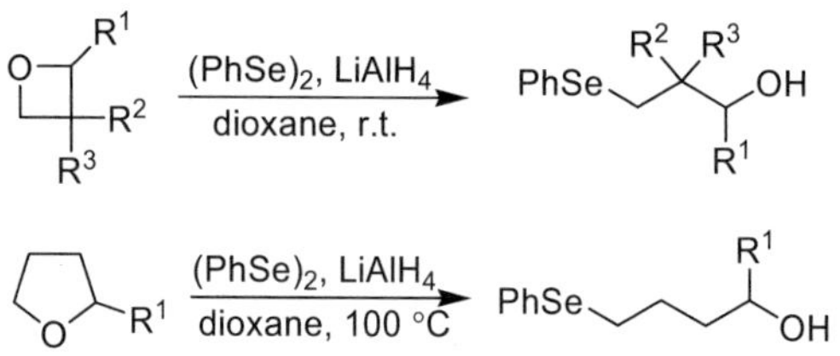

Scheme 3.25

3.3.5.2. Aziridines

The *N*-protected and unprotected aziridines are attacked by nucleophilic selenium species to produce the corresponding *β*-seleno amines. The latter compounds are the building blocks for the synthesis of many bioactive molecules, such as selenocysteine (Se-Cys, **33**), selenocystine and selenolanthionine **34** (Fig. **3.1**) [131]. Additionally, selenoaminoacids are important building blocks for the synthesis of selenoproteins [131c,d] and also for their potential biological activity, for example, in protection against neurodegenerative diseases, such as Alzheimer's and Parkinson's diseases [132].

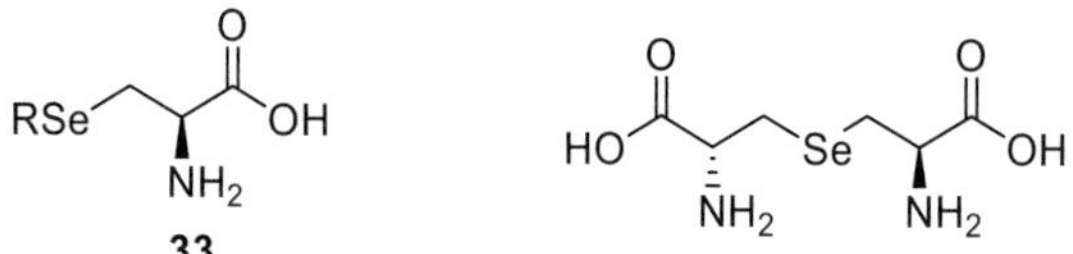

33 **34**

Figure 3.1: Two biologically active β-seleno amines.

The preparation of *β*-Seleno amines by the addition of benzene selenol to aziridines in the presence of triethylamine [133] and nucleophilic ring opening of aziridine derivatives by zinc selenolates, prepared *in situ* from reductive cleavage of diselenides using the $Zn/AlCl_3$ system [31f] or $PhSeSePh/(n\text{-}Bu)_3P$ [134] have been reported.

In the last few years, several protocols have been introduced for the synthesis of chiral *β*-seleno amines by Braga and co-workers (Scheme **3.26**) [135]. Recently, Ganesh and coworkers [136] introduced an efficient procedure for the synthesis of *β*-seleno amines by aziridine ring-opening with selenolate anions derived from diselenides in the presence of Rongalite.

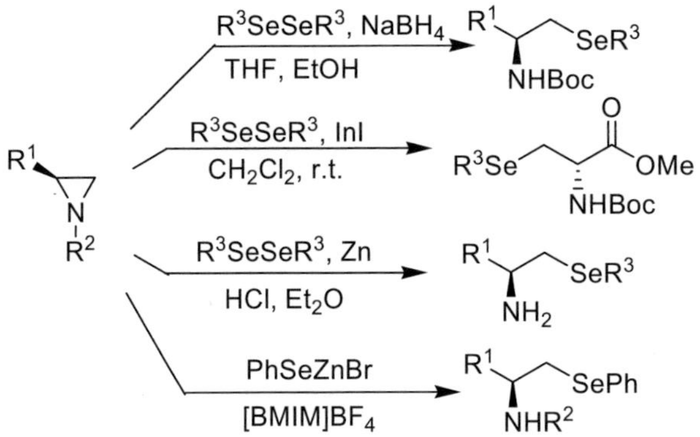

Scheme 3.26

3.3.5.3. Cyclopropanes

The nucleophilic cleavage of electron-deficient cyclopropane derivatives has been widely applied in organic synthesis and is known as the homologous (or 1,5) version of the classical Michael addition, which has found many applications in the total synthesis of natural products [137]. Cyclopropanes which contain, at least, one ester, ketone or cyano substituent are ring-opened by nucleophilic selenolate reagents to provide *γ*-selenides **35** (Scheme **3.27**) [25, 138]. In 2002, Adams and coworkers reported the ring-opening reaction of the chiral imide **36** using sodium phenyl selenolate generated from diphenyl diselenide and $NaBH_4$ (Scheme **3.28**) [139].

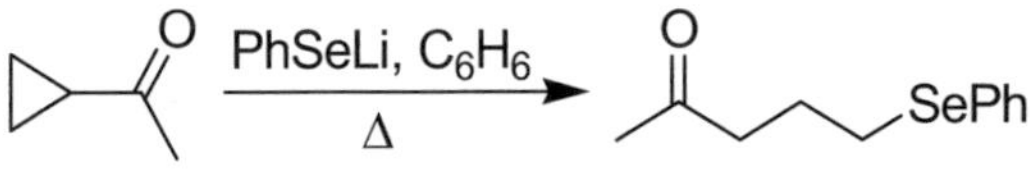

35

Scheme 3.27

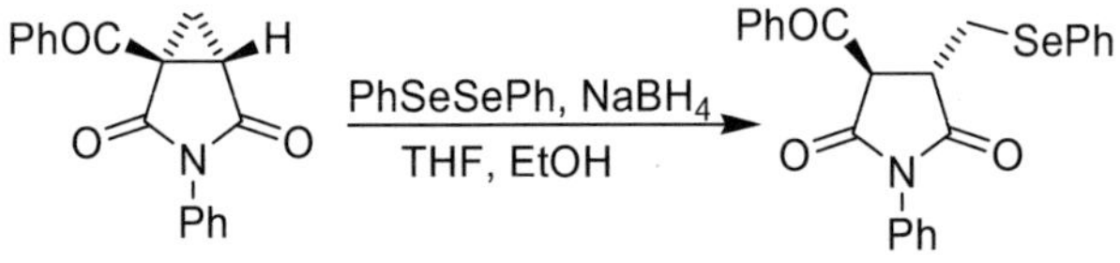

(+)-**36** **37**

Scheme 3.28

Very recently, nucleophilic ring-opening of mono-activated cyclopropanes **38** with zinc selenolates to produce γ-arylselenenyl ketones, acids, esters, and nitriles **39** has been reported (Scheme **3.29**) [31d].

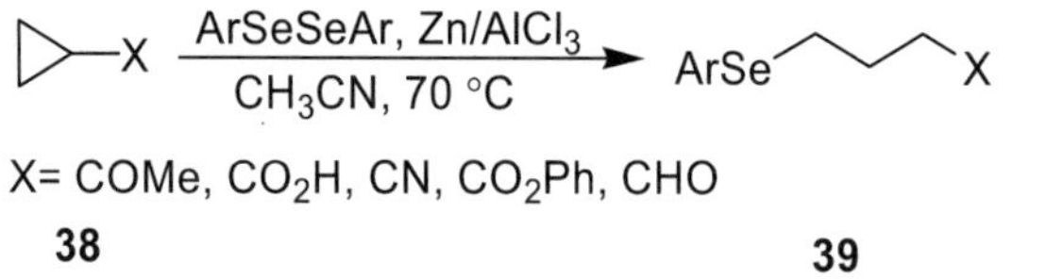

Scheme 3.29

3.3.6. Reactions with Esters and Lactones

Selenolates cleave certain esters and lactones attacking the softer alkyl rather than harder acyl carbon atom, resulting in the displacement of the carboxylate anion (Scheme **3.30**).

Scheme 3.30

Various reagents such as PhSeNa in DMF [138c, 60a] or THF [140] as well as the more reactive potassium salt PhSeK [23] have been utilized. The trimethylsilyl selenide **40** was also used in a similar capacity in the presence of KF and 18-Crown-6 [43] or ZnI_2 [141]. In contrast to the above reagents, dimethylaluminium methanselenolate **41** attacks esters and lactones at the acyl carbon atom to afford high yields of selenoesters [111]. The nucleophilicity of the selenolate anion can be greatly attenuated by varying the counterion and/or the degree of solvation of the anion [3].

In a report by Clive and coworkers [142], bicyclic lactones of type **42** were treated with sodium phenyl selenide and then with diazomethane to produce methyl esters **43** and these give *trans* ring-fused bicyclic compounds **44** when treated with triphenyltin hydride in the presence of a radical initiator (Scheme **3.31**). In order to prepare the pyrromethenone **46**, Jacobi and coworkers [143] reported the synthesis of the carboxylic acid **45** upon S_N2 ring-opening of butyrolactone with *p*-chlorophenyl selenide anion (Scheme **3.32**).

In an elegant report, sodium phenylselenide, prepared from diphenyl diselenide with sodium in THF-HMPA, ring-opened *L*-serine-*β*-lactone **47** to give *N*-Boc-*L*-PhSeAlanine **48** aimed at the synthesis of optically pure *β*-phenylselenoalanine **49** (Scheme **3.33**) [144].

In the report describing the preparation of carboxylic acids **50** and **51**, alternative nucleophilic cleavages of lactones and esters were introduced, utilizing zinc selenolates (Schemes **3.34** and **3.35**) [31a]. It was found that as the size of the substituent at the carbinol carbon of lactones increases, the rate of the alkyl-oxygen cleavage process decreases which is compatible with the S_N2 nature of the reaction. This investigation also showed that upon exposure to phenylselenolate anions, methyl esters underwent excellent alkyl-oxygen cleavage reactions, in spite of steric hinderance around the ester.

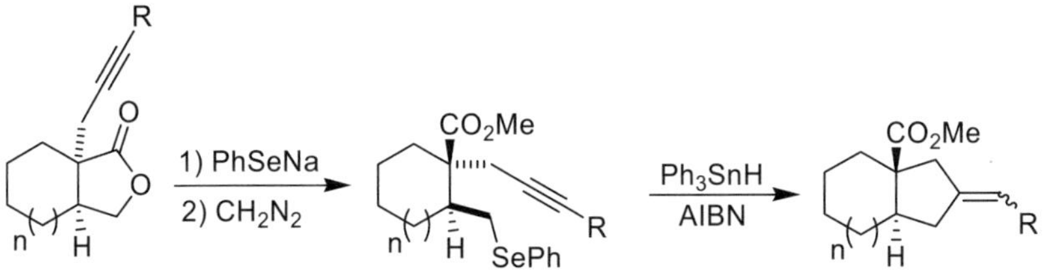

Scheme 3.31

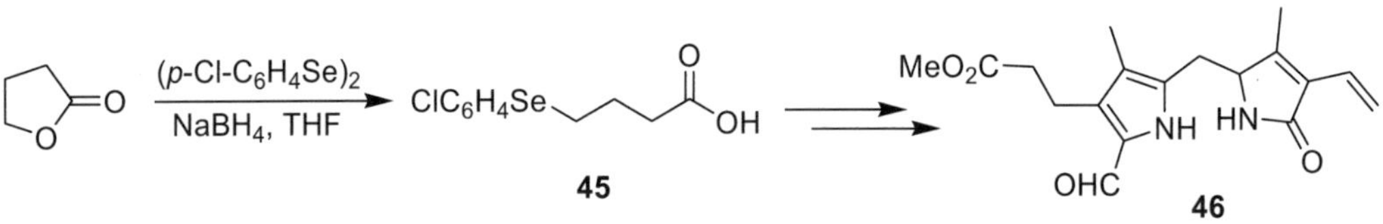

Scheme 3.32

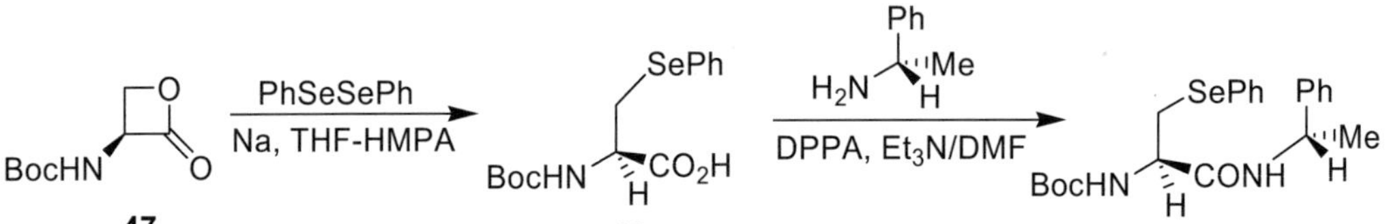

Scheme 3.33

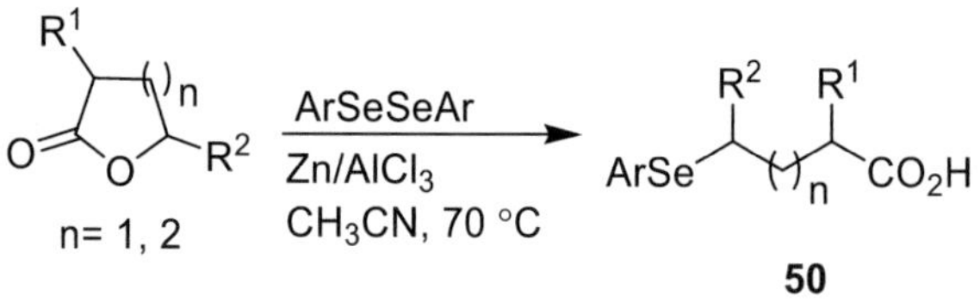

Scheme 3.34

Scheme 3.35

3.3.7. Miscellaneous

Treatment of a wide variety of amines with an equivalent of benzeneselenolates in diglyme in the presence of ruthenium catalyst resulted in formation of the corresponding phenyl selenides, which upon oxidation were transformed into other products (Scheme **36a**, **36b**) [145].

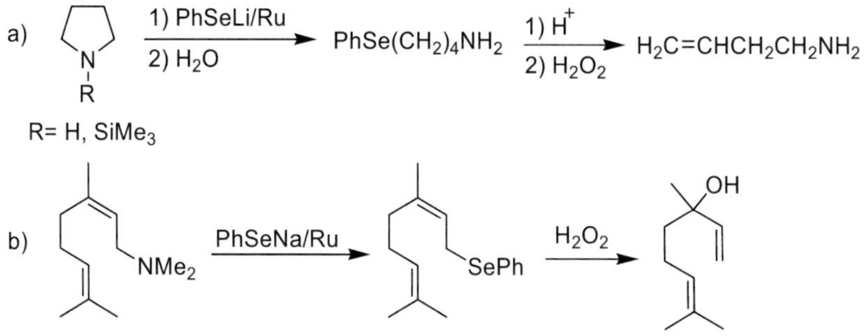

Scheme 3.36

When phenyl trimethylsilyl selenide-$AlBr_3$ system was employed, direct conversion of various benzylic hydroxyl groups into a selenenyl group was accomplished (Scheme **3.37**) [146].

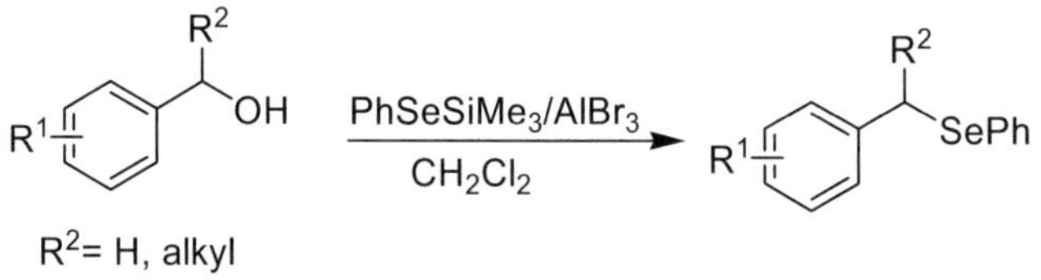

Scheme 3.37

Ranu and coworkers [147] studied the reactions of diphenyl diselenide with aldehydes in the presence of In-TMSCl. Aliphatic aldehydes provided the corresponding selenoacetals, whereas aromatic aldehydes predominantly led to alkyl phenyl selenides under the same reaction conditions (Scheme **3.38**).

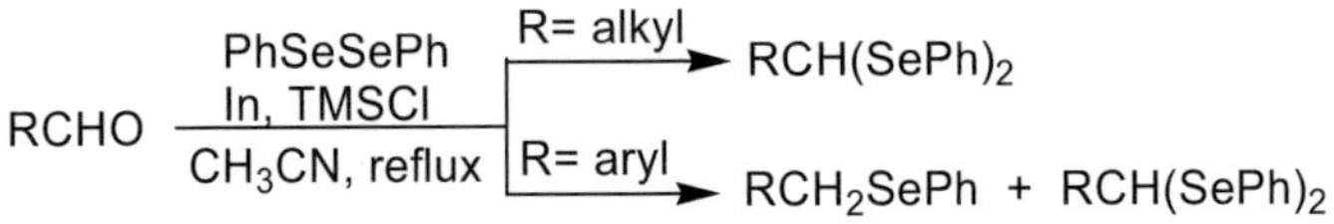

Scheme 3.38

3.4. APPLICATIONS TO SOLID-PHASE SYNTHESIS

Since the first organoselenium resin was reported in 1976 [148], several groups have developed support-bond organoselenium resins as synthetic intermediates and convenient linkers. Ruhland and coworkers described a protocol for the preparation of polystyrene-bound selenium and demonstrated its first application in solid-phase synthesis (SPS) by the synthesis of a library of single aryl alkyl ethers using the Mitsunobu reaction (Scheme **3.39**) [149]. In the same year, Yanada and co-workers [150] utilized a selenide exchange resin, prepared from reduction of selenium with borohydride exchange resin, to give dialkylselenide. Also, a cesium-promoted, one-step coupling of Merrifield's resin with benzeneselenol in the presence of Cs_2CO_3 and TBAI leading to the synthesis of unsymmetrical selenides on solid support was reported [151].

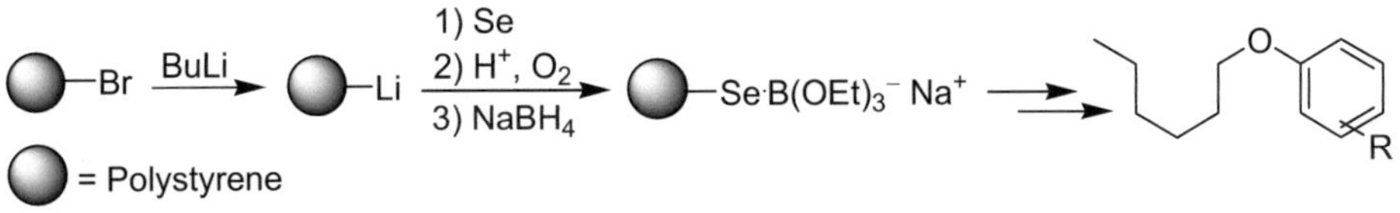

Scheme 3.39

In 2001, Qian and coworkers [152] prepared polystyrene supported selenoesters **52** from the polystyrene supported selenenyl bromide by reaction with sodium borohydride and subsequent acylation with acyl chlorides; this resin was then used as acyl transfer agent to prepare *α,β*-acetylenic ketones **53** by reaction with alkynylcoppers (Scheme **3.40**).

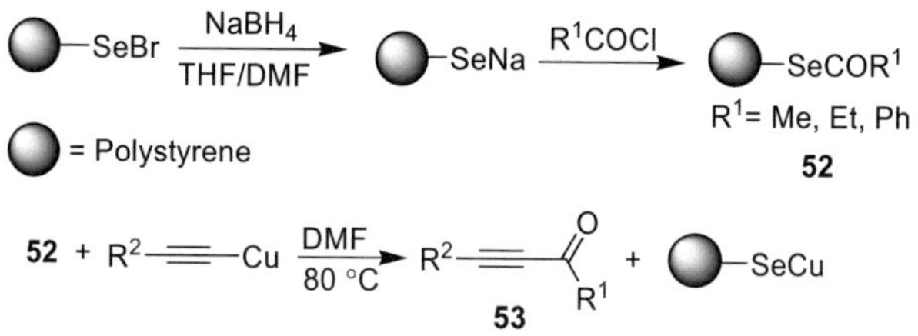

Scheme 3.40

Several years later, applying the same strategy, they developed a novel polystyrene-supported α-seleno acetate **54**; treatment of this resin with LDA, then reaction with aldehydes, followed by the *syn*-elimination of selenoxide giving the β-keto esters **55** (Scheme **3.41**) [153].

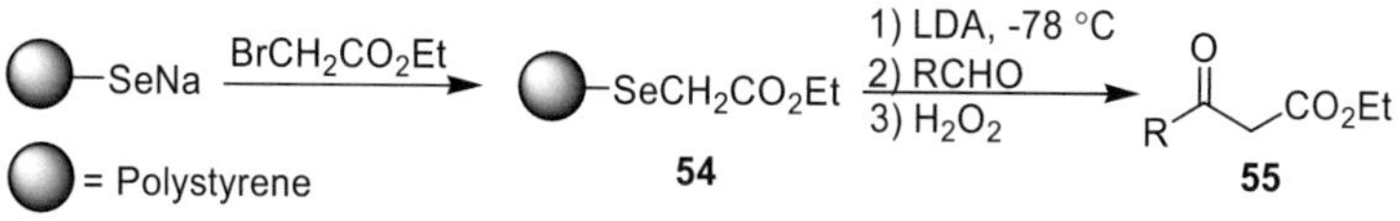

Scheme 3.41

CONCLUSION AND OUTLOOK

In the last forty years or so, our understanding of the reactivity of nucleophilic selenium has advanced to such a level that their use in routine synthesis is now possible. The goal of this chapter was to present the rich chemistry associated with this area of synthetic endeavor in both the distant and the recent past; it is clear that the recent exciting developments in selenium nucleophiles indicate a most promising future for the practitioners in the field in particular and the greater synthetic community in general.

ACKNOWLEDGEMENT

The authors are indebted to Dr. Sogand Noroozizadeh for editing the English content of this chapter.

CONFLICT OF INTEREST

The authors confirm that this chapter contents have no conflict of interest.

REFERENCES

[1] (a) Back, T. G. Organoselenium Chemistry: A Practical Approach; Oxford University Press: Oxford: UK, **1999**; (b) Wirth, T. Organoselenium Chemistry: Synthesis and Reactions; Wiley-VCH, **2012**; (c) Rappoport, Z. Z. The Chemistry of Organic Selenium and Tellurium Compounds; John Wiley & Sons, **2012**, Vol. 3.

[2] Monahan, R.; Brown, D.; Waykole, L.; Liotta, D. In: Organoselenium Chemistry, Liotta, D., Ed.; Wiley: NewYork, **1987**.

[3] Liotta, D.; Sunay, U.; Santiesteban, H.; Markiewicz, W. Phenyl selenide anion, a superior reagent for the S_N2 cleavage of esters and lactones. J. Org. Chem., **1981**, 46, 2605-2610.

[4] Paulmeier, C. Selenium Reagents and Intermediates in Organic Synthesis; Pergamon Press: Oxford, **1986**.

[5] Back, T. G. In The Chemistry of Organic Selenium and Tellurium Compounds, Patai, S., Ed.; Wiley: Chichester, **1987**; Vol. 2, Chapter 3.

[6] Krief, A.; Hevesi, L. Organoselenium Chemistry I: Functional Group Transformations; Springer: Berlin, **1988**.

[7] Iwaoka, M.; Tomoda, S. In Topics in Current Chemistry; Wirth, T., Ed.; Springer-Verlag: Berlin, **2000**, Vol. 208, pp 55-80.

[8] Krief, A. In Comprehensive Organometallic Chemistry II; Abel, E. W.; Stone, F. G. A.; Wilkinson, G.; McKillop, A., Ed.; Pergamon Press: Oxford, **1995**, Vol. 1, pp. 515-569.

[9] Reich, H. J.; Cohen, M. L. Organoselenium Chemistry. Dealkylation of amines with benzeneselenol. J. Org. Chem., **1979**, 44, 3148-3151.

[10] Koketsu, M.; Mizutani, K.; Ogawa, T.; Takahashi, A.; Ishihara, H. Synthesis of 3-Acyl-1-alkyl-2-alkylseleno-1-cyclobutene using alkyneselenolate. J. Org. Chem., **2004**, 69, 8938-8941.

[11] Huang, X.; Xu, X. -H. Insertion of elemental selenium into zinc carbon bond and application in synthesis of α-selenocarbonyl compound. Synth. Commun., **1998**, 28, 807-811.

[12] Klayman, D. L.; Günther, W. H. H. (Eds.), Organic Selenium Compounds: Their Chemistry and Biology, John Wiley & Sons: New York, **1973**.

[13] Brandsma, L.; Wijers, H. E. A simple laboratory scale preparative method for dialkyl sulphides, selenides, tellurides, thiols and selenols. Recl. Trav. Chim., **1963**, 82, 68-74.

[14] (a) Anderson, J. M.; Barker, G. K.; Drake, J. E.; Rodger, M. Methylseleno-derivatives of Group IV. J. Chem. Soc., Dalton Trans., **1973**, 1716-1724; (b) Tiecco, M.; Testaferri, L.; Tingoli, M.; Chianelli, D.; Montanucci, M. Nucleophilic aromatic substitutions of nactivated aryl halides by methyl selenide anions. Synthesis and selective dealkylations of aryl alkyl selenides. J. Org. Chem., **1983**, 48, 4289-4296.

[15] Detty, M. R.; Seidler, M. D. Silyl halides from (phenylseleno)silanes. Reaction with oxiranes and alcohols to give hydrolytically stable silyl ethers. J. Org. Chem., **1981**, 46, 1283-1292.

[16] Kozikowski, A. P.; Ames, A. Dimethylaluminum methylselenolate: a remarkable reagent for the preparation of active acyl-transfer agents. J. Org. Chem., **1978**, 43, 2735-2737.

[17] Gladysz, J. A.; Hornby, J. L.; Garbe, J. E. Convenient one-flask synthesis of dialkyl selenides and diselenides *via* lithium triethylborohydride reduction of sex. J. Org. Chem., **1978**, 43, 1204-1208.

[18] Syper, L.; Mlochowski, J. The convenient syntheses of organoselenium reagents. Synthesis, **1984**, 439-442.

[19] Sekiguchi, M.; Tanaka, H.; Takami, M.; Ogawa, A.; Ryu, I.; Sonoda, N. Reduction of elemental selenium by samarium diiodide: Selective synthesis of diorganyl selenides and diselenides. Heteroatom Chem., **1991**, 2, 427-430.

[20] Tschguaeff, L. Über selenomercaptane und einige derivate derselben. Chem. Ber., **1909**, 42, 49-54.

[21] Schmidt, M.; Block, H. D. Zur kenntnis von tris(alkyl- und -arylseleno)boranen. J. Organomet. Chem., **1970**, 25, 17-21.

[22] Liotta, D. New organoselenium methodology. Acc. Chem. Res., **1984**, 17, 28-34.

[23] Dowd, P.; Kennedy, P. Convenient preparation of sodium and potassium phenyl selenides. Reduction of the selenium-selenium bond by sodium hydride and potassium hydride. Synth.Commun., **1981**,11, 935-941.

[24] (a) Liotta, D.; Paty, P. B.; Johnston, J.; Zima, G. Synthetic applications of phenyltrimethylsilylselenide. Tetrahedron Lett., **1978**, 19, 5091-5094; (b) Miyoshi, N.; Ishii, H.; Kondo, K.; Murai, S.; Sonoda, N. Convenient syntheses of phenyl trimethylsilyl selenide and benzeneselenol. Synthesis, **1979**, 300-301; (c) Detty, M. R.; Wood, G. P. Substitution reactions of thallous thiophenoxide and thallous phenyl selenide with halogen-bearing substrates. J. Org. Chem., **1980**, 45, 80-89.

[25] Smith, A. B. III; Scarborough, R. M. Jr. Nucleophilic ring cleavage of mono-activated cyclopropanes *via* sodium and lithium phenyl selenolate. Tetrahedron Lett., **1978**, 19, 1649-1652.

[26] Bauer, H. Über O-Nitrophenyl-selencyanid und O-Amino-selenophenol. Ber., **1913**, 46, 92-98.

[27] Cravador, A.; Krief, A. Nouvelles méthodes de synthèse des méthylsélénoacétals et cétals. C. R. Acad. Sci. Paris, **1979**, 289c, 267-269.

[28] (a) Bao, W.; Zheng, Y.; Zhang, Y.; Zhou, J. A novel synthesis of allyl and propargyl selenides in aqueous media promoted by indium. Tetrahedron Lett., **1996**, 37, 9333-9334; (b) Zheng, Y.; Bao, W.; Zhang, Y. A novel synthesis of allyl and propargyl selenides promoted by indium. Synth. Commun., **1997**, 27, 79-83.

[29] Santi, C.; Santoro, S.; Testaferri, L.; Tiecco, M. A simple zinc-mediated preparation of selenols. Synlett, **2008**, 1471-1474.

[30] Movassagh, B.; Tatar, A. Zn/$RuCl_3$-promoted cleavage of diselenides: An efficien Michael addition of zinc selenolates to conjugated alkenes in aqueous media. Synlett, **2007**, 1954-1956.

[31] (a) Nazari, M.; Movassagh, B. Nucleophilic cleavage of lactones and esters with zinc selenolates prepared from diselenides in the presence of Zn/$AlCl_3$. Tetrahedron Lett., **2009**, 50, 438-441; (b) Movassagh, B.; Mirshojaei, F. Synthesis of selenol esters from acid chlorides and organic diselenides in the presence of Zn/$AlCl_3$ system. Monat. Chem., **2003**, 134, 831-835; (c) Movassagh, B.; Shamsipour, M. Stereo- and regioselective zinc mediated ring-opening of epoxides with diselenides. Synlett, **2005**, 1316-1318; (d) Nazari, M.; Movassagh, B. Nucleophilic ring opening of mono-activated cyclopropanes with aryl-selenolates generated from diselenides in the presence of a Zn/$AlCl_3$ system. Synlett, **2009**, 1803-1805; (e) Movassagh, B.; Farshbaf, S. Regioselective hydroxyselenation of styrenes by diselenides using the Zn/$AlCl_3$ system. Synthesis, **2010**, 33-35; (f) Movassagh, B.;

Salehi Morovat, E. Regioselective ring-opening of aziridines with diselenides and disulfides using the Zn/$AlCl_3$ system. J. Sulf. Chem., **2011**, 32, 117-122.

[32] Narayanaperumal, S.; Alberto, E. E.; Gul, K.; Rodrigues, E. D.; Braga, A. L. Synthesis of diorganyl selenides mediated by zinc in ionic liquid. J. Org. Chem., **2010**, 75, 3886-3889.

[33] Nishino, T.; Okada, M.; Kuroki, T.; Watanabe, T.; Nishiyama, Y.; Sonoda, N. One-pot synthetic method of unsymmetrical diorganyl selenides: Reaction of diphenyl diselenide with alkyl halides in the presence of lanthanum metal. J. Org. Chem., **2002**, 67, 8696-8698.

[34] Silveira, C.; Braga, A. L.; Larghi, E. L. Synthesis of thiol, selenol, and tellurol esters by the reaction of organochalcogeno mercurials with acid chlorides. Organometallics, **1999**, 18, 5183-5186.

[35] Braga, A. L.; Schneider, P. H.; Paixão, M. W.; Deobald, A. M. Efficient synthesis of diorganyl selenides *via* cleavage of Se-Se bond of diselenides by indium(III) catalyst and zinc. Tetrahedron Lett., **2006**, 47, 7195-7198.

[36] Ranu, B. C.; Mandal, T.; Samanta, S. Indium(I) iodide-mediated cleavage of diphenyl diselenide. An efficient one-pot procedure for the synthesis of unsymmetrical diorganyl selenides. Org. Lett., **2003**, 5, 1439-1441.

[37] Wang, L.; Zhang, Y. Reductive Cleavage of the Se-Se Bond by the Sm-Me_3SiCl-H_2O System: Preparation of Unsymmetrical Phenyl Selenides. J. Chem. Res. (S), **1998**, 598-599.

[38] Schmidt, M.; Block, H. D.; Deng, H.; Chen, J.; Liu, M.; Ding, M.; Wu, H. Notiz zur vereinfachten Darstellung von Alkanselenolen. Chem. Ber., **1970**, 103, 3348-3349.

[39] Dan, W.; Deng, H.; Chen, J.; Liu, M.; Ding, M.; Wu, H. A new odorless one-pot synthesis of thioesters and selenoesters promoted by Rongalite. Tetrahedron, **2010**, 66, 7384-7388.

[40] Fukuzawa, S.; Niimoto, Y.; Fujinami, T.; Sakai, S. Samarium (II) diiodide-induce reductive cleavage of Se-Se and Te-Te bond in diphenyl diselenide and ditelluride leading to a samarium phenylselenolate and tellurolate: Preparation of unsymmetrical alkylphenyl selenides and tellurides. Heteroatom Chem., **1990**, 1, 491-495.

[41] Pedersen, M. L.; Berkowitz, D. B. Formal alpha-vinylation of amino acids. Use of a new benzeneselenolate equivalent. J. Org. Chem., **1993**, 58, 6966-6975.

[42] Sakakibara, M.; Katsumata, K.; Watanabe, Y.; Toru, T.; Veno, Y. A Convenient Procedure for the Preparation of Organic Selenides. Synthesis, **1992**, 377-379.

[43] Detty, M. R. Phenylselenotrimethylsilane. A novel source of phenylselenide anion. Tetrahedron Lett., **1978**, 19, 5087-5090.

[44] Maruoka, K.; Miyazaki, T.; Ando, M.; Matsumura, Y.; Sakane, S.; Hattori, K.; Yamamoto, H. Organoaluminum-promoted Beckmann rearrangement of oxime sulfonates. J. Am. Chem. Soc., **1983**, 105, 2831-2843.

[45] Klayman, D. L. The synthesis of aminoethyl-substituted selenium compounds. J. Org. Chem., **1965**, 30, 2454-2456.

[46] Inokuchi, T.; Kusumoto, M.; Torri, S. Electroreductive ring-opening of α,β-epoxy carbonyl compounds and their homologs through recyclable use of diphenyl diselenide or diphenyl ditelluride as a mediator. J. Org. Chem., **1990**, 55, 1548-1553.

[47] Torii, S.; Inkuchi, T.; Hasegawa, N. A direct arylselenenylation of enones with diaryl diselenides by a electroreductive procedure. Chem. Lett., **1980**, 639-640.

[48] Engman, L.; Stern, D. Thiol/Diselenide exchange for the generation of benzeneselenolate ion. Catalytic reductive ring-opening of α,β-Epoxy Ketones. J. Org. Chem., **1994**, 59, 5179-5183.

[49] Krief, A.; Delmotte, C.; Dumont, W. Chemoselective reduction of organoselenocyanates to diselenides and selenolates. Tetrahedron, **1997**, 53, 12147-12158.

[50] (a) Sharpless, K. B.; Young, M. W. Olefin synthesis. Rate enhancement of the elimination of alkyl aryl selenoxides by electron-withdrawing substituents. J. Org. Chem., **1975**, 40, 947-949; (b) Grieco, P. A.; Noguez, J. A.; Masaki, Y. (±)-Deoxyvernolepin. A cytotoxic vernolepin prototype. J. Org. Chem., **1977**, 42, 495-502.

[51] Tamura, Y.; Adachi, M.; Kawasaki, T.; Kita, Y. Selenocyanation using a combined reagent of triphenylphosphine and selenocyanogen. A new and simple synthesis of alkyl selenocyanates and symmetrical alkyl diselenides from alcohols. Tetrahedron Lett., **1979**, 20, 2251-2252.

[52] Fredga, A. Amides of aliphatic selena-carboxylic acids. Acta Chem. Scand., **1963**, 17, S51-S54.

[53] Logan, G.; Igunbor, C.; Chen, G. -X.; Davis, H.; Simon, A.; Salon, J.; Huang, Z. A simple strategy for incorporation, protection, and deprotection of selenium functionality. Synlett, **2006**, 1554-1558.

[54] Harvey, A. B.; Durig, J. R.; Morrissey, A. C. Vibrational spectra and structure of four-membered ring molecules. XIV. Vibrational analysis and ring puckering vibration of trimethylene selenide and trimethylene selenide-d_4. J. Chem. Phys., **1969**, 50, 4949- 4961.

[55] Günther, W. H. H. Methods in selenium chemistry. II. Bis(methoxymagnesium) diselenide, a novel reagent for the introduction of selenium into organic molecules. J. Org. Chem., **1967**, 32, 3929-3931.

[56] Ścianowski, J. Convenient route to dialkyl diselenides from alkyl tosylates. Synthesis of di(*cis*-myrtanyl) diselenide. Tetrahedron Lett., **2005**, 46, 3331-3334.

[57] Liotta, D.; Saindane, M.; Barnum, C. Reactions involving selenium metal as an electrophile. 2. A general procedure for the preparation of unsaturated β-dicarbonyl compounds. Tetrahedron Lett., **1981**, 22, 3043-3046.

[58] Tiecco, M.; Testaferri, L.; Tingoli, M.; Chianelli, D.; Montanucci, M. Convenient synthesis of lithium methyl selenide. An efficient reagent for the S_N2 cleavage of esters. Synth. Commun. **1983**, 13, 617-620.

[59] (a) Hoye, T. R.; Caruso, A. J. A 1,3-diene-2-carboxylic ester synthesis; Reaction of alkylidene-γ-lactones with potassium and sodium phenyl selenoate. Tetrahedron Lett., **1978**, 19, 4611-4614; (b) Liotta, D.; Markiewicz, W.; Santiesteban, H. The generation of uncomplexe Tetrahedron Lett., **1977**, 18, 4365-4367.

[60] (a) Kametani, T.; Takeuchi, Y.; Yoshii, E. An efficient synthetic route to (±)-nanaomycin A. J. Org. Chem., **1983**, 48, 2630-2632; (b) Arita, M.; Adachi, K.; Ito, Y.; Sawai, H.; Ohno, M. Enantioselective synthesis of the carbocyclic nucleosides (–)-aristeromycin and (–)-neplanocin A by a chemicoenzymatic approach. J. Am. Chem. Soc., **1983**, 105, 4049-4055; (c) Krief, A.; Trabelsi, M.; Dumont, W. Syntheses of alkali selenolates from diorganic diselenides and alkali metal hydrides: Scope and limitations. Synthesis, **1992**, 933-935.

[61] Krief, A.; Derock, M.; Lacroix, D. Zinc-mediated synthesis of tertiary alkyl selenide from tertiary alkyl halides. Synlett, **2005**, 2832-2834.

[62] Zheng, Y.; Bao, W.; Zhang, Y. Synthesis of allyl-type selenides and α-selenoesters in aqueous media promoted by cadmium. Synth. Commun., **2000**, 30, 1731-1736.

[63] (a) Guo, H.; Zheng, Y.; Zhang, Y. Convenient one-pot synthesis of α-selenonitriles, α-selenoesters and asymmetrical selenides by a $Sm/ZnCl_2$ system in $DMF-H_2O$. J. Chem. Res. (S), **2001**, 160-161; (b) Zheng, Y.; Guo, H.; Zhang, Y. Synthesis of allyl selenides

promoted by an Sm/$ZnCl_2$ bimetal system in the presence of water. Chin. J. Chem., **2001**, 19, 530-532

[64] Fukuzawa, S.; Fujinami, T.; Sakai, S. Samarium(II) di-iodide induced synthesis of allylicphenyl selenides from allylic acetates and diphenyl diselenide in the presence of palladium catalyst. Chem. Lett., **1990**, 927-930.

[65] Liao, P.; Bao, W.; Zhang, Y. A novel synthesis of allyl and prop-2-ynyl selenides promoted by tin in the presence of water. J. Chem. Res. (S), **1998**, 150-151.

[66] Su, W.; Zhang, Y.; Zheng, Y.; Li, Y. Facile formation of ytterbium diiodide and its use in the synthesis of allyl selenides. Chin. J. Chem., **2002**, 20, 174-177.

[67] Kundu, A.; Roy, S. Copper(II)/Tin(II) reagent for allylation, propargylation, alkynylation, and benzylation of diselenides: A novel bimetallic reactivity. Organometallics, **2000**, 19, 105-107.

[68] (a) Movassagh, B.; Shamsipour, M. Zinc-mediated cleavage of diselenides: A novel synthesis of unsymmetrical diorganyl selenides in aqueous media. Synlett, **2005**, 121-122; (b) Zhao, X.; Yu, Z.; Yan, S.; Wu, S.; Liu, R.; He. W.; Wang, L. Ruthenium(III) chloride catalyzed efficient synthesis of unsymmetrical diorganyl selenides *via* cleavage of dibenzyl and diphenyl diselenides in the presence of zinc. J. Org. Chem., **2005**, 70, 7338-7341.

[69] Cristau, H. J.; Chabaud, B.; Labaudiniere, R.; Christol, H. Synthesis of aryl phenyl and heteroaryl phenyl selenides by nickel(II)-catalyzed arylation of sodium benzeneselenolate. Organometallics, **1985**, 4, 657-661.

[70] (a) Nishiyama, Y.; Tokunaga, K.; Sonoda, N. New synthetic method of diorganyl selenides: Palladium-catalyzed reaction of $PhSeSnBu_3$ with aryl and alkyl halides. Org. Lett., **1999**, 1, 1725-1727; (b) Beletskaya, I. P.; Sigeev, A. S.; Peregudov, A. S.; Petrovskii, P. V. Copper(I)-catalyzed arylselenylation of aryl bromides and iodides. Tetrahedron Lett., **2003**, 44, 7039-7041.

[71] Movassagh, B.; Fazeli, A. Zinc-mediated synthesis of diaryl selenides from diaryl diselenides and diaryliodonium salts in aqueous media. Z. Naturforsch., **2006**, 61b, 194-196.

[72] (a) Fukuzawa, S.; Tanihara, D.; Kikuchi, S. Palladium-catalyzed coupling reaction of diaryl dichalcogenide with aryl bromide leading to the synthesis of unsymmetrical aryl chalcogenide. Synlett, **2006**, 2145-2147; (b) Ranu, B. C.; Chattopadhyay, K.; Banerjee, S. Indium(I) iodide promoted cleavage of diphenyl diselenide and disulfide and subsequent palladium(0)-catalyzed condensation with vinylic bromides. A simple one-pot synthesis of vinylic selenides and sulfides. J. Org. Chem., **2006**, 71, 423-425.

[73] Wong, Y. -C.; Jayanth, T. T.; Cheng, C. -H. Cobalt-catalyzed aryl-sulfur bond formation. Org. Lett., **2006**, 8, 5613-5616.

[74] (a) Kumar, S.; Engman, L. Microwave-assisted copper-catalyzed preparation of diaryl chalcogenides. J. Org. Chem., **2006**, 71, 5400-5403; (b) Taniguchi, N. Convenient synthesis of unsymmetrical organochalcogenides using organoboronic acids with dichalcogenides *via* cleavage of the S−S, Se−Se, or Te−Te bond by a copper catalyst. J. Org. Chem., **2007**, 72, 1241-1245; (c) Wang, L.; Wang, M.; Huang, F. A simple copper salt-catalyzed synthesis of unsymmetrical diaryl selenides and tellurides from arylboronic acids with diphenyl diselenide and ditelluride. Synlett, **2005**, 2007-2010.

[75] Alves, D.; Santos, C. G.; Paixão, M. W.; Soares, L. C.; deSouza, D.; Rodrigues, O. E. D.; Braga, A. L. CuO nanoparticles: an efficient and recyclable catalyst for cross-coupling

reactions of organic diselenides with aryl boronic acids. Tetrahedron Lett., **2009**, 50, 6635-6638.

[76] Li, Y.; Wang, H.; Li, X.; Chen, T.; Zhao, D. CuS/Fe: a novel and highly efficient catalyst system for coupling reaction of aryl halides with diaryl diselenides. Tetrahedron, **2010**, 66, 8583-8586.

[77] Swapna, K.; Murthy, S. N.; Nageswar, Y. V. D. Magnetically separable and reusable copper ferrite nanoparticles for cross-coupling of aryl halides with diphenyl diselenide. Eur. J. Org. Chem., **2011**, 1940-1946.

[78] Singh, V. P.; Singh, H. B.; Butcher, R. J. Synthesis and glutathione peroxidase-like activities of isoselenazolines. Eur. J. Org. Chem., **2011**, 5485-5497.

[79] Millois, C.; Diaz, P. Solution-phase synthesis of diaryl selenides using polymer-supported borohydride. Org. Lett., **2000**, 2, 1705-1708.

[80] Perin, G.; Lenardão, E. J.; Jacob, R. G.; Panatieri, R. B. Synthesis of vinyl selenides. Chem. Rev., **2009**, 109, 1277-1301.

[81] Tiecco, M.; Testaferri, L.; Tingoli, M.; Chianelli, D.; Montanucci, M. New synthesis of vinyl selenides nucleophilic substitutions of unactivated vinyl halides by selenide anions. Tetrahedron Lett., **1984**, 25, 4975-4978.

[82] (a) Testaferri, L.; Tiecco, M.; Tingoli, M.; Chianelli, D. Stereospecific synthesis of divinyl diselenides from vinyl acetyl selenides. Tetrahedron, **1986**, 42, 4577-4584; (b) Testaferri, L.; Tiecco, M.; Tingoli, M.; Chinelli, D. Stereospecific synthesis of divinyl selenides nucleophilic substitutions of unactivated vinyl halides by vinyl selenide anions. Tetrahedron, **1986**, 42, 63-69.

[83] Chang, D.; Bao, W. Stereoselective synthesis of vinyl selenides by coupling of vinyl bromides with diorgano diselenides catalyzed by copper(I) in [BMIM]BF_4. Synlett, **2006**, 1786-1788.

[84] Yan, J.; Chen, Z. -C. Hypervalent iodine in synthesis 44: Stereoselective synthesis of vinylic selenides by the reaction of sodium selenolates with vinyl (phenyl) iodonium salts. Synth. Commun., **2000**, 30, 1009-1015.

[85] Braga, A. L.; Barcellos, T.; Paixão, M. W.; Deobald, A. M.; Godoi, M.; Stefani, H. A.; Cella, R.; Sharma, A. Copper(I)-catalyzed efficient and stereoselective synthesis of (E)-vinyl selenides and tellurides by the reaction of potassium vinyltrifluoroborates with diphenyl dichalcogenides. Organometallics, **2008**, 27, 4009-4012.

[86] Stefani, H. A.; Comasseto, J. V.; Petragnani, N.; Braga, A. L.; Menezes, P. H.; Gusevskaya, E. V. Nickel(II) catalyzed substitution of halogens in 1-halo-1-chalcogene alkenes by chalcogenate anions. Phosphorous, Sulfur, Silicon, **1997**, 126, 211-222.

[87] Santi, C.; Santoro, S.; Battistelli, B.; Testaferri, L.; Tiecco, M. Preparation of the first bench-stable phenyl selenolate: an interesting "on water" nucleophilic reagent. Eur. J. Org. Chem., **2008**, 5387-5390.

[88] Santoro, S.; Battistelli, B.; Testaferri, L.; Tiecco, M.; Santi, C. Vinylic substitutions promoted by PhSeZnCl: Synthetic and theoretical investigations Eur. J. Org. Chem., **2009**, 4921-4925.

[89] Grieco, P. A.; Yokoyama, Y.; Williams, E. Aryl selenocyanates and aryl thiocyanates: reagents for the preparation of activated esters. J. Org. Chem., **1978**, 43, 1283-1285.

[90] Pfenniger, J.; Heuberger, C.; Graf, W. Steroids and sex hormones. Part 262. The radical induced stannane reduction of selenoesters and selenocarbonates: A new method for the

degradation of carboxylic acids to nor-alkanes and for desoxygenation of alcohols to alkanes Helv. Chim. Acta, **1980**, 63, 2328-2337.

[91] (a) Hiscock, S. M.; Swann, D. A.; Turnbull, J. H. Axial and equatorial selenols: 3α- and 3β-selenyl derivatives of Cholestane and Androstanone. J. Chem. Soc. D, **1970**, 1310a-1310a; (b) Mullen, G. P.; Luthra, N. P.; Dunlap, R. B.; Odom, J. D. Synthesis and multinuclear magnetic resonance study of para-substituted phenyl selenobenzoates. J. Org. Chem., **1985**, 50, 811-816.

[92] (a) Nishiyama, Y.; Kawamatsu, H.; Funato, S.; Tokunaga, K.; Sonoda, N. Phenyl tributylstannyl selenide as a promising reagent for introduction of the phenylseleno group J. Org. Chem., **2003**, 68, 3599-3602; (b) Beletskaya, I. P.; Sigeev, A. S.; Peregudov, A. S.; Petrovskii, P. V. Tributylstannyl aryl selenides as efficient arylselenating agents in the synthesis of seleno esters. Russ. J. Org. Chem., **2001**, 37, 1703-1709.

[93] Lu, G.; Zhang, Y. Facile formation of samarium(III) selenolates from potassium iodide catalyzed reaction of samarium metal and diselenides, and their application in synthesis of selenoesters. Synth. Commun., **1999,** 29, 219-225.

[94] Wang, L.; Zhang, Y. Water accelerated Sm/TMSCl reductive cleavage of the Se-Se bond: Synthesis of selenoesters and selenoformates. Synth. Commun., **1999,** 29, 3107- 3115.

[95] Liu, Y.; Zhang, Y. Reductive cleavage of the Se-Se bond by the Sm/$CrCl_3$ system: A novel one-pot method for the synthesis of selenoesters. Synth. Commun., **1999**, 29, 4043- 4049.

[96] Chen, R.; Zhang, Y. Cleavage of Se-Se Bond by Sm/cat. $CoCl_2$ System: A Novel Metho for the Synthesis of Selenoesters. Synth. Commun., **2000**, 30, 1331-1336.

[97] Munbunjong, W.; Lee, E. H.; Ngernmaneerat, P.; Kim, S. J.; Singh, G.; Chavasiri, W.; Jang, D. O. Indium-mediated cleavage of diphenyl diselenide and diphenyl disulfide: efficient one-pot synthesis of unsymmetrical diorganyl selenides, sulfides, and selenoesters. Tetrahedron, **2009**, 65, 2467-2471.

[98] Movassagh, B.; Shamsipour, M.; Joshaghani, M. Reductive cleavage of the Se-Se bond in the presence of a Zn/$AlCl_3$ system: synthesis of selenol esters. J. Chem. Res., **2004**, 148-149.

[99] (a) Barrett, A. G. M.; Kwon, H.; Wallace, E. M. The conversion of carboxylic acids into isonitriles *via* selenium-phenyl selenocarbamates. J. Chem. Soc., Chem. Commun., **1993**, 1760-1761; (b) Bachi, M. D.; Bosch, E. Synthesis of γ- and δ-lactones by free-radical annelation of Se-phenyl selenocarbonates. J. Org. Chem., **1992**, 57, 4696-4705; (c) Clive, D. L. J.; Yang, W. Preparation of tetrahydrofuran, γ-lactone, chromanol and pyrrolidine systems by sequential 5-exo-digonal radical cyclization, 1,5-hydrogen transfer from silicon, and 5-endo-trigonal cyclization. Chem. Commun., **1996**, 1605-1606; (d) Toyofuku, M.; Fujwara, S. -I.; Shin-ike, T.; Kuniyasu, H.; Kambe, N. Palladium-catalyzed intramolecular selenocarbamoylation of alkynes with carbamoselenoates: Formation of α-alkylidene-β-lactam framework. J. Am. Chem. Soc., **2005**, 127, 9706-9707.

[100] Hartwig, W. Modern methods for the radical deoxygenation of alcohols. Tetrahedron, **1983**, 39, 2609-2645.

[101] Lucas, M. A.; Nguyen, O. T. K.; Schiesser, C. H.; Zheng, S. -L. Preparation of 5-selenopentopyranose sugars from pentose starting materials by samarium(II) iodide or (phenylseleno)formate mediated ring closures. Tetrahedron, **2000**, 56, 3995-4000.

[102] Reinerth, W. A.; Tour, J. M. Protecting groups for organoselenium compounds. J. Org. Chem., **1998**, 63, 2397-2400.

[103] Schiesser, C. H.; Skidmore, M. A. Phenylselenotris(trimethylsilyl)silane and phenyltellurotris((trimethylsilyl)silane: versatile reagents for the preparation of phenylseleno- and phenyltelluro-formates. J. Chem. Soc., Perkin Trans. 1, **1997**, 2689-2690.

[104] Lucas, M. A.; Schiesser, C. H. (Aryltelluro)formates as precursors of alkyl radicals:□ Thermolysis and photolysis of primary and secondary alkyl (aryltelluro)formats. J. Org. Chem., **1996**, 61, 5754-5761.

[105] (a) Movassagh, B.; Fazeli, A. Zinc-mediated cleavage of diselenides: A novel synthesis of selenoformates in aqueous media. Monatsh. Chem., **2007**, 138, 863-865; (b) Movassagh, B. Tatar, A. An efficient, one-pot synthesis of alkyl arylselenoformates using the zinc-ruthenium chloride system in aqueous media. Org. Prep. Proc. Int., **2008**, 40, 477-481.

[106] Rigby, J. H.; Danca, D. M.; Horner, J. H. Carbamoyl radicals from Se-phenylselenocarbamates: Intramolecular additions to alkenes. Tetrahedron Lett., **1998**, 39, 8413-8416.

[107] (a) Ishihara, H.; Koketsu, M.; Fukuta, Y.; Nada, F. Reaction of lithium aluminum hydride with elemental selenium: Its application as a selenating reagent into organic molecules. J. Am. Chem. Soc., **2001**, 123, 8408-8409; (b) Koketsu, M.; Ishida, M.; Takakura, N.; Ishihara, H. Preparation and characterization of N-alkyl-Se- alkylselenocarbamates. J. Org. Chem., **2002**, 67, 486-490.

[108] Sharpless, K. B.; Lauer, R. F. Mild procedure for the conversion of epoxides to allylic alcohols. First organoselenium reagent. J. Am. Chem. Soc., **1973**, 95, 2697-2699.

[109] (a) Reich, H. J.; Chow, F. Selenium stabilized anions. Conversion of β-hydroxy selenides into olefins. J. Chem. Soc., Chem. Commun., **1975**, 790-791; (b) Sevring, M.; Domount, W.; Henesi, L.; Krief, A. Transformation of selenides to alkylhalides new routes for homologization of primary alkylhalides. Tetrahedron Lett., **1976**, 2647-2650; (c) Van Ende, D.; Domount, W.; Krief, A. A new route to epoxides. Angew. Chem. Int. Ed. Engl., **1975**, 14, 700-702; (d) Domount, W.; Krief, A. New synthetic route to di- and trisubstituted epoxides. Angew. Chem. Int. Ed. Engl., **1975**, 14, 350-351.

[110] Cravador, A.; Krief, A. Unusual reactivity of selenoboranes towards epoxides: new selective routes to β-hydroxyselenides and allylalcohols. Tetrahedron Lett., **1981**, 22, 2491-2494.

[111] (a) Kozikowski, A. P.; Ames, A. Dimethylaluminium methaneselenolate-a useful reagent for the preparation of selenoesters. A new Friedel-Crafts acylation procedure promoted by Cu(I). Tetrahedron, **1985**, 41, 4821-4834; (b) Bartlett, P. A.; McQuaid, L. A. Total synthesis of (±)-methyl shikimate and (±)-3-phosphoshikimic acid. J. Am. Chem. Soc., **1984**, 106, 7854-7860.

[112] Posner, G. H.; Rogers, D. Z. Organic reactions at alumina surfaces. Mild and selective opening of epoxides by alcohols, thiols, benzeneselenol, amines, and acetic acid. J. Am. Chem. Soc., **1977**, 99, 8208-8214.

[113] Sridhar, R.; Srinivas, B.; Surendra, K.; Krishnaveni, N. S.; Rao, K. R. Synthesis of β-hydroxy selenides using benzeneselenol and oxiranes under supramolecular catalysis in the presence of β-cyclodextrin in water. Tetrahedron Lett., **2005**, 46, 8837-8839.

[114] Nishiyama, Y.; Ohashi, H.; Itoh, K.; Sonoda, N. Activation of Sn-Se bond: Regioselective ring opening reaction of epoxides with tributylstannyl phenylselenolate in the presence of a Lewis acid. Chem. Lett., **1998**, 2, 159-160.

[115] Sakakibara, M.; Katsumata, K.; Watanabe, Y.; Toru, T.; Veno, Y. A convenient procedure for the preparation of organic selenides. Synthesis, **1992**, 377-379.

[116] Wang, J. -X.; Wu, X.; Huang, D.; Hu, Y. The synthesis of 1-arylseleno-3-alkoxy-2-propanols under microwave irradiation conditions. Synth. Commun., **2002**, 32, 1903- 1908.

[117] Chen, X.; Wu, H.; Su, W.; Xu, R.; Liu, M.; Ding, J. Active metallic indium-mediated ring-opening of epoxides with diphenyl diselenides: a novel one-pot synthesis of β-hydroxy selenides in aqueous media. J. Chem. Res., **2007**, 325-327.

[118] do Rego Barros, O. S.; de Carvalho, A. B.; Lang, E. S.; Peppe, C. Indium(I) iodide-mediated regioselective ring opening of epoxides with diphenyldiselenide: The preparation of β-hydroxy selenides. Lett. Org. Chem., **2004**, 1, 43-46.

[119] Yang, M.; Zhu, C.; Yuan, F.; Huang, Y.; Pan, Y. Enantioselective ring-opening reaction of meso-epoxides with ArSeH catalyzed by heterometallic Ti-Ga-Salen system. Org. Lett., **2005**, 7, 1927-1930.

[120] Tiecco, M.; Testaferri, L.; Marini, F.; Sternativo, S.; Delverme, F.; Santi, C.; Bagnoli, L.; Temperini, A. Synthesis of enantiomerically enriched β-hydroxy selenides by catalytic asymmetric ring opening of meso-epoxides with (phenylseleno)silanes. Tetrahedron, **2008**, 64, 3337-3342.

[121] Nishibayashi, Y.; Singh, J. D.; Fukuzawa, S. -I.; Vemura, S. Synthesis of chiral diferrocenyl dichalcogenides and their application to asymmetric nucleophilic ring opening of meso-epoxides. J. Chem. Soc., Perkin Trans. 1, **1995**, 2871-2876.

[122] Knapp, S.; Zhao, D. Synthesis of the sialidase inhibitor siastatin B. Org. Lett., **2000**, 2, 4037-4040.

[123] Rigby, J. H.; Maharoof, U. S. M.; Mateo, M. E. Studies on the Narciclasine alkaloids: Total synthesis of (+)-Narciclasine and (+)-Pancratistatin. J. Am. Chem. Soc., **2000**, 122, 6624-6628.

[124] Treadwell, E. M.; Neighbors, J. D.; Wiemer, D. F. A cascade cyclization approach to Schweinfurthin B. Org. Lett., **2002**, 4, 3639-3642.

[125] Still, W. C.; Murata, S.; Revial, G.; Yoshihara, K. Synthesis of the cytotoxic germacranolide eucannabinolide. J. Am. Chem. Soc., **1983**, 105, 625-627.

[126] Schlessinger, R. H.; Lopes, A. A total synthesis of racemic senepoxide: formal syntheses of crotepoxide and pipoxide. J. Org. Chem., **1981**, 46, 5252-5253.

[127] Stevens, R. V.; Albizati, K. F. Synthetic approach to the amphilectane diterpenes: the use of nitriles as terminators of carbocation-olefin cyclizations. J. Org. Chem., **1985**, 50, 632-640.

[128] Haraguchi, K.; Tanaka, H.; Miyasaka, T. Preparation of γ- and δ-phenylselenenyl alcohols *via* ring cleavage of oxetane and oxolane. Synthesis, **1989**, 434-436.

[129] Kataoka, T.; Yoshimatsu, M.; Shimizu, H.; Kawase, Y.; Hori, M. The C-O bond cleavage of cyclic ethers with tris(phenylseleno)borane-Lewis acid. Heterocycles, **1990**, 31, 889-893.

[130] Abe,H.; Yamasaki, A.; Fuji, H.; Harayama, T. Convenient transformation of benzyl alcohol into its corresponding selenide using the selenolate anion-aluminium chloride combination system. Chem. Pharm. Bull., **1996**, 44, 2223-2226.

[131] (a) Schneider, A.; Rodrigues, O. E. D.; Paixão, M. W.; Appelt, H. R.; Braga, A. L.; Wessjohann, L. A. Stereoselective synthesis of Boc-protected L-seleno- and tellurolanthionine, L-seleno- and tellurocystine and derivatives. Tetrahedron Lett., **2006**, 47, 1019-1021; (b) Tanner, D.; He, H. M. Enantioselective routes toward 1β-methylcarbapenems from chiral aziridines. Tetrahedron, **1992**, 48, 6079-6086; (c) Moroder, L. Isosteric replacement of sulfur with other chalcogens in peptides and proteins. J. Pept.

Sci., **2005**, 11, 187-214; (d) Pegoraro, S.; Fiori, S.; Cramer, J.; Rudolph-Böhner, S.; Moroder, L. The disulfide-coupled folding pathway of apamin as derived from diselenide-quenched analogs and intermediates. Protein Sci., **1999**, 8, 1605-1613.

[132] (a) Xiong, S.; Markesbery, W. R.; Shao, C.; Lovell, M. A. Seleno-L-methionine protects against β-amyloid and iron/hydrogen peroxide-mediated neuron death. Antiox. Redox Signaling, **2007**, 9, 457-467; (b) Lyons, T. P.; Power, R.; EP 1774972, 2007; Chem. Abstr., **2007**, 146, 428741.

[133] Ide, N. D.; Galonic, D. P.; Van der Donk, W. A.; Gin, D. Y. Conjugation of selenols with aziridine-2-carboxylic acid-containing peptides. Synlett, **2005**, 2011-2014.

[134] Zhang, W. -X.; Ye, K.; Ruan, S.; Chen, Z. -X.; Xia, Q. -H. Tri-n-butylphosphine mediated ring-opening reactions of aziridines or epoxides with diphenyl diselenide. Chin. J. Chem., **2007**, 25, 1758-1761.

[135] Salman, S. M.; Schwab, R. S.; Alberto, E. E.;Vargas, J.; Dornelles, L.; Rodrigues, O. E. D.; Braga, A. L. Efficient ring opening of protected and unprotected aziridines promoted by stable zinc selenolate in ionic liquid. Synlett, **2011**, 69-72; (b) Braga, A. L.; Schwab, R. S.; Alberto, E. E.; Salman, S. M.; Vargas, J.; Azeredo, J. B. Ring opening of unprotected aziridines by zinc selenolates in a biphasic system. Tetrahedron Lett., **2009**, 50, 2309-2311; (c) Braga, A. L.; Schneider, P. H.; Paixão, M. W.; Deobald, A. M.; Peppe, C.; Bottega, D. P. Chiral seleno-amines from indium selenolates. A straightforward synthesis of selenocysteine derivatives. J. Org. Chem., **2006**, 71, 4305-4307; (d) Braga, A. L.; Paixão, M. W.; Marin, G. Seleno-Imine: A new class of versatile, modular N, Se ligands for asymmetric palladium-catalyzed allylic alkylation. Synlett, **2005**, 1675-1678.

[136] Ganesh, V.; Chandresekaran, S. One-pot synthesis of β-amino/β-hydroxy selenides and sulfides from aziridines and epoxides. Synthesis, **2009**, 3267-3278.

[137] (a) Danishefsky, S. Electrophilic cyclopropanes in organic synthesis. Acc. Chem. Res., **1979**, 12, 66-72; (b) Wrobel, T.; Takahashi, K.; Honkan, V.; Lannoye, G.; Cook, T. M.; Bertz, S. H. α-Lithio ketones. 1. Stereocontrolled synthesis of (±)-modhephene *via* the Weiss reaction. J. Org. Chem., **1983**, 48, 139-141; (c) Zutterman, E.; De Wilde, H.; Mijugheer, R.; De Clercq, P.; Vandewalle, H. Evidence for an olefinic intermediate in the configurational inversion accompanying hydrogenolysis of a 7-oxanorbornyl vicinal dibromide. Tetrahedron, **1979**, 35, 2389-2392.

[138] (a) Masamune, S.; Kaiho, T.; Garvey, D. S. Aldol methodology: synthesis of versatile intermediates, 3-hydroxy-2-vinylcarbonyl compounds. J. Am. Chem. Soc., **1982**, 104, 5521-5523; (b) Kocienski, P.; Yeates, C. New synthetic routes to spiroacetals. The 3,4-dihydro-2H-pyran approach to (±)-talaromycin B. J. Chem. Soc., Perkin Trans. 1, **1985**, 1879-1883; (c) Scarborough, R. M.; Toder, B. H.; Smith, A. B. A stereospecific total synthesis of (±)-methylenomycin A and its epimer, (±)-epimethylenomycin A. J. Am. Chem. Soc., **1980**, 102, 3904-3913.

[139] Adams, D. J.; Blake, A. J.; Cooke, P. A.; Gill, C. D.; Simpkins, N. S. Highly enantioselective synthesis of chiral imides and derived products *via* chiral base desymmetrisation. Tetrahedron, **2002**, 58, 4603-4615.

[140] Takano, S.; Tamura, N.; Ogasawara, K. Synthesis of a potential synthon for the chiral synthesis of the corynanthe-type indole alkaloids: enantioselective total synthesis of (–)-antirhine. J. Chem. Soc., Chem. Commun., **1981**, 1155-1156.

[141] Miyoshi, N.; Ishii, H.; Murai, S.; Sonoda, N. Synthesis of alkyl phenyl selenides by the reaction of phenyl trimethylsilyl selenide with acetates and lactones. Chem. Lett., **1979**, 873-876.

[142] Clive, D. L. J.; Manning, H. W.; Boivin, T. L. B.; Postema, M. H. D. Formation of trans ring-fused compounds by an alkylation-radical cyclization sequence. J. Org. Chem., **1993**, 58, 6857-6873.

[143] Jacobi, P. A.; DeSimone, R. W.; Ghosh, I.; Guo, J.; Leung, S. H.; Pippin, D. New syntheses of the C,D-ring pyrromethenones of phytochrome and phycocyanin. J. Org. Chem., **2000**, 65, 8478-8489.

[144] Sakai, M.; Hashimoto, K.; Shirahama, H. Synthesis of optically pure β-phenylselenoalanine through serine-β-lactone: A useful precursor of dehydroalanine. Heterocycles, **1997**, 44, 319-324.

[145] Murahashi, S. -I.; Yano, T. Novel transformation of primary, secondary, and tertiary amines to organoselenides with ruthenium catalyst. J. Am. Chem. Soc., **1980**, 102, 2456-2458.

[146] Abe, H.; Yamasaki, A.; Harayama, T. Direct conversion of a benzylic hydroxy group into a selenenyl group using the phenyl trimethylsilyl selenide-aluminium bromide combination. Chem. Pharm. Bull., **1998**, 46, 1311-1313.

[147] (a) Ranu, B. C.; Mandal, T. An indium-TMSCl promoted reaction of diphenyl diselenides and aldehydes: novel routes to selenoacetals and alkyl phenyl selenides. Tetrahedron Lett., **2006**, 47, 5677-5680; (b) Ranu, B. C.; Saha, A.; Mandal, T. An indium-TMSCl promoted reaction of diphenyl diselenide and diorganyl disulfides with aldehydes: novel routes to selenoacetals, thioacetals and alkyl phenyl selenides. Tetrahedron, **2009**, 65, 2072-2078.

[148] Dörwald, F. Z. In Organic Synthesis in Solid Phase: Supports, Linkers, Reactions; Wiley-VCH: NewYork, **2000**, pp. 287-288.

[149] Ruhland, T.; Andersen, K.; Pedersen, H. Selenium-linking strategy for traceless solid-phase synthesis: Direct loading, aliphatic C-H bond formation upon cleavage and reaction monitoring by gradient MAS NMR spectroscopy. J. Org. Chem., **1998**, 63, 9204-9211.

[150] Yanada, K.; Fujita, T.; Yanada, R. Reduction of selenium with BER in methanol. Application to synthesis of dialkyl selenides. Synlett, **1998**, 971-972.

[151] Cohen, R. J.; Fox, D. L.; Salvatore, R. N. A novel and highly efficient synthetic route to unsymmetrical organoselenides using cesium bases. J. Org. Chem., **2004**, 69, 4265- 4268.

[152] Qian, H.; Shao, L. -X.; Huang, X. Polymer-supported selenol esters as useful acylating reagents. Application to α,β-acetylenic ketones synthesis. Synlett, **2001**, 1571-1572 and references cited there in.

[153] Qian, H.; Huang, X. Preparation of polystyrene-supported α-seleno acetate and application to solid-phase synthesis of β-keto esters. Synlett, **2006**, 1547-1548.

CHAPTER 4

Organoselenium Compounds as Precursor of Radicals

Jarosław Lewkowski*

University of Łódź, Faculty of Chemistry, Department of Organic Chemistry, Tamka 12; 91-403 Łódź, Poland

Abstract: Radical reactions with the participation of selenium compounds have been described here in the point of view of the source of selenium radicals. Consequently, the following groups have been reviewed: alkyl phenyl selenides, acyl phenyl selenides, imidoyl phenyl selenides, benzeneselenol and other selenides as the source of radicals.

The chapter covers the literature from early 90s until 2012.

Keywords: Radicals, Selenides, Stannanes, Silanes, Diselenides, Addition, Cyclization, Coupling, Stereoselectivity, Regioselectivity.

4.1. INTRODUCTION

The act of writing this chapter turned out to be a very difficult task, as it is generally an uneasy challenge to overview the problem, which has been summarised one year earlier. It is to know that W. Russell Bowman has written the excellent and the high-impact chapter [1] about selenium compounds in radical reactions, which appeared in the recently published book. Considering that I was under great intellectual influence of this publication, the reader can imagine my task was twice more difficult. Nevertheless, despite all formal and personal obstacles, I tried to show my view to this problem.

Radical reactions with the participation of selenium compounds have been divided considering the source of selenium radicals. Therefore the chapter is divided into 5 subsections describing consequently:

- Alkyl phenyl selenides as the source of radicals.

***Address correspondence to Jarosław Lewkowski:** University of Łódź, Faculty of Chemistry, Department of Organic Chemistry, Tamka 12; 91-403 Łódź, Poland; Tel/Fax: +48.42.635.57.51;
E-mail: jlewkow@uni.lodz.pl

Claudio Santi (Ed)

- Acyl phenyl selenides as the source of radicals.
- Imidoyl phenyl selenides as the source of radicals.
- Benzeneselenol as the source of radicals and finally.
- Other selenides as the source of radicals.

Because the chapter by Bowmann covered the literature until 2010, I have introduced several papers from last half of 2011 and 2012.

4.2. ALKYL PHENYL SELENIDES AS THE SOURCE OF RADICALS

A phenylselenyl group is an efficient leaving group in some radical conversions, where radicals are generated by the action of tributyltin hydride and tris(trimethylsilyl)silane. They are abstracted by primary alkyl, stannyl or germanyl radicals with the same efficacy as bromine, but much less than iodine. Bromides and iodides are very good leaving groups, so there would not be the necessity of replacing them, if they survived the conditions of certain transformations. But they do not and thus they need to be introduced shortly before the radical reaction. Apart from this, the use of bromides and iodides has some other important disadvantages; they undergo elimination reactions in basic media, nucleophilic substitution, or homolytic cleavage. These reactions do not occur for phenyl selenides. Their stability to most reaction conditions is their huge advantage over halides in SN_2 substitution reactions. Moreover, phenyl selenide group is stable to most common and important reactions, such as: hydrolyses, attack of common nucleophiles and reducing agents ($LiAlH_4$, DIBAL, $NaBH_4$), it is neither susceptible to imine formation catalyzed by acids, nor to Wittig reactions and even nor to metathesis. Literature data [2, 3] prove that this functional group is stable even in conditions of the Swern oxidation.

Another advantage of a phenylselenyl group is that it can be introduced nearly at the beginning of the reaction sequence, is highly stable to a variety of transformations so that it is untacked, and finally it may be removed in the course of a radical reaction. According to Bowmann [1], these selenyl groups are often called "*radical protective groups*", protecting the potential radical center until the moment of its "*deprotection*" performed by the action of a radical reagent.

It is to state that a phenylselenyl group is not attacked by nucleophilic *N*-centers, which are present in many molecules of importance bearing amine or imine groups [2, 3]. An example, where phenylselenyl moiety used in radical reaction demonstrates its chemical stability in various reactional conditions including contact with amines and imines is the synthesis of *N*-cyclopentyl benzylamine from 5-bromopentanoic acid ethyl ester shown in Scheme **4.1** [2]. The phenylselenyl moiety was inserted by the substitution of diselenide on the alkyl bromide, obtained carboxylic acid was undergone the reduction with DIBAL to an aldehyde, which, on reaction with an amine, was transformed to an imine. Further action of tributyltin hydride caused the radical reaction with SePh as a leaving group. In this way, the phenylselenyl moiety has proved to stable against sodium borohydride, DIBAL, amines, imines, and towards acid catalysis.

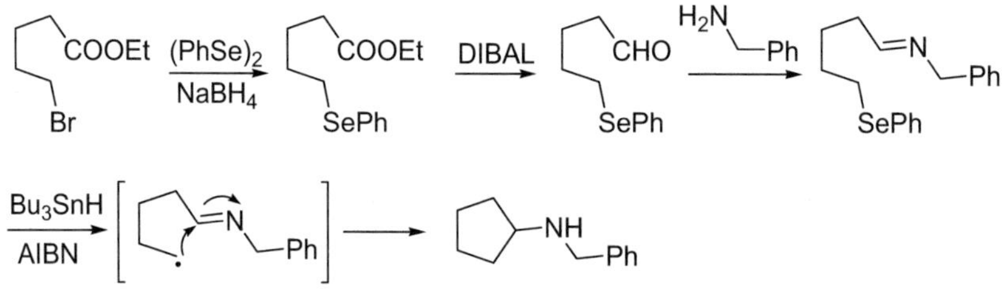

Scheme 4.1

There are more experimental evidences for the stability of the phenylselenyl moiety against the nitrogen nucleophiles - important sources demonstrated that they are not attacked by nucleophilic *N* centers of basic heteroarenes. Numerous examples showed that the phenylselenyl moiety is not susceptible to imidazoles [4, 5], benzimidazoles [4-6], pyrazoles [5, 7], and azines [8]. This fact is of importance because such heteroaromatic species are present in many biologically active molecules. The phenylselenyl moiety is easily introduced into such nitrogen-containing heteroareanes by simple *N*-alkylation, and the ω-(phenylselenyl)alkyl group linked to nitrogen, after radical substitution undergoes the ring closure.

The synthesis of compounds demonstrating promising cytostatic properties is the best example for the methodology described above [9] (Scheme **4.2**). 1,5-Dihydro-benzo[1,2-*d*;4,5-*d'*]diimidazole **1** is easily converted to 1,5-bis-(3-phenylselanyl-propyl)-1,5-dihydro-benzo[1,2-d;4,5-d']diimidazole by alkylation, which undergoes radical aromatic substitution cyclisation to give 2,3,8,9-tetrahydro-1*H*,7*H*-pyrrolo[1,2-*a*]pyrrolo[1',2':1,2]-imidazo[5,4-*f*]benzimidazole (**2**).

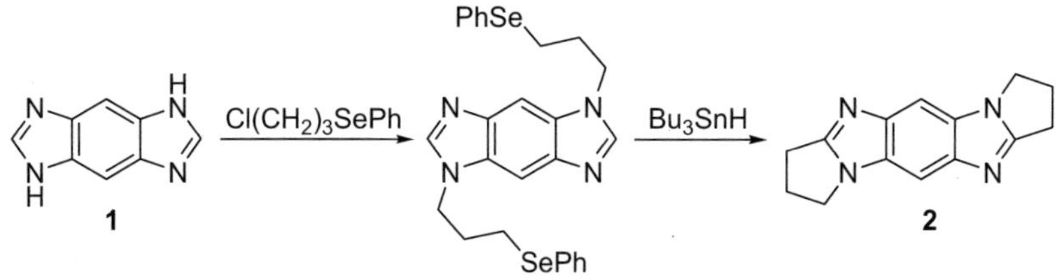

Scheme 4.2

The most commonly used reagent for radical abstraction of the phenylselenyl moiety is tributyltin hydride, but it is not the only one reagent used for this purpose. Tris(trimethylsilyl)silane, tributylgermanium hydride [10], ethylpiperidine hypophosphite and diethylphosphine oxide (DEPO) [11] have been also applied. For example, the cyclisation onto the aromatic ring (aromatic homolytic substitution) of the phenylselenyl derivative of the anilide to indolinone, initiated by the AIBN-DEPO pair is shown to be efficient (Scheme **4.3**) [11]. This example demonstrates that the phenylselenyl moiety is well stable in the presence of trimethylaluminium used in the amide synthesis. Instead of AIBN, triethylborane may be also used as an initiator.

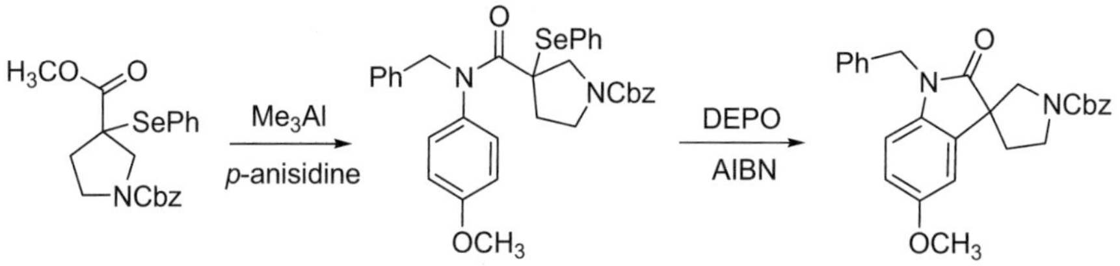

Scheme 4.3

The use of alkyl phenyl selenides provides the answer to the question how to perform the transformation of 2-isopropylidene-1-(1-phenyl-3-phenylselanyl-propyl)-aziridine (**3**) to 5-isopropylidene-2-phenyl-piperidine (**4**), which is shown in Scheme **4.4** [12]. The formed radical **5** yields the 1-aza-bicyclo[3.1.0]hexyl intermediate **6**, which undergoes the ring opening to give the desired product **4**. This report provides the proof for the stability of the phenylselenyl moiety in the presence of sodium amide.

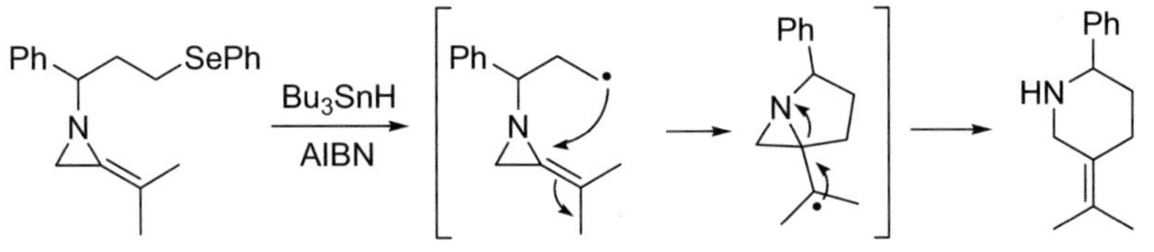

Scheme 4.4

The conversion of phenylselenyl derivative of acetophenone to 3-methyl-5-phenyl-3,4-dihydro-2H-pyrrole is the next example for the use of alkyl selenide precursors to generate radical intermediates required for radical cyclization (Scheme **4.5**) [13, 14]. This is also another example for the stability of the PhSe group in the imine formation conditions.

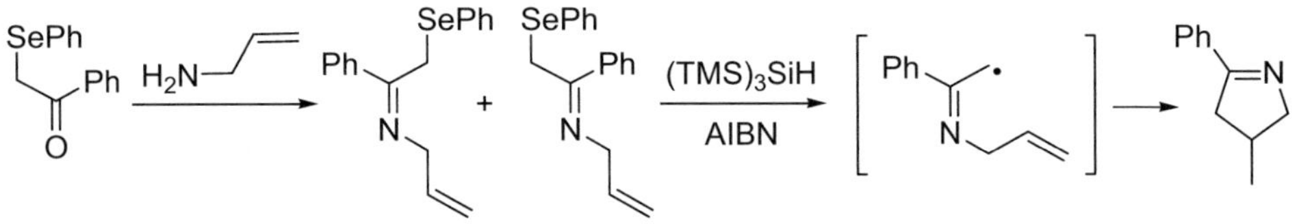

Scheme 4.5

Aldehydes [15] and nitriles [16] bearing phenylselenyl moiety are interesting agents for cyclisation. The "ring-expansion" reaction of hydroxymethyloxirane to (*R,S*) and (*S,S*) isomers of 2-hydroxymethyl-4-hydroxytetrahydrofuran is its distinct example, which is shown in Scheme **4.6** [15]. The phenylselenyl aldehyde **7** was synthesized by cleaving the epoxide ring with diphenyl diselenide and subsequent substitution to bromoacetate, and subsequent reduction with DIBAL. The action of tributyltin hydride and AIBN generated the radical **8**, which undergoes the cyclisation. Final deprotection gave hydroxytetrahydrofuran derivatives.

Cyclization may also be performed to a double bond of α,β-unsaturated carbonyl compounds and two examples may be quoted [17-19]. First is a distinct application of the phenylselenyl moiety to the synthesis of nucleosides. 4-[1-(2,4-Dioxo-3,4-dihydro-2H-pyrimidin-1-yl)-2-phenylselanyl-ethoxy]-5-trityloxy-pent-2-enoic acid methyl ester, the uracyl derivative bearing the phenylselenyl moiety on the action of tributyltin hydride and triethylborane generated the radical **9**, which cyclizes onto a double bond from an α,β-unsaturated ester to yield two diastereoisomers of 3′- substituted 2′,3′-dideoxyribonucleosides in a 96% dr (Scheme **4.7**) [17].

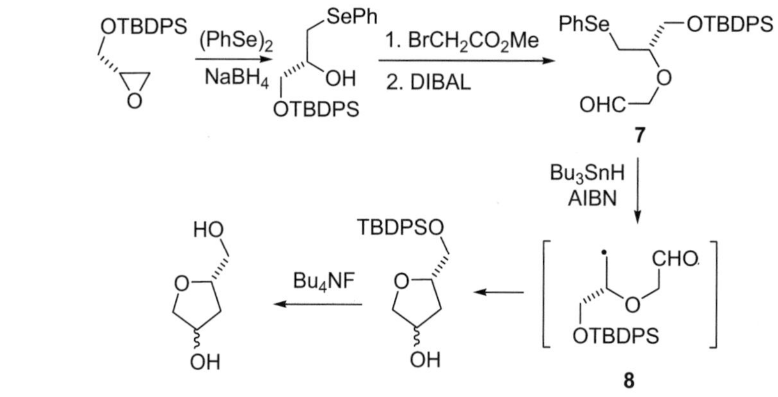

Scheme 4.6

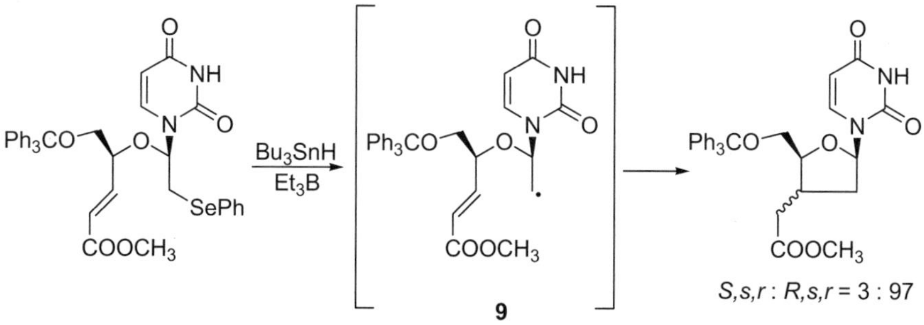

Scheme 4.7

The second example demonstrates generating an alkyl radical **10**, its addition to carbon monoxide to form an intermediate acyl radical **11**, which undergoes cyclization into an α,β-unsaturated carbonyl group to obtain stereoisomers of tetrahydrofuranone in rather high stereoselectivity of *de* exceeding 70%, (Scheme **4.8**) [18].

The reaction leading to similar products is depicted in Scheme **4.9**, where the action of *N*-(phenylseleno)phthalimide (NPSP) on (*S*)-(*E*)-1-phenyl-3,4-but-1-en-3,4-diol causes the formation of two diastereoisomers of 2-phenyl-4-phenylselenyl-3-hydroxytetrahydrofuran. The distinct chiral assistance in this reaction is due to the influence of a hydroxyl group linked to a chiral centre [20]. After removal of a phenylselenyl group with a triphenyltin hydride-AIBN pair, these two isomers gave two diastereomers of 2-phenyl-4-hydroxytetreahydrofuran (Scheme **4.9**).

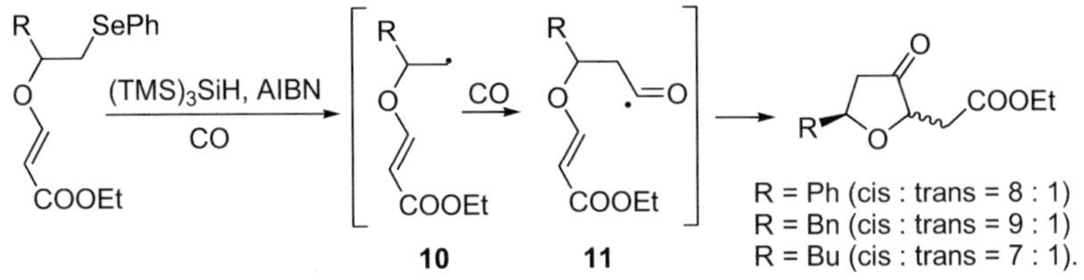

Scheme 4.8

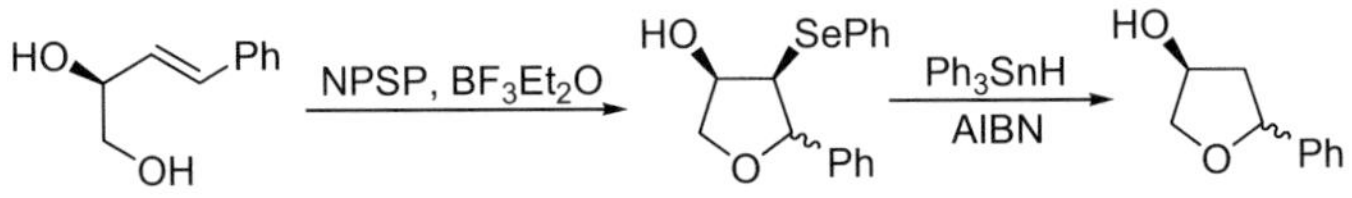

Scheme 4.9

Diels-Alder reactions of cyclic dienes with phenyl vinyl selenide as the dienophile has been well described as the efficient route to substituted bicyclic compounds bearing a phenylselenyl group [21, 22], which, on the action of tributyltin hydride forms radical intermediates undergoing rearrangement [21, 22]. For example, the Diels-Alder reaction of 2-methoxycarbonyl-6,6-dimethoxycyclohexane-1,3-dien-5-on with vinyl selenide yielded the phenylselenyl derivative of bicyclo[2.2.2]octenone [22], as in Scheme **4.10**.

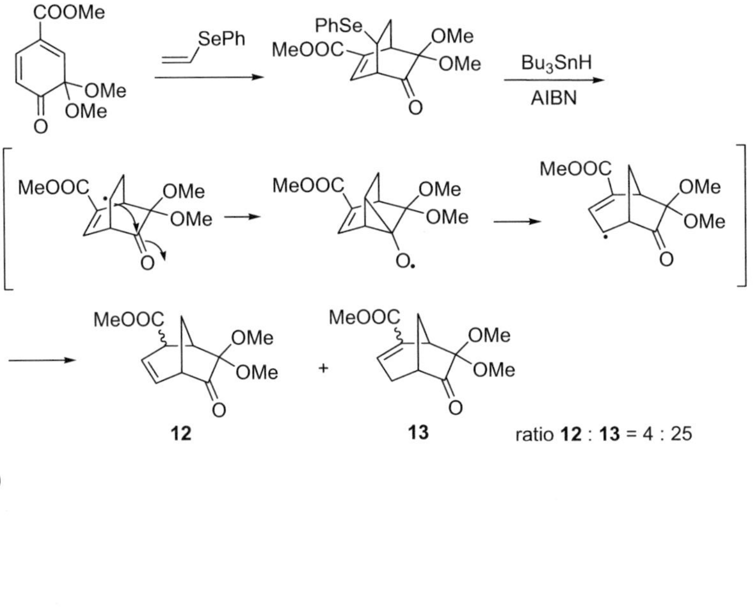

Scheme 4.10

This phenylselenyl derivative, after treatment with tributyltin hydride gave the radical, which rearranged to give bicycle [3.2.1]octenones **12** and **13**. Coming back to the stability of the phenylselenyl moiety, it is worth of mention its compability with the conditions of the Grignard reaction. Therefore, the reaction of 5-phenylselanylpentanal with 1-phenylbut-1-en-4-yl magnesium bromide can occur to give 1-phenyl-9-phenylselanyl-non-1-en-5-ol [3] (Scheme **4.11**). Subsequent oxidation of hydroxyl group followed by the action of benzylamine allowed to obtain the imine **14**, which after the action of tributyltin hydride with AIBN undergoes the subsequent radical cyclization and led to 5,5-spirocyclic amine, *i.e.*, 1,2-dibenzyl-1-aza-spiro[4.4]nonane. The synthesis of other spirocyclic amines was also reported [3].

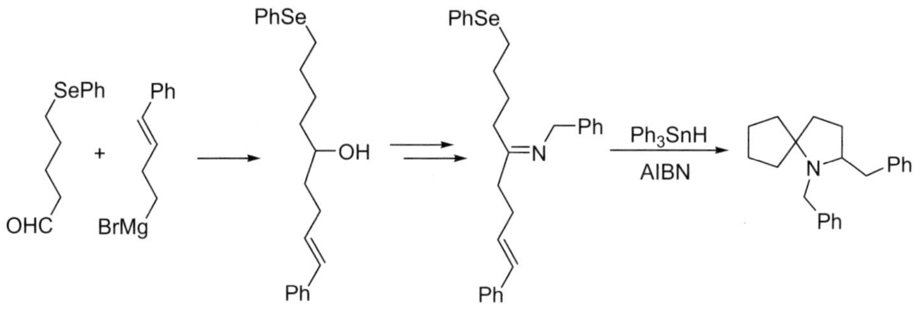

14

Scheme 4.11

The ring closing metathesis is commonly applied to the formation of desired cyclic compounds. To achieve this reaction, the Grignard reaction, similar to example above was performed. The reaction of 4-phenylselanylbutanal with prop-1-en-3-yl magnesium halide occurred to give 1-phenyl-9-phenylselanyl-non-1-en-5-ol (Scheme **4.12**) [23]. Then, under ester formation reaction, but-3-enoic acid 1-phenyl-9-phenylselanyl-non-1-en-5-yl ester occurred, which was converted to 7-(3-phenylselanyl-propyl)-6,7-dihydro-3H-oxepin-2-one under the ring closing metathesis reaction conditions. Its radical cyclization reaction, initiated by a tributyltin hydride-AIBN pair led to the bicyclic derivative, namely 2-oxa-bicyclo[4.3.1]decan-3-one in rather fair yield.

The progress of synthesis performed on solid phase caused certainly that several procedures involving radical selenium chemistry have been developed. The problem of application of solid phase based on selenides has been reviewed [24]. Therefore, several examples must find their place in our chapter.

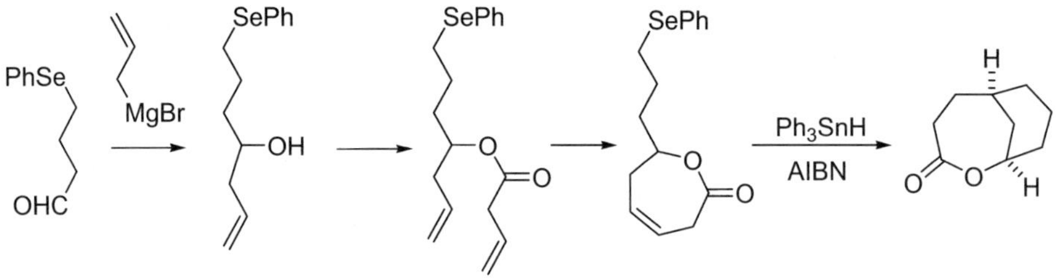

Scheme 4.12

The synthesis of 6,7,8,9-tetrahydro-pyrido[1,2-a]indole, performed by the route of radical cyclization to the indole ring system was carried out on Wang solid phase resin. 2-Chloro-1-(3-phenylselanyl-propyl)-1*H*-indole was linked to the mercaptophenyl moiety to form 1-(3-phenylselanyl-propyl)-2-arylsulfanyl-1*H*-indole fragment, which, upon the action of a tributyltin hydride-AIBN pair generated an alkyl radical. This radical underwent aromatic radical substitution with the solid phase as the leaving group (Scheme **4.13**) [6], resulting in the cyclization reaction to give 6,7,8,9-tetrahydro-pyrido[1,2-a]indole in 60% yield.

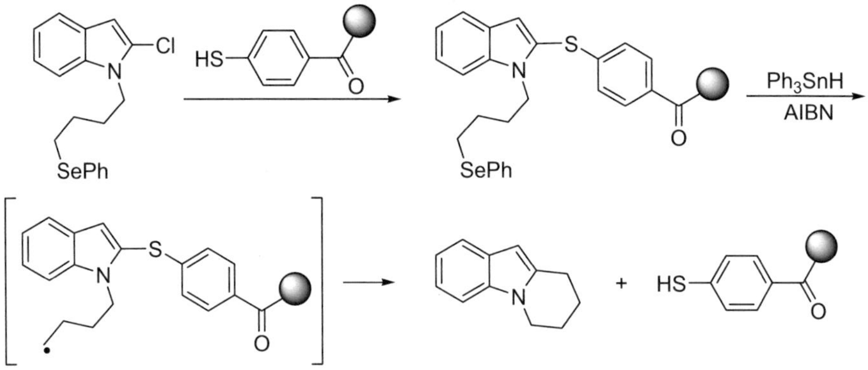

Scheme 4.13

The solid phase synthetic methodology, which uses selenide-assisted radical reactions did not reach exclusively for phenylselenyl derivatives but generally for ones having arylselenyl leaving groups. For example, the arylselenyl bromide linked to the Quadragel® reacted with 1-chloro-4-iodobutane to give 4-chlorobutyl-aryl selenide linked to the resin (Scheme **4.14**) [5].

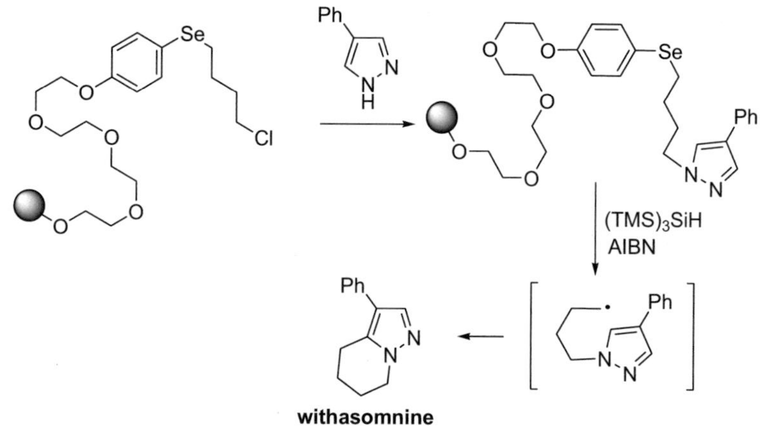

Scheme 4.14

This selenide, on the reaction with pyrazole in DMF in basic condition led to the formation of pyrazolylbutyl-aryl selenide moiety, which under the influence of tris(trimethylsilyl)silane with AIBN generated the intermediate radical, which cyclized to give 3-phenyl-4,5,6,7-tetrahydro-pyrazolo[1,5-a]pyridine - the naturally occurring pyrazole alkaloid withasomnine, which has been isolated from the roots of *Withania somnifera* Dun. (*Solanaceae*), which are used medicinally in India., as it is shown in Scheme **4.14** [5].

Stereoselective or even enantioselective intramolecular radical cyclization with a phenylselenyl leaving group turned out to be possible, when an appropriate catalyst is used [25, 26]. Racemic 5-cyclohex-1-enyl-2-methyl-3-oxo-2-phenylselanyl-pentanoic acid ethyl ester undergoing the action of triethylborane as the radical initiator and magnesium perchlorate with a ligand **15** as an asymmetric catalyst led to the formation of a (1-*R*, 4a-*S*, 8a-*R*) stereoisomer of 1-methyl-2-oxo-4a-phenylselanyl-decahydro-naphthalene-1-carboxylic acid ethyl ester in a 87% *ee* (Scheme **4.15**) [25]. A ligand **15** is coupled with Mg(II) and, thus, they facilitate the enantioselectivity.

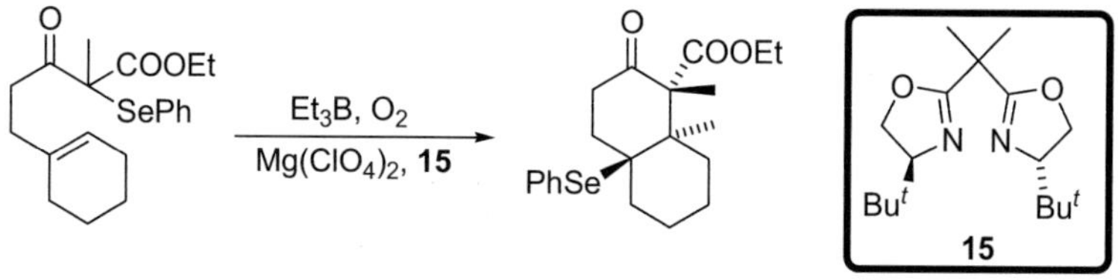

Scheme 4.15

Interesting example of cyclization has been given during last months. Vinylidenecyclopropanes bearing an allene moiety connected with a highly strained cyclopropane have been converted to bicyclic compounds on the action of diphenyldiselenide in the course of radical reaction [27]. For example, reaction of disubstituted diphenylvinylidenecyclopropane with diphenyldiselenide gave the defined stereoisomer of 1-(2,2-diphenyl-1-phenylselanyl-vinyl)-5-phenyl-2-phenylselanylmethyl-bicyclo[3.1.0]hexane (Scheme **4.16**) [27].

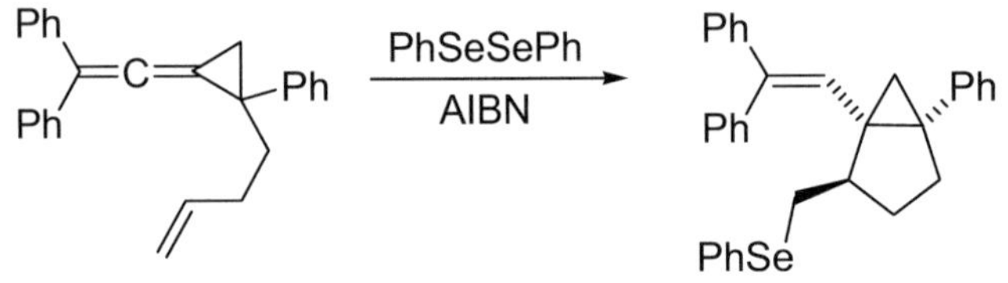

Scheme 4.16

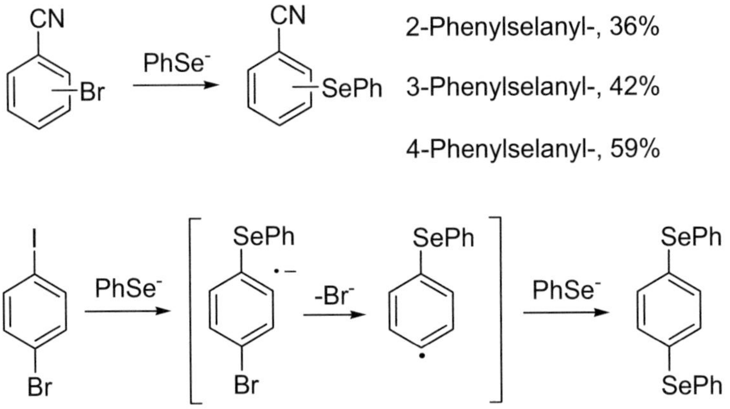

Scheme 4.17

Phenyl selenide anion is able, in reaction with aryl halides to undergo single electron transfer to yield phenyl selenide radicals and radical anions. The mechanism of this reaction has been well reviewed twice and very recently [1, 28], therefore there is no need to repeat it. This radical reaction seems to be useful in converting aryl halides to phenylselenyl arenes. For example *o*-, *m*- and *p*-bromobenznitrile easily react with phenylselenyl anion to give *o*-, *m*- and *p*-(phenylselanyl)-benzonitrile respectively in fair yields (Scheme **4.17**) [29]. And what is important, the reaction with dihalogenoarenes led to the exclusive formation of di-(phenylseleno)benzenes, as it is shown on Scheme **4.17** [29].

4.3. ACYL PHENYL SELENIDES AS THE SOURCE OF RADICALS

Acyl phenyl selenides found their large application in radical reactions [30]. It is due to the fact that the phenylselenyl moiety is a good leaving group, and radicals are easily generated by the action of tributyltin hydride and tris-(trimethylsilyl)silane. In some cases, tributylgermanium hydride may be used [10]. Acyl phenyl selenides win the comparison with acyl halides, due to their relative stability. It is also an easy task to prepare them; the general methodology involves the conversion of carboxylic acids with the use of relatively mild reagents as the source of selenium, which do not attack most of functional groups [23]. According to Bowman [1], benzeneselenol should be avoided as the reagent due to its extremely unpleasant smell. We do agree with him! But there are several other methods for synthesis of acyl phenyl selenides, therefore it is highly possible to avoid it. The simplest but not necessarily the best method involves reaction of the respective acid chloride with the benzeneselenolate anion, which is prepared *in situ* by the reduction of diphenyl diselenide sodium borohydride. But as the use of acyl halides suffers from one important disadvantage, *i.e.* hydrogen halide evolution, it is better to avoid this method too. Especially that, the direct action of diphenyl diselenide or *N*-(phenylselenyl)phthalimide on the carboxylic acid, catalyzed by tributylphosphine is reported to be efficient [1].

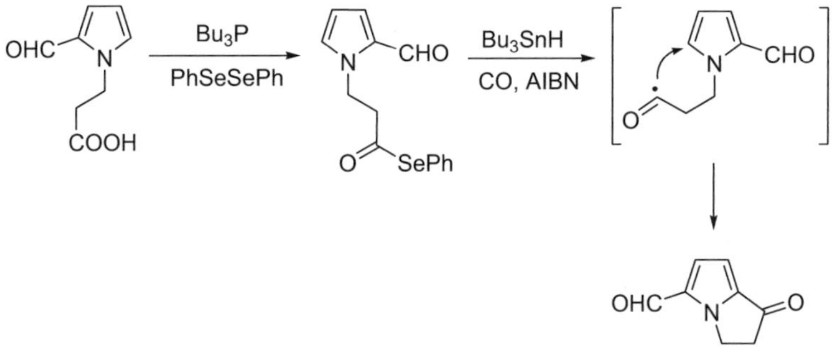

Scheme 4.18

The excellent example of such a protocol is the synthesis of 7-oxo-6,7-dihydro-5*H*-pyrrolizine-3-carbaldehyde by the cyclization of acyl radical onto pyrrole [5, 31] (Scheme **4.18**). The acyl phenyl selenide, namely 3-(2-formyl-pyrrol-1-yl)-selenopropionic acid *Se*-phenyl ester was prepared from 3-(2-formyl-pyrrol-1-yl)-propionic acid. The action of tributyltin hydride with AIBN caused generating a radical, which underwent the cyclization to give the desired fused pyrrole in reasonable yields. However, due to the fact that this cyclization is a relatively slow reaction, the problem of the loss of carbon oxide was observed. Therefore, the reaction must be carried out under an atmosphere of carbon oxide (Scheme **4.18**).

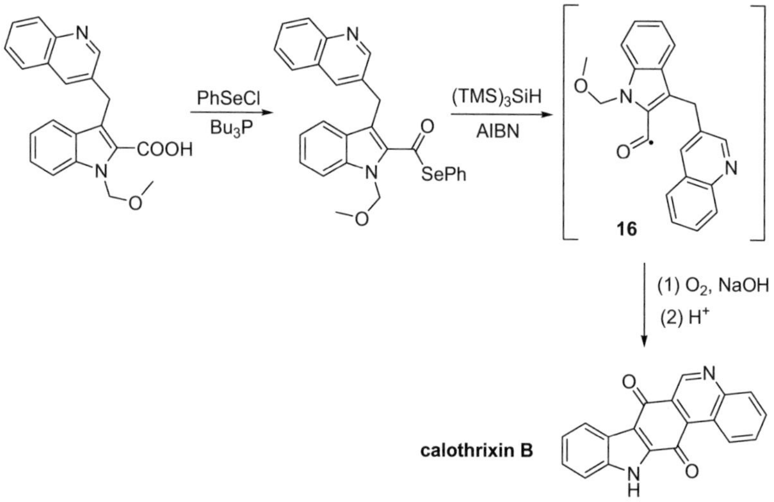

Scheme 4.19

When the acyl group is linked to an aryl moiety, no problem with CO loss occurs, as the carbonyl group is linked to an aryl carbon with much stronger bond [32-36]. The synthesis of 12*H*-5,12-diaza-indeno[1,2-*b*]phenanthrene-7,13-dione, known as calothrixin B - a potent antimalarial alkaloid is a distinct example (Scheme **4.19**) [32]. The *N*-protected 3-quinolin-3-ylmethyl-1H-indole-2-carboselenoic acid *Se*-phenyl ester, on the action of tris-(trimethylsilyl)silane and AIBN generated the radical **16**, which underwent the cyclization to give *N*-protected 7,12-dihydro-5,12-diaza-indeno[1,2-*b*]phenanthren-13-one. Its oxidation with oxygen combined with deprotection led to the formation of the desired alkaloid. The acyl selenide is easily prepared from the corresponding carboxylic acid, namely 3-quinolin-3-ylmethyl-1H-indole-2-carboxylic acid by the action of phenylselanyl chloride and tributylphosphine.

The similar case of a 2-indolylacyl radical cyclization is met in a case of the synthesis of the alkaloid dasycarpidone **17** and its epimer epidasycarpidone **18** (Scheme **4.20**) [34, 35]. 3-(3-Ethyl-1-methyl-1,2,5,6-tetrahydro-pyridin-2-

ylmethyl)-1*H*-indole-2-carboselenoic acid *Se*-phenyl ester, on the action of tributyltin hydride with triethylborane, was converted to the 2-indolylacyl radical, which underwent the 6-*endo* cyclization to give dasycarpidone **17** and epidasycarpidone **18**.

4.4. IMIDOYL PHENYL SELENIDES AS THE SOURCE OF RADICALS

Imidoyl selenides are as efficient source for imidoyl radicals, as acyl selenides are for acyl radicals [37-40]. This methodology has the important advantage in heteroarene synthesis as the use of imidoyl radicals allows to introduce one unsaturated bond.

Synthesis of imidoyl phenyl selenides is easily performed starting from benzoic acid amides, which are converted to imidoyl chlorides. Imidoyl chlorides are transformed to desired imidoyl phenyl selenides by the action of the *in situ* generated phenylselanyl anion. The phenylselanyl anion is generated by the action of the K-selectride in THF on diphenyldiselenide (Scheme **4.21**) [38-40].

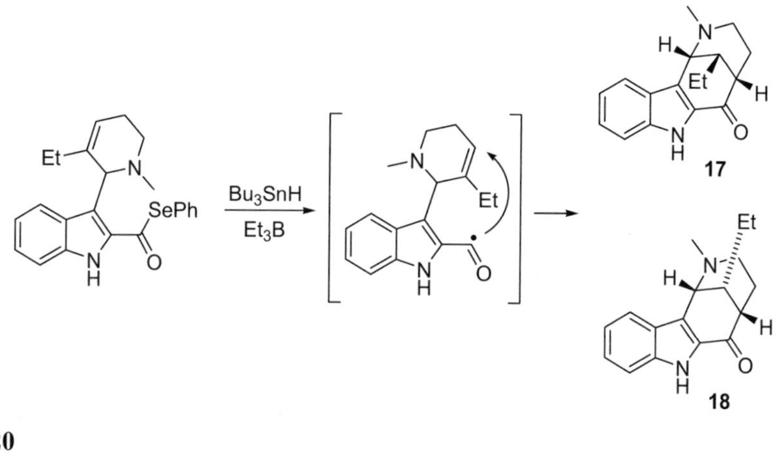

Scheme 4.20

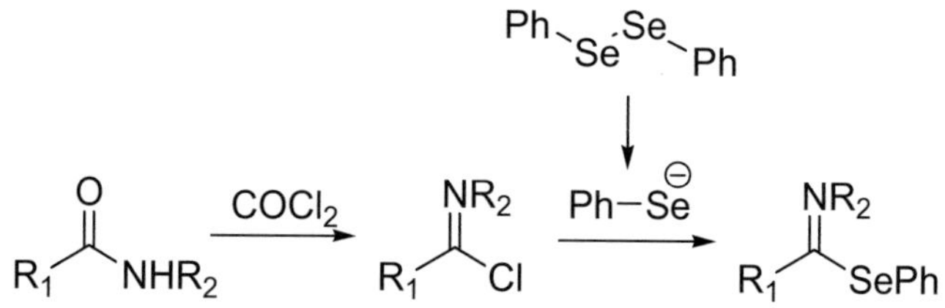

Scheme 4.21

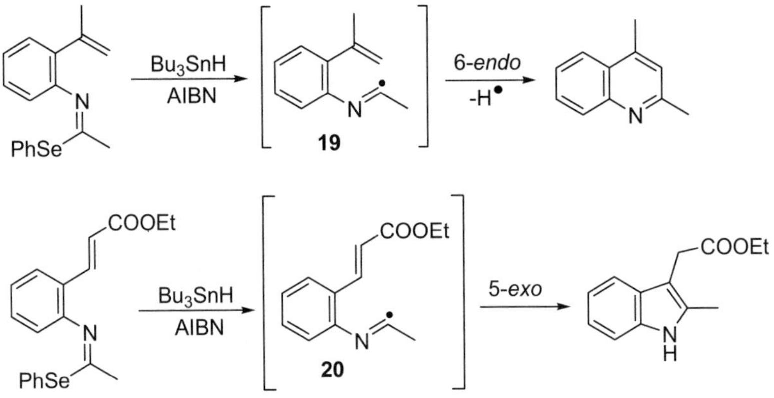

Scheme 4.22

A good and distinct example of imidoyl phenyl selenides is the synthesis of indoles and quinolines (Scheme **4.22**) [39, 40]. *N*-(2-Isopropenyl-phenyl)-selenoacetimidic acid phenyl ester reacted easily with tributyltin hydride to yield the imidoyl radical **19**, which underwent the 6-*endo* cyclization to give 2,4-dimethyl-quinoline in 80% yield. The imidoyl radical **20**, generated from 3-[2-(1-phenylselanyl-ethylideneamino)-phenyl]-acrylic acid ethyl ester underwent the 5-*exo* cyclization to achieve (2-methyl-1*H*-indol-3-yl)-acetic acid ethyl ester.

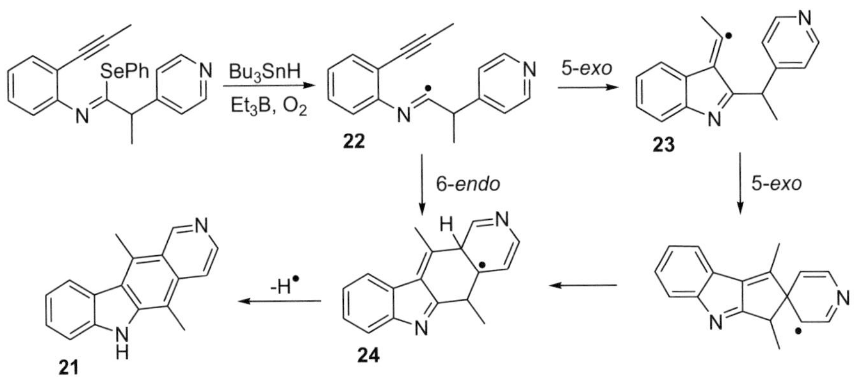

Scheme 4.23

The synthesis of the anticancer alkaloid ellipticine **21** can be considered as the reaction of the same type. From 4-[1-Methyl-2-phenylselanyl-3-(2-prop-1-ynyl-phenyl)-allyl]-pyridine, the imidoyl radical **22** was generated in reaction with tributyltin hydride and triethylborane in the oxygen atmosphere. The radical **22** underwent 5-*exo* cyclization to the alkenyl radical **23** (Scheme **4.23**) [39], which, after the 5-*exo* cyclization, followed by the neophyl rearrangement, was converted to the radical **24**. The alkenyl radical **23** underwent also the 6-*endo* cyclization to give directly the radical **24**. The abstraction of hydrogen followed by the tautomeric conversion allowed to give the anticancer alkaloid ellipticine, which chemical name is 5,11-dimethyl-6*H*-pyrido[4,3-*b*]carbazole **21** [39, 40].

4.5. PHENYLSELENOL AS THE SOURCE OF RADICALS

Benzenoselenol is able to dissociate hydrogen in a homolytical reaction to form a phenylselenyl radical in the course of the reaction with tributyltin hydride. Despite the fact that benzeneselenol is claimed to have one of the most ugly odour, its application in organic synthesis, where it is the source of radicals is rather large.

It is to note, howeverm that diphenyldiselenide is a good precursor for benzeneselenol, since it is easily reduced *in situ* by the action of tributyltin hydride and this allows to avoid handling with benzeneselenol.

Generally, the conversions with benzeneselenol as the source of radicals are best applicable to reactions of aryl halides, especially iodides with any other arene. After abstraction of halogen, a radical is generated, which reacts with a molecule of any other arene to form an arene substituted by a cyclohexadienyl moiety.

The mechanism of such a reaction is presented in Scheme **4.24** [41]. A tributyltin radical reacts with aryl iodide to generate an aryl radical, which subsequently reacts with an arene molecule to form a biaryl radical **25**. On the action of benzeneselenol, the biaryl radical forms cyclohexadienyl arene product.

This reaction turned out to be very effective in reactions with heteroarenes, such as, pyridines and pyrroles [42, 43] and thiophenes and furans [43, 44].

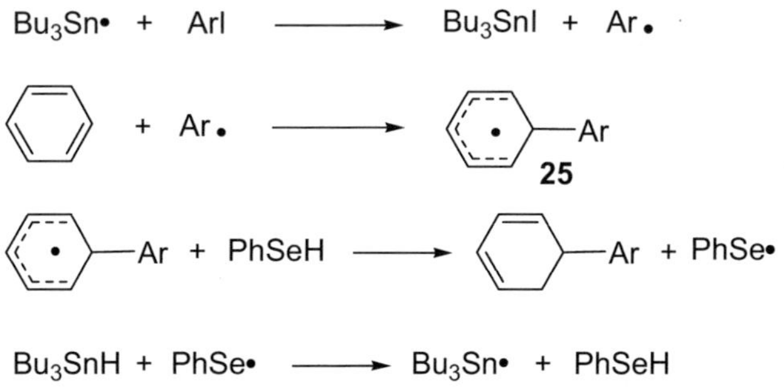

Scheme 4.24

A good and distinct example of the use of this methodology in organic synthesis is the preparation of carbazomycin B, a known antimicrobial antibiotic, which chemically is 3-methoxy-1,2-dimethyl-9*H*-carbazol-4-ol. Its preparation is the sequence of very interessting conversions, which are presented in Scheme **4.25** and includes radical reactions as well as cyclizations of their products [45]. 2-Iodo-6-methoxy-3-methoxycarbonylamino-4,5-dimethyl-phenyl acetate reacted with benzene in the presence of diphenyl diselenide and a tributyltin hydride-AIBN pair to give 2-cyclohexa-2,5-dienyl-6-methoxy-3-methoxycarbonylamino-4,5-dimethyl-phenyl acetate, in the course of a radical reaction. The latter cyclohexadienyl derivative underwent the addition of phenylselenyl group in reaction with phenylselenyl bromide to achieve 5-acetoxy-6-methoxy-7,8-dimethyl-1-phenylselanyl-1,2,4a,9a-tetrahydro-carbazole-9-carboxylic acid methyl ester, which is oxidized with *tert*-butyl hydroperoxide to give the desired carbazomycin B [45].

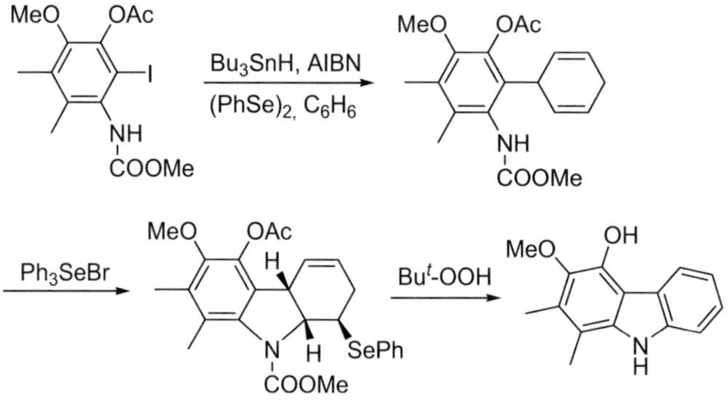

Scheme 4.25

This methodology allowed to force the 5-*exo* cyclization in situation, when occurrence of the 6-*endo* cyclization was normally expected as a sole route. When 2-iodo-*N*-methyl-*N*-phenyl-benzamide underwent the reaction with benzeneselenol in the presence of a tributyltin hydride-AIBN pair, a radical **26** occurred, which underwent the 5-*exo* cyclization to give a spirodienyl radical, which gave, under the action of benzeneselenol, 3,4-benzo-1-methyl-1-aza-spiro[4.5]deca-3,6,9-trien-2-one **27** in 43% yield (Scheme **4.26**) [46]. The cyclization following the 6-*endo* mechanism also occurred and gave 5-methyl-4a,10b-dihydro-5*H*-phenanthridin-6-one **28** but in a twice less yield (22%) [46].

The cyclization of *O*-trityl oximes with benzeneselenol as the source of radicals should be considered as the useful method for synthesis of cyclic oximes [47]. Diphenyl diselenide was rapidly reduced by tributyltin hydride to benzeneselenol, which reacting with 3-(2-iodo-cyclohexyl)-propionaldehyde O-trityl-oxime generated a corresponding cyclohexyl radical. The radical underwent the cyclization with elimination of trityl moiety to give hexahydro-benzofuran-3-one oxime (Scheme **4.27**) [47].

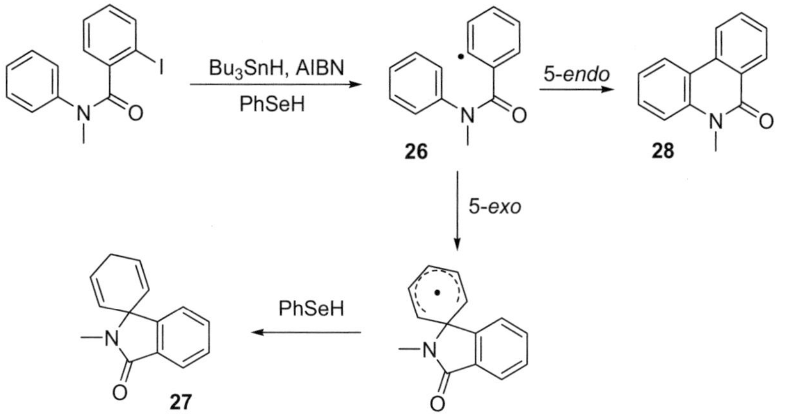

Scheme 4.26

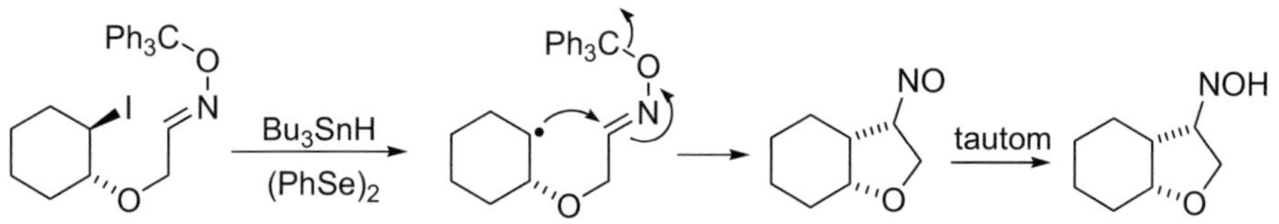

Scheme 4.27

4.6. OTHER SELENIDES AS THE SOURCE OF RADICALS

As usually, while creating the systematic division of any phenomenon, one reaches the point, where the group called "other items" is necessary to be introduced. This term is a some kind of the bag of a great capacity, where one can place all cases not fitting to previous ones.

Our topic is not the exception and a group named "*other selenides as the source of radicals*" must be introduced. But these "other selenides" lie not far away from these mentioned above, since phenylselenyl formates can be considered as a special case of acyl selenides and benzyl selenides are, in some extend relative to phenylselenyl derivatives.

Oxycarbonyl radicals are generated from (phenylselenyl)formates [48-50]. A good example for its use is the synthesis of a γ-lactone moiety of juruenolide C (**29**), which is presented in Scheme **4.28** [48]. The (phenylselenyl)formate derivative **30** is prepared from the carboxylic acid using the common method and subsequently underwent the action of the tributyltin hydride-AIBN system to generate the oxycarbonyl radical **31**. The latter radical, after a series of rearrangements led to the subproduct bearing a γ-lactone moiety (**32**).

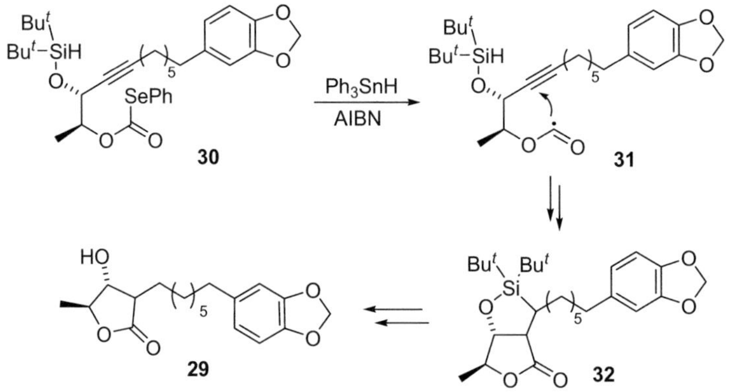

Scheme 4.28

Benzyl aryl selenides are easily synthesized by the reduction of dibenzyl diselenide with sodium borohydride and the nucleophilic substitution to any aryl halide. They are used for the organic synthesis in the same way as phenyl selenides but in a smaller degree.

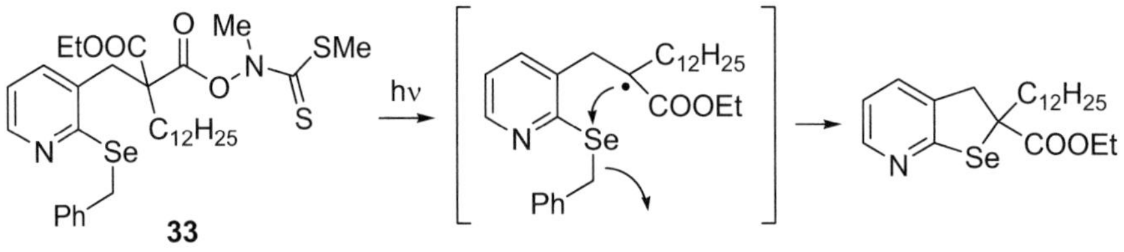

Scheme 4.29

A good example is the conversion of dithiocarbamoyl derivative of 2-(2-benzylselanyl-pyridin-3-ylmethyl)-2-dodecyl-malonic acid ethyl ester (**33**) to give

2-dodecyl-2,3-dihydro-selenolo[2,3-b]pyridine-2-carboxylic acid methyl ester. After the photochemical decarboxylation and subsequent generation of a radical, the desired product has been obtained (Scheme **4.29**) [8].

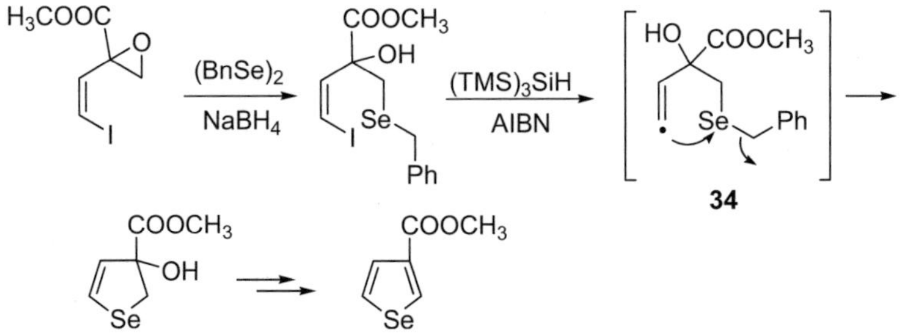

Scheme 4.30

Benzyl alkyl selenides bearing a C=C double bond in the molecule are able to generate vinyl radicals, which undergo the cyclization reactions. For example, the synthesis of methyl 3-selenophane carboxylate is performed starting from 2-(2-iodo-vinyl)-oxirane-2-carboxylic acid methyl ester, which underwent the reaction with diphenyl diselenide and sodium borohydride to give 2-benzylselanylmethyl-2-hydroxy-4-iodo-but-3-enoic acid methyl ester. The vinyl radical **34** was formed by the action of a $(TMS)_3SiH$-AIBN pair and underwent the cyclization to yield 3-hydroxy-2,3-dihydro-selenophene-3-carboxylic acid methyl ester, which was then converted to the desired methyl 3-selenophane carboxylate (Scheme **4.30**) [51].

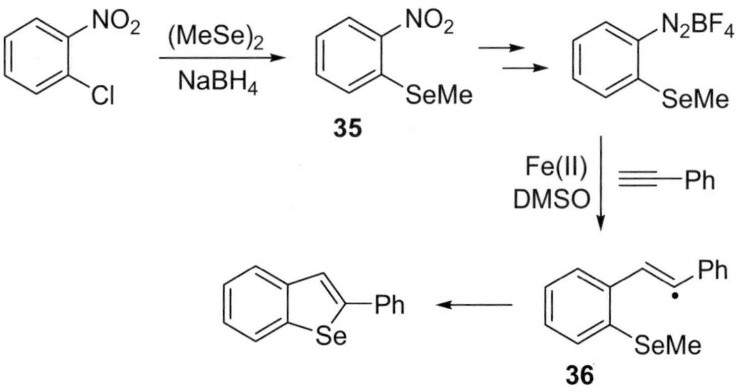

Scheme 4.31

The synthesis of 2-phenylbenzoselenophene was performed *via* methylselanyl derivative. The reaction of 1-chloro-2-nitrobenzene with dimethyl diselenide and sodium borohydride gave 1-nitro-2-methylselanyl-benzene (**35**). This methylselanyl derivative **35**, after the reduction and diazotization resulted in diazo salt, which underwent the radical addition to phenylacetylene in the presence of iron (II). The resulting alkenyl radical **36** cyclised easily to give 2-phenyl-benzo[*b*]selenophene in 30% yield (Scheme **4.31**) [52].

Solid phase selenyl bromide has been also used as the source of radicals [52-54]. Its reaction with 2-allyl-aniline followed by acylation resulted in the formation of dihydroindole derivative **37**, which, after abstraction of selenyl moiety by the action of tributyltin hydride was converted to indol methyl radical **38**. The latter underwent the radical cyclization to obtain (6a-*R*,10-*R*,11a-*R*)-7,8,9,10,10a,11,11a,12-octahydro-6aH-indolo[1,2-*b*]isoquinolin-6-one **39** (Scheme **4.32**) [52]. This methodology was demonstrated to be important as a tool for the stereoselective synthesis of fused-ring indoles.

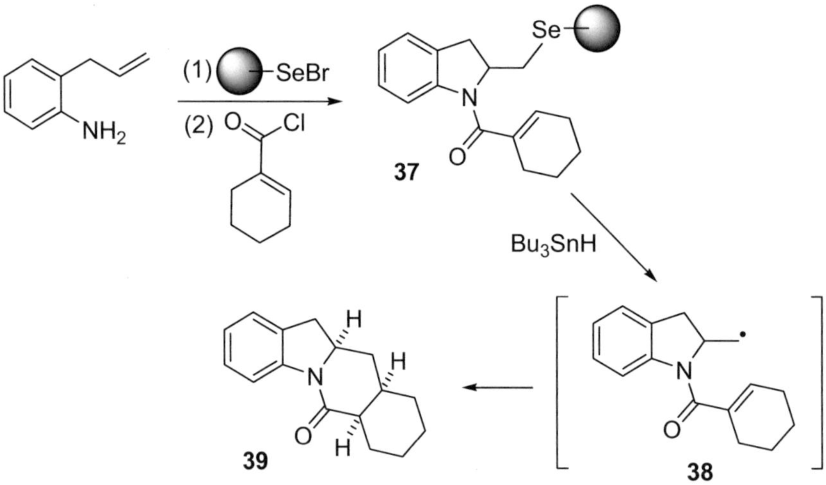

Scheme 4.32

Derivatives of 6-oxo-6,9-dihydro[1,2,5]selenadiazolo[3,4-*h*]quinoline may also be considered as the source of radicals, since it was established that upon their electrochemical anodic oxidation in alkaline solutions, the selenadiazole ring is replaced, forming instead the paramagnetic species analogous to the *ortho* semiquinone radical anions, which was confirmed by the ESR spectroscopy (Scheme **4.33**) [55].

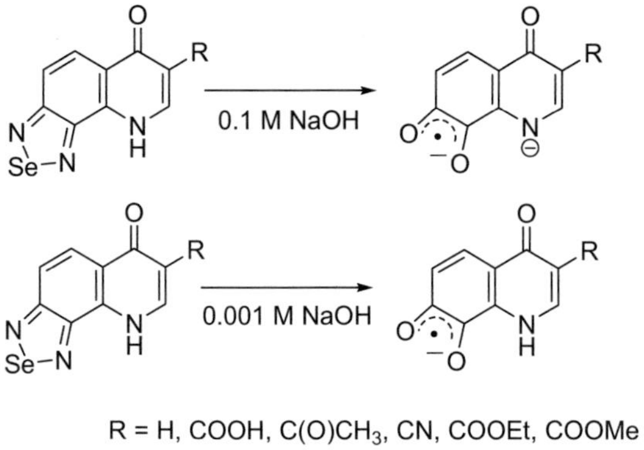

Scheme 4.33

Dependingly on the concentration of sodium hydroxide, the product occurred in the deprotonated (for 0.1 M aqueous NaOH) or the protonated form (for 0.001 M aqueous NaOH).

It is difficult to predict the synthetic application for this reaction of a radical formation, however, one can imagine its potency in preparation of fused-ring heterocycles.

ACKNOWLEDGEMENT

Declared none.

CONFLICT OF INTEREST

The authors confirm that this chapter contents have no conflict of interest.

REFERENCES

[1] Bowman, W. R. In *Organoselenium Chemistry. Synthesis and Reactions*; T. Wirth ed., Wiley VCH, **2012**; pp. 111-146.

[2] (a) Bowman, W.R.; Stephenson, P.T.; Terrett, N.K.; Young, A.R. Cyclisation of carbinyl radicals onto imines and hydrazones *Tetrahedron Lett.*, **1994**, *35*, 6369-6372; (b) Bowman, W.R.; Stephenson, P.T.; Terrett, N.K.; Young, A.R. Radical cyclisations of imines and hydrazones *Tetrahedron*, **1995**, *51*, 7959-7980.

[3] (a) Bowman, W.R.; Stephenson, P.T.; Young, A.R. Synthesis of nitrogen heterocycles using tandem radical cyclisation of imines *Tetrahedron Lett.*, **1995**, *36*, 5623-5626; (b) Bowman, W.R.; Stephenson, P.T.; Young, A.R. Cascade radical cyclisations of imines *Tetrahedron*, **1996**, *52*, 11445 - 11462.

[4] (a) Aldabbagh, F.; Bowman, W.R. Radical cyclisation onto imidazoles and benzimidazoles *Tetrahedron Lett.*, **1997**, *38*, 3793-3794; (b) Aldabbagh, F.; Bowman, W.R.; Mann, E. Oxidative radical cyclisations onto imidazoles and pyrroles using Bu_3SnH *Tetrahedron Lett.* **1997**, *38*, 7937-7940; (c) Aldabbagh, F.; Bowman, W.R. Radical cyclisation onto imidazoles and benzimidazoles *Tetrahedron*, **1999**, *55*, 4109-4122; (d) Aldabbagh, F.; Bowman, W.R.; Mann, E.; Slawin, A.M.Z. Bu_3SnH mediated oxidative radical cyclisation onto imidazoles and pyrroles *Tetrahedron*, **1999**, *55*, 8111-8128.

[5] Allin, S.M.; Barton, W.R.S.; Bowman, W.R.; Bridge, E.; Elsegood, M.R.J.; McInally, T.; McKee, V. Bu_3SnH-mediated radical cyclisation onto azoles *Tetrahedron*, **2008**, *64*, 7745-7758.

[6] Allin, S.M.; Bowman, W.R.; Karim, R.; Rahman, S.S. Aromatic homolytic substitution using solid phase synthesis *Tetrahedron*, **2006**, *62*, 4306-4316.

[7] Allin, S.M.; Barton, W.R.S.; Bowman, W.R.; McInally, T. Radical cyclisation onto pyrazoles: synthesis of withasomnine *Tetrahedron Lett.*, **2002**, *43*, 4191-4193.

[8] Fenner, T.; White, K.M.; Schiesser, C.H. Preparation of 2,3-dihydroselenolo[2,3-*b*]pyridines and related compounds by free-radical means *Org. Biomol. Chem.*, **2006**, *4*, 466-474.

[9] Fagan, V.; Bonham, S.; Carty, M.B.; Aldabbagh, F. One-pot double intramolecular homolytic aromatic substitution routes to dialicyclic ring fused imidazobenzimidazole-quinones and preliminary analysis of anticancer activity *Org. Biomol. Chem.* **2010**, *8*, 3149-3156.

[10] Bowman, W.R.; Krintel, S.L.; Schilling, M.B. Tributylgermanium hydride as a replacement for tributyltin hydride in radical reactions *Org. Biomol. Chem.*, **2004**, *2*, 585-592.

[11] Murphy, J.A.; Tripoli, R.; Khan, T.A.; Mali, U.W. Novel Phosphorus Radical-Based Routes to Horsfiline *Org. Lett.* **2005**, *7*, 3287-3289.

[12] Prévost, N.; Shipman, M. Synthesis of substituted piperidines, decahydroquinolines and octahydroindolizines by radical rearrangement reactions of 2-alkylideneaziridines *Tetrahedron,* **2002**, *58*, 7165-7175.

[13] Srivastava, P.; Engman, L. A radical cyclization route to cyclic imines *Tetrahedron Lett.*, **2010**, *51*, 1149-1151.

[14] Bowman, W.R.; Storey, J.M.B. Synthesis using aromatic homolytic substitution - recent advances *Chem. Soc. Rev.*, **2007**, *36*, 1803-1822.

[15] Tiecco, M.; Testaferri, L.; Marini, F.; Sternativo, S.; Santi, C.; Bagnoli, L.; Temperini, A. Intramolecular addition of carbon radicals to aldehydes: synthesis of enantiopure tetrahydrofuran-3-ols *Tetrahedron* **2007**, *63*, 5482-5489.

[16] Bowman, W.R.; Bridge, C.F.; Brookes, P. Radical cyclisation onto nitriles *Tetrahedron Lett.* **2000**, *41*, 8989-8995.

[17] Kumamoto, H.; Ogamino, J.; Tanaka, H.; Suzuki, H.; Haraguchi, K.; Miyasaka, T.; Yokomatsu, T.; Shibuya, S. Radical-mediated furanose ring reconstruction from 2′,3′-*seco*-uridine *Tetrahedron,* **2001**, *57*, 3331-3341.

[18] Berlin, S.; Ericcson, C.; Engman, L. Radical Carbonylation/Reductive Cyclization for the Construction of Tetrahydrofuran-3-ones and Pyrrolidin-3-ones *J. Org. Chem.*, **2003**, *68*, 8386-8398.

[19] Berlin, S.; Ericcson, C.; Engman, L. Construction of Tetrahydrofuran-3-ones from Readily Available Organochalcogen Precursors *via* Radical Carbonylation/Reductive Cyclization *Org. Lett.*, **2002**, *4*, 3-6.

[20] Tiecco, M.; Testaferri, L.; Bagnoli, L.; Terlizzi, R.; Temperini, A.; Marini, F.; Santi, C.; Scarponi, C. Synthesis of enantiomerically pure perhydrofuro[3,4-*b*]pyrans and perhydrofuro[3,4-*b*]furans *Tetrahedron Assymmetry* **2004**, *15*, 1949-1955.

[21] Markó, I.E.; Warriner, S.L.; Augustyns, B. Radical-Initiated, Skeletal Rearrangements of Bicyclo[2.2.2] Lactones *Org. Lett.,* **2000**, *2*, 3123-3125.

[22] Gao, S.Y.; Chittimalla, S.K.; Chuang, G.J.; Liao, C.- C. Efficient Synthesis and Subsequent Transformations of Phenylsulfanylbicyclo[2.2.2]octenones and Phenylselenylbicyclo [2.2.2]-octenones *J. Org. Chem.,* **2009**, *74*, 1632-1639.

[23] Clive, D.L.J.; Cheng, H. Tandem ring-closing metathesis-radical cyclization based on 4-(phenylseleno)butanal and methyl 3-(phenylseleno)propanoate - a route to bicyclic compounds *Chem. Commun.,* **2001**, 605 - 606.

[24] McAllister, L.A.; McCormick, R.A.; Procter, D.J. Sulfide- and selenide-based linkers in phase tag-assisted synthesis *Tetrahedron,* **2005**, *61*, 11527-11576.

[25] Yang, D.; Zheng, B.-F.; Gao, Q.; Gu, S.; Zhu, N.-Y. Enantioselective PhSe-Group-Transfer Tandem Radical Cyclization Reactions Catalyzed by a Chiral Lewis Acid *Angew. Chem. Int. Ed.,* **2006**, *45*, 255-258.

[26] Yang, D.; Gao, Q.; Lee, O.-Y. Lewis Acid Promoted Phenylseleno Group Transfer Tandem Radical Cyclization Reactions *Org. Lett.* **2002**, *4*, 1239-1241.

[27] Yuan, W.; Wei, Y.; Shi, M.; Li, Y. Reactions of Vinylidenecyclopropanes with Diphenyl Diselenide in the Presence of AIBN and Thermally-Induced Further Transformations *Chem. Eur. J.*, **2012**, *18*, 1280-1285.

[28] Rossi, R.A.; Pierini, A.B.; Penénory, A.B. Nucleophilic Substitution Reactions by Electron Transfer *Chem. Rev.,* **2003**, *103*, 71-167.

[29] Palacios, S.M.; Alonso, R.A.; Rossi, R.A. Photostimulated reactions of 1-iodoadamantane and iodobenzene with thiolate, selenate, and tellurate ions *Tetrahedron,* **1985**, *41*, 4147-4156.

[30] (a) Chatgilialoglu, C.; Crich, D.; Komatsu, M.; Ryu, I. Chemistry of Acyl Radicals *Chem. Rev.* **1999**, *99*, 1991-2069. (b) Yet, L. Free radicals in the synthesis of medium-sized rings *Tetrahedron,* **1999**, *55*, 9349-9403.

[31] Allin, S.M.; Barton, W.R.S.; Bowman, W.R.; McInally, T. Acyl radical cyclisation onto pyrroles *Tetrahedron Lett.,* **2001**, *42*, 7887-7890.

[32] Bennasar, M.L.; Roca, T.; Ferrando, F. A New Radical-Based Route to Calothrixin B *Org. Lett.* **2006**, *8*, 561-564.

[33] Bennasar, M.L.; Roca, T.; Ferrando, F. Regioselective Intramolecular Reactions of 2-Indolylacyl Radicals with Pyridines: A Direct Synthetic Entry to Ellipticine Quinones *J. Org. Chem.,* **2005**, *70*, 9077-9080.

[34] Bennasar, M.L.; Roca, T.; Garcia -Diaz, D. A New Acyl Radical-based Route to the 1,5-Methanoazocino[4,3-*b*]indole Framework of Uleine and Strychnos Alkaloids *J. Org. Chem.,* **2008**, *73*, 9033-9039.

[35] (a) Bennasar, M.L.; Roca, T.; Garcia -Diaz, D. 6-*Exo* Cyclizations of 2-Indolylacyl Radicals: Access to the Uleine Alkaloid Skeleton *Synlett* **2008**, 1487-1490; (b) Bennasar, M.L.; Roca, T.; Garcia -Diaz, D. Novel 7- and 8-Endo 2-Indolylacyl Radical Cyclizations: Efficient Construction of Azepino- and Azocinoindoles *J. Org. Chem.,* **2007**, *72*, 4562-4565.

[36] (a) Bennasar, M.L.; Roca, T.; Griero, R.; Bassa, M.; Bosch, J. New Cascade 2-Indolylacyl Radical Addition−Cyclization Reactions *J. Org. Chem.,* **2001**, *66*, 7547-7551; (b)

Bennasar, M.L.; Roca, T.; Griero, R.; Bassa, M.; Bosch, J. Generation and Intermolecular Reactions of 3-Indolylacyl Radicals *J. Org. Chem.,* **2002**, *67*, 6268-6271.

[37] (a) Bachi, M.D.; Denenmark, D. Cyclizations of ene radicals. Imidoyl radicals as intermediates in the synthesis of heterocyclic compounds *J. Am. Chem. Soc.,* **1989**, *111*, 1886-1888 ; (b) Bachi, M.D.; Balanov, A.; Bar-Ner, N.; Bosch, E.; Denenmark, D.; Mizhiritskii, M. Synthesis of non-aromatic heterocyclic compounds through free-radical reactions *Pure Appl. Chem.* **1993**, *65*, 595-601.

[38] Bowman, W.R.; Fletcher, A.J.; Lovell, P.J.; Pedersen, J.M. Synthesis of Indoles Using Cyclization of Imidoyl Radicals *Synlett,* **2004**, 1905-1908.

[39] Pedersen, J.M.; Bowman, W.R.; Elsegood, M.R.J.; Fletcher, A.J.; Lovell, P.J. Synthesis of Ellipticine: A Radical Cascade Protocol to Aryl- and Heteroaryl-Annulated[*b*]carbazoles *J. Org. Chem.,* **2005**, *70*, 10615-10618.

[40] Bowman, W.R.; Fletcher, A.J.; Pedersen, J.M.; Lovell, P.J.; Elsegood, M.R.J.; Hernandez Lopez, E.; McKee, V.; Potts, G.B.S. Amides as precursors of imidoyl radicals in cyclisation reactions *Tetrahedron,* **2007**, *63*, 191-203.

[41] Crich, D.; Grant, D.; Krishnamurthy, V.; Patel, M. Catalysis of Stannane-Mediated Radical Chain Reactions by Benzeneselenol *Acc. Chem. Res.,* **2007**, *40*, 453-463.

[42] Crich, D.; Patel, M. Radical dearomatization of arenes and heteroarenes *Tetrahedron,* **2006**, *62*, 7824-7837.

[43] Crich, D.; Patel, M. Direct Synthesis of Heterobiaryls by Radical Addition to Pyridine: Epeditious Synthesis of Chelating Ligands *Heterocycles,* **2004**, *64*, 499-504.

[44] Crich, D.; Patel, M. Facile Dearomatizing Radical Arylation of Furan and Thiophene *Org. Lett.,* **2005**, *7*, 3625-3626.

[45] Crich, D.; Rumthao, S. Synthesis of carbazomycin B by radical arylation of benzene *Tetrahedron,* **2004**, *60*, 1513-1516.

[46] Crich, D.; Hwang, J.-T. Stannane-Mediated Radical Addition to Arenes. Generation of Cyclohexadienyl Radicals and Increased Propagation Efficiency in the Presence of Catalytic Benzeneselenol *J. Org. Chem.,* **1998**, *63*, 2765-2770.

[47] Clive, D.L.J.; Pham, M.P.; Subedi, R. Carbocyclization by Radical Closure onto *O*-Trityl Oximes: Dramatic Effect of Diphenyl Diselenide *J. Am. Chem. Soc.,* **2007**, *129*, 2713-2717.

[48] Clive, D.L.J.; Ardelean, E.S. Synthesis of (+)-Juruenolide C: Use of Sequential 5-Exo-Digonal Radical Cyclization, 1,5-Intramolecular Hydrogen Transfer, and 5-Endo-Trigonal Cyclization *J. Org. Chem.,* **2001**, *66*, 4841-4844.

[49] Lucas, M.A.; Nguyen, O.T.K.; Schiesser, C.H.; Zheng, S.-L. Preparation of 5-Selenopentopyranose Sugars from Pentose Starting Materials by Samarium(II) Iodide or (Phenylseleno)formate Mediated Ring Closures *Tetrahedron,* **2000**, *56*, 3995-4000.

[50] (a) Singh, A.K.; Bakshi, R.K.; Corey, E.J. Total synthesis of (±)-atractyligenin *J. Am. Chem. Soc.,* **1987**, *109*, 6187-6189; (b) Bachi, M.D.; Bosch, E. Synthesis of gamma- and delta-lactones by free-radical annelation of Se-phenyl selenocarbonates *J. Org. Chem.,* **1992**, *57*,4696-4705.

[51] Grange, R.L.; Ziogas, J.; North, A.J.; Angus, J.A.; Schiesser, C.H. Selenosartans: Novel selenophene analogues of milfasartan and eprosartan *Bioorg. Med. Chem. Lett.,* **2008**, *18*, 1241-1244.

[52] Staples, M.K.; Grange, R.L.; Angus, J.A.; Ziogas, J.; Tan, N.P.H.; Taylor, M.K.; Schiesser, C.H. Tandem free-radical addition/substitution chemistry and its application to the preparation of novel AT_1 receptor antagonists *Org. Biomol. Chem.*, **2011**, *9*, 473-479.

[53] Nicolaou, K.C.; Roecker, J.; Pfefferkorn, J.A.; Cao, G.-Q. A Novel Strategy for the Solid-Phase Synthesis of Substituted Indolines *J. Am. Chem. Soc.,* **2000**, *122*, 2966-2967.

[54] (a) Nicolaou, K.C.; Pfefferkorn, J.A.; Cao, G.-Q.; Kim, S.; Kessabi, J. A Facile Method for the Solution and Solid-Phase Synthesis of Substituted [3.3.1] Bicycles *Org. Lett.,* **1999**, *1*, 807-810; (b) Ruhland, T.; Andersen, K.; Pedersen, H. Selenium-Linking Strategy for Traceless Solid-Phase Synthesis: Direct Loading, Aliphatic C–H Bond Formation upon Cleavage and Reaction Monitoring by Gradient MAS NMR Spectroscopy *J. Org. Chem.,* **1998**, *63*, 9204-9211.

[55] Stasko, A.; Zalibera, M.; Barbierikova, Z.; Rimarcík, J.; Lukes, V.; Bella, M.; Milata, V.; Brezova, V. Anodic oxidation of selenadiazoloquinolones in alkaline media *Magn. Reson. Chem.* **2011**, *49*, 168-174.

CHAPTER 5

Hypervalent Selenium Derivatives

Józef Drabowicz[1,*], Piotr Kiełbasiński[2], Adrian Zając[2] and Patrycja Pokora-Sobczak[2]

[1]Centre of Molecular and Macromolecular Studies, Polish Academy of Sciences, Sienkiewicza 112, Lodz 90-363, Poland; [1]Jan Dlugosz University in Czestochowa, Department of Chemistry, Environment Protection and Biotechnology, 42-200-Czestochowa, Armii Krajowej Ave. 13/15, Poland and [2]Centre of Molecular and Macromolecular Studies, Polish Academy of Sciences, Sienkiewicza 112, Lodz 90-363, Poland

Abstract: Selenium forms a very large number of inorganic and organic compounds showing different reactivity and structural properties. Among them there are hypervalent compounds, selenuranes - the derivatives in which the central selenium atom has expanded its valence shell from 8 to 10 or 12 electrons. Such compounds are presented in the chapter using the criterion of classification which is based on the formal oxidation state and number (N) of ligands bonded to selenium. For each class, the methods of synthesis, complemented with the mechanistic and stereochemical studies, are comprehensively discussed.

Keywords: Hypervalent Selenium, Hypervalent Bonding, Selenium Trihalides, Selenuranes, Selenurane Oxides, Perselenuranes, Spiroselenuranes, Dihaloselenuranes, Bicyclic selenuranes, Chirality.

5.1. INTRODUCTION

Inorganic and organic compounds of selenium constitute a large family of derivatives showing various reactivities and structural properties. To classify them different criteria can be used. Among that, the criterion based on simultaneous consideration of formal oxidation state and number (N) of ligands bonded to selenium atom can be considered as the most convenient one. Based on it, the derivatives in which the central selenium atom has expanded its valency shell

*Address correspondence to Józef Drabowicz:** Centre of Molecular and Macromolecular Studies, Polish Academy of Sciences, Sienkiewicza 112, Lodz 90-363, Poland; Tel: +48 42 680 32 34; Fax: +48 42 6803261; E-mail: draj@cbmm.lodz.pl

Claudio Santi (Ed)

from 8 to 10 or 12 electrons can be considered as hypervalent compounds (the concept was for the first time proposed by Musher in 1969) [1]. Using a general systematic scheme proposed by Martin and coworkers [2], such derivatives can be assigned to the following subclasses:

a) trivalent, tricoordinated negatively charged selenium ate-complexes (selenuranide anions, considered as 10-Se-3 species **1**);

b) tetravalent, tetracoordinated selenuranes (considered as 10-Se-4 derivatives **2**);

c) hexavalent, pentacoordinated selenurane oxides (considered as 12-Se-5 species **3**);

d) hexavalent, hexacoordinated perselenuranes (viewed as 12-Se-6 derivatives **4**);

e) compounds having the formal oxidation state and the numbers of ligands higher than 6 (considered as the 2n-Se-n (N> 6) group **5**) (Fig. **5.1**).

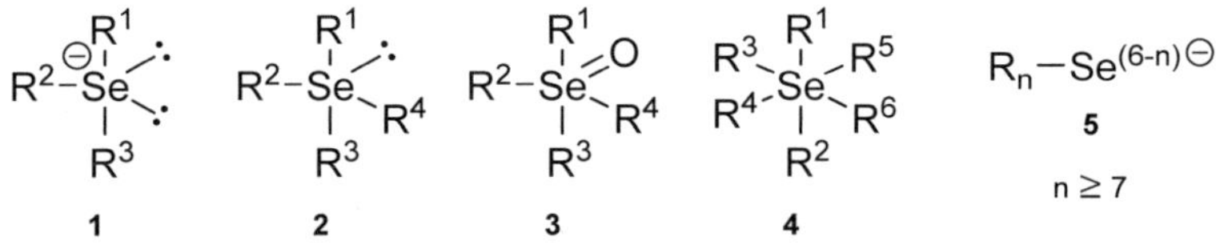

Figure 5.1: Classification of hypervalent selenium species.

The definition of hypervalent selenium derivatives is also formally fulfilled by the corresponding selenium ylides (in which a selenium atom has expanded its formal valence shell from 8 to 10). When one considers that the important resonance structures of these compounds **6** are polarized from selenium to carbon, it becomes evident that they represent 8-Se-3 species (Fig. **5.2**). Therefore they do not meet the condition of hypervalency and their chemistry is not discussed in this chapter.

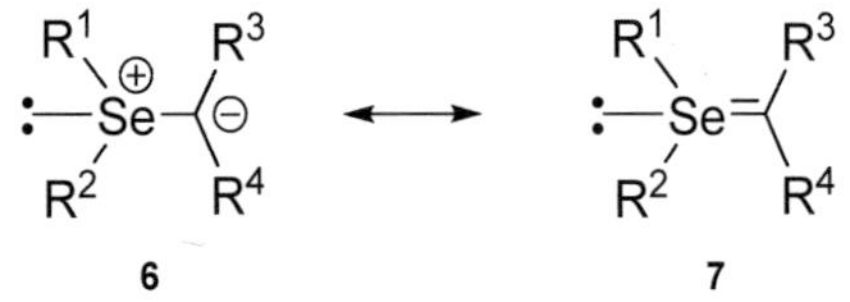

Figure 5.2: Resonance structures of selenium ylides.

The first comprehensive review devoted to advances in the chemistry of these selenium derivatives constitutes the chapter entitled "Tetra- and higher-valent (hypervalent) derivatives of selenium and tellurium" included to the first volume of "The Chemistry of Organic Selenium and Tellurium Compounds", edited by Patai and Rappoport and published in 1986 [3a]. The similar chapter included to the third volume of "The Chemistry of Organic Selenium and Tellurium Compounds", edited by Rappoport and published in 2012 [3b] covers the advances in most recent years (usually after 1998). In 1999 another excellent monographic chapter devoted to this topic, co-authored by Furukawa and Sato [4], included in "The Chemistry of Hypervalent Compounds" edited by Akiba covered the developments before 1997. Hypervalent selenium compounds are also discussed in many other textbook compilations and review papers [5].

The nature of bonding and the molecular geometry of hypervalent derivatives, including selenium compounds, is presented in details in a few chapters of "The Chemistry of Hypervalent Compounds" edited by Akiba [6,7a,b]. In this context, it is interesting to note that the nature of chemical bonding in a numbers of hypervalent molecules (including selenium) with a variety of the ligands (F, Cl, O, CH_3, and CH_2) studied recently using the topological analysis of the electron localization function (ELF) [8], led to a conclusion that "*the bonds in hypervalent molecules are very similar to those in the corresponding nonhypervalent compounds*". Therefore, in the author's opinion the term hypervalent has little significance except for the indication that an atom in a molecule forms more than four electron pair bonds.

Accepting the existence of hypervalent bonding, it should be expected that 10-Se-4 and 10-Se-5 derivatives will always form a trigonal bipyramidal structure (TBP). On the other hand, perselenuranes 12-Se-6 species and 12-Se-5 (bearing free electron pair), will have a square bipyramidal structure (SBP) (Fig. **5.3**) [7a].

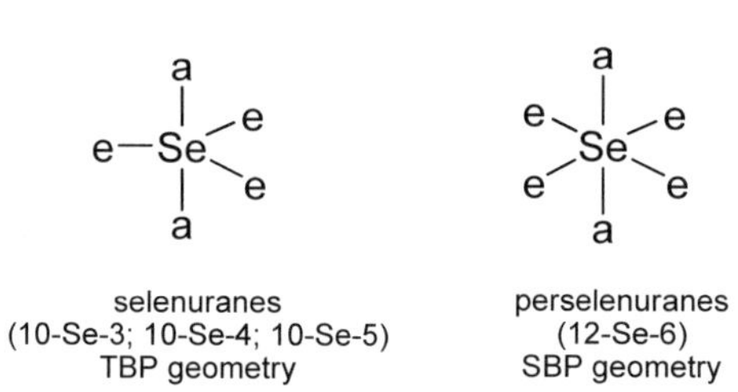

Figure 5.3: Geometry of hypervalent selenium compounds.

The Mutterties rule [9] (the more electronegative ligands tend to occupy the apical positions whereas the lone electron pair should be located at the equatorial position) is fulfilled by hypervalent selenium compounds having TBP geometry. Moreover, five- and six-membered rings, which stabilize the hypervalent molecules, span both the axial and equatorial positions [2a]. In full accord with a very early molecular orbital calculations [10] supported by the valence bond treatment of Shaik and coworkers [11], selenuranes have been found to be less stable than telluranes whereas sulfuranes are least stable.

In hypervalent 10-Se-4 and 10-Se-5 species the apical and equatorial ligands can be interchanged with each other according to the Berry pseudorotation mechanism [12] (Scheme **5.1**) or by the "turnstile" rotation mechanism proposed by Ugi and Ramirez [13] (Scheme **5.2**).

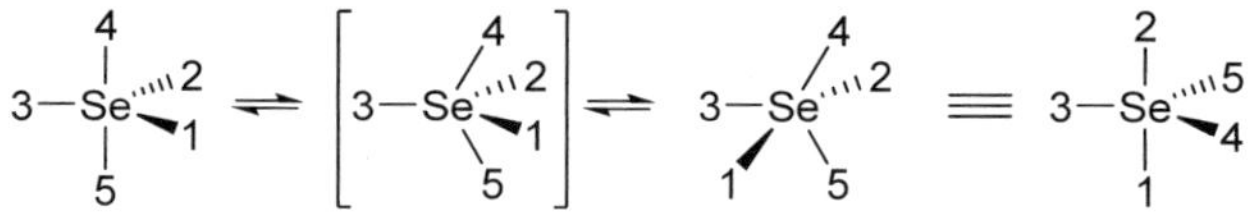

Scheme 5.1

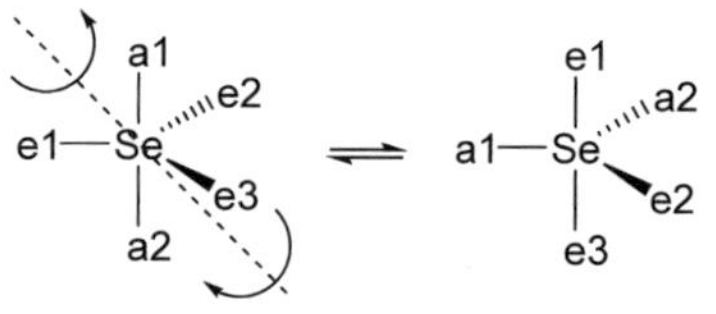

Scheme 5.2

In the molecules, which can be described as 12-Se-6 species, the apical and equatorial ligands can be interchanged with each other very slowly according to a nondissociative mechanism (the so called Bailar twist) (Scheme **5.3**) [5e,7a].

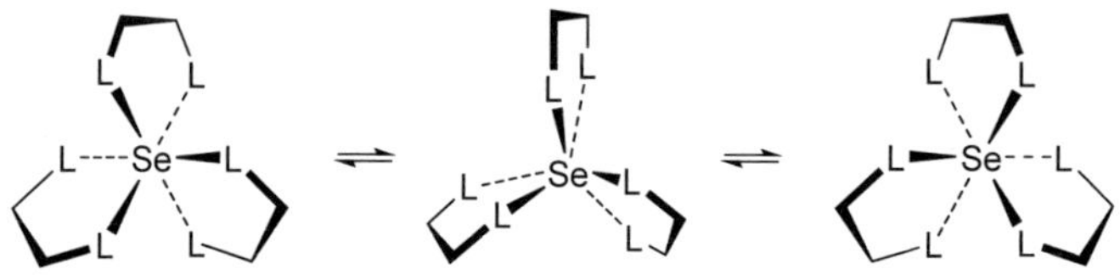

Scheme 5.3

For hypervalent selenium compounds of 10-Se-4 or 10-Se-5 structures the two most common modes constitute the ligand exchange reaction (LER) and the ligand coupling reaction (LCR).[7a] It results from the tendency of the central selenium atom to return to the normal valency by extruding either a ligand bearing pair of electrons or a pair of ligands that afford stable compounds having an octet around the selenium atom (Scheme **5.4**).

$$R_2Se + R{-}R \longleftarrow R_4Se{-}: \longrightarrow R_3Se^{\oplus} + R^{\ominus}$$

Scheme 5.4

5.2. REACTIVE INTERMEDIATES CONTAINING A HYPERVALENT SELENIUM ATOM

The selenium ate-complex **9** was generated in the organoselenium - organolithium exchange reaction between cyclic selenide **8** and *n*-butyllithium (Scheme **5.5**) [14].

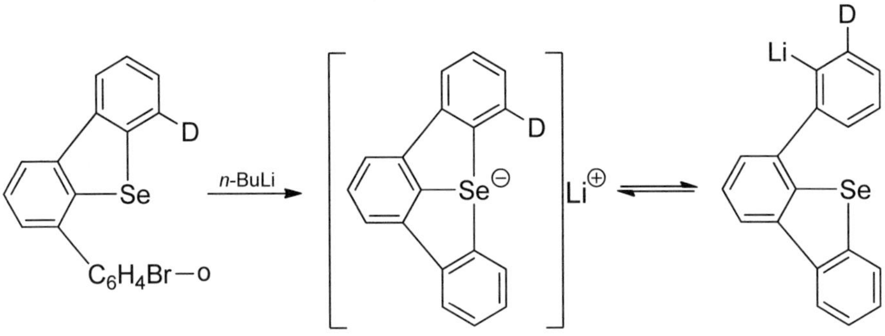

8 **9** **10**

Scheme 5.5

Tetraphenylselenurane **c11** [15] and 2,2'-biphenylylene-diphenylselenurane **12** [16], were observed by low-temperature NMR spectroscopies when one or two equivalents of phenyllithium were added to the corresponding substrate (Scheme **5.6**).

Similarly, the formation of 2,2'-biphenylylenedimethylselenurane **14** [17], (which afforded the corresponding dibenzoselenophene **15** quantitatively at room temperature) was observed by the ^{1}H, ^{13}C, and ^{77}Se NMR spectroscopies at low temperature, in the reactions of dibenzoselenophene Se-oxide **13** with methyllithium (Scheme **5.7**).

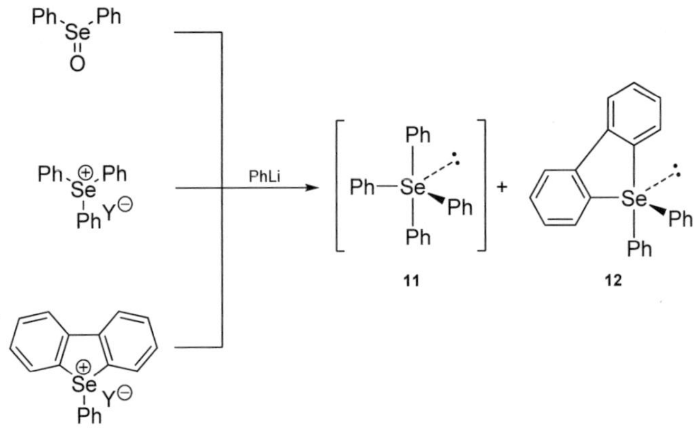

11 **12**

Scheme 5.6

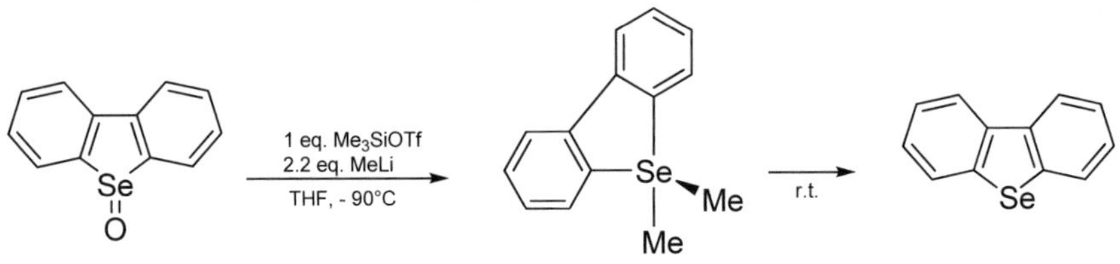

13 **14** **15**

Scheme 5.7

The generation of selenuranes **17a-c** as intermediates (Scheme **5.8**) [18] was suggested to explain the stereospecific oxidation of hydrazine into *cis*-diimide and the catalytic disproportionation of hydrogen peroxide effected by selenoxides **16a-b**.

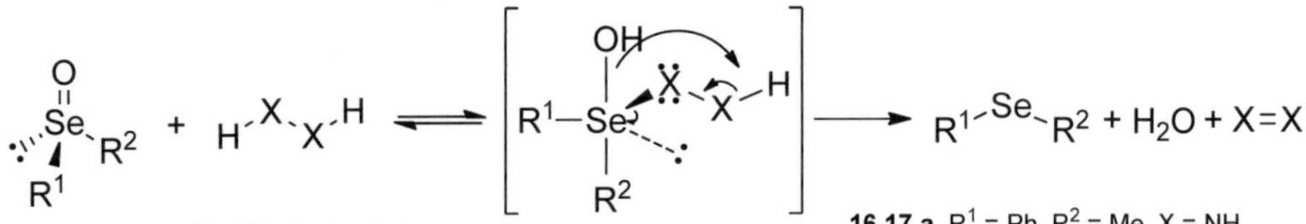

16a,b

X = NH - hydrazine
X = O - hydrogen peroxide

17a-c

16,17 a, R^1 = Ph, R^2 = Me, X = NH
16,17 b, R^1 = Ph, R^2 = *p*-$MeOC_6H_4$, X = NH
16,17 c, R^1 = Ph, R^2 = Me, X = O
16,17 d, R^1 = Ph, R^2 = *p*-$MeOC_6H_4$, X = O

Scheme 5.8

An acid–base equilibrium between the Se–N selenurane oxides **20a,b** and the corresponding selenoxides **19a,b** were detected by ^{1}H and ^{77}Se NMR during the oxidation of selenomethionines **18a,b** using hydrogen peroxide in aqueous solutions (Scheme **5.9**). The theoretically calculated ^{77}Se NMR chemical shifts (which were found to be in good agreement with experimental results for both **19a** and **20a**). Fully support experimental data (Scheme **5.9**) [19].

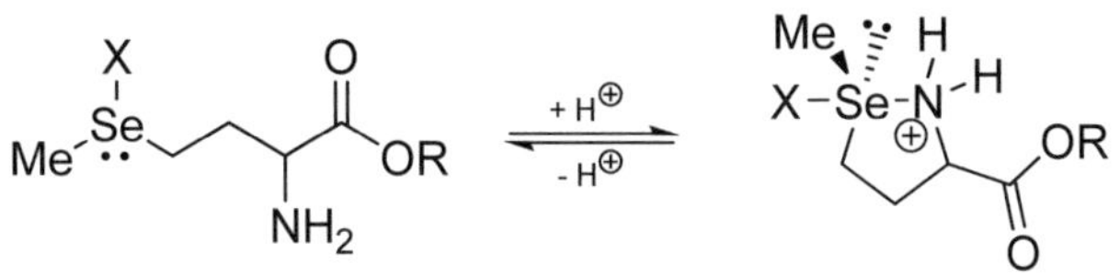

18a,b **18a**, R = H, X = :
18a,b **18b**, R = Me, X = :
19a,b **19a**, R = H, X = O
19a,b **19b**, R = Me, X = O

20a,b **20a**, R = H
20b, R = Me

Scheme 5.9

To explain an enhanced *para* selectivity observed during the electrophilic chlorination of toluene using *bis*(4-chlorophenyl) selenide/Lewis acids as catalysts [20], the intermediacy of a *bis*(4-chlorophenyl)selenium dichloride/Lewis acid complex **21** was suggested. Its existence was supported by ^{1}H NMR studies (Fig. **5.4**).

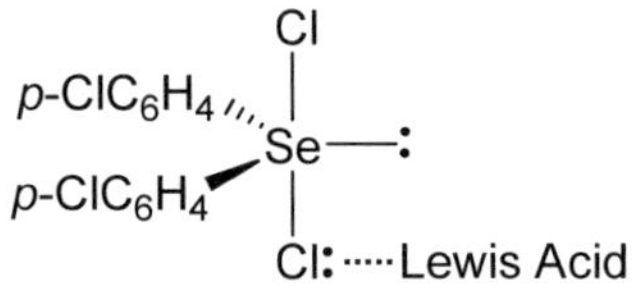

21

Figure 5.4: *bis*(4-Chlorophenyl)selenium dichloride/Lewis acid complex.

Short-lived selenuranes **23a-b** (formed by addition of thiol to the seleninic acid) were suggested by density-functional theory (DFT) calculations used to model the thiol reduction of methane- and benzeneseleninic acids **22a-b** into the corresponding selenenic acids **25a-b** (Scheme **5.10**) [21].

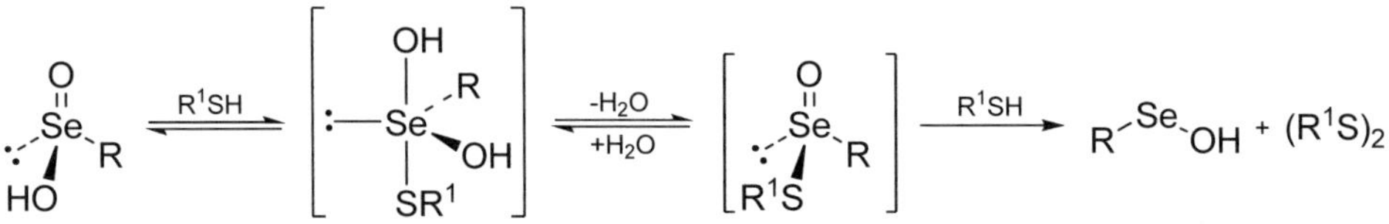

22a,b **23a,b** **24a,b** **25a,b**

22,23,24,25 a, R = Me
22,23,24,25 b, R = Ph

Scheme 5.10

An equilibrium between the optically active selenoxide **27b** (generated *in situ* by treatment of selenide **26** with the optically active Davis oxaziridine) and an optically active cyclic selenurane **27a** was suggested to explain low asymmetric induction in the selenoxide **27**-catalyzed oxidation of sulfides **28** to sulfoxides **29** (Scheme **5.11**) [22].

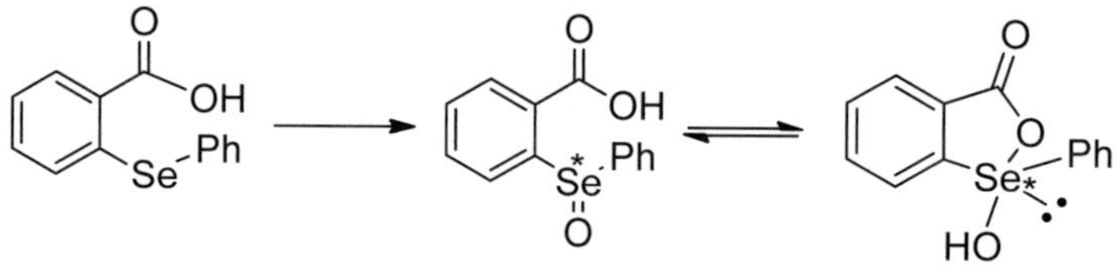

26 **27b** **27a**

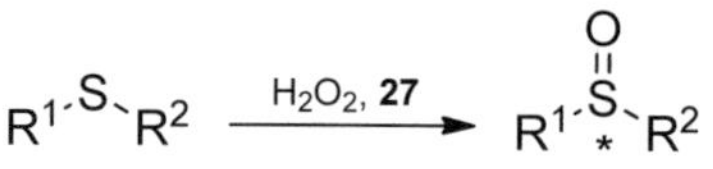

28 **29**

29		
R^1	R^2	Yield [%]
p-Tol	Me	75
CH_2Ph	Me	69

Scheme 5.11

5.3. ISOLABLE SELENURANES

A relatively rich number of isolable hypervalent selenium derivatives has been well documented in the chapters by Bergman and coworkers [3a], Furukawa and Sato [4] and Drabowicz, Kielbasinski and Zając [3b]. Due to this fact only the most representative examples of isolable hypervalent derivatives will be presented. They are divided in accord with the commonly accepted Martin's N-Se-L (A_nB_m) coding system, in which N stands for the number of valence electrons associated formally with a central selenium atom and L shows the number of ligands (A and B stand for the bonding element) [23].

5.3.1. 10-Se-3 Selenuranes

Selected isolable π-hypervalent selenium derivatives having the general structures **30** [24,25], **31** [26] and the single bisaryl analogue **32** [27] are shown in Fig. **5.5**.

Similarly, 1,6-dioxa-6a-selena(IV)-2,5-diazapentalenes **34** were prepared by treatment of *bis*-oximes **33** with selenium dioxide (Scheme **5.12**) [28].

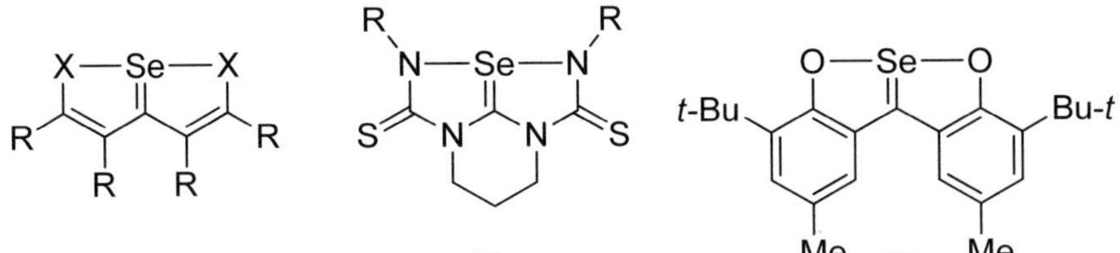

30
R = H or alkyl
X = O or S

31
R = Me or Et

32

Figure 5.5: General structures of selected isolable hypervalent selenium derivatives.

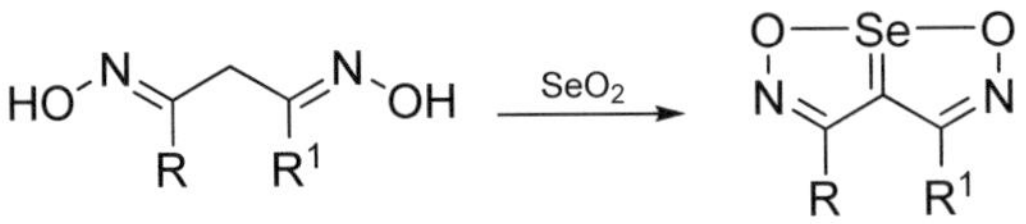

33 **34**

R = H, Me, CO_2Et
R^1 = H, Ph, 4-MeC_6H_4,4-ClC_6H_4, Me
R-R^1 = -$(CH_2)_3$-, -$(CH_2)_4$-

Scheme 5.12

Selenium-containing hypervalent cation **36** was prepared starting from the corresponding divalent precursor **35** as shown in Scheme **5.13** [4,29].

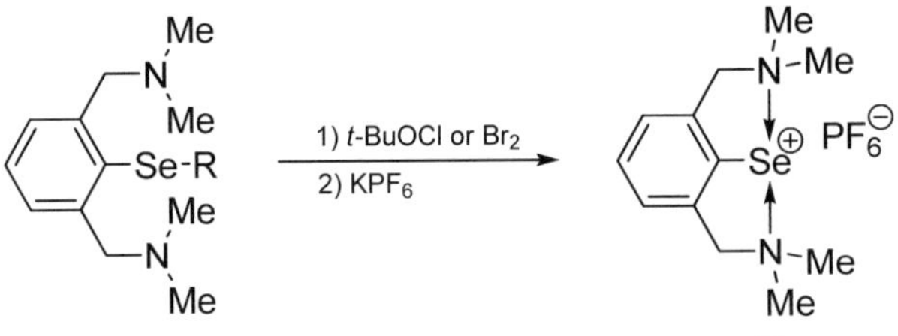

35 **36**

Scheme 5.13

Also 1,2-oxaselenolyl-1-ium chlorides **37a,b**, with partial covalent bonding between Se and the more electronegative ligand atoms, can be considered as hypervalent compounds [30a]. While treated with *p*-nitrobenzoyl chloride or *p*-anisoyl chloride they give the appropriate dioxaselenapentalenes **38a-c** in modest yield (Scheme **5.14**) [30b].

5.3.2. 10-Se-4 Selenuranes

This class of selenuranes has been regarded for years as intermediates in many organic reactions involving tetravalent, tricoordinated selenium compounds. More recently, attention has been focused on studies devoted to their synthetic utility and to experiments aimed to find and explain their biological activity [as compounds that mimic the properties of the selenium-containing enzyme (GSH-Px), possessing different levels of immunomodulating activity or showing good activity to catalyze the reaction of hyperoxide with thiols] [31].

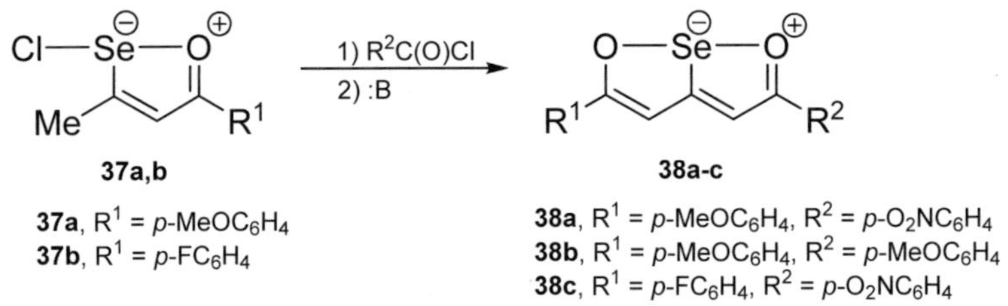

Scheme 5.14

The formation of 10-Se-4 selenuranes has been proposed as the key process in these reactions. The most important findings are briefly presented below while the synthesis, structure and reactivity of particular groups of isolable 10-Se-4 selenuranes are discussed.

Metathetical reactions between selenium tetrachloride and silver sulfonates **39** in 1:1 stoichiometry yielded 10-Se-4 selenuranes **40a-c**, which turned out to be highly sensitive to moisture and to fume in moist air. All the three compounds were found to be nonconducting in nitromethane or acetonitrile (as milimolar solutions), thus suggesting that the sulphonate group is covalently bonded to the selenium atom (Scheme **5.15**) [32]. It should be noted that earlier it was reported that selenium tetrachloride did not yield any defined compounds with trifluoromethanesulfonic acid [33].

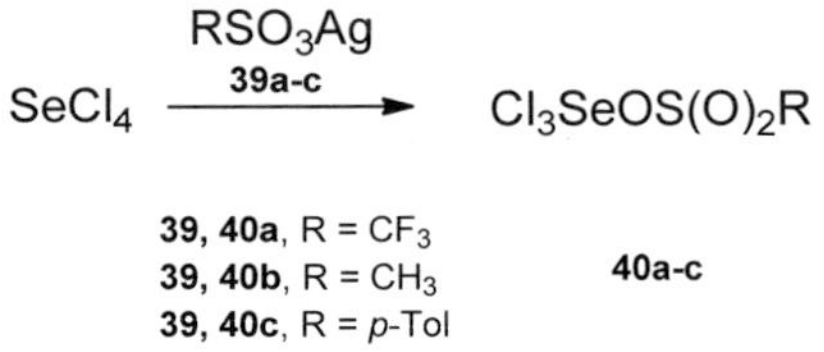

Scheme 5.15

A large number of organic chalcogenium trihalides have been prepared since the first members of this family were unambiguously characterized in the middle of the nineteen twenties [35,36] and they are listed in a review [37] and in a book [38].

The oldest and still the most general procedure for the preparation of selenium trihalides **42** and **43** is based on treatment of the corresponding diselenides **41** with halogens (Scheme **5.16**) [34-36, 39]. Recently, phenylselenium trichloride **42a** was synthesized in the reaction of diphenyl diselenide **41a** with three molar equivalents of sulfuryl chloride [39].

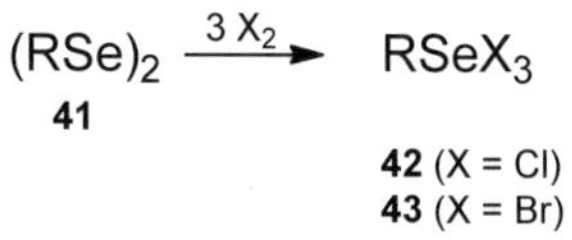

Scheme 5.16

Upon treatment of the diselenides **41a-h** with 3 equivalents of xenon difluoride the corresponding organoselenium trifluorides **44a-h** were isolated (Scheme **5.17**) [40a,b,d]. Interestingly, the chloride ion facilitates this reaction [40b]. They are much less stable than the trichlorides **42a** and tribromides **43a**, and their decomposition starts at +4 °C within days.

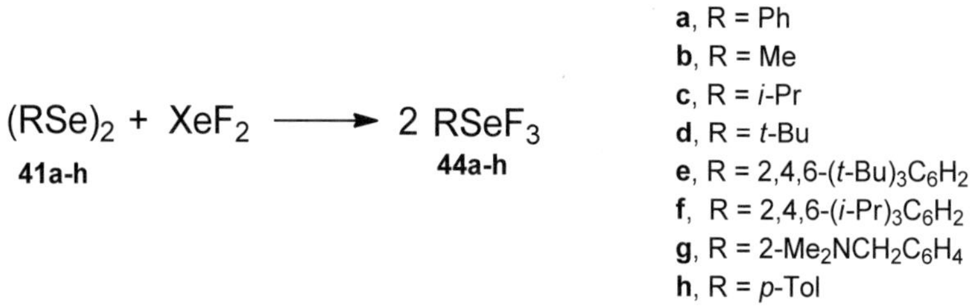

Scheme 5.17

Organoselenium trihalides have found use in the introduction of selenium into organic molecules. A review article devoted to phenylselenium trichloride **42a** is included in the Encyclopedia of Reagents for Organic Synthesis [41].

The reaction of the organoselenium trifluorides **44a-c,i-k** (generated *in situ* according to Scheme **5.17**) with 3 equivalents of Me_3SiN_3 at low temperatures furnished the corresponding organoselenium triazides **45a-c,i-k** (Scheme **5.18**)

[40a]. They are unstable even at -50 °C and, when slowly warmed up, decompose immediately to the corresponding diselenide and dinitrogen (Scheme **5.18**).

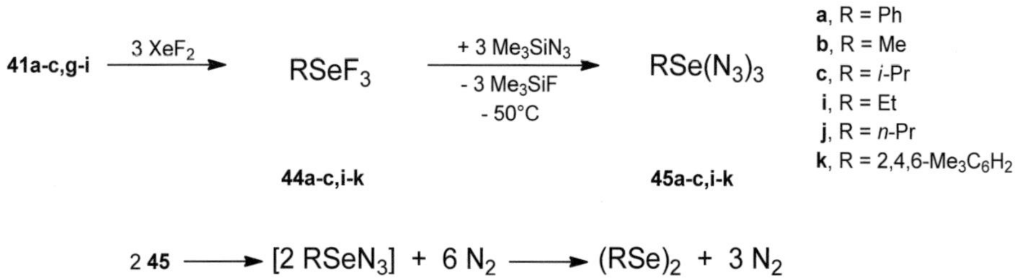

Scheme 5.18

The preparation of dihalogenoselenuranes based on the oxidative halogenation of acyclic and cyclic selenides with elemental halogens is very convenient from the experimental point of view. Thus, chlorination of perfluorinated diphenyl selenide **46a** in liquid chlorine gave very unstable dichloride **47a** which, upon removal of excess chlorine, immediately re-forms the selenide **46a** (Scheme **5.19**) [42].

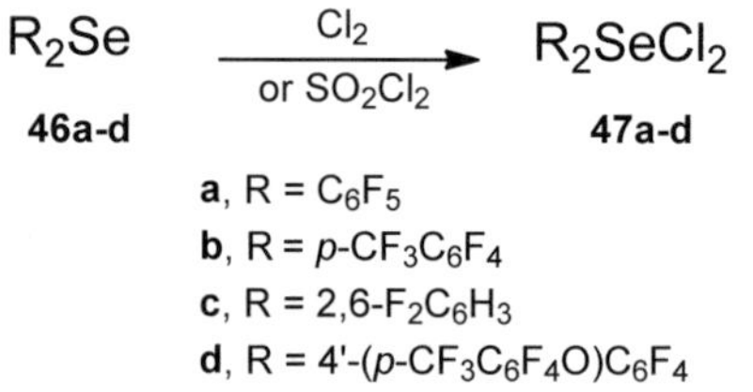

Scheme 5.19

A few dihaloselenuranes were found to be useful halogenating agents. For example, the reaction of optically active benzyl alcohol **49** with dichlorodimethyl selenurane **48**, used in the presence of triphenylphosphine, afforded (α-trifluoromethyl)benzyl chloride **50** with inversion of configuration (Scheme **5.20**). Thiophosphate **51** underwent the similar reaction giving thiophosphorochloridite **52** (Scheme **5.21**) [22,43].

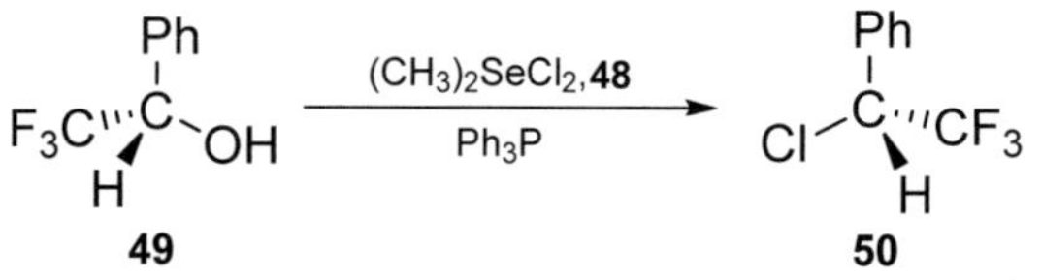

Scheme 5.20

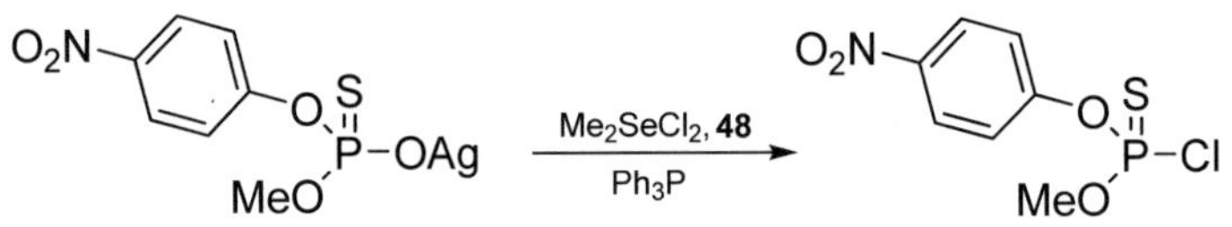

51 **52**

Scheme 5.21

The reaction of ferrocenyl-*n*-heptyl selenium dibromide **53** with 4-pentenoic acid led to the formation of 5-ferrocenylselanylmethyl-dihydrofuran-2-one **54** and *n*-heptyl bromide **55** (Scheme **5.22**) [44].

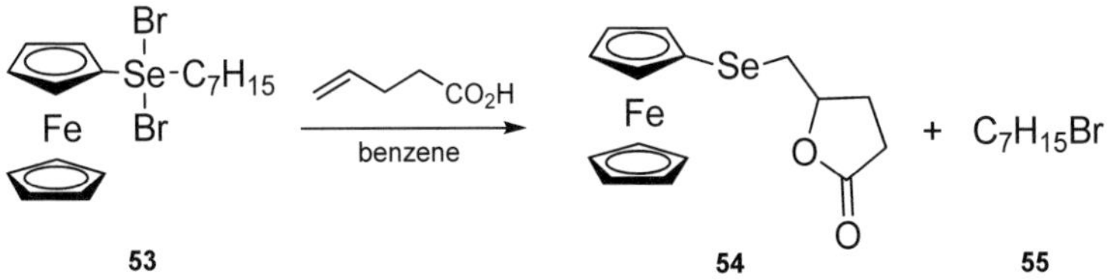

53 **54** **55**

Scheme 5.22

A very large group of diorganodifluoro selenuranes **56a-l** were prepared by oxidative fluorination of the selenides **46** with xenon difluoride (Scheme **5.23**) [40a,42,45,46]. Isolated compounds are usually pale-yellow liquids or colorless solids. They are, like already discussed trifluorides, extremely sensitive towards moisture and are storable only at +4 °C [in perfluoroalkoxy-copolymer (PFA) vessel] over a period of about 2 weeks at maximum.

The nucleophilic exchange of fluorine atoms in difluoroselenuranes **56a-l** with trimethylsilyl azide afforded the corresponding organoselenium diazides **57a-l** (Scheme **5.24**) [40a,c,45,46,47].

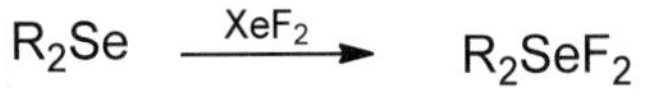

46a-l **56a-l**

a, R = C_6F_5
b, R = p-$CF_3C_6F_4$
c, R = 2,6-$F_2C_6H_3$
d, R = 4'-(p-$CF_3C_6F_4O$)C_6F_4
e, R = Me
f, R = Et
g, R = *i*-Pr
h, R = *n*-Pr
i, R = Ph
j, R = 2,4,6-$Me_3C_6H_2$
k, R = 2,4,6-(*t*-Bu)$_3C_6H_2$
l, R = *o*-(Me_2NCH_2)C_6H_4

Scheme 5.23

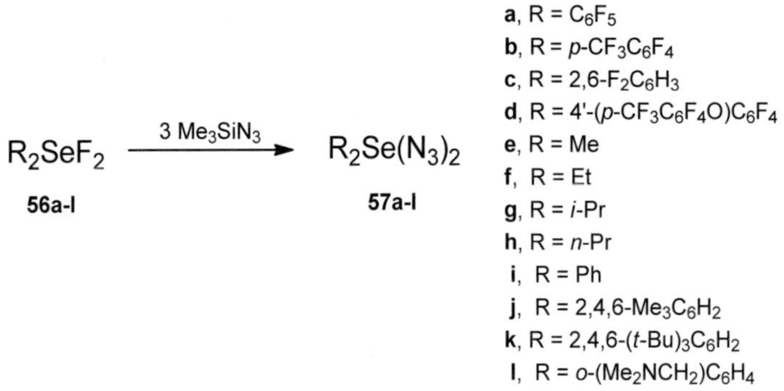

Scheme 5.24

They were stable only at temperatures around -50 °C and decomposed at increased temperatures under vigorous formation of dinitrogen to the corresponding monoselenides.

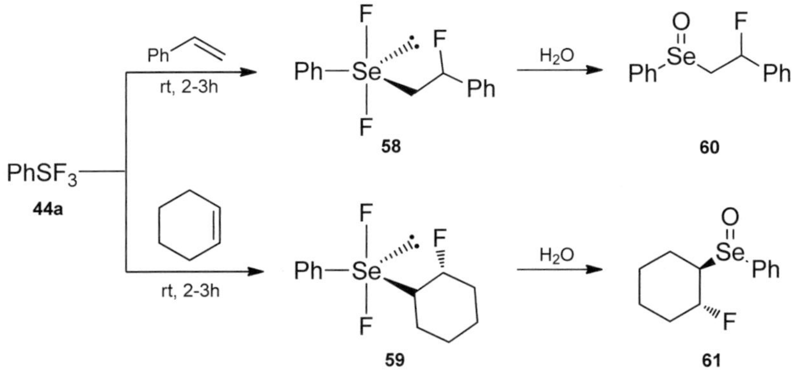

Scheme 5.25

The unstable difluoroselenuranes **58** and **59** were formed by treatment of olefins (styrene and cyclohexene) with phenyltrifluoroselenurane **44a**. They could not be isolated due to their extreme hydrolytic susceptibility. After treatment with aqueous $NaHCO_3$ or on silica gel, they were converted quantitatively into more stable selenoxides **60** or **61** (Scheme **5.25**) [40d].

The chloroselenuranes **64a-c** were isolated upon treatment of suitably functionalized ω-hydroxyalkyl selenides **62a-c** with *t*-butyl hypochlorite (Scheme **5.26**) [48]. Chlorophenylselenurane **64a** (which has a slightly distorted trigonal bipyramidal geometry around the central selenium atom) reacted with tellurides

63a and **63c** in a methylene chloride solution at room temperature giving selenide **62a** and telluranes **65a** and **65c** in quantitative yield (Scheme **5.26**). The similar oxidation of ethyl selenide **62c** needed a longer time. Moreover, ethylselenurane **64c** showed a lower reactivity compared with phenylselenurane **64a**, and phenyltellurane **65a** was able to oxidize only alkyl selenide **62c** [48].

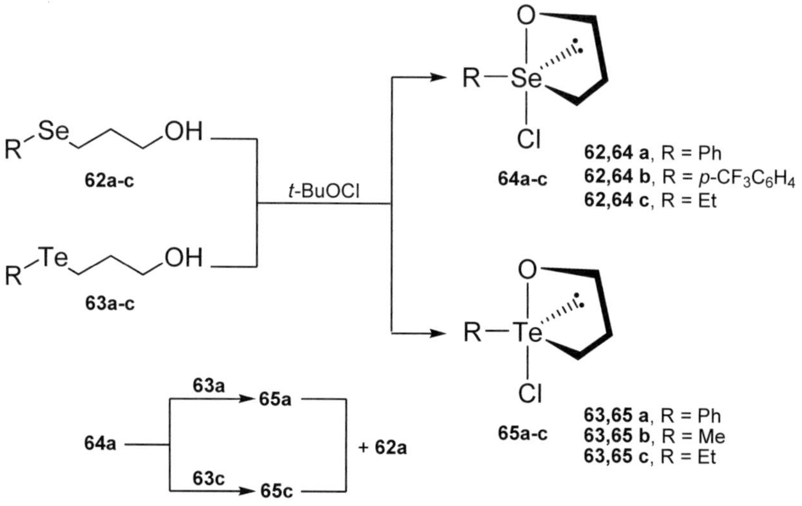

Scheme 5.26

Sulfides **66** were selectively converted to the corresponding sulfoxides **67** upon the reaction with chloroselenuranes **64a** and **64c** with the simultaneous formation of selenides **62a,c** (Scheme **5.27**) [48].

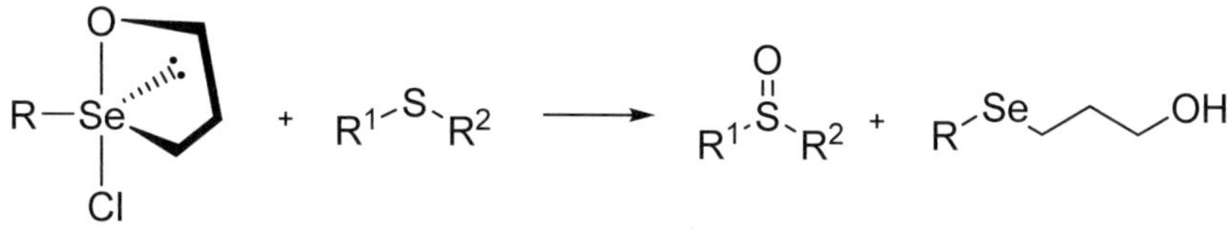

64a,c **66** **67** **62a-c**

	67		
selenurane	R^1	R^2	Yield [%]
64a	Me	Me	100
64a	*n*-Bu	*n*-Bu	95
64a	*c*-C_6H_{11}	Me	92
64c	*p*-$MeOC_6H_4$	Et	50

Scheme 5.27

Diasteromeric chloroselenuranes **69a-d** (X = Cl) were formed rapidly (10 min at 0 °C) and as single stereoisomers (89-100% yield) upon the reaction of bicyclic hydroxy selenides containing the bornyl moiety **68a-d** with *t*-BuOCl (Scheme **5.28**) [49]. Their instantaneous hydrolysis at 0 °C resulted in the selenoxides **74a-d** again as single diastereomers (90-100% yield). When the selenoxide **74a** was treated with HCl the starting chloroselenurane **69a** was recovered as a single diastereomer (100% yield). A similar treatment of **74a** with HBr (96% yield) gave bromoselenurane **70**. A few other diastereomerically pure selenuranes **71-73** were also prepared by the reaction of the selenoxide **74a** with 3,5-dinitrobenzoic acid, *p*-toluenesulfonic acid and trifluoromethanesulfonic acid in the presence of $MgSO_4$ (88- 91% yield) (Scheme **5.28**) [49].

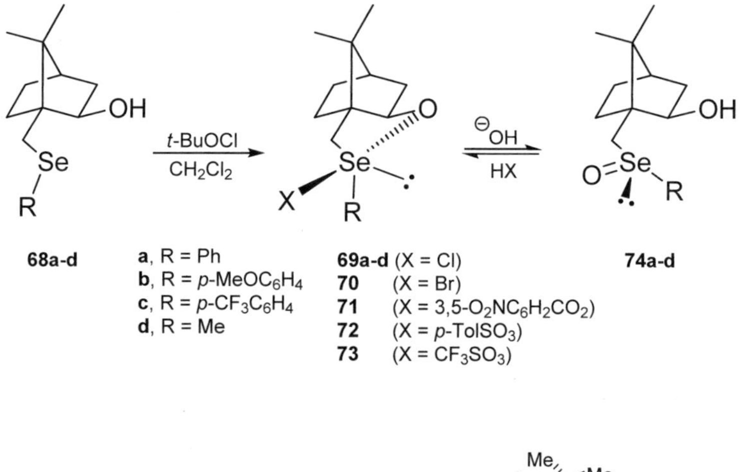

68a-d **a**, R = Ph; **b**, R = *p*-MeOC$_6$H$_4$; **c**, R = *p*-CF$_3$C$_6$H$_4$; **d**, R = Me

69a-d (X = Cl)
70 (X = Br)
71 (X = 3,5-O$_2$NC$_6$H$_2$CO$_2$)
72 (X = *p*-TolSO$_3$)
73 (X = CF$_3$SO$_3$)

74a-d

Scheme 5.28

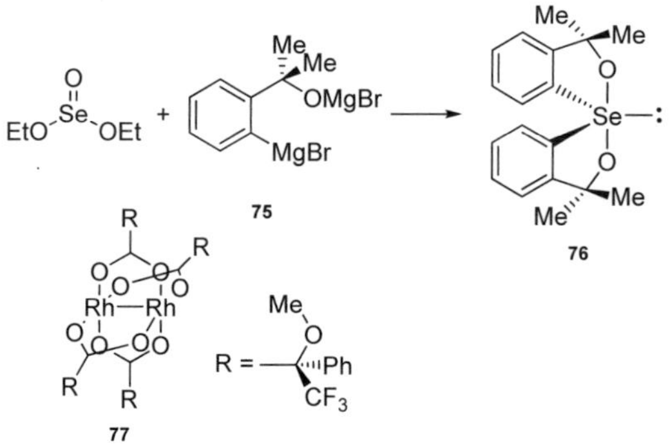

Scheme 5.29

Spiroselenuranes containing a hypervalent selenium atom with trigonal bipyramidal geometry exhibit axial chirality even if both ligand arms are equal.

Recently, one example of this type of spiroselenurane, namely *C*2-symmetric 3,3,3'3'-tetramethyl-1,1'-spirobi[3*H*,2,1]-benzoxaselenole **76**, available earlier in the racemic form by a route involving a few steps [50], was synthesized in one step from diethyl selenite and the Grignard reagent **75** derived from *o*-bromocumyl alcohol (Scheme **5.29**) [51].

The stereogenic character of the selenium atom in these compounds is clearly indicated by the presence of the two well-separated methyl singlets in the ^{1}H NMR spectrum at 1.59 and 1.63 ppm and in the ^{13}C NMR spectrum at 32.89 and 33.89 ppm. A semi-preparative separation of the racemate on the chiral column gave a sample of each enantiomer fully characterized by means of polarimetry, ^{1}H NMR and CD spectroscopy. The individual adduct species formed between spiroselenurane **76** and a chiral dirhodium tetracarboxylate complex **77** were characterized by low-temperature NMR spectroscopy [52]. Moreover, the experimental vibrational circular dichroism (VCD) spectra were obtained for both enantiomers. The theoretical VCD spectra were obtained by means of density functional theoretical calculations with the B3LYP density functional. By comparing experimental and theoretical VCD spectra, the absolute configuration of an enantiomer with positive specific rotation in methylene chloride at 589 nm was determined to be *R*. This conclusion was ultimately verified by comparing the results of experimental optical rotatory dispersion (ORD) and electronic circular dichroism (ECD) with predictions of the same properties using the B3LYP functional [53].

A glutathione peroxidase mimic, a very stable diazaselenurane **79**, was obtained by the oxidation of selenide **78** with hydrogen peroxide (Scheme **5.30**) [54]. It is interesting to note that its crystallization is accompanied by a spontaneous optical resolution leading to the isolation of optically active samples.

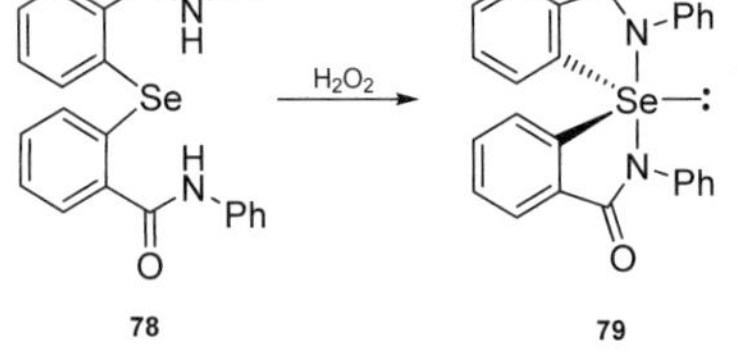

Scheme 5.30

Similarly, monoselenide **80**, upon reaction with H_2O_2, afforded bicyclic diazaselenurane **27**, presumably *via* the selenoxide intermediate (**81**) (Scheme **5.31**) [55].

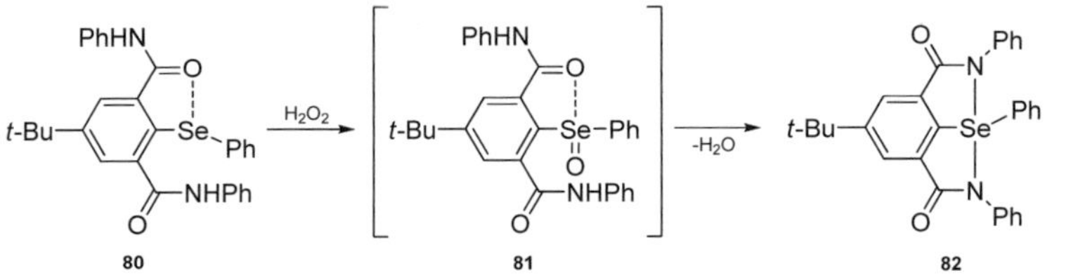

Scheme 5.31

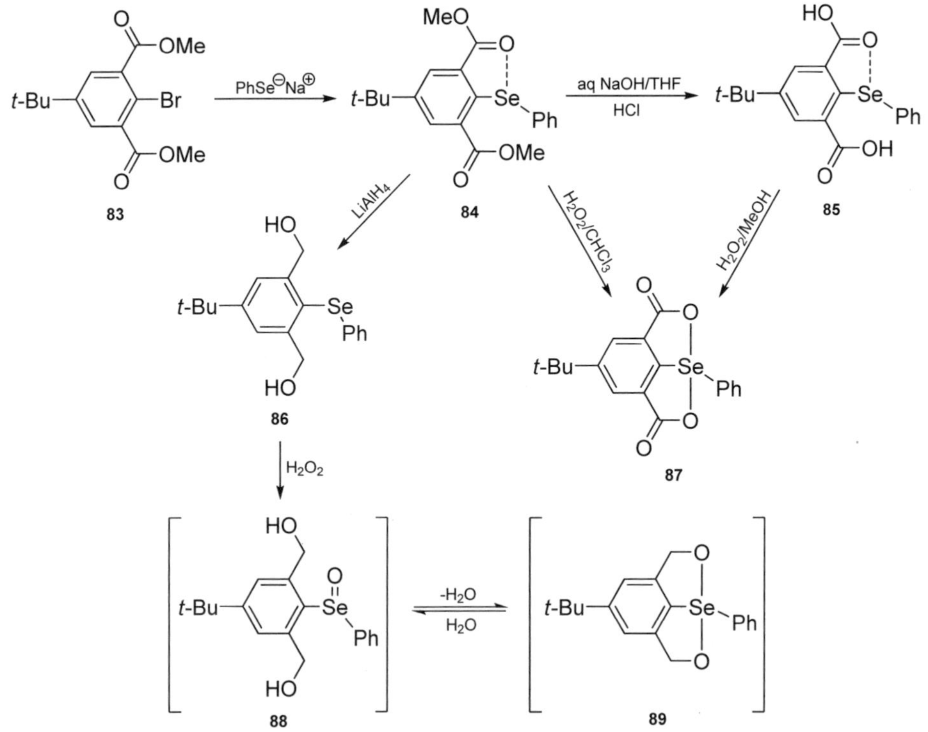

Scheme 5.32

This approach was also applied in the preparation of selenuranes **87** and **89** (Scheme **5.32**). Thus, the oxidation of monoselenide **85** with H_2O_2 resulted in the formation of bicyclic selenurane **87** in three hours (Scheme **5.32**). However, reaction of **84** with H_2O_2 resulted in the formation of **87** after three days. On the other hand, the reaction of monoselenide **86** with an excess of hydrogen peroxide resulted in the formation of a mixture of selenoxide **88** and dialkoxyselenurane **89** (Scheme **5.32**) (based on ^{77}Se NMR spectroscopy). However, it was difficult to separate the bicylic selenurane **89** from the mixture [55].

Three tetraarylic mono- and bicyclic selenuranes **11-12** bearing four carbon ligands, which were prepared before 1998 and discussed by Furukawa and Sato [4], are shown in Fig. **5.6**.

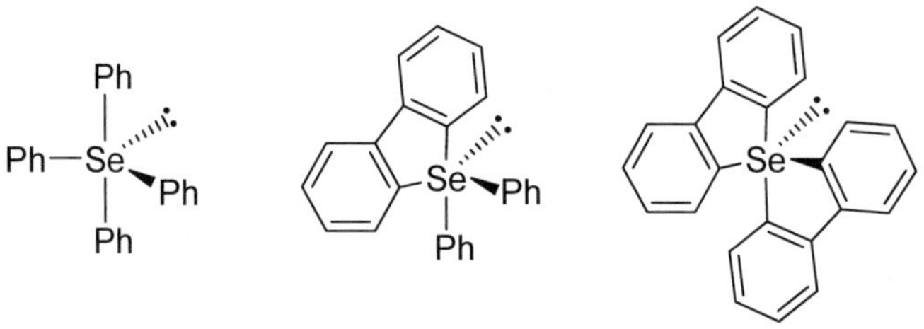

11 12 90

Figure 5.6: Stable tetraarylic selenuranes.

5.3.3. Isolable Selenurane Oxides (10-Se-5 Species)

The chemistry of selenurane oxides has been less extensively studied if compared with the chemistry of tetracoordinated selenuranes. Therefore, in the last few decades only a few papers dealing with this topic have been published. Among them there is a report on the first successful isolation of the enantiomers of C2-symmetric 3,3,3',3'-tetramethyl-1,1'-spirobi[3*H*,2,1]-benzoxaselenole oxide **91** *via* liquid chromatography of the racemate [prepared by oxidation of the parent selenurane **76** with *meta*-chloroperbenzoic acid (Scheme **5.33**)] using a column with a chiral stationary phase or by its spontaneous optical resolution, taking place during slow evaporation of its acetonitrile solution [56].

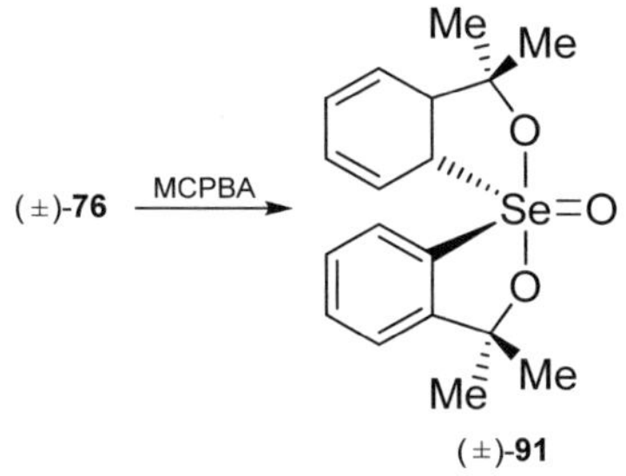

Scheme 5.33

The reactivity of the selenurane oxide is in a sharp contrast to that of its sulfur analogue. For example, it is reduced to the parent selenurane **76** in the presence of HCl and converted into the symmetrical *bis*-hydroxyalkyl selenide **92** by the action of two equivalents of triphenylphosphine in the presence of water, with simultaneous formation of triphenylphosphine oxide (Scheme **5.34**) [22,56].

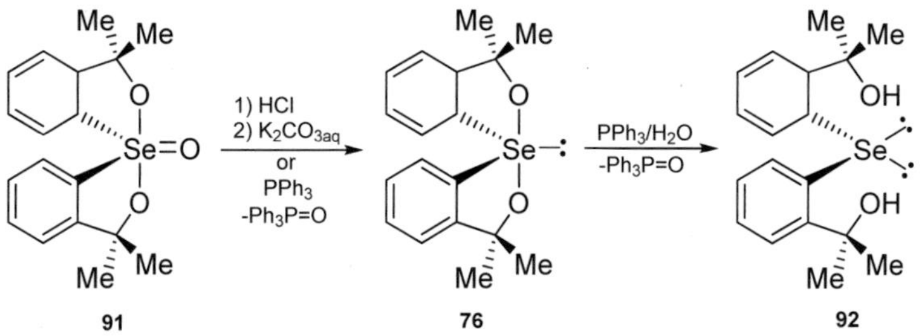

Scheme 5.34

5.4. ISOLABLE 12-Se-6 PERSELENURANES

Hexacoordinated selenium compounds are quite rare. Due to the involvement of 12 electrons to bond 6 ligands to the central selenium atom, perselenuranes are considered to have three sets of 3c-4e bonds which are perpendicular to each other. Therefore, they should have an octahedral geometry. Considering that, according to expanded Rundle - Musher model [4], the nonbonding orbital splits into bonding and antibonding orbitals, the haxacoordinated selenium species are more stable than the corresponding tetravalent, tetracoordinated selenuranes [4].

Phenylselenopentafluoride **93** was recently prepared by the room temperature reaction of diphenyl diselenide **41a** with xenon difluoride (Scheme **5.35**) [40b,57]. Their reaction with olefins afforded the corresponding 1,2-difluorides as the main products (Scheme **5.35**) [57, 58d].

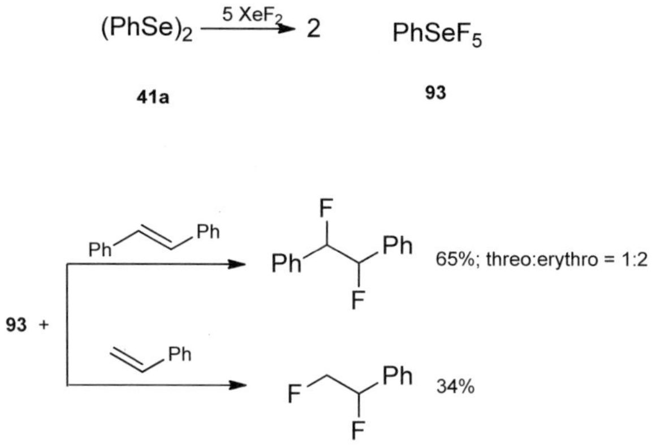

Scheme 5.35

Experiments devoted to the preparation and structural studies of difluoro derivative **94** (Scheme **5.36**) were very comprehensively discussed by Furukawa and Sato in their book chapter [4] and in more recent papers [58a,b].

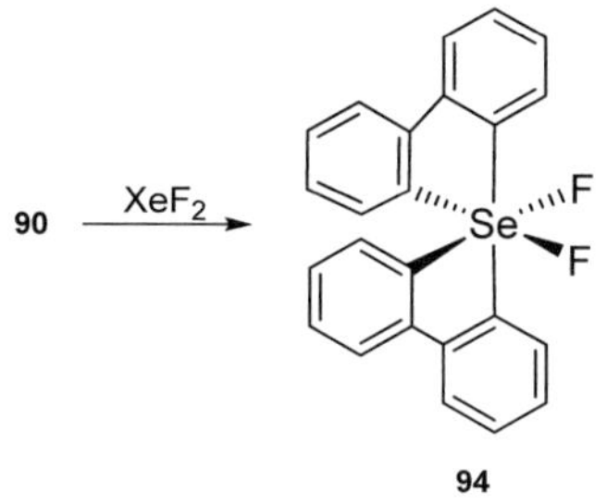

Scheme 5.36

5.5. ISOLABLE 12-Se-5 PERSELENURANES

Perselenurane **95** (in which the formal three center, four-electron bond, typical for hypervalent perchalcogenuranes can also be identified) was isolated by the addition of chlorine to 1,2-oxaselenolyl-l-ium chloride **37a** (Scheme **5.37**) [30a].

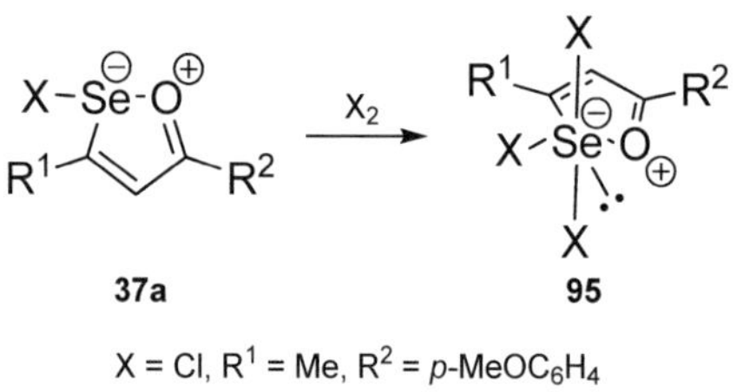

Scheme 5.37

ACKNOWLEDGEMENTS

This paper was financially supported by the National Science Center fund awarded based on the decision UMO-2011/01/B/ST5/06304 UMO-2012/06/A/ST5/00227 (to JD).

CONFLICT OF INTEREST

The authors confirm that this chapter contents have no conflict of interest.

REFERENCES

[1] Musher, J. I. The Chemistry of Hypervalent Molecules† *Angew. Chem. Intern. Ed. Engl.*, **1969**, *8*, 54-68.

[2] (a) Perkins, C. W.; Martin, J. C.; Arduengo, A. J.; Lau, W.; Alergia, A.; Kochi, K. J. An Electrically Neutral σ-Sulfuranyl Radical from the Homolysis of a Perester with Neighboring Sulfenyl Sulfur: 9-S-3 Species *J. Am. Chem. Soc.*, **1980**, *102*, 7753-7759. (b) Hayes, R.A.; Martin, J. C. In: *Organic Sulfur Chemistry: Theoretical and Experimental Advances*;. Bernardi, F.; Csizmadia, J.G.; Mangini, A. Eds.; Elsevier, Amsterdam, **1985**; Chapter *8*, pp. 408-458.

[3] (a) Bergman, J.; Engman, L.; and Siden, J. In: *The Chemistry of Organic Selenium and Tellurium Compounds*; Patai, S.; Rappoport, Z. Eds.; John Wiley and Sons, Chichester, **1986**, Chapter *14*, pp. 517-558. (b) Drabowicz, J.; Kielbasinski, P.; Zajac, A. In: *The Chemistry of Organic Selenium and Tellurium Compounds*; Rappoport, Z. Ed.; John Wiley and Sons, Chichester, **2012**, Vol. *3*, Part *2*, Chapter *15*, pp. 891-948.

[4] Furukawa, N; Sato, S. In: *Chemistry of Hypervalent Compounds*; Akiba, K. Ed., Wiley-VCH, New York, **1999**, Chapter *8*, pp. 241-278.

[5] (a) Paulmier, C. Ed.; In: *Selenium Reagents and Intermediates in Organic Synthesis*, Pergamon Press, Oxford, **1986**; (b) Liotta, D. Ed.; In: *Organoselenium Chemistry*, John Wiley & Sons, New York, **1987**; (c) Back, T. G. Ed.; In: *Organoselenium Chemistry: A Practical Approach*, Oxford University Press, Oxford, **1999**; (d) Drabowicz, J.; Mikołajczyk, M. Selenium at Higher Oxidation State *Top. Curr. Chem.*, **2000,** *208*, 143-176; (e) Drabowicz, J.; Halaba, G. In: Stereochemical aspect of the chemistry of hypervalent chalcogen compounds *Rev. Heteroatom Chem.*, **2000**, *22*, 1-32; (f) Drabowicz, J. Unexpected Conversions of Selected Heteroorganic Compounds *Heteroatom Chem.*, **2002**, *13*, 437-442; (g) Allenmark, S. Recent Advances in Spirochalcogenurane Stereochemistry - A Mini Review *Chirality*, **2008**, *20*, 544 -551; (h) Wirth, T. In: *Comprehensive Organometallic Chemistry III*; Crabtree, R. H.; Mingos, D. M. P. Eds.; Elsevier, Oxford, **2007**, Vol. *9*, pp. 457-500; (i) Wirth, T. Organoselenium Chemistry in Stereoselective Reactions *Angew. Chem. Int. Ed.*, **2000**, *39*, 3740-3749; (j) Zhu, C.; Huang, Y. Asymmetric Synthesis of Chiral Organoselenium Compounds *Curr. Org. Chem.*, **2006**, *10*, 1905-1920; (k) Braga, A. L. ; Lüdtke, D. S.; Vargas, F. Enantioselective Synthesis Mediated by Catalytic Chiral Organoselenium Compounds *Curr. Org. Chem.*, **2006**, 10,

1921 -1938; (l) Murai, T.; Kimura, T. Syntheses and Properties of Phosphinoselenoic Chlorides, Acids, and Their Salts *Curr. Org. Chem.*, **2006**, *10*, 1963-1973; (m) Coles, M. P. Utilization of Nonbonded Interactions Involving Organoselenium Compounds *Curr. Org. Chem.*, **2006**, *10*, 1993-2005; (n) Freudendahl, D. M.; Shahzad, S. A.; Wirth, T. Recent Advances in Organoselenium Chemistry *Eur. J. Org. Chem.*, **2009**, *48*, 1649-1664.

[6] Drabowicz, J. In: *Chemistry of Hypervalent Compounds* Akiba, K. Ed.; Wiley-VCH, New York, **1999**, Chapter *7*, pp. 211-240.

[7] (a) Akiba, K. In: *Chemistry of Hypervalent Compounds* Akiba, K. Ed.; Wiley-VCH, New York, **1999**, Chapter *1*, pp. 1-8; (b) Chapter *2*, pp. 9- 47.

[8] Noury, S.; Silvi, B; Gillespie, J.R.; Chemical Bonding in Hypervalent Molecules: Is the Octet Rule Relevant? *Inorg. Chem.*, **2002**, *41*, 2164-2172 (2002).

[9] Mutterties, E. L; Schunn, R. A. Pentaco-Ordination *Quart. Rev.*, **1966**, *20*, 245-299.

[10] Chen, M. M. L.; Hoffmann, R. Sulfuranes. Theoretical aspects of bonding, substituent site preferences, and geometrical distortions *J. Am. Chem. Soc.*, **1976**, *98*, 1647-1653.

[11] Sini, G.; Ohanessian, G.; Hiberty, P. C.; Shaik, S. S. Why is SiH_5- a stable intermediate while CH_5- is a transition state? A quantitative curve crossing valence bond study *J. Am. Chem. Soc.*, **1990**, *112*, 1407-1413.

[12] Berry, R. S. Correlation of Rates of Intramolecular Tunneling Processes, with Application to Some Group V Compounds *J. Chem. Phys.*, **1960**, *32*, 933-938.

[13] Ugi, I.; Marquarding, D.; Klusacek, H.; Gillespie, P.; Ramirez, F. Berry pseudorotation and turnstile rotation *Acc. Chem. Res.*, **1971**, *4*, 288-296.

[14] Reich, H. J.; Gudmundson, B. O.; Dykstra, D. R. The first organoselenium ate complex *J. Am. Chem. Soc.*, **1992**, *114*, 7937-1938.

[15] Ogawa, S.; Sato, S.; Erata, T.; Furukawa, N. Reactions of triaryl selenonium salts and diaryl selenoxides with phenyllithium: First detection of tetraaryl, selenuranes [10-SE-4(C4)] by ^{77}SE, 1H, and ^{13}C $_{NMR}$ *Tetrahedron Lett.*, 1991, **32**, 3179-3182.

[16] Sato, S.; Furukawa, N. First detection of 2,2′-biphenylylenediphenylsulfurane and -selenurane [10-M-4(C4), M = S, Se] by low temperature NMR experiments and isolation of the tellurane *Tetrahedron Lett.*, **1995**, *36*, 2803-2806.

[17] Sato, S.; Matsuo, M.; Nakahodo, T.; Furukawa, N.; Nabeshima, T. Generation and detection of tellurane [10–Te–4(C4)] and selenurane [10–Se–4(C4)] having alkyl and aryl ligands *Tetrahedron Lett.*, **2005**, *46*, 8091-8093.

[18] Henriksen, L.; Jakobsen, S. Dissociative cycloelimination, a new selenium based pericyclic reaction *Chem. Commun.*, **2001**, 2448-2449.

[19] Ritchey, J. A.; Davis, B. M.; Pleban, P. A.; Bayse, C. A. Experimental and theoretical evidence for cyclic selenurane formation during selenomethionine oxidation *Org. Biomol. Chem.*, **2005**, *3*, 4337-4342.

[20] Graham, J. C.; Feng, C. H.; Orticochea, M.; Ahmed, G. The use of bis(4-chlorophenyl) selenide/Lewis acid catalysts in the electrophilic chlorination of toluene *J. Org. Chem.*, **1990**, *55*, 4102-4109.

[21] Bayse, C. A. Model mechanisms of sulfhydryl oxidation by methyl- and benzeneseleninic acid, inhibitors of zinc-finger transcription factors *J. Inorg. Biochem.*, **2010**, 104, 1-8.

[22] Drabowicz, J.; Łuczak, J.; Łyżwa, P.; Kiełbasiński, P.; Mikołajczyk, M.; Yamamoto, Y.; Matsukawa, S.; Akiba, K.; Wang, F.; Polavarapu, P. L.; Wieczorek, M.W. Some Chiral and Achiral Low- and High-Coordinated Selenium and Tellurium Compounds: Syntheses, Structural Determination, and Selected Synthetic Applications *Phosphorus Sulfur Silicon*, **2005**, *180*, 741-753.

[23] Perkins, C. W.; Martin, J. C.; Arduengo, A. J.; Lau, W.; Algeria, A.; Kochi, J. K. An electrically neutral.sigma.-sulfuranyl radical from the homolysis of a perester with neighboring sulfenyl sulfur: 9-S-3 species *J. Am. Chem. Soc.*, **1980**, *102*, 7753-7759.

[24] Reid D. H.; Webster, R. G.; Heterocyclic compounds. XVII. Synthesis of 1,6-dioxa-6a-thia- and 1,6-dioxa-6a-selenapentalenes *J. Chem. Soc. Perkin Trans.1,* **1975**, *8*, 775-780.

[25] Reid, D. H.; Webster, R. G. Heterocyclic compounds. X. 6a-Selenathiophthenes *J. Chem. Soc.***1971**, *19*, 3187-3190.

[26] (a) Iwasaki, F.; Murakami, H.; Yamazaki, N.; Yasui, M.; Tomura, M.; Matsumura, N. Structures of 6a-thia/selena-1,3,4,6-tetraazapentalene derivatives: hypervalent S/Se-N bonds *Acta Cryst.*, **1991**, *C47*, 998-1003; (b) Matsumura, N.; Tomura, M.; Inazu, K.; Inoue, H.; Yamazaki, N.; Iwasaki, F. Synthesis and characterization of 10-Se-3 type selenatetraazapentalene derivatives with a hypervalent selenium *Phosphorus Sulfur Silicon,* **1992**, *67*, 135-140.

[27] Foroughifar, N.; Synthesis and thermal rearrangement of oxygen heterocyclic spirans *J. Sci., Islamic Republic of Iran*, **1993**, *4*, 273-290.

[28] Perrier, M.; Vialle, J. Aluminum nitride thin films *Bull. Soc. Chem. Fr*, **1979**, 199-214.

[29] Fujihara, H.; Mima, H.; Furukawa, N. First Isolation and Crystal Structure of Heavier Chalcogenenium Cations (RSe^+, RTe^+) Stabilized by Two Neighboring Amino Groups *J. Am. Chem. Soc.,* **1995**, *117*, 10153-10154.

[30] (a) Detty, M. R.; Perlstein, J. Radical anion stabilities in 1,6-dioxa-6a-tellurapentalenes and related compounds. The quantum efficiency of photoconduction as a measure of E.degree *Organometallics,* **1987**, *6*, 1597-1606; (b) Detty, M. R.; Murray, B. Cyclization of 3-(arylchalcogeno)propenoyl chlorides. 2. Chalcogen and substituent control in the regiochemistry of intramolecular acylation. Preparation of benzo[b]telluropyrones *J. Am. Chem. Soc.,* **1983**, *105*, 883 -890.

[31] (a) Mauri, P.; Bennazi, L.; Floh′e, L.; Maiorino, M.; Pietta, P. G.; Pilawa, S.; Roveri, A.; Ursini, F. Versatility of Selenium Catalysis in PHGPx Unraveled by LC/ESI-MS/MS *Biol. Chem.*, **2003**, *384*, 575-588; (b) Ren, B.; Huang, W.; Akesson, B.; Ladenstein, R. The Crystal Structure of Seleno-Glutathione Peroxidase from Human Plasma at 2.9 Å Resolution *J. Mol. Biol.*, **1997**, *268*, 869-885; (c) Mugesh, G.; Singh, H. G. Synthetic organoselenium compounds as antioxidants: glutathione peroxidase activity *Chem. Soc. Rev.,* **2000**, *29*, 347-457); (d) Bhabak, K. P.; Mugesh, G. Functional Mimics of Glutathione Peroxidase: Bioinspired Synthetic Antioxidants *Acc. Chem. Res.,* **2010**, *43*, 1408-1419.

[32] Kapoor, R.; Wadhawanv, P.; Katyal, V.; Sood, V. R.; Kaapoor, P. Preparation, reactions, and characterization of trichloroselenium(IV) and trichlorotellurium(IV) sulfonates: $SeCl_3(XSO_3)$ and $TeCl_3(XSO_3)$ with X = CF_3, CH_3, and p-$CH_3C_6H_4$ *Can. J. Chem.,* **1989**, *67*, 1760-1764.

[33] Howell, R. D.; McCow, J. D. Trifluoromethanesulfonic acid a nd derivatives *Chem. Rev.*, **1977**, *77*, 69-92.

[34] (a) Irgolic, K. J. In: *Houben-Weyl* Klamann, D. Ed.; Thieme, Sttutgart, **1990**, Vol. *E12b* pp. 71-73 and 75-76 and 84-118; (b) Irgolic, K. J In: *Houben-Weyl* Klamann, Ed. D.; Thieme, Sttutgart, **1990**, Vol. *E12b*, pp.119-150.

[35] Shaw Jr. E. H.; Reid, E. E. Some derivatives of ethyl selenomercaptan *J. Am. Chem. Soc.,* **1926**, *48*, 520-528.

[36] (a) Morgan, G. T.; Kellet, J. CXL.-Interactions of tellurium tetrachloride and aryl alkyl ethers. Part II *J. Chem. Soc.,* **1926**, 1080-1088; (b) Morgan, G. T.; Drew H. D. K.

CCCXV.—Interactions of tellurium tetrachloride and aryl alkyl ethers. Part I *J. Chem. Soc., Trans.,* **1925**, 2307-2315.

[37] (a) Bergman, J. Tellurium in organic chemistry—I : A novel synthesis of biaryls *Tetrahedron*, **1972**, *28*, 3323-3331; (b) Sadekov, I. D.; Bushkov, A. Y.; Minkin, V. I. The Synthesis and Structure of Telluranes *Russ. Chem. Rev.*, **1979**, *48*, 343-362.

[38] Irgolic, K. J. In: *The Organic Chemistry of Tellurium*, Gordon and Breach, New York, 1974.

[39] Barnes, N. A.; Godfrey, S. M.; Halton R. T. A.; Pritchard, R. G. A comparison of the solid state structures of the selenium(IV) compounds $PhSeX_3$ (X = Cl, Br) *J. Chem. Soc. Dalton Trans.*, **2005**, 1759-1761.

[40] (a) Klapötke, T. M.; Krumm, B.; Scherr, M. Studies on the Properties of Organoselenium(IV) Fluorides and Azides *Inorg. Chem.*, **2008**, *47*, 4712-4722; (b) Ou, X., Janzen, A. F. Oxidative fluorination of S, Se and Te compounds *J. Fluorine Chem.*, **2000**, *101*, 279-283; (c) Klapötke, T. M.; Krumm, B.; Mayer, P.; Naumann, D.; Schwab, I. Fluorinated tellurium(IV) azides and their precursors *J. Fluorine Chem.*, **2004**, *125*, 997-1005; (d) Lermontov, S. A.; Zavorin, S. I.; Bakhtin, I. V.; Pushin, A. N.; Zefirov, N. S.; Stang, P. J.; Fluorination of olefins with $PhSeF_3$, $PhSeF_5$ and $PhTeF_5$ *J. Fluorine Chem.*, **1998**, *87*, 75-83.

[41] Engman, L. In: *Encyclopedia of Reagents for Organic Synthesis* Paquette L. A. Ed; John Wiley and Sons, Chichester, **1995**, Vol. *6*, pp. 4014-4015.

[42] Klapötke, T. M.; Krumm, B.; Mayer, P.; Polborn, K.; Ruscitti,O. P. New aspects in the chemistry of aromatic and fluoroaromatic selenium and tellurium compounds: similarities and diversities *Posphorus Sulfur Silicon*, **2001**, *171-172*, 373-382.

[43] Drabowicz, J; Łuczak, J. unpublished results.

[44] Kumar, S.; Helt, J.-C. P.; Autschbach, J.; Detty, M. R. A New Reaction for Organoselenium Compounds: Alkyl Transfer from Diorganoselenium(IV) Dibromides to Alkenoic Acids To Give γ- and δ-Lactones *Organometallics*, **2009**, *28*, 3426-3436.

[45] Klapötke, T. H.; Krumm, B.; Mayer, P.; Piotrowski, H.; Ruscitti, O. P.; Schiller, A. Novel Organotellurium(IV) Diazides and Triazides *Inorg. Chem.*, 2002, **41**, 1184-1193.

[46] Hammerel, A.; Klapotke, T. M.; Krumm, B.; Scherr, M. Tellurium(IV) Fluorides and Azides containing the Nitrogen Donor Substituent R = $2\text{-}Me_2NCH_2C_6H_4$; Crystal Structure of $RTeF_3$ and of an Unusual Tellurium(VI) Fluoride Salt *Z. Anorg. Allg. Chem.*, **2007**, *633*, 1618-1626.

[47] Klapötke, T. M.; Krumm, B.; Mayer, P.; Ruscitti, O. P. First Synthesis and Structures of Aryltellurium(IV) Diazides *Inorg. Chem.*, **2000**, *39*, 5426-5427.

[48] Zhang. Z.; Koizumi, T. Reactivity of chlorooxachalcogenuranes: oxidation of sulfides to sulfoxides using chlorooxaselenuranes *Synth. Commun.*, **2000**, *30*, 979-987.

[49] Kurose, N.; Takahashi, T.; Koizumi, T. First synthesis of optically pure selenuranes and stereoselective alkaline hydrolysis. Their application to asymmetric [2,3] sigmatropic rearrangement of allylic selenoxides Tetrahedron, **1997**, *53*, 12115-12129.

[50] Reich, H. J.; Organoselenium stereochemistry. Configurational stability of dialkoxydiarylselenium compounds *J. Am. Chem. Soc.*, **1973**, *95*, 964-966.

[51] (a) Drabowicz, J.; Łuczak, J.; Mikołajczyk, M, Yamamato, Y.; Matsukawa, S.; Akiba, K. The first enantiomerically pure, C_2-symmetric spiroselenurane: 3,3,3′,3′-tetramethyl-1,1′-spirobi[3H,2,1]-benzoxaselenole *Tetrahedron: Asymm.*, **2002**, *13*, 2079-2082; (b) Drabowicz J. unpublished results.

[52] Gati, T.; Toth, T.; Drabowicz, J.; Moeller, S.; Hofer, E.; Polavarapu, P.; Duddeck, H. Effective enantiodifferentiation of spirochalcogenuranes by the dirhodium method: Towards the determination of absolute configurations? *Chirality*, **2005**, *17*, S40-S47.

[53] Petrovic, A. G.; Polavarapu, P. L.; Drabowicz, J.; Zhang, Y.; McConnell, O. J.; Duddeck, H. Absolute Configuration of C_2-Symmetric Spiroselenurane: 3,3,3′,3′-Tetramethyl-1,1′-spirobi[3 H,2,1]Benzoxaselenole *Chem. Eur. J.*, **2005**, 11, 4257-4262.

[54] Sarma, B. K.; Manna, D.; Minoura, M.; Mugesh, G. Synthesis, Structure, Spirocyclization Mechanism, and Glutathione Peroxidase-like Antioxidant Activity of Stable Spirodiazaselenurane and Spirodiazatellurane *J. Am. Chem. Soc.*, **2010**, *132*, 5364-5374.

[55] Selvakumar, K.; Singh, H. B.; Goel, N.; Singh, U. P.; Butcher, R. J.; Synthesis and structural characterization of pincer type bicyclic diacyloxy- and diazaselenuranes Dalton Trans., **2011**, *40*, 9858-9867.

[56] Drabowicz, J.; Łuczak, J.; Mikołajczyk, M.; Yamamoto, Y.; Matsukawa, S.; Akiba, K. First optically active selenurane oxide: Resolution of C_2-symmetric 3,3,3′,3′-tetramethyl-1,1′-spirobi [3h,2,1]-benzoxaselenole oxide *Chirality*, **2004**, *16*, 598-601.

[57] Lermontov, S. A.; Zavorin, S. I.; Bakhtin, I. V, Zefirov, N. S.; Stang, P. J. Fluorinating properties of $PhTeF_5$ and $PhSeF_5$ towards C=C BOND *Phosphorus, Sulfur Silicon*, **1995**, *102*, 283-286.

[58] (a) Furukawa, N.; Sato, S. Chalcogenuranyl dications bearing unusual bonds and charges *Heteroatom Chem.*, **2002**, *13*, 406-413. (b) Sato, S.; Yamashita, T.; Horn, E.; Takahashi, O.; Furukawa, N. Isolation and structure of bis(2,2′-biphenylylene)dichloro- and -difluoropertelluranes, [12-Te-6(C4X2), X = Cl, F] (λ6-tellane) *Tetrahedron*, **1997**, *53*, 12183-12194.

Organoselenium Chemistry Between Synthesis and Biochemistry, **2014**, **146-165**

CHAPTER 6

Selenoamides, Synthetic Methods and Recent Progress on their Synthetic Applications

Toshiaki Murai*

Department of Chemistry, Faculty of Engineering, Gifu University, Yanagido, Gifu 501-1193, Japan

Abstract: This chapter has overviewed the stability of selenocarbonyl compounds that have been classified based on the substituents attached to their selenocarbonyl carbon atom. Focus has then been laid on the synthetic methods for selenoamides, whose chemistry has grown much more than that of other selenocarbonyl compounds in the last ten years. As starting materials leading to selenoamides, ordinary amides, nitriles and terminal acetylenes are mainly used. As a final part of this review, recent application of selenoamides to synthetic reactions has been introduced. Nucleophilic characters of selenium and nitrogen atoms in selenoamides have enabled several types of intermolecular cyclizations. The carbon atom of selenoamides can accept nucleophilic attack and selenium atom can formally work as a leaving group.

Keywords: Selenocarbonyl compounds, Selenoamides, Amides, Nitriles, Acetylenes, Cyclization, Selenoiminium salts, Copper (0), Complexes, Photoinduction.

6.1. INTRODUCTION

The replacement of an oxygen atom of a range of carbonyl compounds and their derivatives with a selenium atom leads to a wide variety of selenocarbonyl (C=Se) compounds [1]. On the basis of the electronegativity of oxygen and selenium atoms, the formed C=Se bond becomes less polar than C=O. The C=Se are longer double bonds and in the π-bonds, the overlap between a 2p atomic orbital of the carbon atom and 4p atomic orbital of the selenium atom is less efficient. Additionally, energy gap between HOMO and LUMO is smaller than that of carbonyl and thiocarbonyl groups. The visualized scheme of p-orbital has shown

***Address correspondence to Toshiaki Murai:** Department of Chemistry, Faculty of Engineering, Gifu University, Yanagido, Gifu 501-1193, Japan; Tel: +81 58 293 2614; Fax: +81 58 293 2614; E-mail: mtoshi@gifu-u.ac.jp that HOMO as well as LUMO spread over carbon and selenium atoms and are rather localized at the selenium atom. On the basis of these fundamental aspects, selenocarbonyl compounds can be predicted to be less stable and more reactive than carbonyl and thiocarbonyl compounds. Moreover, both carbon and selenium atoms can be electrophilic and nucleophilic centers.

Selenocarbonyl compounds are classified into several categories on the basis of the substituents at the carbon atom of C=Se group. The first category includes compounds with only hydrogen and/or carbon atoms attached to the carbon, that have been called selenoaldehydes and selenoketones [2]. In particular, selenoaldehydes are highly reactive and cannot be isolated unless sterically bulky substituents are introduced to the carbon atom of C=Se [3]. Some selenoketones are isolated, but most of these selenocarbonyl compounds are generated *in situ* and trapped mainly *via* hetero-Diels-Alder reaction. A very recent example of generation of selenoaldehydes is the thermal cycloreversion of [3+2] cycloadducts **2** between selenoaldehydes **1** generated from aldehydes and *bis*-(dimethylaluminum) selenide ($Me_2AlSeAlMe_2$) and anthracene (Scheme **6.1**) [4]. The selenoaldehydes **2** were generated in the presence of dienes and 1,3-dipoles such as **3** and underwent cycloaddition reactions to give **4**.

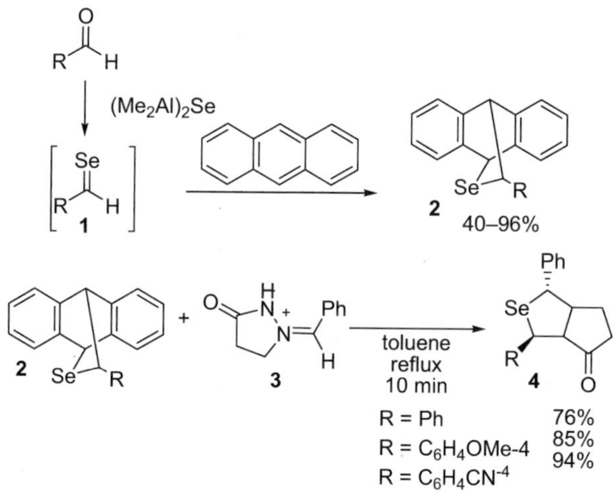

Scheme 6.1

In the second category, one of hydrogen or carbon atom of the compounds in category 1 is replaced with a heteroatom-containing functional group (Scheme **6.2**).

In fact, a range of selenoamides **5** [5], selenoic acid *O*-esters **6** [6], selenothioic acid *S*-esters **7** [7] and diselenoic acid esters **8** [7] have been prepared and isolated, but examples of selenotelluroic acid *Te*-esters **9** have yet to be known. On the basis of the notion that the introduction of heavier elements destabilizes the compounds, the esters **7** and **8** seem to be less readily handled than selenoamides **5** and selenoesters **6**. However the handling of esters **6** is also cumbersome and chemistry of **6** has been less explored compared with that of **7** in recent years. In contrast, increasing attention has been paid to fundamental aspects, synthesis and applications of selenoamides **5** not only from a synthetic point of view but also from biological applications. In this review, focus has been laid on the selenoamides **5** and in particular on their synthetic methods, also some recent examples of their reactions have been shown.

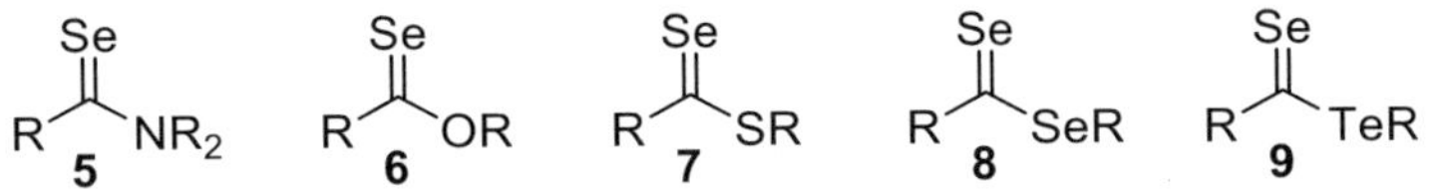

Scheme 6.2

6.2. SELENOAMIDES: SYNTHETIC METHODS

6.2.1. Amides to Selenoamides

One of the most straightforward methods leading to selenoamides is the direct conversion of ordinary amides to selenoamides. To effect the reaction, phosphorus selenide (P_2Se_5) has been used (Scheme **6.3**). With this method, aliphatic secondary and tertiary selenoamides were prepared although in low yields [8]. Selenoamides were obtained in improved yields by adding barium carbonate to the reaction of aliphatic amides and P_2Se_5 in boiling xylene [9].

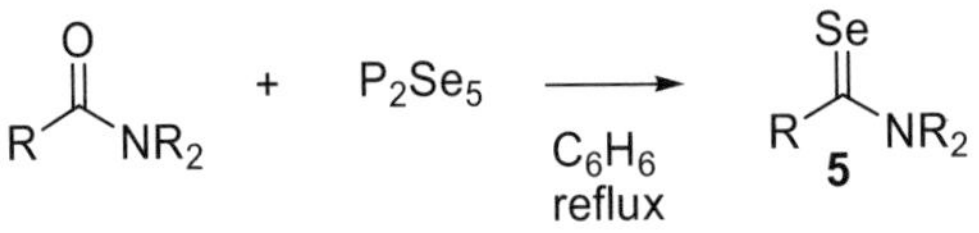

Scheme 6.3

As more readily available phophorus selenides, 2,4-*bis*(phenyl)-1,3-diselenadiphosphetane-2,4-diselenide (**10**) has been used. The compound **10** was originally reported by Woollins and coworkers in 1988 [10a] and was applied to the selenation of benzamides in 2001 (Scheme **6.4**) [10b]. The reaction was carried out in toluene and chromatographic purification gave selenoamides **5** in pure form without the contamination of phosphorus being compound **10** insoluble in the solvent. The yields of tertiary selenoamides were about 70% while secondary and primary selenoamides were obtained in moderate to low yields. Selenation of ε-caprolactam successfully proceeded to give the corresponding selenolactam **11** in 44% of yield. The compound **10** is now commercially available and has been called Woollins' reagent (WR). The conversion of tertiary amides into the corresponding selenoamides with **10** was also performed in C_6H_6 at room temperature [10c] and with this variation, selenoformamides and aliphatic selenoamides were also obtained in moderate to high yields. A wide applicability of **10** was further proved by the selenation of monopeptide [10d], and *N*-Boc protected dipeptide **12** leading to **13** in 37% yield [10e]. The methoxycarbonyl group in **12** remained intact in the product **13**.

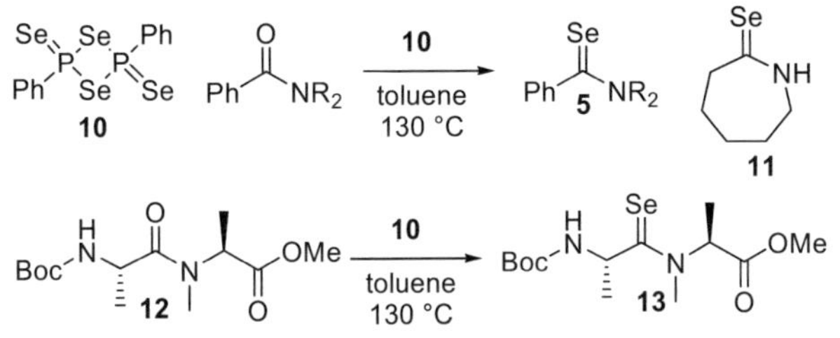

Scheme 6.4

Hydrogen selenide can be used as a selenating agent of amides, but it needs highest care in handling. Moreover, the combination of *bis*(trimethylsilyl)selenide ($Me_3SiSeSiMe_3$) and $BF_3{\bullet}Et_2O$ was used (Scheme **6.5**), although *bis*(trimethylsilyl)selenide had to be separately prepared. The reaction enables the conversion of tertiary and secondary α,β-unsaturated amides and formamides to the corresponding selenoamides, whereas primary selenoamides are not obtained [11].

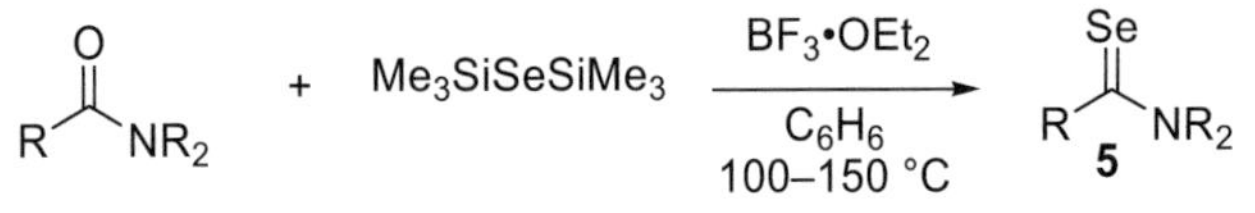

Scheme 6.5

The Se^{2-} species can be basically used as selenating agents of amides. To generate this kind of species, elemental selenium was reduced with diisobutylaluminum hydride (DBAL-H) and after that amides were added to the reaction mixture (Scheme **6.6**) [12]. This reaction resulted to be suitable for the preparation of tertiary selenoformamides **14**.

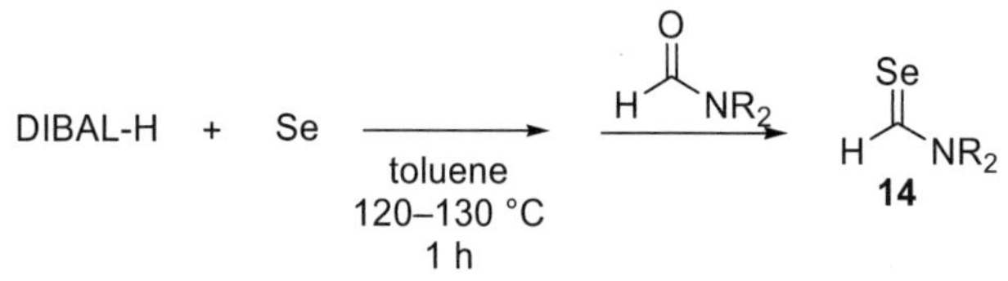

Scheme 6.6

Alternatively, hydrochlorosilanes were used for the reduction of elemental selenium in the presence of a base (Scheme **6.7**) [13]. Using DMAP, the efficient conversion of formamides and *N*-benzyl acetamide into the corresponding selenoformamides **14** and selenoamide **15** was achieved. The choice of the base is substantial in the reaction. In the case of the conversion of acetyl protected sialic acid derivative under almost identical conditions, DABCO gave the corresponding selenoacetamide **15** in moderate yield and acetoxy and methoxycarbonyl groups tolerated the reaction conditions.

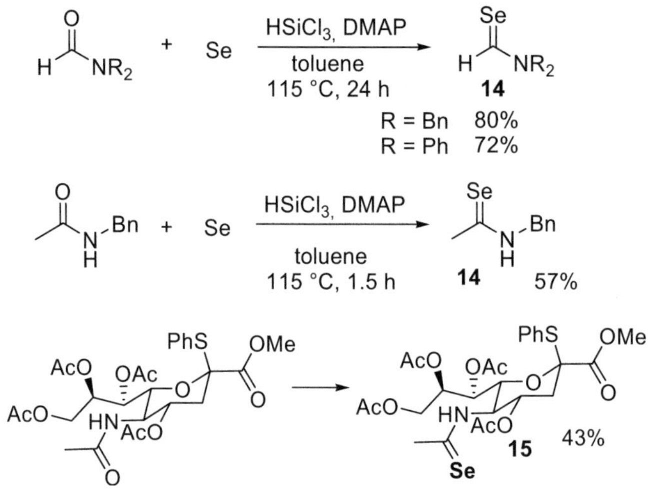

Scheme 6.7

Selenation of amides *via* chloro iminium salts is also available for the preparation of selenoamides. For example, oxalyl chloride reacted with aromatic amides to form iminium salts **16**, which were then treated with a selenating agents generated *in situ* from lithium aluminum hydride ($LiAlH_4$) and elemental selenium in THF to give the corresponding aromatic selenoamides in good yields (Scheme **6.8**) [14].

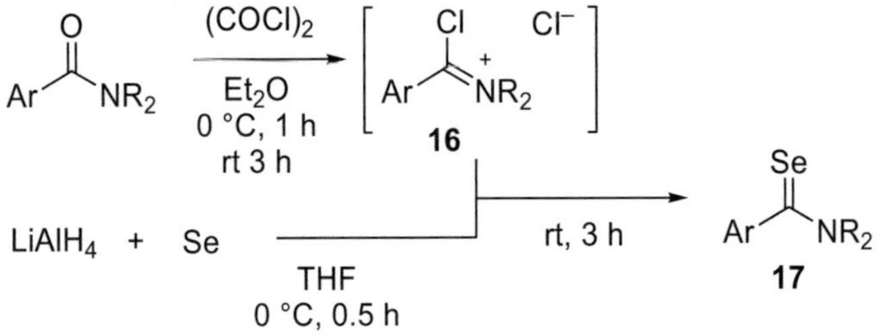

Scheme 6.8

The combination of $LiAlH_4$ and elemental selenium was applied also to the selenation of di- and tripeptide (Scheme **6.9**) [14c]. For example, dipeptide **18** was converted into the corresponding selenoxo dipepdtidomimetics **19** in good yields. Chlorination of **18** was examined with several chlorinating agents. Among them, the use of PCl_5 and 0.3 equiv of DMF with respect to PCl_5 led to **19** with high efficiency. Tetraethylammonium tetraselenotungstate ($(Et_4N)_2WSe_4$) was also used for the selenaton of chloro iminium salts derived from tertiary and secondary amides and *N*-methyl lactams [15].

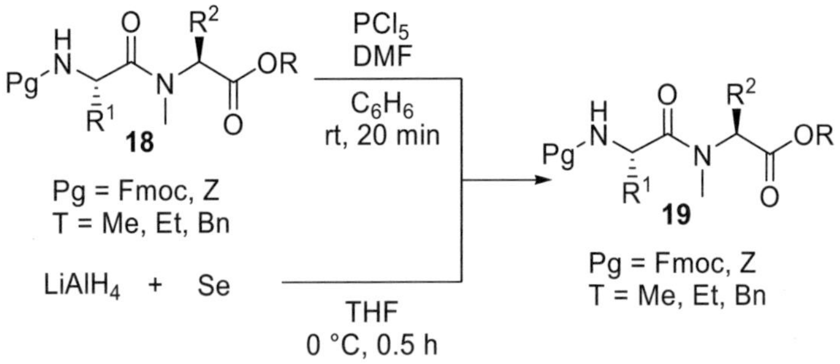

Scheme 6.9

Aminocarbene complexes of chromium and tungsten are also used as starting materials leading to selenoamides (Scheme **6.10**). The complexes **20** reacted with NaHSe generated *in situ* by the reduction of elemental selenium with $NaBH_4$ to give pyrimidineselones **21** along with the formation of chromium or tungsten complexes **22** where selenium atom is coordinated to these metals [16a]. Biscarbene complex **23** was also subjected to selenation with carbonyl selenide derived from CO and elemental selenium in the presence of Et_3N to give the product **24**. In this reaction, both Cr=C bonds in **23** were converted into Se=C bond. Under milder reaction conditions, selenation of Cr=C attached to an ethoxy group selectively proceeded while Cr=C bond next to an amino group did not react at all [16b]. Notably, chromium complexes have been used as a starting materials leading to ferrocenylselenoamides (Scheme **6.11**), that have shown potential utility as anti-cancer drugs [16c]. For this activity, selenium atom is substantial and sulfur isologues have shown to be much less powerful. Initially, Fischer carbene complexes **25** reacted with 1,2-aminoalcohols to give aminocarbene complexes **26**, which were then selenated with NaHSe derived from $Se/NaBH_4$. The reaction leading to **26** was complete within 60 min at room temperature and the yields of **26** are generally high, except for the preparation of tertiary selenoamides.

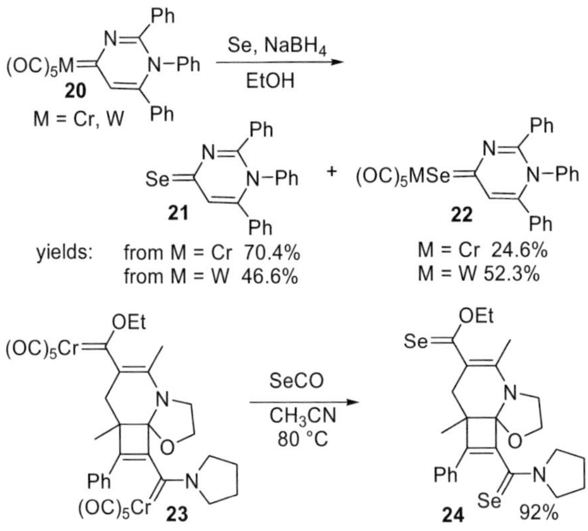

Scheme 6.10

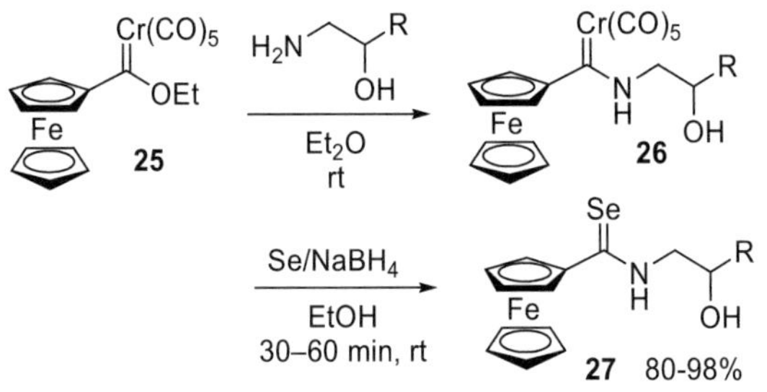

Scheme 6.11

6.2.2. Nitriles to Selenoamides

Addition of hydrogen selenide and its surrogates to nitriles gives primary selenoamides, although aliphatic primary selenoamides are highly labile and not isolated in pure form. As a selenium source, aluminum selenide (Al_2Se_3) was used with pyridine and triethylamine and the obtained primary selenoamides were isolated by recrystallization (Scheme **6.12**) [17]. The reduction of elemental selenium with $NaBH_4$ leads to NaHSe, that was added to nitriles to give primary selenoamides [18].

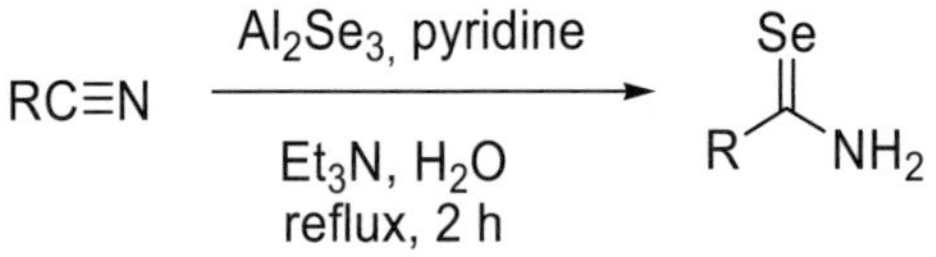

Scheme 6.12

Carbon monoxide can also reduce elemental selenium in the presence of a base in water generating H_2Se, which also reacted with nitriles to give primary selenoamides (Scheme **6.13**) [19]. Recently, this reaction has been carried out under atmospheric pressure of CO [19c]. The amine exchange reaction of primary selenoamides with primary amines led to secondary selenoamides [19b]. Alternatively, these latter have been obtained by addition of selenoic acid potassium salts [20] and Woollins' reagent [21] to nitriles.

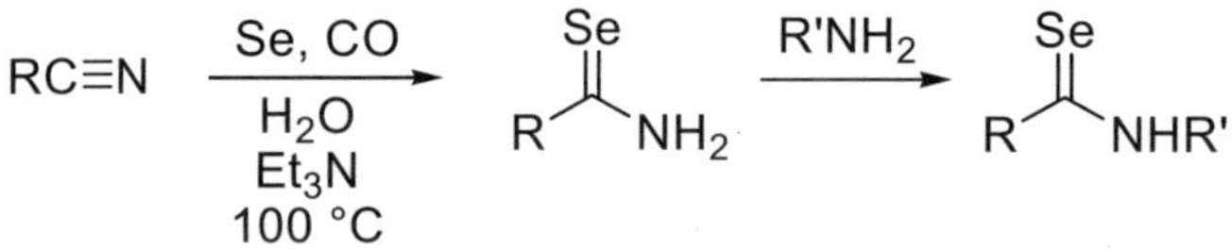

Scheme 6.13

6.2.3. Acetylenes to Selenoamides

Deprotonation of terminal alkynes gives lithium acetylides, which reacts with elemental selenium to generate lithium alkyneselenolates **28** (Scheme **6.14**). The protonation of **28** can lead to selenoketenes **29** as a reactive intermediates. However, amines are not acidic enough to protonate **28**. Therefore, amines have been used as a solvent to eventually give α-monosubstituted selenoamides **30** [22].

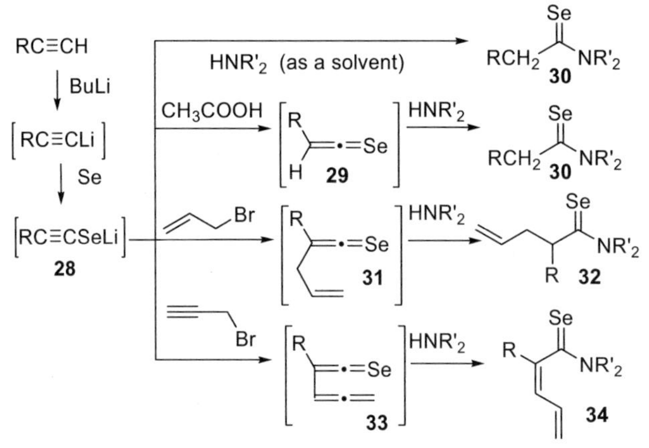

Scheme 6.14

Instead, the protonation of **28** with acetic acid readily proceeded to generate **29** [23]. In fact, at low temperatures, the reaction mixture of **28** and acetic acid was blue, which was indicative of the formation of selenoketenes **29**. One equivalent of amines was then added to give **30** in high yields. A wide range of primary and secondary amines could be used. Allylation of **28** initially takes place at the selenium atom of **28** to form allylic alkynyl selenides, which undergo a seleno-Claisen rearrangement to give selenoketene intermediates **31** [24]. Selenoketenes **31** are highly labile, but can readily accept the nucleophilic attack of amines. Therefore, the allylation in the presence of amines gives γ,δ-unsaturated selenoamides **32** as the four-component coupling products with high efficiency [24b,c]. Likewise, propargylation of **28** takes place at the selenium atom followed by [3,3]-sigmatropic rearrangement to form 1,2-propadienylselenoketeones **33** [25]. The addition of amines to **33** mainly gives α,β,γ,δ-unsaturated selenoamides **34**. Extrusion of nitrogen gas from 1,2,3-selenadiazoles in the presence of potassium hydroxide or under photoirradiation conditions ~~to~~ generates potassium alkyneselenolates or selenoketenes [26]. In the former case, large excess amounts of amines were added to give selenoamides **30**.

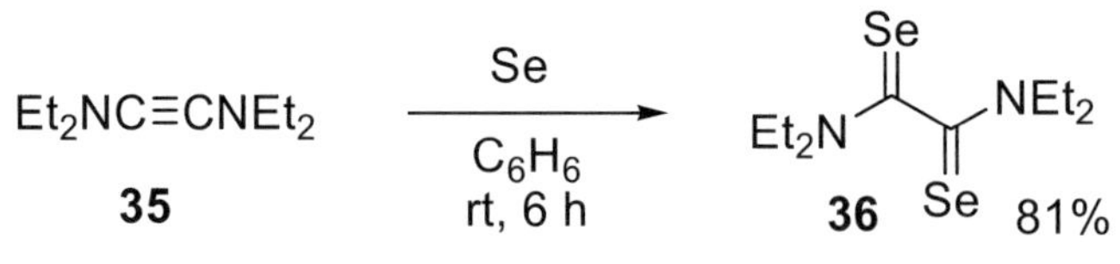

Scheme 6.15

Ethynediamine **35** was reacted with elemental selenium at room temperature to give diselenoamide **36** (Scheme **6.15**) [27].

6.2.4. Miscellaneous

Carbanions generated from benzylphosphonate [28] and dihaloalkanes [29] react with aminoselanes and elemental selenium and then with secondary amines to give tertiary selenoamides. For example, the treatment of dichloroacetic acid ethyl ester with NaH, selenium and Et_2NH in HMPA gave ethoxycarbonyl selenoamide **37** (Scheme **6.16**).

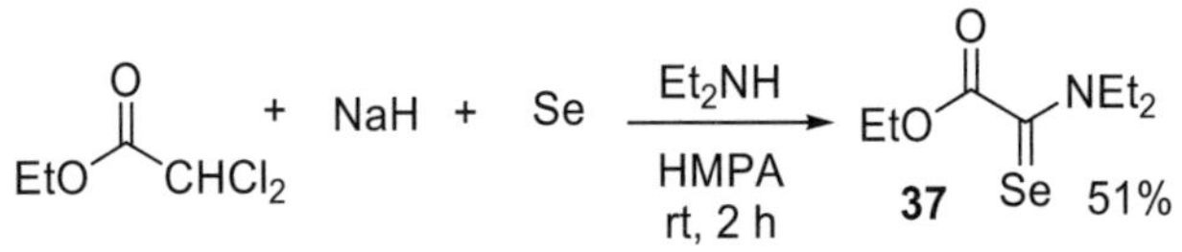

Scheme 6.16

Amination of dibenzyl triselenocarbomate [30], selenoic acid *O*-esters **6** [31] and diselenoic acid esters **8** [32] leads to selenoamides.

Finally, *bis*(2-cyanoethyl)diselenide is used as a good precursor to introduce selenocarbonyl group [33]. The *in situ* generated $NaSeCH_2CH_2CN$ can be introduced into the iminocarbon atom to form **38**. The solid phase conversion enabled decyanoethylation of **38** gives the compound **39** having a selenocarbamoyl group (Scheme **6.17**).

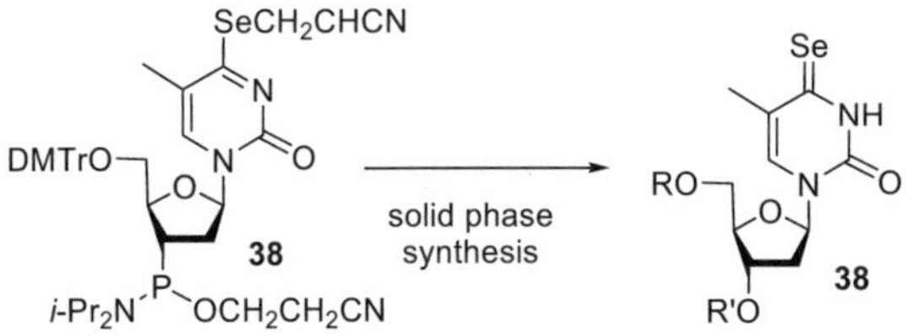

Scheme 6.17

6.3. Selenoamides: Recent Progress on Their Synthetic Applications

6.3.1. Intermolecular Cyclization

Synthesis of nitrogen- and/or selenium-containing heterocycles using selanoamides have been continuously developed [34]. The treatment of primary aromatic selenoamides with DMAD in EtOH gave 4,5-dihydro-1,3-selenazoles **39** in good to high yields (Scheme **6.18**) [35].

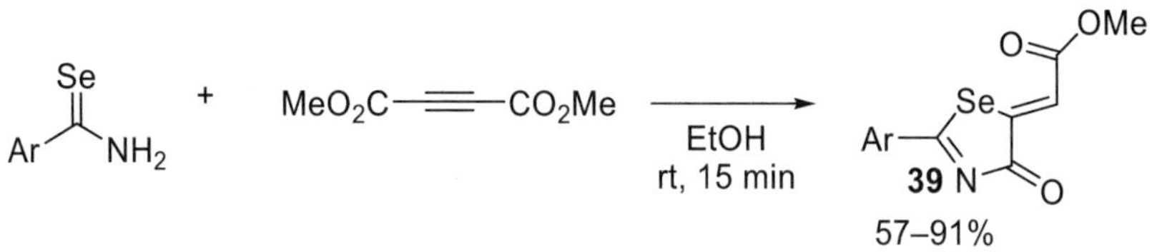

Scheme 6.18

The inner salts **40** reacted with allenes **41** bearing an electron-withdrawing group under reflux in THF to result in the formation of selenophene derivatives **42** [36]. Initially, [3+2] cycloaddition reaction between **40** and **41** proceeds to form spiro-compounds **43**, which then undergo skeletal rearrangement to give **42** *via* **44** (Scheme **6.19**).

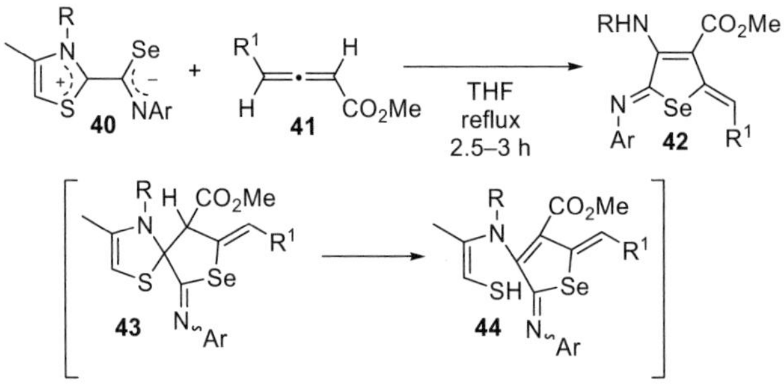

Scheme 6.19

The reaction of propargyl alcohols **45** with selenobenzamide in the presence of catalytic amounts of scandium triflate and ammonium sulfate proceeded in a mixed solvent of nitromethane and water to give selenazoles **46** (Scheme **6.20**) [37].

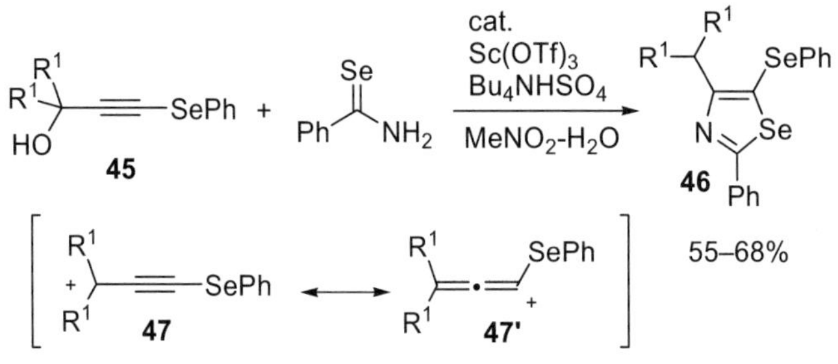

Scheme 6.20

In the initial step, dehydroxylation of **45** may take place to form propadienyl cation **47**, which then undergoes [3+2] cycloaddition reaction with selenobenzamide to reach the final products **46**. α-Cyanoselenoacetamide ($NCCH_2C(Se)NH_2$) has been used for the synthesis of a variety of heterocycles involving selenazoles [38].

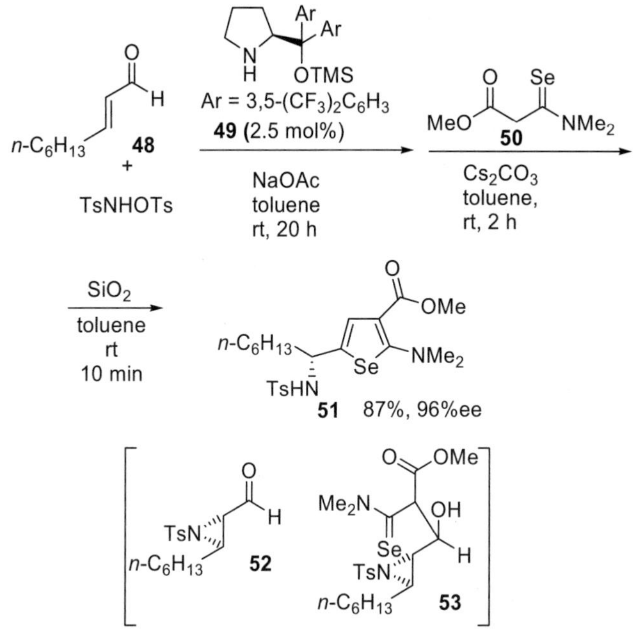

Scheme 6.21

Proline-derived organocatalyst **49** catalyzed the reaction of α,β-unsaturated aldehyde **48** with *N*-tosyl tosyloxyamine, followed by the addition of β-alkoxycarbonyl selenoamide **50** to give selenophene **51** with high efficiency and high enantiomeric excess (Scheme **6.21**) [39]. Initially, aziridine formation takes place in a highly enantioselective manner to form **52**. The enolate generated from **50** may attack the carbonyl carbon atom to form **53**. Intramolecular attack of the selenium atom to cleave the aziridine ring in **53** may take place to form five-membered heterocycle followed by the aromatization to lead to **51**.

6.3.2. Selenoiminium Salts

The successful inter- and intramolecular cyclization of selenoamides is ascribed to the high nucleophilicity of the selenium atom in selenoamides. Therefore, alkylation of selenoamides should proceed smoothly. Even so, about a day was necessary for the methylation of selenoamides with methyl iodide. Instead, methylation of selenoamides with methyl triflate was complete within 30 sec [40]. The reactivity of selenoiminium salts has also been elucidated (Scheme **6.22**) [41].

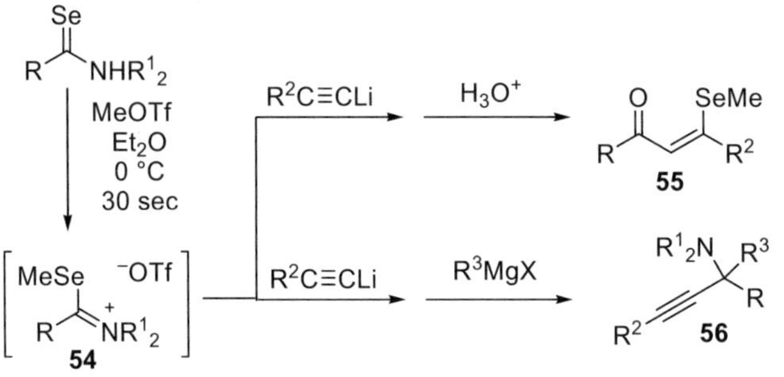

Scheme 6.22

The addition of lithium acetylides to selenoiminium salts **54** followed by hydrolysis gives β-methylselanyl-α,β-unsaturated ketones as a product. In the reaction, the MeSe- group undergoes 1,3-rearrangement. In contrast, prior to the hydrolysis of the reaction mixture of **54** and lithium acetylides, the further addition of Grignard reagents afforded propargylamines **55** with a tetrasubstituted carbon atom. A similar reaction proceeds with thioiminium salts [42], but lower amounts of Grignard reagents are necessary for the reaction with selenoiminium salts **54**. In the reaction in Scheme **6.22**, the order of the addition of two different organometallic reagents is substantial. When Grignard reagents were added to **54** prior to the addition of organolithium reagents, the products where two equiv of Grignard reagents were added as a major product. Tellurium nucleophile generated from $LiAlH_4$ and tellurium powder can add to the iminium carbon atom of selenoiminium salts **54** to form telluroamides **55** in low to good yields (Scheme **6.23**) [43]. The method provides tertiary and secondary telluroamides, but α-monosubstituted telluroamides are fairly labile, ~~and~~ thus were obtained only in 8% of yield.

6.3.3. Copper(0)-Induced Reaction of Selenoamides

Self-dimerization of selenoamides was achieved by reatment with Cu(0) to give 1,2-enediamines **58** (Scheme **6.24**) [44]. This has implied the generation of α-aminocarbene species **59** and/or cupper carbenoids. The putative species were trapped with alkenes with electron-withdrawing groups to form cyclopropylamines **60** as a product. C-H bond insertion of terminal alkynes to **59** was further achieved to afford propargylamines **61**. A variety of terminal acetylenes were used, although terminal acetylenes had to be used as a solvent.

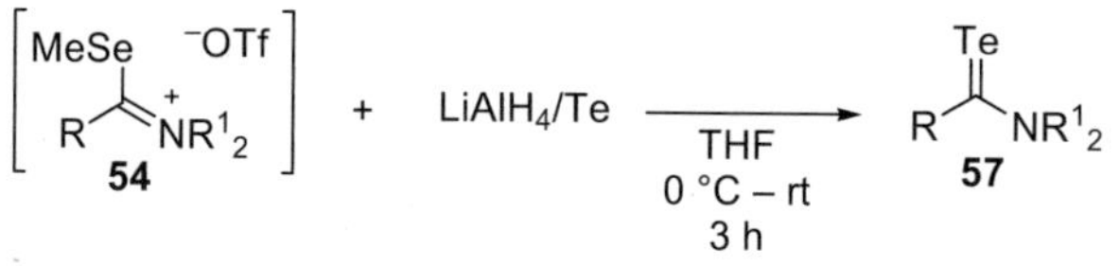

Scheme 6.23

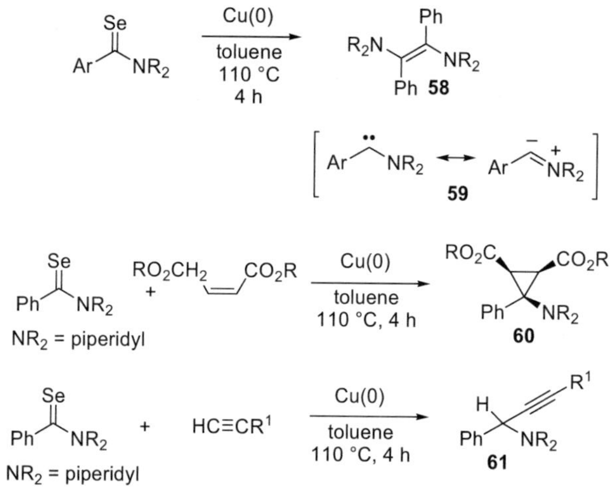

Scheme 6.24

6.3.4. Deprotonation of Selenoamides

Similarly to ordinary amides and thioamides, deprotonation of selenoamides at α to the selenocarbonyl group readily takes place to form metal eneselenolates **62** [45]. As latest examples, allylation of **62** generates allylic vinyl selenides, which readily undergo seleno-Claisen rearrangement (Scheme **6.25**) [46].

The use of geranyl and neryl bromides gave the corresponding products **64** and **65**, where quaternary carbon atoms were present at β to the selenocarbonyl group, with high diastereoselectivity. In contrast, lithium eneselenolate derived from selenoacetamide reacted with methyl choroformate to give β-methoxycarbonyl selenoamide **66** in good yields (Scheme **6.26**) [39].

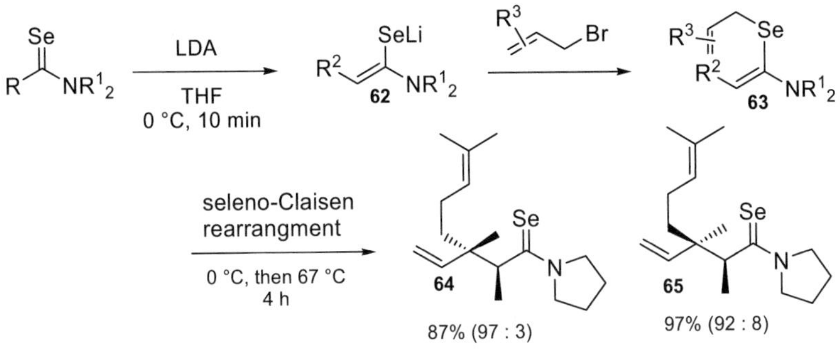

Scheme 6.25

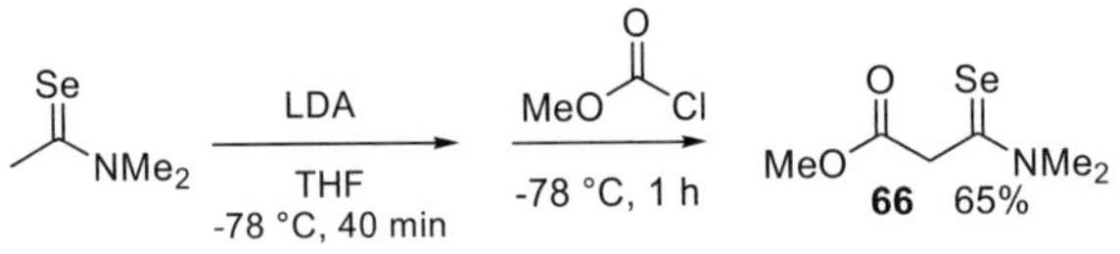

Scheme 6.26

Deprotonation of selenoformamides **67** with LDA was shown to take place at the selenocarbonyl carbon atom to generate (selenocarbamoyl)lithum **68**, which was then trapped with chlorotrimethylsilane to give the corresponding product **69** for the first time (Scheme **6.27**) [47].

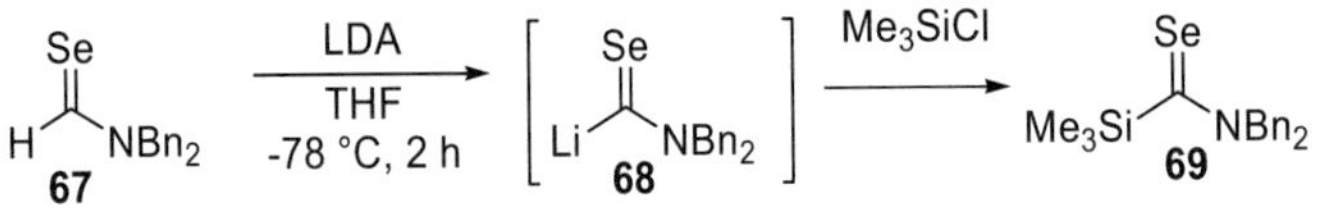

Scheme 6.27

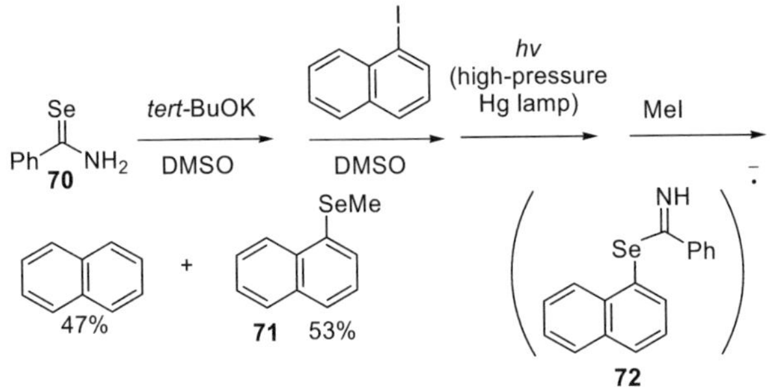

Scheme 6.28

6.3.5. Photoinduced Reaction

The reaction of selenobenzamide **70** with potassium *tert*-butoxide and iodonaphthalene under photo-irradiation conditions, followed by the methylation, gives methyl naphthyl selenide **71** along with naphthalene (Scheme **6.28**) [48]. In the reaction, naphthyl radical may react with selenocarbamate ion derived from **70** and *tert*-BuOK to form *Se*-naphthyl selenoimide radical anion **72**. The elimination of *N*-phenylimino radical from **72** may form naphthaneselenolate, which was methylated to give **71**.

CONCLUSION AND OUTLOOK

In summary, this review has briefly overviewed properties of selenocarbonyl compounds, focusing on the synthetic methods for selenoamides and their recent use in synthetic reactions since in the last several years, new progress has been seen mainly on the chemistry of selenoamides. However, unique properties of all the selenocarbonyl compounds should be more deeply disclosed and some of them can be used in the fields of material and biological sciences. In future, further developments are expected in these fields.

ACKNOWLEDGEMENT

Declared none.

CONFLICT OF INTEREST

The authors confirm that this chapter contents have no conflict of interest.

REFERENCES

[1] (a) Murai, T. In: *Topics in Current Chemistry*; Wirth, T. Ed.; Springer-Verlag, Berlin, **2000**, Vol. *208*; pp. 177–199. (b) Murai, T. In: *Organoselenium Chemistry*; Wirth, T. Ed.; Wiley-VCH, Weinheim **2012**; pp. 257-285.

[2] (a) Guziec, L.J.; Guziec, F.S.Jr. In: *Comprehensive Organic Functional Group Transformations II*; Katritzky, A.R.; Taylor, R.J.K.; Jones, K. Ed.; Elsevier PERGAMON, Amsterdam, **2005**; Vol. *3*, pp. 397–417. (b) Kuhn, N.; Verani, G. In: *Handbook of Chalcogen Chemistry*; Devillanova, F.A. Ed.; RSC Publishing, Cambridge, **2007**; pp. 107–144.

[3] (a) Okazaki, R.; Kumon, N.; Inamoto, N. 2,4,6-Tri-tert-butylselenobenzaldehyde, the first stable selenoaldehyde. *J. Am. Chem. Soc.* **1989**, *111*, 5949-5951. (b) Takeda, N.; Tokitoh, N.; Okazaki, R. The first rotational isomers of stable selenoaldehydes and their η^1-tungsten complexes. *Angew. Chem. Int. Ed. Engl.* **1996**, *35*, 660-662.

[4] Maeda, H.; Takamizawa, Y.; Segi, M. [3+2] Ccycloaddition of seleno- and thioaldehdyes with cyclic azomethine imines. *Heterocycles* **2012**, *84*, 393-400.

[5] (a) Murai, T. In: *Topics in Current Chemistry*; Kato, S. Ed.; Springer-Verlag, Berlin, **2005**; Vol. *251*, pp. 247–272. (b) Moore, A. J. In: *Comprehensive Organic Functional Group Transformations II*, Katritzky, A.R.; Taylor, R.J.K.; Jones, K. Ed.; Elsevier PERGAMON, Amsterdam, **2005**; Vol. *5*, p. 519–570. c) Flynn, C.; Haughton, L. In: *Comprehensive Organic Functional Group Transformations II*; Katritzky, A.R.; Taylor, R.J.K.; Jones, K. Ed.; Elsevier PERGAMON, Amsterdam, **2005**; Vol. *5*, p. 571–581.

[6] Ishii, A.; Nakayama, J. In: *Comprehensive Organic Functional Group Transformations II*; Katritzky, A.R.; Taylor, R.J.K.; Jones, K. Ed.; Elsevier PERGAMON, Amsterdam, **2005**; Vol. *5*, pp 459–491.

[7] (a) Murai, T. Synthesis and reactions of selenothioic acid *S*-esters and diselenoic acid esters. *Synlett* **2005**, 1509–1520. (b) Ishii, A.; Nakayama J. In: *Topics in Current Chemistry,* Kato, S. Ed.; Springer-Verlag, Berlin, **2005**; Vol. *251*, pp. 227–246. c) Murai, T. In: *Comprehensive Organic Functional Group Transformations II*, Katritzky, A.R.; Taylor, R.J.K.; Jones, K. Ed.; Elsevier PERGAMON, Amsterdam, **2005**; Vol. *5*, pp. 493–518.

[8] Collard-Charon, C.; Renson, M. Étude Infra-Rouge Des Substances Possédant une Liaison C=Se Adjacente à; un ou Plusieurs Atomes D'azote III. Sélénoamides. *Bull. Soc. Chim. Belg.* **1963**, *72*, 304-315

[9] Rae, I.D.; Wade, M.J. The preparation of some secondary and tertiary selenoamides and study of their spectra. *Int. J. Sulfur Chem.* **1976**, *8*, 519–528.

[10] (a) Wood, P. T.; Woollins, J. D. Phosphorus–selenium heterocycles. *J. Chem. Soc., Chem. Commun.* **1988**, 1190-1191. (b) Bhattacharyya, P.; Woollins, J. D. Selenocarbonyl synthesis using Woollins reagent. *Tetrahedron Lett.* **2001**, *42*, 5949-5951. (c) Bethke, J.; Karaghiosoff, K.; Wessjohann, L.A. Synthesis of *N,N*-disubstituted selenoamides by O/Se-exchange with selenium-Lawesson's reagent. *Tetrahedron Lett.* **2003**, *44*, 6911–6913. (d) Huhtiniemi, T.; Suuronen, T.; Lahtela-Kakkonen, M.; Bruijn, T.; Jääskeläinen, S.; Poso, A.; Salminen, A.; Leppänen, J.; Jarho, E. *N*ε-Modified lysine containing inhibitors for SIRT1 and SIRT2. *Bioorg. Med. Chem.* **2010**, *18*, 5616–5625. (e) Huang, Y.; Jahreis, G.; Lücke, C.; Wildemann, D.; Fischer, G. Modulation of the Peptide Backbone Conformation by the Selenoxo Photoswitch. *J. Am. Chem. Soc.* **2010**, *132*, 7578-7579.

[11] Takikawa, Y.; Watanabe, H.; Sasaki, R.; Shimada, K. Conversion of amides, *N,N,N',N'*-tetramethylurea, and esters to the corresponding selenoxo compounds by treatment with bis(trimethylsilyl) selenide and $BF_3{\cdot}OEt_2$. *Bull. Chem. Soc. Jpn.* **1994**, *67*, 876–878.

[12] Li, G. M.; Zingaro, R. A. Reaction of diisobutylaluminum hydride with selenium and tellurium: new reagents for the synthesis of seleno- and telluro-amides. *J. Chem. Soc., Prekin Trans 1* **1998**, 647–650.

[13] Shibahara, F.; Sugiura, R.; Murai, T. Direct thionation and selenation of amides using elemental sulfur and selenium and hydrochlorosilanes in the presence of amines. *Org. Lett.* **2009**, *11*, 3064–3067.

[14] (a) Ishihara, H.; Koketsu, M.; Fukuda, Y.; Nada, F. Reaction of lithium aluminum hydride with elemental selenium: its application as a selenating reagent into organic molecules. *J. Am. Chem. Soc.* **2001**, *123*, 8408–8409. (b) Koketsu, M.; Okayama, Y.; Aoki, H. Ishihara, H. Facile synthesis of *N,N*-dialkylselenoamides from amides. *Heteroatom Chem.* **2002**, *1*, 195–198. (c) Vishwanatha, T.M.; Narendra, N.; Chattopadhyay, B.; Mukherjee, M.; Sureshbabu, V.V. Synthesis of selenoxo peptides and oligoselenoxo peptides employing LiAlHSeH. *J. Org. Chem.* **2012**, *77*, 2689-2702.

[15] Saravanan, V.; Mukherjee, C.; Das, S.; Chandrasekaran, S. Facile conversion of amides and lactams to selenoamides and selenolactams using tetraethylammonium tetraselenotungstate. *Tetrahedron Lett.* **2004**, *45*, 681–683.

[16] (a) Zheng, Z.; Chen, J.; Yu, Z.; Han, X. Oxidative demetalation of Fischer alkoxy carbene complexes with stoichiometric pyridine *N*-oxide and $NaBH_4$-promoted demetalation of Fischer iminocarbene complexes with sulfur and selenium. *J. Organomet. Chem.* **2006**, *691*, 3679–3692. (b) Zheng, Z.; Chen, J.; Luo, N.; Yu, Z.; Han, X. Efficient demetalation of Fischer alkoxy- and amino-biscarbene complexes of chromium *via* CO-promoted sulfuration and selenylation. *Organometallics* **2006**, *25*, 5301–5310. (c) Gutiérrez-Hernáandez, A.I.; López-Cortés, J.G.; Ortega-Alfaro, M.C.; Ramirez-Apan, M.T.; Cázares-Marinero, J.de J.; Toscano, R.A. Ferrocenylselenoamides: synthesis, characterization and cytotoxic properties. *J. Med. Chem.* **2012**, *55*, 4652–4663.

[17] Cohen, V.I. Synthesis of unsubstituted aromatic and heterocyclic selenocarboxamides. *Synthesis* **1978**, 668–669.

[18] (a) Lai, L.-L.; Reid, D.H. Synthesis of primary selenocarboxamides and conversion of alkyl selenocarboxamides into selenazoles. *Synthesis* **1993**, 870–872. (b) Zhao, X.-R.; Ruan, M.-D.; Fan, W.-Q.; Zhou, X.-J. Convenient synthesis of primary selenoamides from aryl nitriles, selenium and sodium borohydride. *Synth. Comun.* **1994**, *24*, 1761–1765. (c) Ruan, M.-D.; Zhang, P.-F.; Tao, Y.; Fan, W.-Q. A new synthetic method for *N*-mono- and *N,N*-disubstituted selenoamides. *Synth. Commun.* **1996**, *26*, 2617–2623.

[19] (a) Ogawa, A.; Miyake, J.; Karasaki, Y.; Murai, S.; Sonoda, N. Synthesis utilizing reducing ability of carbon monoxide. A new method for synthesis of selenocarboxamides: reaction of nitriles with selenium, carbon monoxide, and water. *J. Org. Chem.* **1985**, *50*, 384–386. (b) Ogawa, A.; Miyake, J.; Kambe, N.; Murai, S.; Sonoda, N. Synthesis utilizing the reducing ability of carbon monoxide. New methods for synthesis of *N*-substituted selenoamides. *Bull. Chem. Soc. Jpn.* **1985**, *58*, 1448–1451. (c) Chen, Y.; Tian, F.; Song, M.; Lu, S. Facile synthesis of selenocarboxamides from nitriles using $Se/CO/H_2O$ under atmospheric pressure. *Heteroatom Chem.* **2008**, *19*, 211–214.

[20] Ishihara, H.; Yosimi, K.; Kouketsu, M. A facile preparation of aliphatic and aromatic primary selenoamides using 4-methylselenobenzoate as a new selenating reagent. *Chem. Lett.* **1998**, 1287–1788.

[21] Hua, G.; Li, Y.; Slawin, A. M. Z.; Woollins, J. D. Synthesis of primary arylselenoamides by reaction of aryl nitriles with Woollins' reagent. *Org. Lett.* **2006**, *8*, 5251–5254.

[22] Sukuhai, R.S.; Jong, R.; Brandsma, L. A convenient method for the preparation of substituted selenoamides and thioamides. *Synthesis* **1977**, 888–889.

[23] Murai, T.; Ezaka, T.; Niwa, N.; Kanda, T.; Kato, S. An efficient synthetic method of aliphatic selenoamides *via* selenoketene intermediates. *Synlett* **1996**, 865–866.

[24] (a) Schaumann, E.; Grabley, F.-F. Stable selenoketenes *via* selena-Cope rearrangement. *Tetrahedron Lett.* **1980**, *21*, 4251–4254. (b) Murai, T.; Ezaka, T.; Kanda, T.; Kato, S. Highly efficient one-pot synthesis of α,α-disubstituted selenoamides and their first reduction to amines. *Chem. Commun.* **1996**, 1809–1810. (c) Murai, T.; Ezaka, T.; Kato, S. An efficient synthetic method of 4-penteneselenoamides: four-component coupling reaction of terminal acetylenes, selenium, amines, and allylic bromides. *Bull. Chem. Soc. Jpn.* **1998**, *71*, 1193–1200.

[25] (a) Shimada, K.; Akimoto, S.; Itoh, H.; Nakamura, H.; Takikawa, Y. *Chem. Lett.* **1994**, 1743–1746. (b) Koketsu, M.; Kanoh, M.; Itoh, E.; Ishihara, H. Reaction of allenyl

selenoketene, generated by [3,3] sigmatropic rearrangement, with amines. *J. Org. Chem.* **2001**, *66*, 4099–4101.

[26] (a) Malek-Yazdi, F.; Yalpani, M. A facile synthesis of substituted seleno and thio amides. *Synthesis* **1977**, 328–330. (b) Harrit, N.; Rosenkilde, S.; Larsen, B.D.; Holm, A. *J. Chem. Soc. Perkin Trans. 1* **1985**, 907–911. (c) Petrov, M.L.; Lyapunova, A.G.; Androsov, D.A. New synthesis of 2-aminobenzo[b]selenophenes from 4-(2-chloroaryl)-1,2,3-selenadiazoles. *Russ. J. Org. Chem.* **2012**, *48*, 147–149.

[27] Nakayama, J.; Mizumura, A.; Akiyama, I.; Nishio, T.; Iida, I. Synthesis and characterization of tetraethylethanediselenoamide. *Chem. Lett.* **1994**, 77–80.

[28] Okuma, K.; Ikari, K.; Ohta, H. Reactions of phosphonate carbanions with selenium or sulfur in the presence of amines. Synthesis of seleno- and thioamides. *Chem. Lett.* **1992**, 131–134.

[29] (a) Takikawa, Y.; Yamaguchi, M.; Sasaki, T.; Ohnishi; Shimada, K. Convenient syntheses of *N,N*-dialkylselenoamides and *N,N,N',N'*-tetraalkylselenoureas by treating terminal gemdihaloalkanes, chloroform, or sodium trichloroacetate with a base, elemental selenium, and amines. *Chem. Lett.* **1994**, 2105–2108. (b) Shimada, K.; Yamaguchi, M.; Sasaki, T.; Ohnishi, K.; Takikawa, Y. A Willgerodt-Kindler type selenation of dihalomethane derivatives, chloroform, and sodium trichloroacetate by treating with a base, elemental selenium, and an amine. *Bull. Chem. Soc. Jpn.* **1996**, *69*, 2235–2242.

[30] Henriksen, L. Preparation of selenoformamides by selenol reduction of dibenzyl triselenocarbonate. *Synthesis* **1974**, 501-503.

[31] Cohen, V.I. Convenient and general method for aliphatic and aromatic selenonesters and *N*-mono- and *N,N*-disubstituted selenoamides synthesis. *J. Org. Chem.* **1977**, *42*, 2645–2647.

[32] Murai, T.; Mizutani, T.; Kanda, T.; Kato, S. Reactions of diselenoic acid esters with amines and x-ray crystal structure analyses of aromatic selenoamides. *Heteroatom Chem.* **1995**, *6*, 241–246.

[33] Salon, J.; Sheng, J.; Jiang, J.; Chen, G.; Caton-Williams, J.; Huang, Z. Oxygen replacement with selenium at the thymidine 4-position for the Se base pairing and crystal structure studies. *J. Am. Chem. Soc.* **2007**, *129*, 4862–4863.

[34] Koketsu, M.; Ishihara, H. In; *Handbook of Chalcogen Chemistry*; Devillanova, F. A. Ed.; RSC Publishing, Cambridge, **2007**, pp. 145–194.

[35] Koketsu, M.; Sasaki, T.; Ando, H.; Ishihara, H. Preparation of 4,5-dihydro-1,3-selenazoles by reaction of aromatic primary selenoamides with acetylenedicarboxylate. *J. Heterocyclic Chem.* **2007**, *44,* 231–232.

[36] Zhang, J.-H.; Cheng, Y. The [3+2] cycloaddition reaction of thiazole carbene-derived C-C-Se 1,3-dipoles: A concise and highly efficient strategy for the construction of multifunctional dihydroselenophenes and selenopheno[2,3-b]pyrazines. *Org. Biomol. Chem.* **2009**, *7*, 3264–3270.

[37] (a) Yoshimatsu, M.; Yamamoto, T.; Sawa, A.; Kato, T.; Tanabe, G.; Muraoka, O. α-Sulfanyl and α-Selanyl Propadienyl Cations: Regioselective Generations and Cycloadditions with Thioamides and Selenoamides Controlled by $MeNO_2$-H_2O System. *Org. Lett.* **2009**, *11*, 2952–2955. (b) Yoshimatsu, M.; Matsui, M.; Yamamoto, T.; Sawa, A. Convenient preparation of 4-arylmethyl- and 4-hetarylmethyl thiazoles by regioselective cycloaddition reactions of 3-sulfanyl- and selanylpropargyl alcohols. *Tetrahedron* **2010**, *66*, 7975–7987.

[38] (a) Frolov, K.A.; Krivokolysko, S.G.; Dotsenko, V.V.; Litvinov, V.P. Cascade synthesis of pyrimido[4,3-*b*][1,3,5]selenadiazine derivatives. *Chem. Heterocycl. Comp.* **2009**, *45*, 255–

256. (b) Dyachenko, V.D.; Sukach, S.M. Synthesis of 7-acetyl-8-aryl(hetaryl)-6-hydroxy-1,6-dimethyl-3-selenoxo-2,3,5,6,7,8-hexahydroisoquinoline-4-carbonitriles by the condensation of 2,4-diacetyl-3-aryl(hetaryl)-5-hydroxy-5-methyl-cyclohexanones with cyanoselenoacetamide. *Chem. Heterocycl. Comp.* **2011**, *48*, 1467–1470. (c) Dyachenko, V.D.; Roman, S.V. Unexpected formation of ethyl 5-cyano-6-[cyano(4'-phenyl-1,3'-selenazol-2-yl)methyl]-4-(furan-2'-yl)-2-methyl-1,4-dihydropyridine-3-carboxylate. *Russ. J. Gen. Chem.* **2011**, *81*, 442–443. (d) Frolov, K.A.; Dotsenko, V.V.; Krivokolysko, S.G.; Litvinov, V.P. Synthesis and alkylation of triethylammonium 3,5,5-tricyano-4,6-di(2-thienyl)-1,4,5,6-tetrahydropyridine-2-selenolate. *Chem. Heterocycl. Comp.* **2012**, *48*, 442–446. (e) Dyachenko, V.D.; Pugach, Y.Y. Synthesis of 2-(4'-Morpholin-4''-yl-5H-chromeno-[2,3-d]pyrimidin-2'-yl)phenol from salicylaldehyde and substituted acrylonitriles. *Russ. J. Gen. Chem.* **2012**, *82*, 921–926.

[39] Ransborg, L.K.; Albrecht, L.; Weise, C.F.; Bak, J.R.; Jorgensen, K.A. Optically active thiophenes *via* an organocatalytic one-pot methodology. *Org. Lett.* **2012**, *14*, 724–727.

[40] Mutoh, Y.; Murai, T. Synthesis and properties of selenoiminium salts derived from secondary selenoamides. *Organometallics* **2004**, *23*, 3907–3913.

[41] (a) Murai, T.; Mutoh, Y.; Kato, S. MeOTf-mediated alkynylation of selenoamides leading to β-methylselenenyl α,β-unsaturated ketones and their characterization. *Org. Lett.* **2001**, *3*, 1993–1995. (b) Murai, T.; Nogawa, S.; Mutoh, Y. Sequential addition reactions of lithium acetylides and grignard reagents to selenoiminium salts leading to 2-propynyl tertiary amines bearing a tetrasubstituted carbon center. *Bull. Chem. Soc. Jpn.* **2007**, *80*, 2220–2225.

[42] (a) Murai, T. Mutoh, Y.; Ohta, Y.; Murakami, M. Synthesis of tertiary propargylamines by sequential reactions of *in situ* generated thioiminium salts with organolithium and -magnesium reagents. *J. Am. Chem. Soc.* **2004**, *126*, 5968–5969. (b) Murai, T.; Mutoh, Y. Thioamides and thioformamides for sequential reactions with organolithium and Grignard reagents. *Chem. Lett.* **2012**, *41*, 2–8.

[43] Mutoh, Y.; Murai, T.; Yamago, S. Telluration of seleno- and chloroiminium salts leading to various telluroamides, and their structure and NMR properties. *J. Organomet. Chem.* **2007**, *692*, 129–135.

[44] (a) Mitamura, T.; Nomoto, A.; Sonoda, M.; Ogawa, A. Copper(0)-induced aminocyclopropanation of olefins *via* deselenation of *N,N*-disubstituted aromatic selenoamides. *Tetrahedron* **2008**, *64*, 9983–9988. (b) Mitamura, T.; Ogawa, A. Copper(0)-Induced deselenative insertion of *N,N*-disubstituted selenoamides into acetylenic C-H Bond leading to propargylamines. *Org. Lett.* **2009**, *11*, 2045–2048.

[45] Murai, T. Generation and reactions of selenium isologues of enolate ions. *Mini-Rev. Org. Chem.* **2004**, *1* 279–290.

[46] Murai, T.; Ezaka, T.; Kato, S. Synthesis of γ,δ-unsaturated selenoamides *via* seleno-Claisen rearrangement of *in situ* generated allylic vinyl selenides from selenoamides and allylic bromides. *Synthesis* **2012**, *44*, 3197–3201.

[47] Murai, T.; Hori, R.; Maruyama, T.; Shibahara, F. (Selenocarbamoyl)silanes and -germanes: their synthesis using (selenocarbamoyl)lithium and characterization. *Organometallics* **2010**, *29*, 2400–2402.

[48] Bouchet, L. M.; Penénory, A. B.; Argüello, J. E. Synthesis of arylselenide ethers by photoinduced reactions of selenobenzamide, selenourea and selenocyanate anions with aryl halides. *Tetrahedron Lett.* **2011**, *52*, 969–972.

Send Orders for Reprints to reprints@benthamscience.net
166 *Organoselenium Chemistry Between Synthesis and Biochemistry*, 2014, 166-196

CHAPTER 7

Enantioselective Catalysis for the Preparation of Organoselenium Compounds and Applications

Francesca Marini*, Luana Bagnoli and Silvia Sternativo

Department of Pharmaceutical Sciences, University of Perugia, Via del Liceo 1- 06100 Perugia, Italy

Abstract: The peculiar reactivity and the interesting biological properties of chiral organoselenium compounds inspired organic chemists to develop new methodologies for their synthesis. In recent years, interesting catalytic protocols have been found for the asymmetric synthesis of organoselenium compounds starting from prochiral or racemic starting materials. This chapter summarizes these novel strategies based on the use of chiral organometallic catalysts, organocatalysts and biocatalysts.

Keywords: Asymmetric synthesis, Organocatalysis, Organometallic catalysts, Chiral Selenides, Desymmetrizations, [2,3]-sigmatropic rearrangements, Cyclopropanes, Spirolactones, Seleno aldehydes, Biocatalysts.

7.1. INTRODUCTION

Selenium chemistry has developed rapidly over the past years [1]. Commercially available or easily accessible electrophilic or nucleophilic reagents can introduce a selenium group into a variety of substrates in high yield and with a predictable chemo-, regio- and stereo-selectivity. This insertion is concomitant with the introduction of new functional groups and stereocenters or with ring formation. Once incorporated, the selenium moiety can be used for further chemical elaborations, such as substitution or elimination reactions, or it can be simply removed. In certain aspects structure and reactivity of organoselenium compounds are similar to those of sulfur analogues, however, the introduction of the heteroatom, the chemical manipulation of the products and the removal of the selenium functionalities are greatly facilitated and occur under milder reaction

***Address correspondence to Francesca Marini:** Department of Pharmaceutical Sciences, University of Perugia, *Via* del Liceo 1- 06100 Perugia, Italy; Tel: + 39 0755855105; Fax: +39 0755855116; E-mail: francesca.marini@unipg.it conditions. These peculiar properties of organoselenium compounds are particularly appreciated in topical areas of modern organic chemistry, such as asymmetric synthesis. Demand for single enantiomers is widely raised primarily in pharmaceuticals, but also in other areas such as flavour and aroma chemicals, agricultural or material chemistry, and consequently, the search of novel and efficient strategies for the asymmetric synthesis of chiral compounds has been the object of intense investigations. In this context, the easy installation and manipulation of selenium groups made these compounds attractive intermediates in the asymmetric construction of natural products, heterocycles and compounds of pharmaceutical interest [2]. However, the recent growing interest in the preparation of chiral organoselenium compounds is not only due to their applications as synthetic intermediates [1,2] or ligands in asymmetric transformations [3] but also to their potential biological properties. Intense investigations have clarified the important role played by selenium compounds in biological redox processes and opened new perspectives for their use, mainly as antioxidants and chemo-protectors, but also as antitumor, antiviral and antimicrobial agents [4]. Highly enantioenriched selenium compounds have been prepared by reagent-controlled asymmetric synthesis with enantiopure selenium reagents and prochiral substrates or by substrate-controlled asymmetric synthesis with enantiopure substrates and achiral selenium reagents [1,2]. Representative examples of these synthetic strategies are described in chapters 2 and 3. Usually the stoichiometric chiral source, *i.e.* the chiral reagent or the auxiliary, is derived from the natural chiral pool through expensive and time consuming multi-step procedures. The proposal of this chapter is to illustrate recent progress in the synthesis of organoselenium compounds by catalytic asymmetric methods. Asymmetric catalysis consists in the conversion of prochiral or racemic starting materials into enantioenriched products by effect of a chiral catalyst. In recent years this approach has acquired great popularity, due to its cost effectiveness and environmental benefits. The reported reactions demonstrate that the use of privileged catalysts of broad applicability can be successfully extended to selenium chemistry. Novel, mechanistically intriguing and synthetically valuable methods, suitable for the preparation of enantioenriched molecules containing or not selenium, are described. The selected examples are arranged in three sections depending on the nature of the chiral catalyst: metal-organic ligand complex, organocatalyst or biocatalyst.

7.2. SYNTHESIS OF CHIRAL ORGANOSELENIUM COMPOUNDS BY ORGANOMETALLIC CATALYSIS AND APPLICATIONS

Metal catalyzed asymmetric ring opening reactions of epoxides with various nucleophiles represents an efficient strategy for the enantioselective synthesis of versatile 1,2-functionalized building-blocks [5]. Reactions with nucleophilic selenium reagents can give access to 1,2-seleno alcohols that are useful intermediates in the preparation of enantiomerically enriched allylic alcohols or heterocycles [1,6]. Yang and co-workers reported in 2005 the first enantioselective ring-opening reaction of *meso*-epoxides (desymmetrization) with arylselenols using the Gallium-Titanium-Salen complex **3** (Scheme **7.1**) [7]. Before this study, only reagent-controlled methods for the diastereoselective opening of *meso*-epoxides were disposable [8]. Catalytic desymmetrizations of *meso* epoxides typically involve a chiral Lewis acid which activates the epoxide. The use of a heterobimetallic catalyst with two different Lewis centres provides an efficient stereocontrol by effect of a double activation of both the substrate and the nucleophile within the same chiral scaffold. In fact the hard Lewis acid titanium is able to coordinate with the epoxide, whereas the gallium coordinates with the soft Lewis base arylselenol and directs the nucleophilic attack toward one of the epoxide termini.

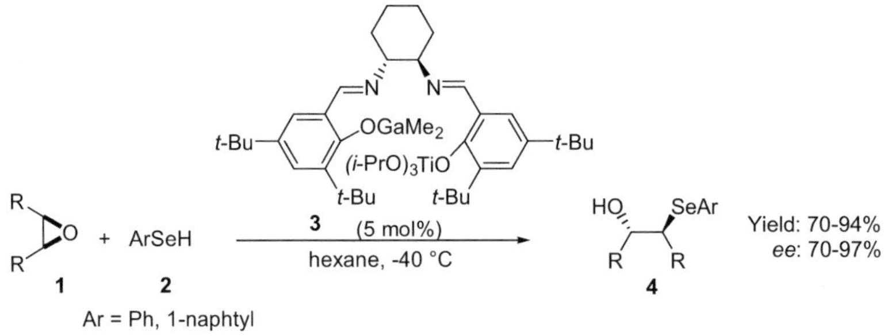

Scheme 7.1

A broad range of aromatic, aliphatic and cyclic epoxides were readily opened with arylselenols in very good yields and 70–97% *ee*. Interestingly, benzenselenol gave better enantioselectivity than the corresponding thiol when reacted with the same epoxide [7b].

In 2008 Tiecco and Marini [9] described a new desymmetrization protocol using commercially or easily available Chromium-Salen complexes **5** and *tert*-butyl(dimethyl)(phenylseleno)silane **6** as the nucleophile. In the same year, Schneider developed a scandium-bipyridine catalyzed desymmetrization with benzenselenol and aromatic *meso* epoxides [10].

Table 7.1: Asymmetric ring opening reactions of *meso* epoxides with selenium nucleophiles and chiral metal complexes

Catalytic system and reaction conditions	Selenium Nucleophile	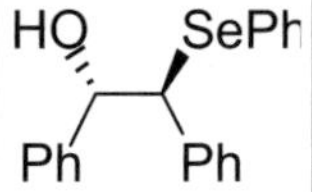	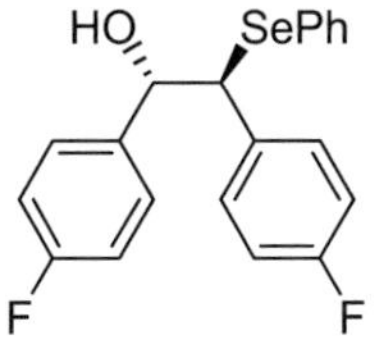	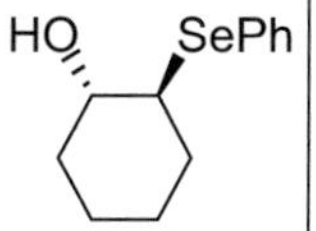
		4a	**4b**	**4c**
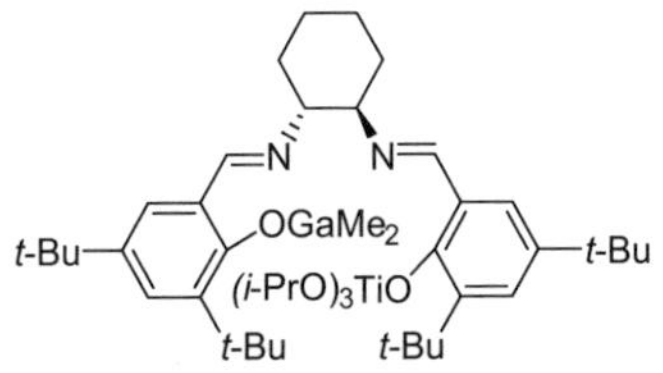	PhSeH **2a**	70% 72% *ee*	-	94% 97% *ee*
3 (5 mol%), hexane, -40 °C				
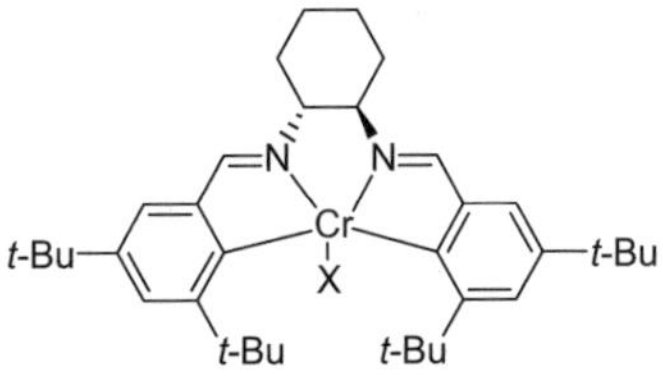	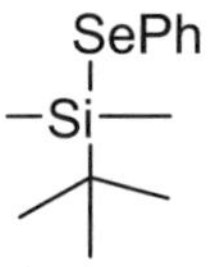	92% 92% *ee*	50% 70% *ee*	88% 62% *ee*
5 (5 mol%), TBME, -10 °C	**6**	X = BF_4,	X = BF_4	X = Cl
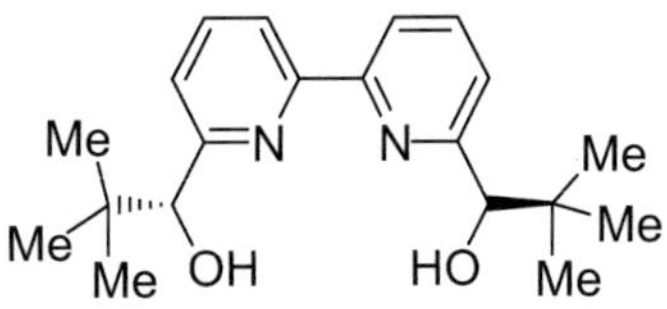 $Sc(OTf)_3$	PhSeH **2a**	*ent*-**4a** 77% 93% *ee*	*ent*-**4b** 51% 93% *ee*	*ent*-**4c** 71% 24% *ee*
7 (10 mol%), CH_2Cl_2, rt				

Selected results of the three methods are compared in Table **7.1**. The methods are complementary since the first protocol gave better results with aliphatic and cyclic epoxides but resulted less efficient in the desymmetrization of aromatic epoxides.

[2,3]-sigmatropic rearrangements of allylic selenoxides and selenimides are well established methods for the preparation of allylic alcohols or amines with several applications in natural product synthesis [1,11]. A typical [2,3]-sigmatropic rearrangement involves the transfer of oxygen or nitrogen from the selenium atom to the 3-carbon atom through cyclic transition states (TS). The following hydrolysis of the selenenate ester or amide leads to allylic alcohols or the amines [12,13]. Selenium based rearrangements favourably compare with reactions of the sulfur analogues, since they proceed at a faster rate and under milder reaction conditions due to lower activation energies. Moreover the equilibrium favours the selenenate in the selenium-based method and the sulfoxide in the sulfur series.

Stoichiometric methods have been widely employed for the preparation of optically active selenoxides or selenimides that are required for an asymmetric rearrangement. Chiral allylic selenoxides (Scheme **7.2**, X=O) were generated either by diastereoselective oxidation of selenides bearing a chiral auxiliary at the selenium atom or by enantioselective oxidation with chiral oxidants [13]. Chiral selenimides are accessible by treatment of chiral non racemic selenides with chloramine T (TsNClNa), N-(p-tolylsulfonyl)-imino(phenyl)-λ^3-iodane (TsN=IPh) or alternatively with a mixture of N-chlorosuccinimide (NCS) and a benzyloxy (Cbz) or *tert*-butoxy (Boc) carbammate [14]. On the other hand, processes employing a catalytic chiral source for the enantioselective formation of selenoxides or selenimides starting from achiral substrates are, up to date, rare.

Very recently a ligand-based strategy for the vanadium-catalyzed selenide oxidation with *in situ* asymmetric [2,3]sigmatropic rearrangement failed. In fact, several chiral Shiff bases, previously employed for the asymmetric oxidation of sulfides, afforded racemic allylic alcohols [13a].

In 1998 Uemura investigated the enantioselective imidation of prochiral selenides with catalytic amounts of CuOTf and the (*R*)-bisoxazoline **11** as chiral ligand [15]. Some examples are reported in Scheme **7.3**. Allylic N-tosylamides **10a-c** were obtained with moderate to good yields and low enantioselectivity *via* [2,3] sigmatropic rearrangement of the initially formed selenimides **9a-c**. In the absence of 3Å MS molecular sieves no enantioselectivity was observed in the formation of the selenimide. They are used to prevent the rapid selenimide-selenoxide equilibrium by removal of water.

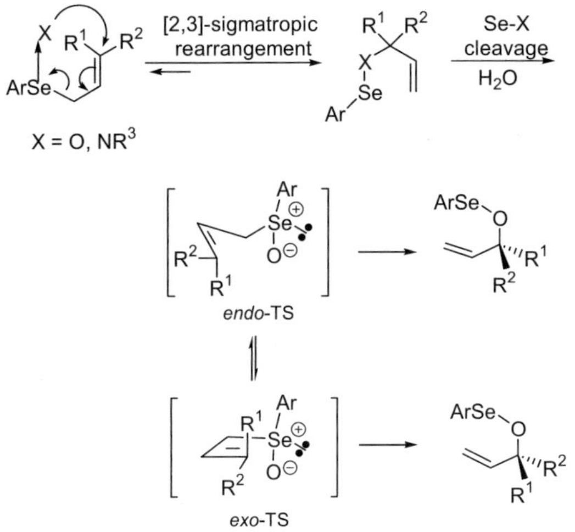

Scheme 7.2

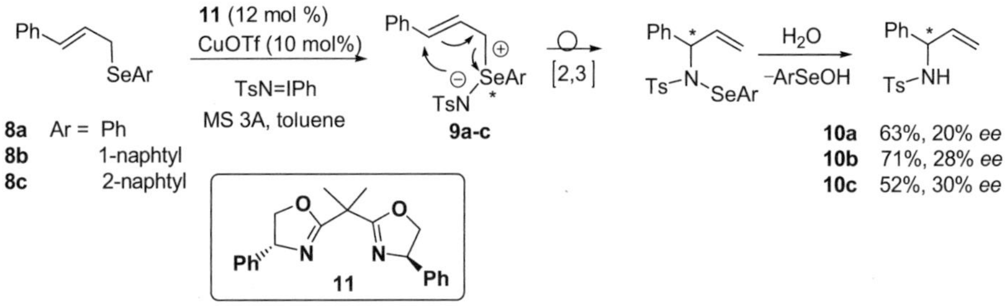

Scheme 7.3

Allylic selenium ylides **13** generated from cinnamyl selenide and ethyl diazoacetate (EDA) in the presence of enantiopure organometal catalysts are also suitable substrates for asymmetric [2,3]-sigmatropic rearrangements [16]. The plausible catalytic cycle is shown in Scheme **7.4**. The chiral copper triflate complex derived from the bis-oxazoline **15** or the Doyle rhodium (II) catalyst, (Rh_2(5*S*-MEPY)$_4$) **16** are competent catalysts for the enantioselective synthesis and subsequent rearrangement of selenium ylides. The metal-catalyzed decomposition of the ethyl diazoacetate generates the chiral metal carbene species **12** which adds to the selenium atom of the phenyl cinnammyl selenide to form the ylide **13**.

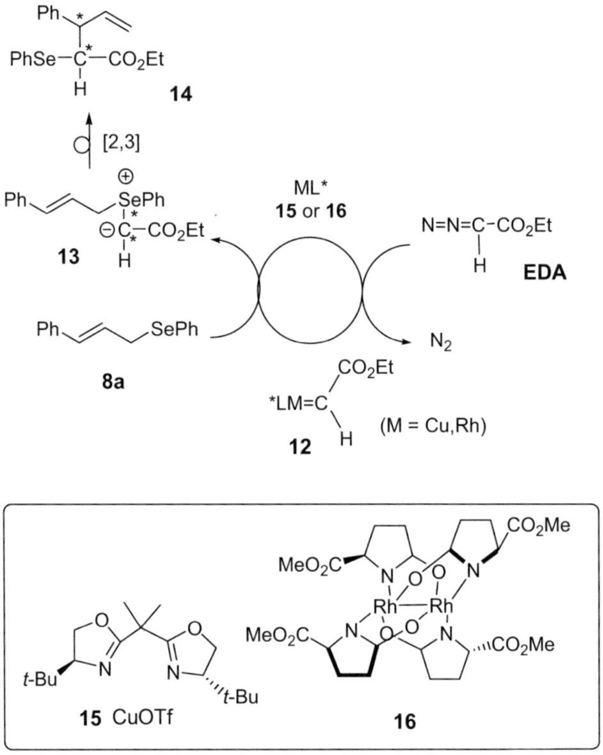

Scheme 7.4

The [2,3]sigmatropic rearrangement generates a carbon-carbon bond and two new vicinal stereocenters and leads to a diastereomeric mixture of esters with moderate enantioselectivity and poor diastereoselectivity (Scheme **7.5**). Considering that usually [2,3]-sigmatropic rearrangements proceed without loss of enantiomeric purity and racemizations at chiral selenium ylides do not occur, it was suggested that the observed enantioselectivity depends on the carbenoid addition to the selenium atom of the prochiral selenide. The results are still not satisfactory, but the selectivity may be improved through an optimization of the catalyst design.

Finally highly enantioenriched allyl selenides generated by palladium-catalyzed decarboxylative selenenylations of cyclohexenone-derived selenocarbonates have been employed in stereospecific [2,3]-sigmatropic rearrangements [17]. Scheme **7.6** shows reagents, catalyst, reaction conditions, yields and enantiomeric excesses.

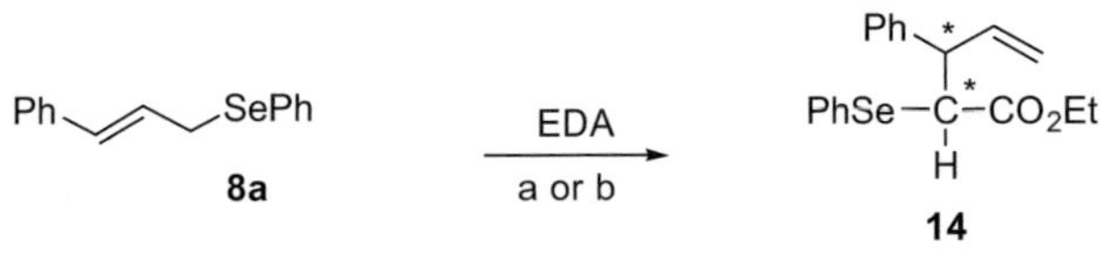

	yield	*dr*	*ee*
a. **15** (5 mol%) + CuOTf (5 mol%), $CHCl_3$, 0 °C	71%	66:34	22, 32
b. **16** (1 mol%), CH_2Cl_2, 40 °C	65%	58:42	25, 41

Scheme 7.5

The method is based on the kinetic resolution of the racemic selenocarbonate **17** with tris(dibenzylideneacetone) dipalladium (Pd_2dba_3) and the (*S,S*)-naphthyl-Trost ligand **20**. Both the unreacted selenocarbonate (*S,S*)-**17** and the allyl selenide (*R,R*)-**18** were isolated with high enantiopurity, showing that an efficient kinetic resolution took place in the first step. The next amination reaction provides the desired amine in good yield and almost complete conservation of the chirality. Other rearrangements with chirality transfer of highly enantioenriched allylic selenoxides or selenimides generated by organocatalytic methods are discussed in section 7.3.

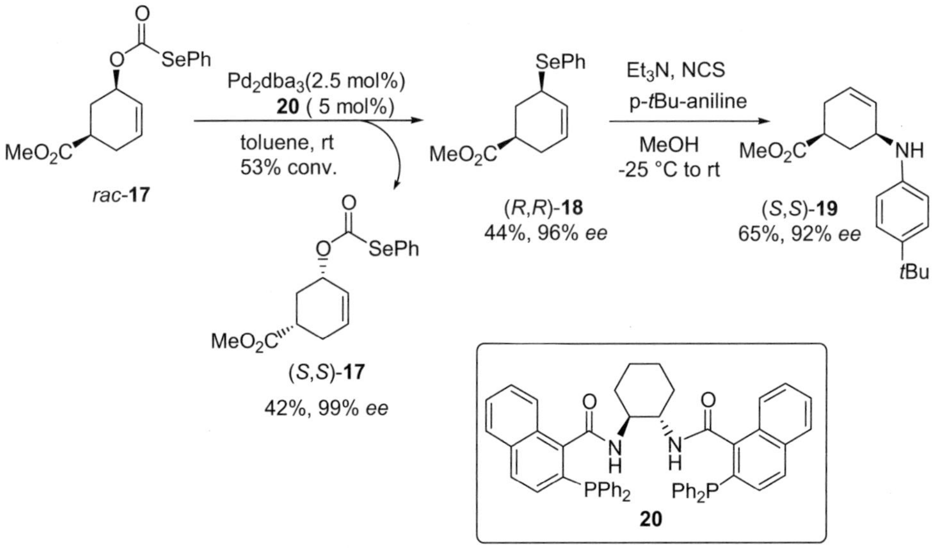

Scheme 7.6

7.3. SYNTHESIS OF CHIRAL ORGANOSELENIUM COMPOUNDS BY ORGANOCATALYSIS AND APPLICATIONS

Asymmetric organocatalysis, *i.e.* the use of relatively small chiral organic molecules to accelerate and control the stereochemical course of a variety of fundamental synthetic processes, has grown explosively during the last years [18]. Some advantages of organocatalytic methods are the use of metal-free conditions, the favourable environmental impact, the low cost and the operational simplicity. Cinchona alkaloid derivatives, proline or other amino acids and peptide-derived compounds are typical organocatalysts. Common activation modes involve the formation of transient intermediates, such as enamines or iminium ions (covalent catalysis) [18a], or weak interactions such as hydrogen bonding or ion pairing formation (non covalent catalysis) [18b]. Wynberg in 1979 introduced cinchonidine **22** as a catalyst for the 1,4-conjugate addition of arylselenols to cyclohexenone (Scheme **7.7**) obtaining 1,4-adducts with excellent yields, but poor enantioselectivities [19].

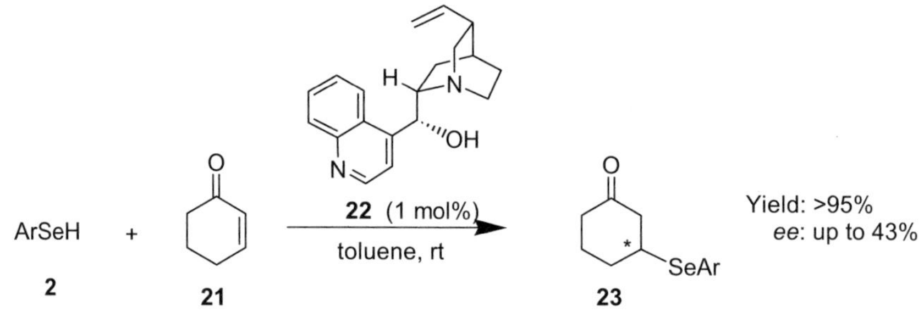

Scheme 7.7

After this pioneering work, only recently other organocatalytic methods based on selenium reagents appeared in the literature. In particular several 1,4-conjugate additions of carbon-centered nucleophiles to vinyl selenones have been investigated with the aim of assembling densely functionalized compounds containing all carbon quaternary stereocenters. The construction of such molecules is nowadays, one of the most dynamic areas in asymmetric synthesis being the enantioselective formation of these stereocenters very challenging. First results were obtained with racemic α-substituted cyanoacetates **24** and the unsubstituted vinyl selenone **25** in the presence of the thioureidic catalyst **26** [20] bearing the (*R*,*R*)-1,2-diphenyl-1,2-ethandiamine as the chiral scaffold (scheme **7.8**). Chiral bifunctional thioureas are powerful organocatalysts which are able to activate simultaneously both the Michael donor and the acceptor and promote the formation of highly organized transition states [18b-c]. The tertiary amine group may activate the carbon nucleophile acting as a base to produce the corresponding enolate, while the thiourea group interacts with the electrophilic alkene by a double H-bonding interaction.

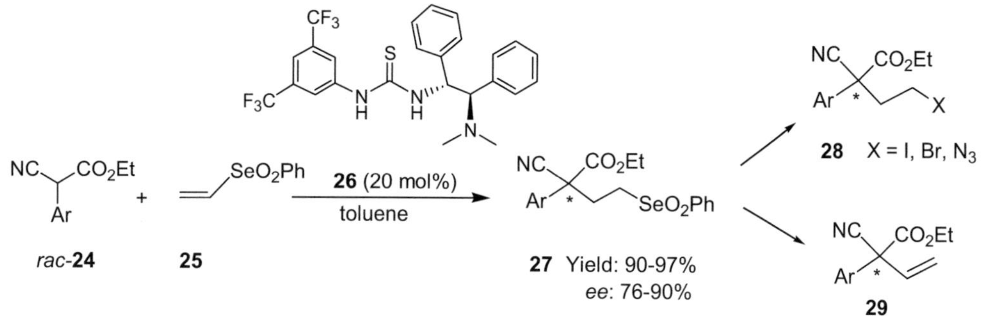

Scheme 7.8

The synergistic effect of thiourea and tertiary amine functionalities is responsible of the high levels of efficiency (90-97% yield) and enantioselectivity (76-90% *ee*). Moreover, the Michael adducts **27** have been converted, without loss of enantiomeric purity, into other multifunctional building-blocks by taking advantage of the excellent leaving group properties of the phenylselenonyl moiety.

Cinchona based thiourea catalysts were used as efficient catalyst for the Michael addition of racemic 3-alkyl or 3-aryl substituted 2-oxindoles in ionic liquids [21]. Stereoselective approaches to oxindole scaffolds have considerable synthetic interest due to their occurrence in diverse natural products and biologically active drug candidates. The 3,3-disubstituted selenium containing oxindoles were generated in good yields and high enantioselectivities (Scheme **7.9**).

Moreover highly enantioenriched alkyl selenones generated by organocatalytic Michael addition of carbon nucleophiles to vinyl selenones have been used as key intermediates for the synthesis of several cyclic or heterocyclic compounds. In these sequential domino processes the starting vinyl selenone acts as a bis-electrophile, since the phenylselenonyl plays a dual role as an activating electron-withdrawing group during the addition step and as a leaving group during the following cyclization.

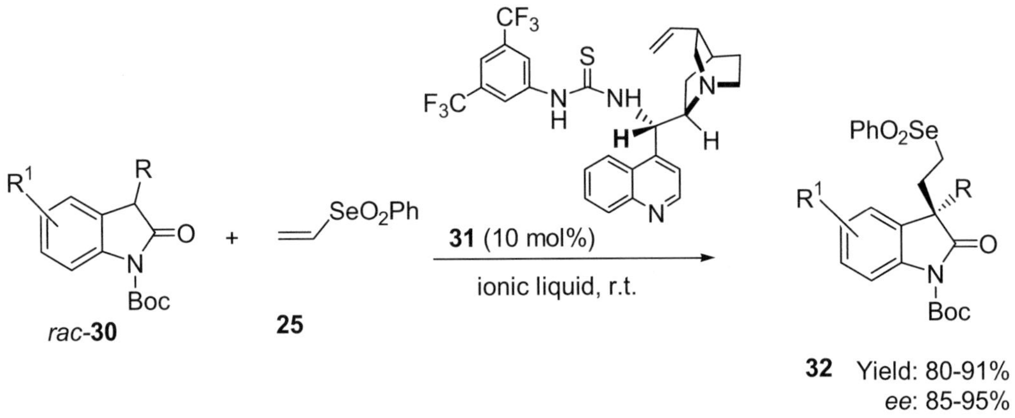

Scheme 7.9

A simple *one-pot* procedure for the diastereo and enantioselective preparation of cyclopropanes is described in Scheme **7.10** [22]. Michael intermediates, generated during the addition of cyanoacetates to easily accessible β-substituted vinyl selenones catalyzed by a hydroquinine derived urea, smoothly cyclize by a Krapcho-type de-ethoxycarbonylation (Method A) followed by an intramolecular nucleophilic substitution of the selenonyl moiety by the enolate intermediate.

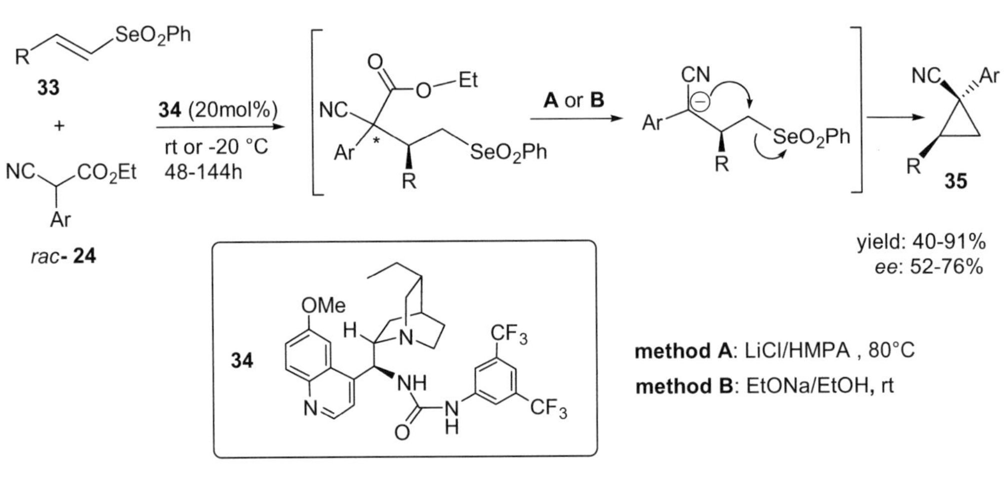

Scheme 7.10

Cyclopropanes having adjacent tertiary and quaternary stereocentres as structural motif have been generated as single isomers in moderate to high yields, complete diastereoselectivity and good enantiomeric excesses. In the presence of β-aryl substituted selenones, the de-ethoxycarbonylation and cyclization steps occur with comparable yields and enantioselectivities also by treatment with EtONa in EtOH (Method B).

Similar procedures provide a rapid access to synthetically and biologically appealing polycyclic structures. As described in scheme **7.11**, a number of optically active spirolactones were obtained *via* a Michael addition/cyclization sequence starting from racemic indanone- or cyclopentanone-derived *tert*-butyl β-ketoesters **36** and the vinyl selenone **25** [23]. Michael adducts were smoothly converted into the corresponding spirolactones by simple stirring with silica gel. The construction of the sterically constrained quaternary spiro centre was achieved with excellent levels of enantioselectivity by using bifunctional quinine or quinidine derivatives bearing a phenolic residue as hydrogen bond donor. The pseudoenantiomeric catalysts **37** (C6'OH-QD) and **38** (C6'OH-Q) derived from quinidine and quinine, respectively and with opposite configuration at C-(8) and C-(9) provided spirolactones **40** with opposite enantioselection and comparable yields and enantiomeric excesses.

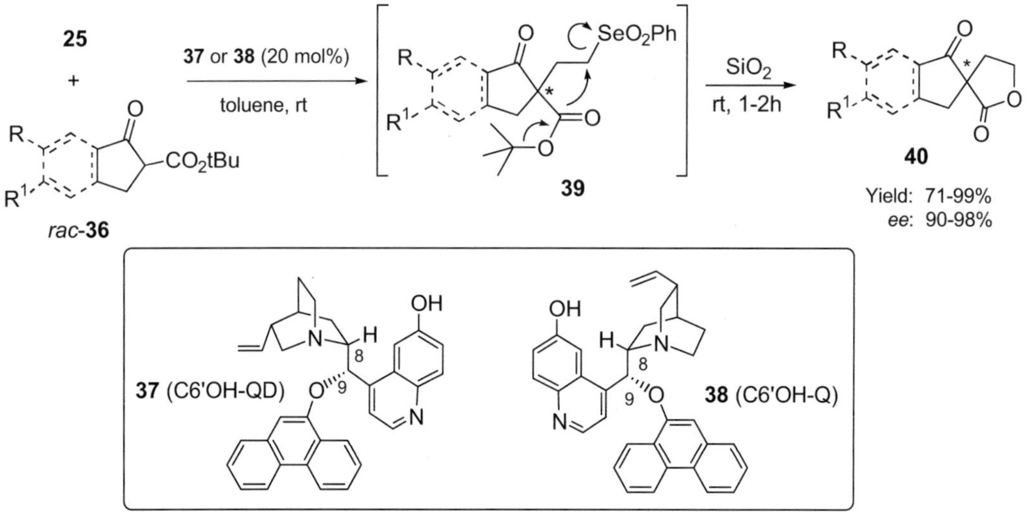

Scheme 7.11

Stereochemical models (Scheme **7.12**) previously developed for other conjugate additions catalyzed by C6'-OH cinchona derivatives can be used to rationalise the stereochemical outcome [18c]. The catalitically active conformer of **38** simultaneously activates and orients the Michael donor and the acceptor by a network of hydrogen-bonding interactions. Catalyst **38** revealed as an efficient catalyst for a stereodivergent reaction on a racemic mixture [23]. This is an asymmetric synthetic strategy based on the use of a single chiral catalyst or reagent [24] to transform the two enantiomers of a racemic substrate into a couple of enantioenriched diastereoisomers. In the present case the spiro carbon center is formed with high enantioselectivity by catalyst control, independently of the pre-existing stereocenter (Scheme **7.13**).

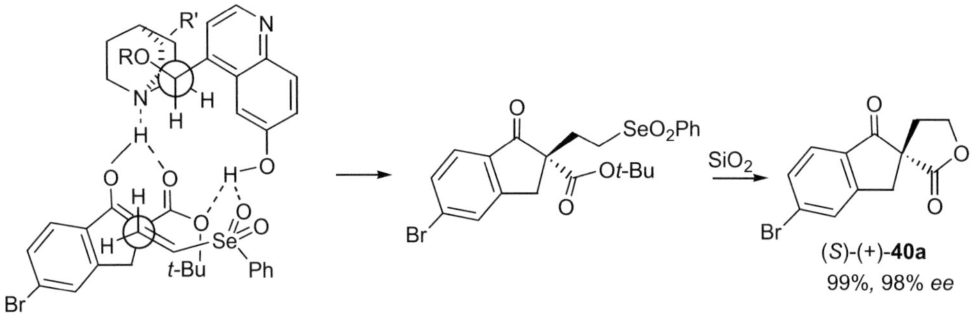

Scheme 7.12

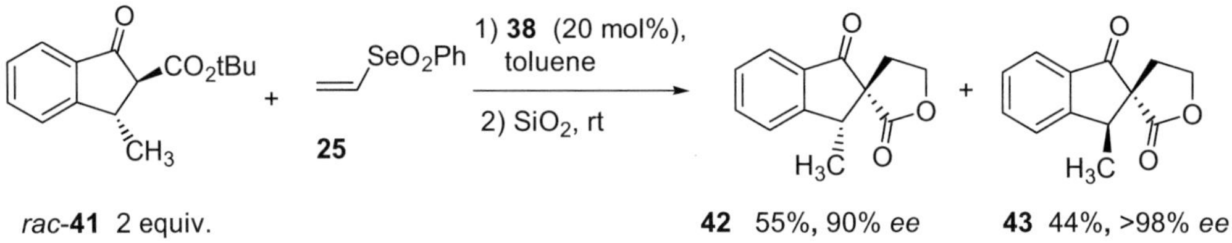

Scheme 7.13

Michael adducts **39** generated by organocatalyzed addition of indanone derived β-ketoester to vinyl selenone are also useful intermediates for the synthesis of conformationally constrained polycyclic pyrrolidines bearing a β-aminoester motif [25]. These drug-like structures containing synthetically challenging contiguous tertiary and quaternary stereocenters were formed with excellent diastereo and enantiocontrol. The operationally simple, multi reaction sequence is described in Scheme **7.14** and involves the following steps:

a) the highly enantioselective addition of an indanone derived β-ketoester to the vinyl selenone catalyzed by C6'OH quinidine or quinine derivatives. A catalyst loading of only 5 mol% was used without compromising the chemical and optical yields.

b) the formation of an azide by nucleophilic substitution of the phenylselenenonyl group,

c) the cyclization reaction *via* a Staudinger/aza-Wittig sequence,

d) the diastereoselective reduction with sodium borohydride in MeOH.

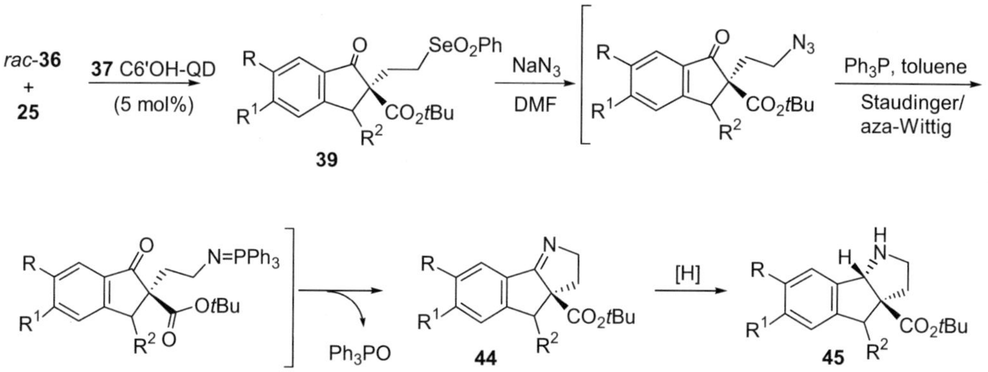

Scheme 7.14

The first three steps were carried out sequentially, excluding purification processes with yields ranging from 60 to 85% and enantiomeric excesses of 93-98%. The following reduction gave the *cis* fused isomer as single product in acceptable yields.

Organocatalyzed Michael addition of α-seleno carbonyl compounds to nitrostyrenes represent an efficient access to a variety of α-seleno ketones, esters or amides containing a selenium substituted quaternary stereocenter [26].

Bifunctional cinchona derived thiourea or squaramide catalysts gave excellent levels of diastereo and enantioselectivity in some cases with a catalytic loading of only 3 mol% (Scheme **7.15**).

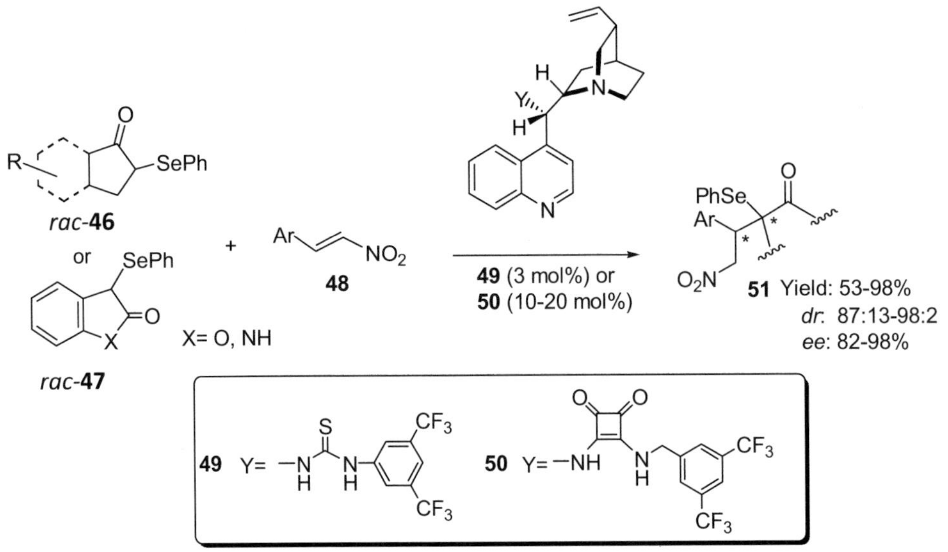

Scheme 7.15

Other important classes of organoselenium compounds such as α-selenoaldehydes are accessible by organocatalytic methods. The first organocatalytic protocol concerning the α-selenenylation of aldehydes using L-prolinamide as catalyst appeared in 2004, but the enantioselectivities were very low [27]. In 2007, two independent protocols described the highly enantioselective synthesis of α-selenoaldehydes with sterically demanding silyl ethers of diarylprolinols and N-(phenylseleno) phthalimide **53** (N-PSP) as the electrophilic selenium source [28,29]. In both cases in order to facilitate work-up and avoid racemisations during the chromatographic purification step, the reaction products were isolated as the corresponding alcohols by *in situ* reduction.

The commercially available prolinol derivative **54** is a particularly efficient catalyst (Scheme **7.16**) for the α-selenenylation of aldehydes with alkyl, alkenyl, and heteroatom-containing residues [28]. High yields and an excellent enantiocontrol were obtained with 5 mol% of catalyst. Acid additives were found to affect both the enantioselectivities and the yields of the products.

The catalytic cycle proposed for secondary amine catalyzed α-selenenylations is described in Scheme **7.17** [29]. The configuration of the final products is controlled by the bulky aryl groups of the amine catalyst which guides the approach of the electrophile from the less hindered face of the generated (*E*)-enamine intermediate. The catalytic activity of recyclable resin-supported prolinol or imidazolidinone derivatives **57** and **58** (Fig. **7.1**) have been evaluated [30]. **57** gave lower levels of reactivity and selectivity in respect to the free catalyst, but the supported imidazolidinone-derived catalyst **58** worked nicely in the α-selenenylation of propanal. The catalyst was recovered by filtration and reused for at least four cycles without loss of catalytic activity.

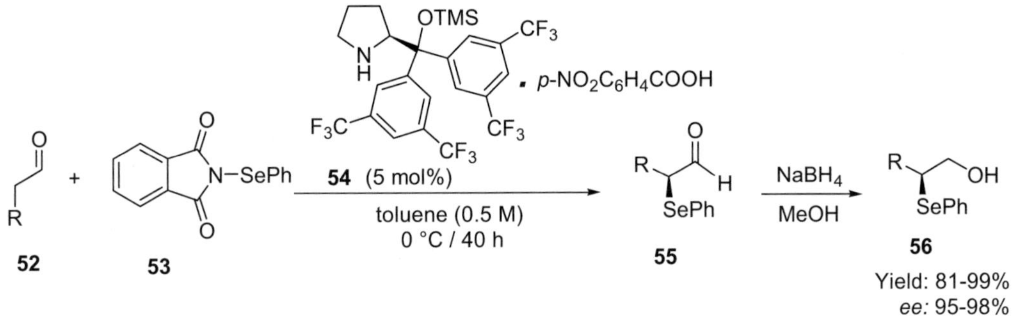

Scheme 7.16

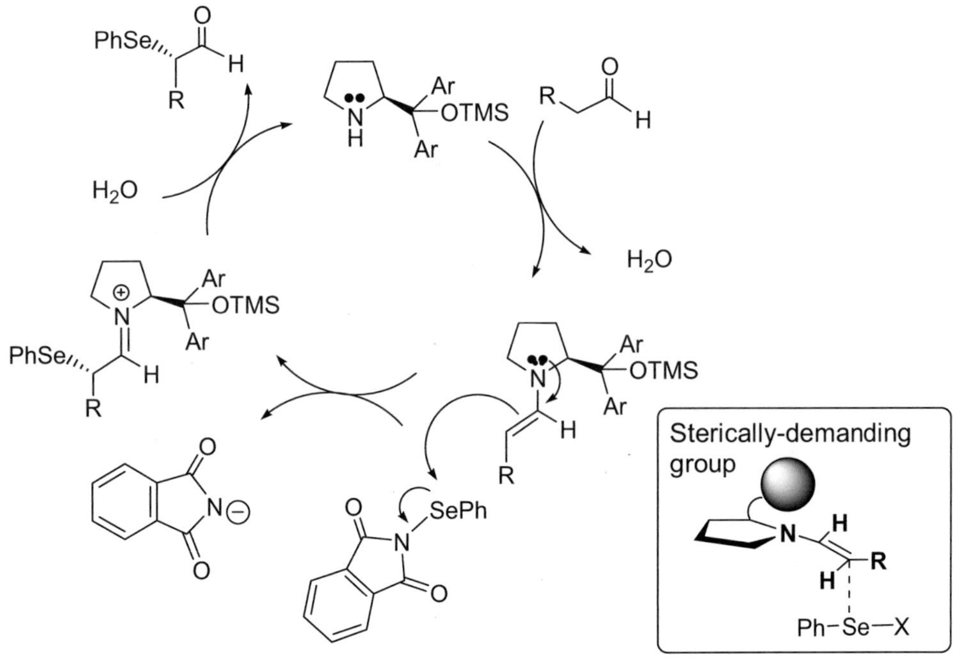

Scheme 7.17

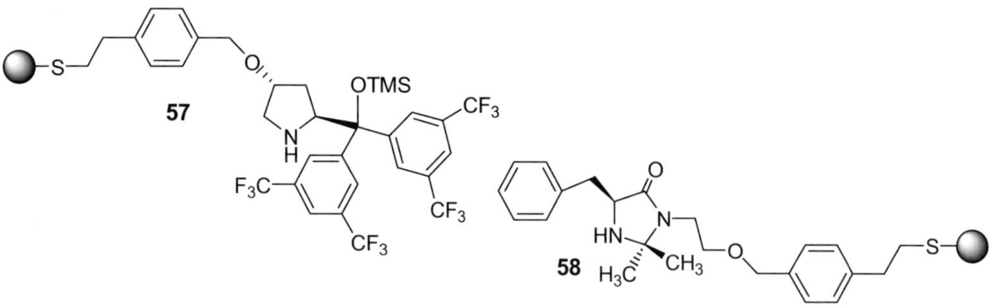

Figure 7.1: Polymer supported organocatalysts.

α-Seleno aldehydes are versatile building blocks for the synthesis of polyfunctionalized compounds and natural products. For example, they can undergo an aldol-type reactions with an electron-withdrawing-stabilized carbanion (Scheme **7.18**). Then, a three-step *one-pot* process of selenide oxidation/epoxide formation and epoxide opening generates γ-hydroxy-(*E*)-α,β-unsaturated sulfones or esters with acceptable overall yields and excellent control of absolute stereochemistry (*ee* values ≥95%) [31]. This procedure has been applied in the concise formal synthesis of the biologically active alkaloid (+)-α-conhydrine **62**. The intermediate γ-hydroxy protected-(*E*)-α,β-unsaturated ester **61** was advantageously prepared in four steps.

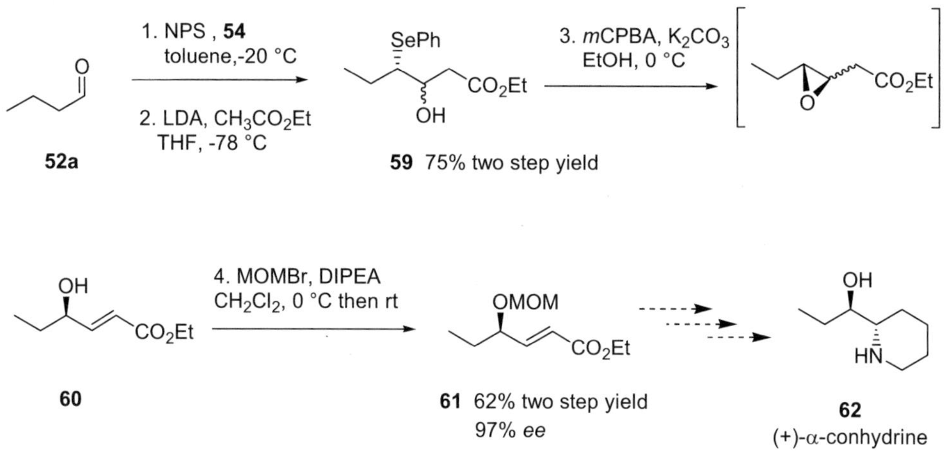

Scheme 7.18

Scheme **7.19** describes two nice protocols for the asymmetric synthesis of α-hydroxy-(*E*)-β,γ- unsaturated esters **64** [32] and α-amino acids **65** [33]. The methods combines the enantioselective amine-catalyzed α-selenenenylation with a Wittig or Horner-Wadsworth-Emmons (HWE) type olefination to yield in *one-pot* the highly enantioenriched γ-phenylseleno-α,β-unsaturated esters **63**. Then, the oxidation to selenoxide or the N-chlorosuccinimmide (NCS)-mediated formation of a selenimide followed by a [2,3]-sigmatropic rearrangement produced the target products in high yields and excellent enantiomeric excesses. As previously reported these rearrangement proceed through cyclic transition states which facilitate an efficient transfer of chirality. The selenium-based protocols described in Scheme **7.19** have some advantages over the corresponding sulfur methods. In fact the synthesis of the starting α-selenoaldehydes occurs from a stable and commercially available selenium reagent and the seleno aldehydes are more resistant to racemization during olefination reactions than the sulfur derivatives. Moreover, because of the scission of weak selenium bonds, [2,3]-sigmatropic rearrangements occur under milder conditions.

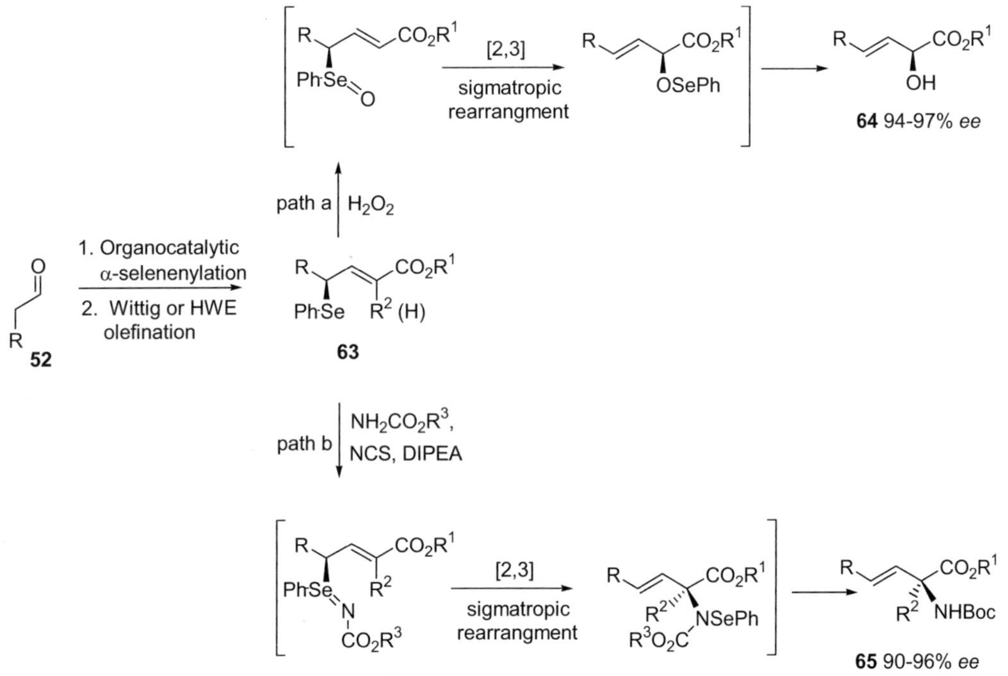

Scheme 7.19

Chiral γ-seleno-α,β-unsaturated esters **63** (Scheme **7.20**) when treated with sulfuryl chloride and ethyl vinyl ether were converted into the corresponding α-chloro-α,β-unsaturated esters **66** with 1,3-*syn* transfer of chirality [34]. Reactions on multigram scale of these allylic chlorides with carbon or nitrogen nucleophiles such as methylcuprate or sodium azide generated γ-substitution products and vinylogous aminoesters with *anti*-transfer of chirality.

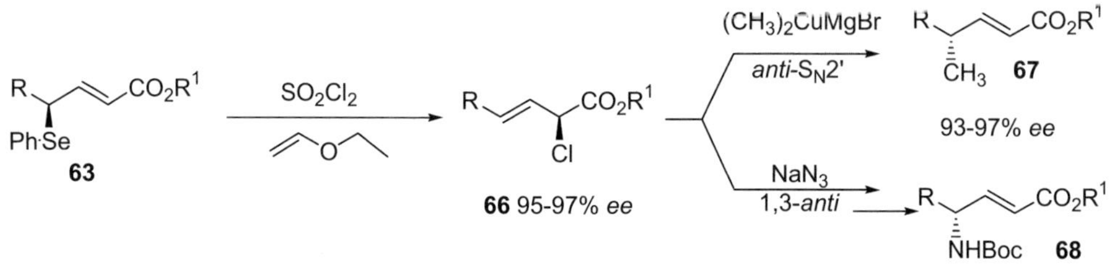

Scheme 7.20

Chiral α-seleno aldehydes are also useful intermediates in the synthesis of heterocycles.

Scheme **7.21** describes the preparation of enantioenriched 1,3-oxazolidinones. The β-phenylseleno alcohol, generated by direct reduction of the aldehyde 55b, was converted into the corresponding carbamate and then oxidized *in situ* to generate a selenonyl group. The ring closure reaction occurs by stereospecific intramolecular nucleophilic displacement of this group by the nitrogen atom [28].

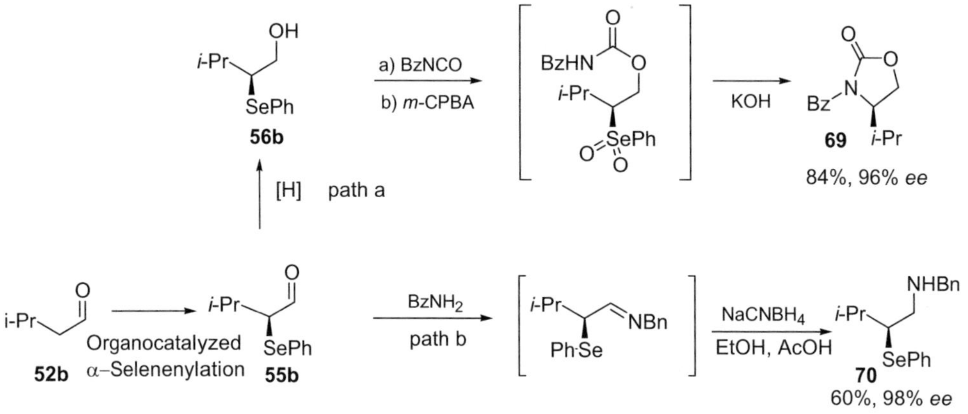

Scheme 7.21

Finally α-seleno aldehydes were converted into 1,2-selenoamines in good yield and without loss of enantiomeric purity (Scheme **7.21**, path b) [28].

Enantiomerically pure 1,2-selenamines are not only appreciated ligands or intermediates for enantioselective reactions and heterocyclic synthesis, but also promising pharmacological agents [35]. An alternative asymmetric catalytic approach to vicinal selenoamines is based on the desymmetrization of *meso* 3,5-dinitrobenzoyl aziridines. The ring opening reaction with a 1:2 mixture of (phenylseleno)trimethylsilane/benzenselenol as the nucleophile and VAPOL - hydrogen phosphate **72** as the chiral catalyst generated products with good to excellent enantioselectivities (Scheme **7.22**) [36]. Aziridine substrate can be varied with small changes in yield and selectivities. Further studies demonstrated that commercially available phosphoric acids as well as those synthesized by established procedures may contain variable amounts of metallic phosphate impurities and that the presence of these metal species is essential to achieve high levels of catalytic activity and enantioselectivity [37]. New experiments showed that the cooperative effect of a 1:1 mixture of calcium and magnesium phosphate salts maximizes yields and enantiomeric excesses.

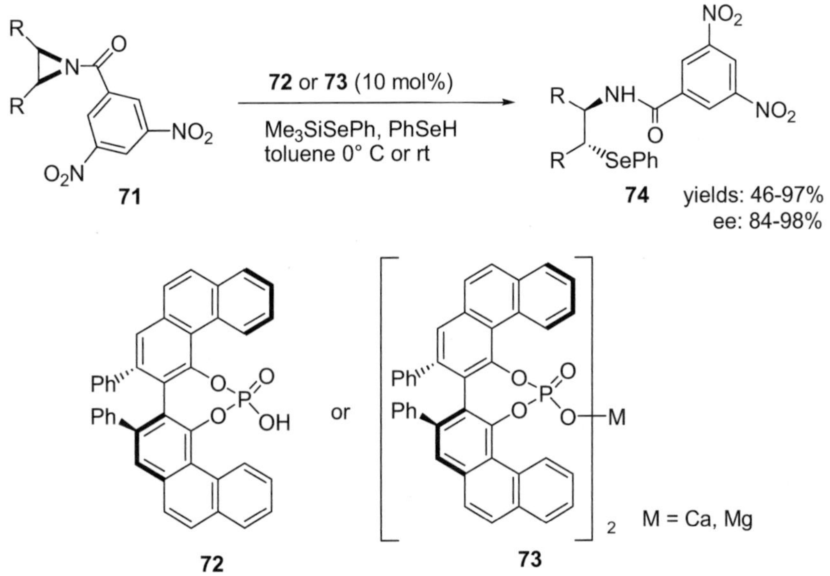

Scheme 7.22

In 2010 Denmark [38] reported the first catalytic enantioselective selenoetherification reaction of unsaturated alcohols (Scheme **7.23**). The method is based on the concept of “Lewis base activation of Lewis acids.” As depicted in

Fig. **7.2** the weak selenium(II) electrophile N-arylselenophthalimide **75** combines with the chiral Lewis base 1,1'-binaphtalene-2,2'-diamine-derived (BINAM) thiophosphoroamide **77**. The resultant ionic strong electrophile reacts with prochiral alkenes to afford enantioenriched seleniranium ions. The ring closure reactions by the suitably positioned oxygen atom release the products **78** and regenerate the Lewis base catalyst.

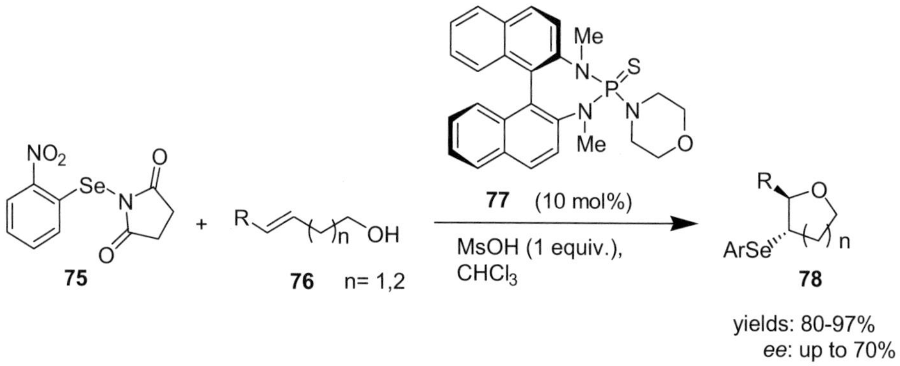

Scheme 7.23

Exo selectivity in ring closure reactions is not always observed. The generation of configurationally stable seleniranium ions is crucial for the enantioselectivity. On the basis of extensive studies, authors suggested that erosion in enantiomeric purity of seleniranium ions is possible either by attack of a nucleophile on selenium atom regenerating the double bond along with an achiral selenium species, or by non-stereocontrolled transfer of the selenium ion between alkenes (alkene to alkene transfer). The use of an electrophilic reagent with a 2-nitrophenyl group linked to the selenium atom significantly reduces the racemization of the seleniranium ion intermediates with beneficial effects on the enantioselectivity of the process. Computational modelling and a careful design of catalysts might further improve the enantioselectivity.

Chiral Bronsted acid catalyzed enantioselective carboxyselenenylation of alkenes has also been recently explored (Scheme **7.24**). Although mechanistic details are not clarified, this process is feasible with the binaphtol-derived N-triflyl phosphoramide **80** and N-phenylselenophtalimide. Cyclic alkenes gave better results than acyclic ones with enantiomeric excesses up to 84% [39].

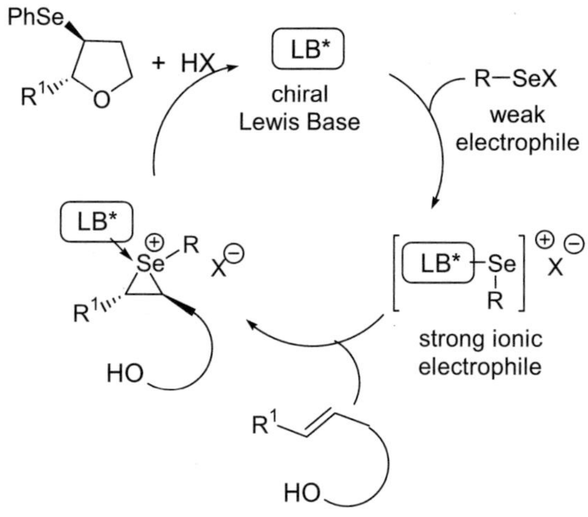

Figure 7.2: Catalytic cycle for asymmetric selenocyclizations.

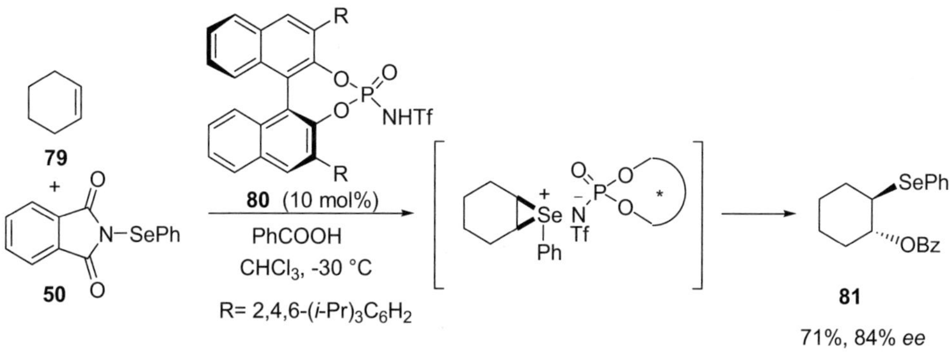

Scheme 7.24

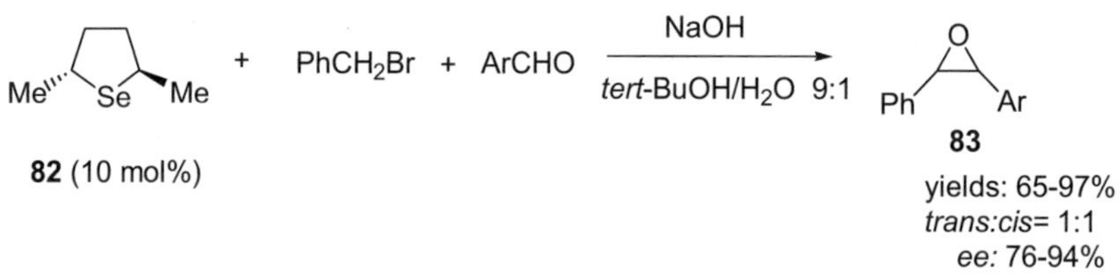

Scheme 7.25

Enantiopure selenium ylides are generated by addition of benzyl bromide to a catalytic amount of the C_2 symmetric (2*R*,5*R*)-2,5-dimethylselenolane **82** in the presence of NaOH [40]. These ylides react in *one-pot* with a variety of aldehydes to give oxiranes in high yields and excellent enantiomeric excesses but no diastereoselectivity (Scheme **7.25**). The plausible catalytic cycle is reported in

Fig. **7.3**. A catalyst loading of 10 mol % was required. Comparison between selenium and sulfur analogues revealed that selenium series leads to enhanced reactivity and better asymmetric induction [41].

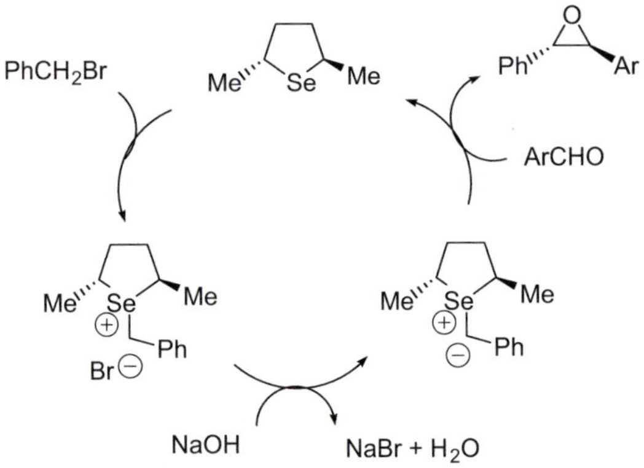

Figure 7.3: Catalytic cycle for asymmetric conversion of aldehydes into epoxides.

7.4. SYNTHESIS OF ORGANOSELENIUM COMPOUNDS BY BIOCATALYSIS

Biocatalysis, *i.e.* the use of enzymes or whole cell systems for performing chemical transformations, offers nowadays an alternative and ecological way to generate enantiopure organic compounds. In fact high levels of chemo-, enantio- and regioselectivity can be obtained in mild reaction conditions of pH and temperature. Few applications of biocatalysis to the preparation of selenium compounds were published to date [42]. Among them the enzymatic kinetic resolutions of racemic β-phenylseleno alcohols by commercially available lipases in organic solvents.

Figure 7.4: Schematic representation of Kazlauskas rule.

Cyclic or acyclic secondary alcohols were resolved through transesterification with vinyl acetate and immobilized lipases such as CALB (*Candida Antarctica* lipase type b [43] or PS-C II Amano lypase [44] in non polar solvents. Both products and unreacted starting materials were recovered with high enantiomeric excesses, except when bulky substituents were present at the carbinol carbon. Selected results of both methods are shown in Table **7.2**. The empirical "Kazlauskas rule" (Fig. **7.4**) correctly predicts the observed stereopreference for the acetylation of the *(R)* and the *(RR)* isomers.

Table 7.2: Lipase-catalyzed resolution of β-hydroxy selenides

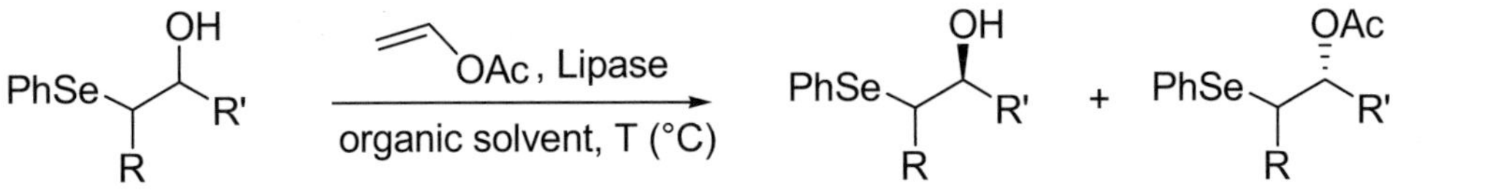

rac-**84a-f** (*S*) or (*S*,*S*)-**84a-f** (*R*) or (*R*,*R*)-**85a-f**

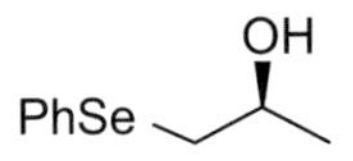	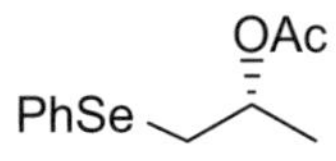	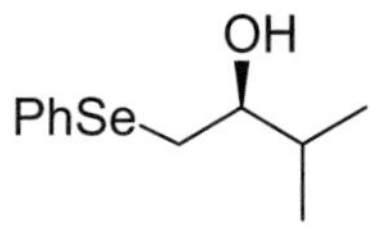	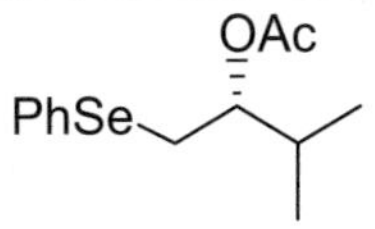
(*S*)-**84a** >99% *ee*[a]	(*R*)-**85a** 99% *ee*[a]	(*S*)-**84b** 69% *ee*[a]	(*R*)-**85b** 86% *ee*[a]
Method A (c= 50%)		Method A (c= 45%)	
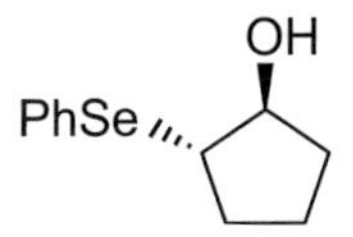	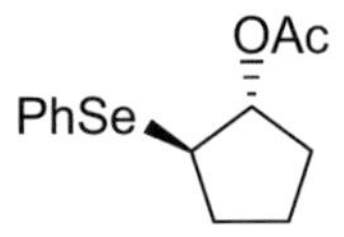	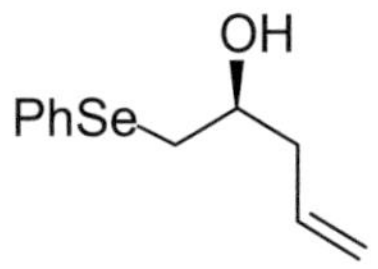	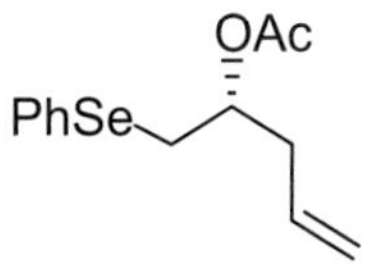
(*S*,*S*)-**84c** >99% *ee*[b]	(*R*,*R*)-**85c**>99% *ee*[b]	(*S*)-**84d** >99% *ee*[b]	(*R*)-**85d** 94% *ee*
Method B (c= 50%)		Method B (c= 51%)	
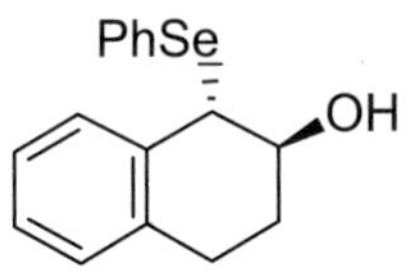	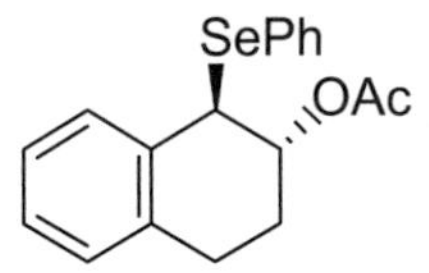	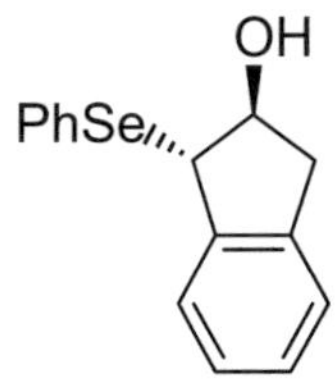	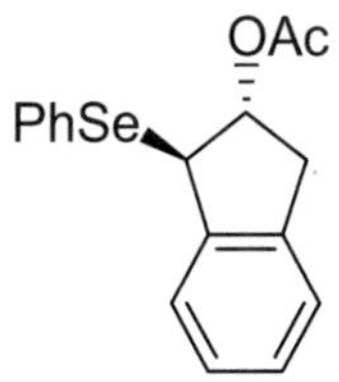
(S,*S*)-**84e** 82% *ee*[b]	(*R*,*R*)-**85e**>99% *ee*[b]	(*S*,*S*)-**84f** >99% *ee*[b]	(*R*,*R*)-**85f** >99% *ee*[b]
Method B (c= 45%)		Method B (c= 50%)	

[a] Method A: CALB lipase, hexane, T= 32°C or 40°C.
[b] Method B: PS-CII lipase, toluene, T= 30°C.

Acetylation mediated by lipases have also been employed for the kinetic resolution of organoselenium amines [45]. More recently, chiral organoselenium-1-phenylethanamines were synthesized with high yields (up to 87%) and excellent enantioselectivities through a chemoenzymatic dynamic kinetic resolution (DKR) [46]. Such kind of protocols overcome the limitation of classical kinetic resolutions allowing for the complete transformation of a racemate into a single enantiomeric product with a theoretical maximum yield of 100% as a consequence of the continuous *in situ* racemization of the less reactive enantiomer (Scheme **7.26**).

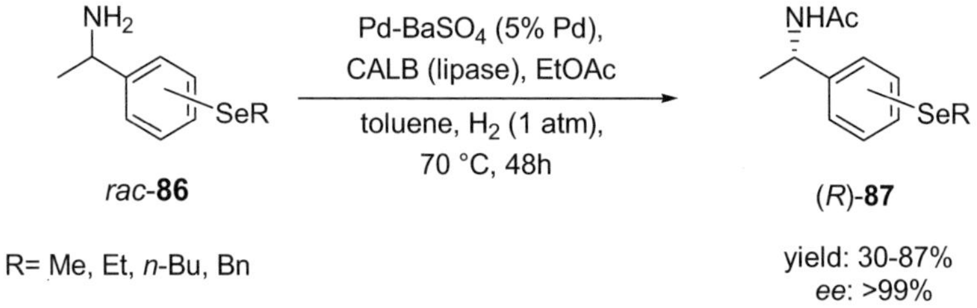

Scheme 7.26

The protocol combines in one reaction vessel the CALB lipase, as the resolution biocatalyst, and palladium on $BaSO_4$, as the racemization chemocatalyst. The selenium functionality is leaved intact in the final amides obtained with enantiomeric excesses up to 99%.

ω-Transaminases have been recently evaluated for the kinetic resolution of selenium containing amines [47]. Both enantiomers of the amines are accessible in high enantiomeric excesses (up to 99%) by using complementary enzymes.

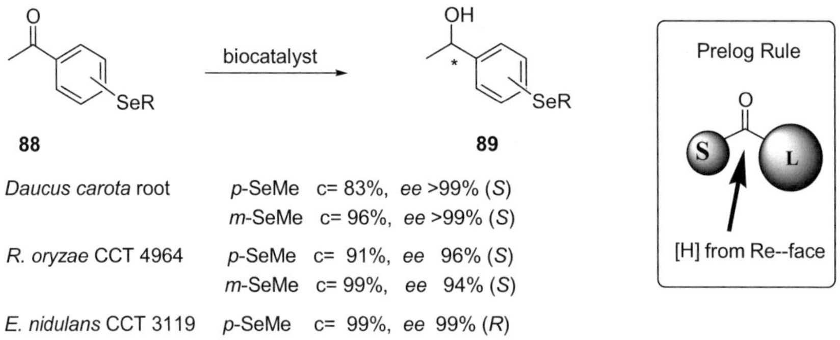

Scheme 7.27

Almost enantiopure organoselenium-1-arylethanols **89** have been prepared not only by enzymatic kinetic resolution with *Candida Antarctica* lipase [48], but also by enantioselective reduction of selenosubstituted acetophenones catalyzed by whole fungal cells of the genera *Rhizopus oryzae*, *Aspergillus terreus* and *Emericella nidulans* [49] or by vegetables such as *Daucus carota* root [50]. Reductions follow the Prelog's rule and occur from the *Re*-face, except for the reaction catalyzed by *Emericella nidulans* which shows opposite enantioselectivity (Scheme **7.27**). The efficiency and enantioselectivity of the process is strictly dependent from steric effects. In fact *orto* selenosubstituted acetophenones are unreactive toward all the biocatalysts employed.

7.5. CONCLUSIONS

Asymmetric catalytic methods with privileged chiral metal-complexes, organocatalysts and biocatalysts are emerging as effective tools for the enantioselective synyhesis of important classes of organoselenium compounds. Although the reactions developed are yet limited, in many cases excellent results in terms of reactivity and stereoselectivity have been achieved under mild and operationally simple conditions. The peculiar reactivity of selenium containing intermediates generated by asymmetric catalysis have been exploited in practical synthetic sequences for the construction of highly enantioenriched and densely functionalized compounds.

ACKNOWLEDGEMENT

Declared none.

CONFLICT OF INTEREST

The authors confirm that this chapter contents have no conflict of interest.

REFERENCES

[1] For books on this area see: (a) Organoselenium Chemistry: Synthesis and Reactions Ed. Wirth, T. Wiley-VCH, Weinheim, **2012**. (b) Organoselenium Chemistry: Modern Developments in Organic Synthesis Ed. Wirth, T. in Top. Curr. Chem.; Springer, Berlin, **2000**. (c) Organoselenium Chemistry-A Practical Approach Ed. Back, T. G. Oxford

University Press, Oxford, **1999**. (d) Organoselenium Chemistry Ed. Liotta, D.C. John Wiley & Sons, Inc., New York, **1987** (e) Selenium Reagents and Intermediates in Organic Synthesis Ed. Paulmier, C. Pergamon Press, Oxford, **1986**.

[2] (a) Zhu, C.; Huang, Y. Asymmetric Synthesis of Chiral Organoselenium Compounds. Curr. Org. Chem. **2006**, 10 1905-1920. (b) Tiecco, M.; Testaferri, L.; Bagnoli, L.; Marini, F.; Santi, C.; Temperini, A.; Scarponi, G.; Sternativo, S.; Terlizzi, R.; Tomassini, C. Enantioselective Synthesis of Heterocyclic Compounds Mediated by Organoselenium Reagents. Arkivoc **2006**, (vii), 186-206. (c) Tiecco, M.; Testaferri, L.; Marini, F.; Bagnoli, L.; Santi, C.; Temperini, A.; Sternativo, S.; Tomassini, C. Asymmetric Syntheses Promoted by Organoselenium Reagents. Phosphorus, Sulfur and Silicon **2005**, 180, 729-740. (d) Wirth, T. Organoselenium Chemistry in Stereoselective Reactions. Angew. Chem., Int. Ed. **2000**, 39, 3740-3749.

[3] For reviews (a) Godoi, M.; Paixao, M. W.; Braga, A. L. Chiral Organoselenium Transition-metal Catalysts in Asymmetric Transformations. Dalton Trans., **2011**, 40, 11347-11355. (b) Braga, A. L.; Ludtke, D. S.; Vargas, F.; Braga, R. C. Catalytic Applications of Chiral Organoselenium Compounds in Asymmetric Synthesis. Synlett. **2006**, 1453-1466.

[4] For reviews see (a) Nogueira, C. W.; Zeni, G.; Rocha, J. B. T. Organoselenium and Organotellurium Compounds: Toxicology and Pharmacology. Chem. Rev. **2004**, 104, 6255-6285. (b) Mugesh, G.; du Mont, W-W.; Sies, H. Chemistry of Biologically Important Synthetic Organoselenium Compounds. Chem. Rev. **2001**, 101, 2125-2179. (c) Soriano-Garcia, M. Organoselenium Compounds as Potential Therapeutic and Chemopreventive Agents: a Review. Curr. Med Chem. **2004**, 11, 1657-1669. For recent examples see also: (a) Straightforward Synthesis of Non-Natural Selenium Containing Amino Acid Derivatives and Peptides. Braga, A.; Lüdtke, D. S. Paixao, M. W.; Alberto, E. E.; Stefani, H. A; Juliano, L. Eur. J. Org. Chem. **2005**, 4260-4264. (b) Garud, D., R.; Ando, H.; Kawai, Y.; Ishihara, H.; Koketsu, M. Synthesis of Novel Selenapenams, Selenacephems, and Selenazepines Using a 2-(Trimethylsilyl)ethyl Protection Approach Org Lett. **2007**, 9, 4455-4458. (c) Jeong, L. S.; Tosh, D. K.; Kim, H. O.; Wang, T.; Hou, X.; Yun, H. S.; Kwon, Y.; Lee, S. K.; Choi, J. ; Zhao L. X. First Synthesis of 4'- Selenonucleosides Showing Unusual Southern Conformation. Org. Lett. **2008**, 10, 209- 212. (d) Garud, D., R.; Koketsu, M. Synthesis of 3-Selena-1-Dethiacephems and Selenazepines *via* Iodocyclization. Org. Lett. **2008**, 10, 3319-3322. (e) Garud, D., R.; Garud, D. D.; Koketsu, M. Synthesis of Selenium-Containing Bicyclic β-lactams *via* Alkene Metathesis Org. Biomol. Chem., **2009**, 7, 2591-2598. (f) Maza, S. Lopez, O; Martos, S. Maya, I. Fernández-Bolaños J. G. Synthesis of the First Selenium-Containing Acyclic Nucleosides and Anomeric Spironucleosides from Carbohydrate Precursors. Eur. J. Org. Chem. **2009**, 5239-5246. (g) Schwab, R. S.; Schneider, P. H. Straightforward Synthesis of Non-Natural Chalcogen Peptides *via* Ring Opening of Aziridines. Tetrahedron, **2012**, 68, 10449-10455. (h) Affeldt, R.; Braga, H. C.; Baldassari, L. L.; Lüdtke, D. S. Synthesis of Selenium-Linked Neoglycoconjugates and Pseudodisaccharides. Tetrahedron **2012**, 68, 10470-10475.

[5] Carreira, E. M.; Kvaerno, L. Classics in Stereoselective Synthesis. Wiley-VCH, Weinheim, **2009**.

[6] For some examples see (a) Ericsson, C.; Engman, L. Diastereocontrol by Trialkylaluminiums in the Synthesis of Tetrahydrofurans *via* Radical Cyclization. Org. Lett. **2001**, 3, 3459-3462. (b) Gupta, V.; Engman, L. Tetrahydrofuran Derivatives from Epoxides *via* Group Transfer Cyclization or Reductive Radical Cyclization of

Organotellurium and Organoselenium Intermediates. J. Org. Chem. **1997**, 62, 157. (c) Tiecco, M.; Testaferri, L.; Bagnoli, L.; Purgatorio, V.; Temperini, A.; Marini, F.; Santi, C. Synthesis of Enantiomerically Pure Substituted Tetrahydrofurans from Epoxides and Phenylselenium Reagents. Tetrahedron: Asymmetry **2004**, 15, 405-412. (d) Tiecco, M.; Testaferri, L.; Temperini, A.; Bagnoli, L.; Marini, F.; Santi, C. Ring-Closure Reactions through Intramolecular Displacement of the Phenylselenonyl Group by Nitrogen Nucleophiles: a New Stereospecific Synthesis of N-Tosyl and N-Benzoyl-1,3-oxazolidin-2-ones from β-Hydroxyalkyl Phenyl Selenides. Chem. Eur. J. **2004**, 10, 1752-1764.

[7] (a) Yang, M.; Zhu, C.; Yuan, F.; Huang, Y.; Pan, Y. Enantioselective Ring-Opening Reaction of meso-Epoxides with ArSeH Catalyzed by Heterometallic Ti-Ga-Salen System. Org. Lett. **2005**, 7, 1927-1930. (b) Sun, J.; Yang, M.; Yuan, F. ; Jia, X.; Yang, X.; Pan, Y.; Zhu, C. Catalytic Asymmetric Ring-Opening Reaction of meso-Epoxides with Aryl Selenols and Thiols Catalyzed by a Heterobimetallic Gallium-Titanium-Salen Complex. Adv. Synth. Catal. **2009**, 351, 920–930.

[8] (a) Nishibayashi, Y.; Singh, J. D.; Fukuzawa, S. I.; Uemura, S. Synthesis of Chiral Diferrocenyl Dichalcogenides and Their Application to Asymmetric Nucleophilic Ring Opening of meso-Epoxides. J. Chem. Soc., Perkin Trans. 1 **1995**, 2871-2876. (b) Tomoda, S.; Iwaoka, M, Synthesis of Selenium-containing Binaphthyls and Their Applications to the Asymmetric Ring-opening of Cyclohexene Oxide. J. Chem. Soc., Chem. Commun. **1998**, 1283-1284.

[9] Tiecco, M.; Testaferri, L.; Marini, F.; Sternativo, S.; Del Verme, F.; Santi, C.; Bagnoli, L.; Temperini, A. Enantiomerically Enriched β-Hydroxy selenides by Catalytic Asymmetric Ring Opening of meso-Epoxides with (Phenylseleno)silanes. Tetrahedron **2008**, 64, 3337-3342.

[10] Tschop, A.; Nandakumar, M. V.; Pavlyuk, O.; Schneider C. Scandium–Bipyridine-Catalyzed, Enantioselective Selenol Addition to Aromatic meso-Epoxides. Tetrahedron Lett. **2008**, 49, 1030–1033.

[11] For recent examples see: (a) Davoren, E.; Harcken, C.; Martin, S. F. Studies toward the Enantioselective Syntheses of Oxylipins: Total Synthesis and Structure Revision of Solandelactone E. J. Org. Chem. **2008**, 73, 391–402. (b) Zanoni, G.; Castronovo, F.; Perani, E.; Vidari, G. A →J Prostaglandin Swap: A New Tactic for Cyclopentenone Prostaglandin Synthesis. J. Org. Chem. **2003**, 68, 6803-6805. (c) Bradshaw, B.; Extebarria-Jardì, G.; Bonjoch, J. Total Synthesis of (-)-Anominine. J. Am. Chem. Soc. **2010**, 132, 5966-5967.

[12] (a) Reich, H. J.; Yelm, K. E.; Wollowitz, S. Kinetics, Thermodynamics, and Stereochemistry of the Allyl Sulfoxide-Sulfenate and Selenoxide-Selenenate [2,3] sigmatropic rearrangements. J. Am. Chem.Soc. **1983**, 105, 2503-2504. See also: (b) Bayse, C. A.; Antony, S. Theoretical Studies of [2,3]-Sigmatropic Rearrangements of Allylic Selenoxides and Selenimides. Molecules **2009**, 14, 3229-3236.

[13] For selected examples see: (a) Bourland, T. C.; Carter, R. G.; Yokochi, A. F. T. Vanadium-Catalyzed Selenide Oxidation with *in Situ* [2,3]Sigmatropic Rearrangement (SOS Reaction): Scope and Asymmetric Applications. Org Biomol Chem. **2004**, 2, 1315-1329. (b) Nishibayashi, Y.; Singh, J. D.; Fukuzawa, S.; Uemura, S. Synthesis of [R,S;R,S]- and [S,R;S,R]-Bis[2-[1-(dimethylamino)ethyl]ferrocenyl] Diselenides and Their Application to Asymmetric Selenoxide Elimination and [2,3] Sigmatropic Rearrangement. J. Org. Chem. **1995**, 60, 4114-4120. (c) Davis, F.A.; Reddy, R. T. Asymmetric Oxidation of Simple Selenides to Selenoxides in High Selenenate Rearrangement Enantiopurity. Stereochemical

Aspects of the Allyl Selenoxide/Allyl Selenenate Rearrangement. J. Org. Chem. **1992**, 57, 2599-2605. (d) Komatsu, N.; Nishibayashi, Y.; Uemura S. Asymmetric [2,3] Sigmatropic Rearrangement *via* Chiral Selenoxide with Sharpless Oxidants. Tetrahedron Lett. **1993**, 34, 2339-2342. (e) Reich, H. J.; Yelm, K. E. Asymmetric Induction in the Oxidation of [2.2]Paracyclophane-Substituted Selenides. Application of Chirality Transfer in the Selenoxide [2,3] Sigmatropic Rearrangement. J. Org. Chem. **1991**, 56, 5672-5679.

[14] For selected examples see: (a) Shea, R. G.; Fitzner, J. N.; Fankhauser, J. E.; Spaltenstein, A.; Carpino, P. A.; Peevey, R. M.; Pratt, D. V.; Tenge, B. J.; Hopkins P. B. Allylic Selenides in Organic Synthesis: New methods for the Synthesis of Allylic Amines. J. Org. Chem. **1986**, 51, 5243–5252. (b) Nishibayashi, Y.; Chiba, T.; Ohe K.; Uemura, S. High Chirality Transfer in Chiral Selenimides *via* [2,3]Sigmatropic Rearrangement. J. Chem. Soc., Chem. Commun., **1995**, 1243-1244.

[15] (a) Takada, H.; Oda, M.; Miyake, Y.; Ohe, K.; Uemura, S. Catalytic Asymmetric Imidation of Selenides into Selenimides. J. Chem. Soc., Chem. Commun, **1998**, 1557-1558. (b) Miyake, Y.; Oda, M.; Oyamada, A.; Takada, H.; Ohe, K.; Uemura, S. Asymmetric Imidation of Organic Selenides into Selenimides. J. Organomet. Chem. **2000**, 611, 475-487.

[16] Nishibayashi, Y.; Ohe, K.; Uemura S. The First Example of Enantioselective Carbenoid Addition to Organochalcogen Atoms: Application to [2,3]Sigmatropic Rearrangement of Allylic Chalcogen Ylides. J. Chem. Soc., Chem. Commun. **1995**, 1245-1246.

[17] J.Waetzig, S. R.; Tunge, J. A. Synthesis of Allyl Selenides by Palladium-Catalyzed Decarboxylative Coupling. Chem. Commun., **2008**, 3311–3313.

[18] [(a) MacMillan, D. W. C. The advent and Development of Organocatalysis. Nature **2008**, 455, 304-308. (b) Pellissier, H. Asymmetric Organocatalysis. Tetrahedron **2007**, 63. 9267–9331. (c) Yeboah, E. M.O.; Yeboah, S. O.; Singh G. S. Recent Applications of Cinchona Alkaloids and their Derivatives as Catalysts in Metal-free Asymmetric Synthesis. Tetrahedron **2011**, 671725-1762. (d) Alemán, J.; Parra, A.; Jiang, H.; Jørgensen, K. A. Squaramides: Bridging from Molecular Recognition to Bifunctional Organocatalysis. Chem. Eur. J. **2011**, 17, 6890-6899. (e) Mielgo, A.; Palomo C. α,α-Diarylprolinol Ethers: New Tools for Functionalization of Carbonyl Compounds Chem. Asian J. **2008**, 3, 922-948.

[19] Pluim, H.; Wynberg, H. Alkaloid Catalysed Asymmetric synthesis. The Addition of Selenophenols to 2-Cyclohexen-1-ones and Conversion to Optically Active allylic Alcohols. Tetrahedron Lett., **1979**, 20, 1251-1254.

[20] Marini, F.; Sternativo, S.; Del Verme, F.; Testaferri, L.; Tiecco M. Enantioselective Organocatalytic Michael Addition of α-Substituted Cyanoacetates to α,β-Unsaturated Selenones. Adv. Synth. Catal. **2009**, 351, 103-106.

[21] Zhang, T; Cheng L.; Hameed, S.; Liu, L.; Wang, D.; Chen, Y.-J. Highly Enantioselective Michael Addition of 2-Oxindoles to Vinyl Selenone in RTILs Catalyzed by a Cinchona Alkaloid-Based Thiourea. Chem. Commun. **2011**, 47, 6644-46.

[22] Marini, F.; Sternativo, S.; Del Verme, F.; Testaferri, L.; Tiecco M. A New Stereoselective Synthesis of Cyclopropanes Containing Quaternary Stereocentres *via* Organocatalytic Michael Addition to Vinyl Selenones. Adv. Synth. Catal. **2009**, 351, 1801-1806.

[23] Sternativo, S.; Calandriello, A.; Costantino, F.; Testaferri, L.; Tiecco, M.; Marini F. A Highly Enantioselective One-pot Synthesis of Spirolactones by an Organocatalyzed Michael Addition/Cyclization Sequence. Angew. Chem. Int. Ed. **2011**, 50, 9382-9385.

[24] Miller, L. C.; Sarpong, R. Divergent Reactions on Racemic Mixtures. Chem.Soc. Rev. **2011**, 40, 4550-4562.

[25] Sternativo, S.; Walczak, O.; Battistelli, B.; Testaferri, L.; Marini F. Organocatalytic Michael Addition of Indanone Carboxylates to Vinyl Selenone for the Asymmetric Synthesis of Polycyclic Pyrrolidines. Tetrahedron **2012**, 68, 10536-10541.

[26] Marcos, V.; Aleman, J.; Garcıa Ruano, J. L.; Marini F.; Tiecco, M. Asymmetric Synthesis of α-Alkyl α-Seleno Carbonyl Compounds Catalyzed by Bifunctional Organocatalysts. Org. Lett., **2011**, 13, 3052–3055.

[27] Wang, W.; Wang, J.; Li, H. A Simple and Efficient L-Prolinamide-Catalyzed α-Selenenylation Reaction of Aldehydes. Org Lett. **2004**, 6, 2817-20. See also Wang. J.; Li, H.; Mei, Y.; Lou, B.; Xu, D.; Xie, D.; Guo, H.; Wang, W. Direct, Facile Aldehyde and Ketone α-Selenenylation Reactions Promoted by L-Prolinamide and Pyrrolidine Sulfonamide Organocatalysts. J. Org. Chem. **2005**, 70, 5678-5687.

[28] Tiecco, M.; Carlone, A.; Sternativo, S.; Marini, F.; Bartoli, G.; Melchiorre, P. Organocatalytic Asymmetric α-Selenenylation of Aldehydes. Angew. Chem., Int. Ed. **2007**, 46, 6882-6885.

[29] Sundén, H.; Rios, R.; Córdova A. Organocatalytic Highly Enantioselective α-Selenenylation of Aldehydes. Tetrahedron Lett. **2007**, 48, 7865-7869.

[30] Giacalone, F.; Gruttadauria, M.; Agrigento, P.; Campisciano, V.; Noto, R. Polystyrene-Supported Organocatalysts for α-Selenenylation and Michael Reactions. A Common Post-modification Approach for Catalytic Differentiation. Catal. Commun. **2011**, 16, 75-80.

[31] Petersen, K. S.; Posner, G. H. Asymmetric Organocatalytic, Three-Step Synthesis of γ-Hydroxy-(E)-α,β-Unsaturated Sulfones and Esters. Org Lett. **2008**, 10, 4685-4687.

[32] Hess, L. C.; Posner, G. H. Asymmetric Organocatalytic, Three-Step Synthesis of α-Hydroxy-(E)-β,γ-Unsaturated Esters. Org. Lett. **2010**, 12, 2120-2122.

[33] Armstrong, A.; Emmerson, D. P. G. Enantioselective Synthesis of α-Alkyl, α-Vinyl Amino Acids *via* [2,3]-Sigmatropic Rearrangement of Selenimides. Org. Lett. **2011**, 13, 1040-1043.

[34] Genna, D. T.; Hencken, C. P.; Siegler, M. A.; Posner, G. H. α-Chloro, β,γ-Ethylenic Esters: Enantiocontrolled Synthesis and Substitutions. Org. Lett. **2010**, 12, 4694-4697.

[35] For biological properties of chiral 1,2-selenoamines see: (a) de Souza Prestes, A.; Terra Stefanello, S.; Salman, S. M.; Martini Pazini, A.; Schwab, R. S.; Braga, A. L.; de Vargas Barbosa, N. B.; Rocha, J. B. T. Antioxidant Activity of β-Selenoamines and Their Capacity to Mimic Different Enzymes. Mol. Cell. Biochem. **2012**, 365, 85-92. (b) Woznichak, M. M.; Overcast, J. D.; Robertson, K.; Neumann, H. M.; May, S. W. Reaction of Phenylaminoethyl Selenides with Peroxynitrite and Hydrogen Peroxide. Arch. Biochem. Biophys., **2000**, 379, 314-320; (c) May, S.W.; Wang, L.; Gill-Woznichak, M. M.; Browner, R. F.; Ogonowski, A. A.; Smith J. B.; Pollock, S. H. An Orally Active Selenium-Based Antihypertensive Agent with Restricted CNS Permeability. J. Pharm. Exp. Ther., **1997**, 283, 470-477.

[36] Senatore, M.; Lattanzi, A.; Santoro, S;. Santi, C. Della Sala, G. A General Phosphoric Acid-Catalyzed Desymmetrization of meso-Aziridines With Silylated Selenium Nucleophiles. Org. Biomol. Chem., **2011**, 9, 6205-6207.

[37] Della Sala, G. Studies on the true catalyst in the phosphate-promoted desymmetrization of meso-aziridines with silylated nucleophiles. Tetrahedron, **2013**, 63, 50-56.

[38] Denmark, S. E.; Kalyani, D.; Collins, W. R. Preparative and Mechanistic Studies Toward the Rational Development of Catalytic, Enantioselective Selenoetherification Reactions. J. Am. Chem. Soc. **2010**, 132, 15752-15765. For an asymmetric catalytic sulfenylation of alkenes using a chiral selenophosphoramide catalyst see: Denmark, S. E.; Kornfilt, D. J. P. Vogler, T. Catalytic Asymmetric Thiofuctionalization of Unactivated Alkenes. J. Am. Chem. Soc. **2011**, 133, 15308-15311.

[39] Guan H.; Wang, H.; Huang, D; Shi, Y. Enantioselective oxysulfenylation and oxyselenenylation of olefins catalyzed by chiral Brønsted acids. Tetrahedron **2012**, 68, 2728-2735.

[40] Takada, H.; Metzner, P.; Philouze, C. First Chiral Selenium Ylides Used for Asymmetric Conversion of Aldehydes into Epoxides. Chem. Commun. **2001**, 2350-2351.

[41] Zanardi, J.; Leriverend, C.; Aubert, D.; Julienne, K.; Metzner, P. A Catalytic Cycle for the Asymmetric Synthesis of Epoxide Using Sulfur Ylides. J. Org. Chem. **2001**, 66, 5620-5623.

[42] For a review see: Comasseto, J. V.; Gariani, R. A. Biotransformations on Organic Selenides and Tellurides: Synthetic Applications. Tetrahedron **2009**, 65, 8447-8459.

[43] Costa, C. E.; Clososki, G. C.; Barchesi, H. B.; Zanotto, S. P.; Nascimento, M. G.; Comasseto J. V. Enzymatic Resolution of (RS)-β-Hydroxy Selenides in Organic Media. Tetrahedron: Asymmetry **2004**, 15, 3945.

[44] Gruttadauria, M.; Lo Meo, P.; Riela, S.; D'Anna, F.; Noto, R. Lypase-catalyzed Resolution of β-Hydroxy Selenides. Tetrahedron: Asymmetry **2006**, 17, 2713.

[45] Andrade, L. H.; Silva, A. V. First Chemoenzymatic Synthesis of Organoselenium Amines and Amides. Tetrahedron: Asymmetry, **2008**, 19, 1175-1181.

[46] Andrade, L. H.; Silva, A. V.; Pedrozo, E. C. First Dynamic Kinetic Resolution of Selenium-containing Chiral Amines Catalyzed by Palladium ($Pd/BaSO_4$) and Candida Antarctica Lipasi (CALB). Tetrahedron Lett. **2009**, 50, 4331-4334.

[47] Andrade, L. H.; Silva, A. V; Milani, P; Koszelewski, D; Kroutil, W. ω-Transaminases as Efficient Biocatalysts to Obtain Novel Chiral Selenium-amine Ligands for Pd-Catalysis Org. Biomol. Chem. **2010**, 8, 2043-2051.

[48] Omori, A. T.; Assis, L. F.; Andrade, L. H.; Comasseto J. V.; Porto A. L. M. Enantiomerically Pure Organoseleno-1-Arylethanols by Enzymatic Resolution with Candida Antarctica Lipase: Novozym 435. Tetrahedron: Asymmetry, **2007**, 18, 1048–1053.

[49] Andrade, L. H.; Omori, A. T.; Porto, A. L. M.; Comasseto, J. V. Asymmetric Synthesis of Arylselenoalcohols by Means of the Reduction of Organoseleno Acetophenones by Whole Fungal Cells J. Mol. Catal. B: Enzym. **2004**, 29, 47-54.

[50] Comasseto, J. V.; Omori, A. T.; Porto, A. L. M.; Andrade, L. H Preparation of Chiral Organochalcogeno-α-Methylbenzyl Alcohols *via* Biocatalysis. The role of Daucus Carota Root Tetrahedron Lett. **2004**, 45, 473- 476.

Organoselenium Chemistry Between Synthesis and Biochemistry, 2014, 197-267

CHAPTER 8

Eco-Friendly Access and Application of Organoselenium Reagents: Advances Toward Green Chemistry

Antonio L. Braga[1,*], Ricardo S. Schwab[2,*] and Oscar E.D. Rodrigues[3]

[1]Federal University of Santa Catarina, Florianópolis, SC, Brazil; [2]Federal University of São Carlos, São Carlos, SP, Brazil and [3]Federal University of Santa Maria, Santa Maria, RS, Brazil

Abstract: In recent years, an increasingly broad array of applications has been found for new eco-friendly protocols related to the synthesis of organoselenium compounds. One of the most important challenges in this field is the development of new procedures that involve the use of alternative and/or recyclable solvents or solvent-free conditions. In this context, ionic liquids, water, glycerol, ethanol and polyethylene glycol have been used as 'green' solvents. The advantages of using these solvents will be discussed, mainly with regard to their reusable and easily separable systems. Solvent-free methodologies will also be described in detail. Additionally, catalytic amounts of selenium-containing organic molecules can be applied to effect several chemical transformations, leading to their use as organocatalysts within a "greener perspective" and this probably represents the most important recent advance in this field. Other important selective contributions in this field will also be addressed.

Keywords: Green chemistry, Selenium, Ionic liquids, Glycerol, Solvent Free, Microwave, Hydrogen Peroxide, Oxidation, One-Pot Reaction, Atom Economy.

8.1. INTRODUCTION

Organoselenium chemistry began to gain considerable attention in the early 1970s after the discovery of selenoxide elimination, although the element selenium was discovered back in 1818 by Berzelius [1]. New insights are continually being gained into the biological and medicinal roles of selenium and organoselenium compounds and various selenoproteins which are involved in a wide number of physiological processes in mammals, including antioxidant defense, thyroid

***Address correspondence to Antonio L. Braga:** Federal University of Santa Catarina, Florianópolis, SC, Brazil; Tel: +55 48 3721-6427; E-mail: braga.antonio@ufsc.br
Ricardo S. Schwab: Federal University of São Carlos, São Carlos, SP, Brazil; Tel: +55 16 3351-8081; Email: rschwab@ufscar.br

Claudio Santi (Ed)

hormone production and immune responses, have been identified [2]. The synthesis and reactivity of organoselenium compounds, and related areas, thus represents an important field in organic chemistry. In particular, the development of stable organoselenium compounds has gained increasing attention, along with their application in asymmetric catalysis [3], green chemistry [4] and biological studies [5].

In recent decades, the several global signs regarding climate change have led to criticism regarding the sustainability of the planet and the role of human activity in this scenario. As a consequence, it is widely agreed that an array of actions and efforts in all spheres has to be implemented to try to revert the critical environmental situation of the planet. Interestingly, this concern for the care of the earth is not new. In 1972, the Stockholm Conference [6], held by the United Nations, discussed this issue and provided some directions to diminish anthropogenic impact and to protect and improve the human environment while preserving the quality of the planet. In this regard, Principle 6 of this conference describes "the discharge of toxic substances or of other substances and the release of heat, in such quantities or concentrations as to exceed the capacity of the environment to render them harmless, must be halted in order to ensure that serious or irreversible damage is not inflicted upon ecosystems. The just struggle of the peoples of ill countries against pollution should be supported". This concept highlights the strong connection with industry and, in some respects, with research in chemistry, in terms of the search for more environmentally-friendly protocols. In the 1990s, Paul Anastas moved forward the concept of Green Chemistry [7] and, in collaboration with John Warner, he developed the 12 Principles of Green Chemistry: ***i.*** Prevention; ***ii.*** Atom Economy; ***iii.*** Less Hazardous Chemical Syntheses; ***iv.*** Designing Safer Chemicals; **v.** Safer Solvents and Auxiliaries; ***vi.*** Design for Energy Efficiency; ***vii.*** Use of Renewable Feedstocks; ***viii.*** Reduce Derivatives; ***ix.*** Catalysis; ***x.*** Design for Degradation; ***xi.*** Real-time analysis for Pollution Prevention; ***xii.*** Inherently Safer Chemistry for Accident Prevention. With these concepts launched, several efforts were made in the search for more benign protocols to address these principles. The chapter presented herein covers the scientific literature (excluding the patent literature) which addresses green approaches to the synthesis of organoselenium compounds as well as their synthetic applications involving green processes. There is considerable information available on the chemistry of selenium compounds and this chapter does not address areas which are not directly related to the development of eco-friendly processes. Regrettable, this means that several important and interesting studies on selenium compounds are not cited herein.

8.2. GREEN APPROACHES TO THE SYNTHESIS OF ORGANO-SELENIUM COMPOUNDS

In terms of reaction processes, solvents are generally in the top of the list of damaging chemicals because they are often used in huge amounts and are generally volatile liquids that are difficult to contain [8]. In green approaches to the synthesis of organoselenium compounds different classes of these compounds have been described in the literature. Also, the development of more environmentally-friendly protocols has gained considerable attention, particularly procedures involving the use of alternative and/or recyclable solvents or solvent-free conditions.

8.2.1. Ionic Liquids as Recyclable Solvents

The use of ionic liquids as versatile reaction media for organic transformations has resulted in many diverse and flexible tools establishing highly effective, reusable and easily separable systems. Room temperature ionic liquids are noted for a number of unique properties which have been shown to have a large number of applications [9]. Due to the special design of these molecules, they can act as a solvent only or they can be a ligand and solvent simultaneously. Thus, the development of ionic liquid-mediated organic reactions is gaining prominence and in the area of organoselenium chemistry a variety of transformations employing ionic liquids as solvents has been reported in the literature.

A series of selenoesters **3** has been prepared from acid chlorides **1** and diaryl diselenides **2** employing a variety of ionic liquid (IL) as recyclable solvents (Scheme **8.1**). For instance, the use of metallic indium allowed the preparation of the desired compounds on gentle heating to 50 ^{o}C in good yields [10]. In terms of electronic effects, the authors observed that the reactions were not influenced by electron-donating or electron-withdrawing groups of the acid chlorides, but the insertion of substituents in the diselenide moiety decreased the reaction yield. The recyclability of ILs was studied and after four runs the selenoester was obtained with shortly variation in the reaction. Additionally, selenoesters were also obtained using CuO nanopowder and ionic liquids, with yields ranging from 57 to 91 % [11]. Bases, such as KOH and Cs_2CO_3, showed to be efficient on the synthesis of selenoesters and [bmim]PF_6 afforded higher yields compared with [bmim]BF_4 and [bmim]NTf_2. In terms of catalytic activity, it has been observed that 5 mol% was the best amount and the reaction shows sensitivity, in terms of electronic effects, for the acid chloride and diaryl diselenide. The recyclability was studied and the ionic liquid efficiency was not altered after four successive reaction cycles.

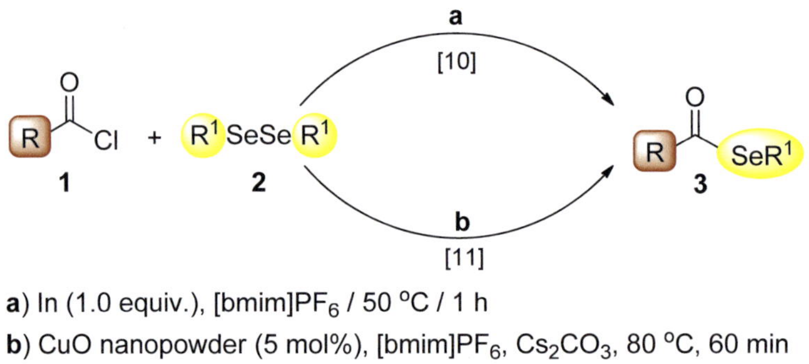

Scheme 8.1: Preparation of selenoesters.

A series of diaryl selenides **7** has been prepared in metal-free and catalyst-free synthesis, employing electrophilic selenium reagents **4** and aryl boronic acids **5** or potassium aryltrifluoroborate **6** at room temperature [12]. A small variety of ionic liquids were tested and [bmim]PF_6 was more efficient than [bmim]BF_4, [bmim]NTf_2, or [bmmim]BF_4. In general, the sensitivity of the reactions to the electronic effects of the aryl boronic moiety was low, affording slightly higher yields for those bearing electron-donating groups compared with electron-withdrawing or electron-neutral groups. The reactions were also performed under MW irradiation, furnishing the selected diaryl selenides in similar yields to those associated with conventional conditions but in shorter reaction times. The recoverability and reuse of the [bmim]PF_6 was evaluated and after 5 successive runs the respective ILs showed the same level of efficiency (Scheme **8.2**).

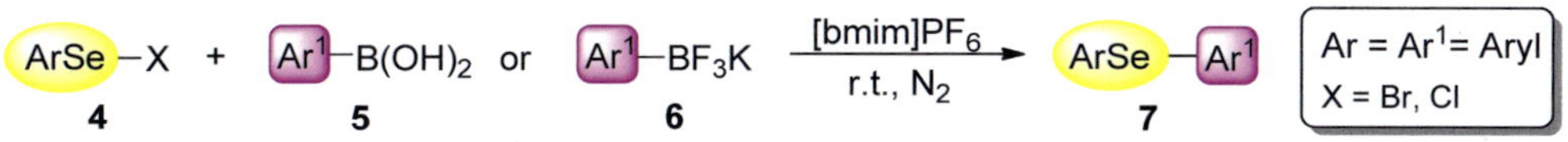

Scheme 8.2: Synthesis of diaryl selenides.

Protocols for the preparation of diaryl and/or dialkyl selenides using nucleophilic selenolate species are also described in the literature. For instance, the use of Zn dust and catalytic ZnO nanoparticles allowed the preparation of a variety of chiral selenoamines **10** from *N*-protect aminomesylates **8** using [bmim]BF_4 as the solvent and diaryl diselenides **9** as the selenium source as depicted in Scheme **8.3** [13]. The possibility for the reuse of the ionic liquid were investigated and was possible verify the activity of the IL after 4 successive runs.

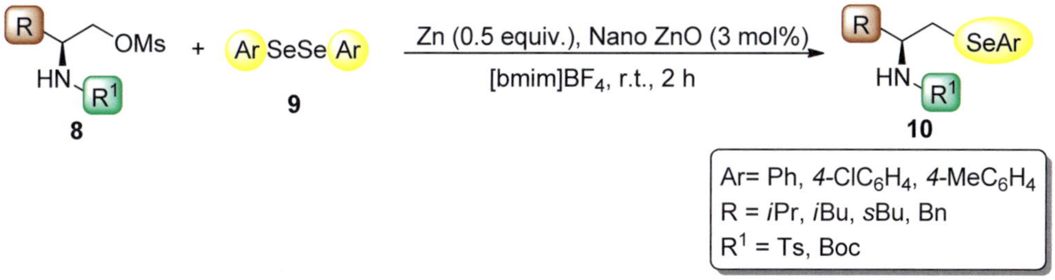

Scheme 8.3: Preparation of chiral β-selenoamines.

Zn dust and InI have also been employed for the synthesis of diorganyl selenides **12** from diselenides **2** and alkyl halides **11** in different ionic liquids (Scheme **8.4**). Firstly, using [bmim]BF_4 as the solvent, Zn dust and different diselenides the corresponing products were obtained in good yields [14]. Four ionic liquids were investigated and it was observed that [bmim]BF_4 afforded better yields than [bmim]PF_6, [bmim]NTf_2 and [bmmim]BF_4. The hydrogen-bonding parameter (HBD) was used to explain the greater efficiency of [bmim]BF_4 and [bmim]PF_6 compared with other ionic liquids such as [bPy]BF_4 and [bmmim]BF_4. The ionic liquid was reused for another 5 runs without loss of effectivity. InI has also been employed for the synthesis of unsymmetrical diorganyl selenides and [bmim]BF_4 was more active than [bmim]PF_6, [bmim]NTf_2 and [bPy]BF_4 [15]. The reactions proceeded under mild conditions and in short reaction times. The procedure shows significant sensitivity to steric influence, and the *para*-substitution in the diselenide or organic halide moiety afforded better yields than the *meta* or *ortho* position. This is in agreement with the proposed SN_2 mechanism associated with this methodology.

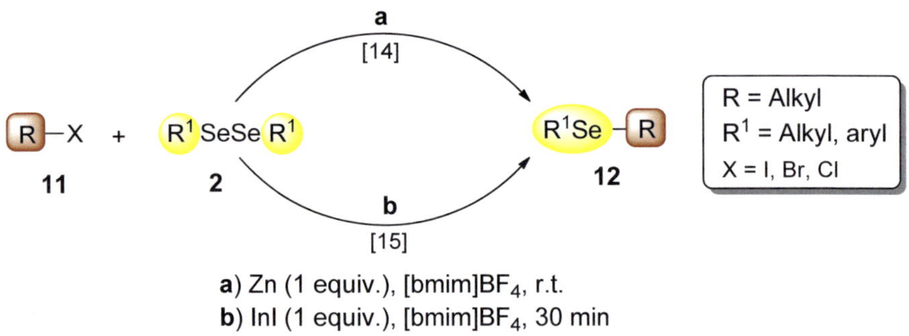

Scheme 8.4: Preparation of diorganyl selenides.

Zinc dust has also been used for the preparation of a series selenoesters **3** from a variety acid chlorides **1** and diorganoyl diselenides **2** with yields ranging from 39 to 95 %. [bmim]PF_6 was found to be a more effective ionic liquid compared with [bPy]BF_4, [bmim]BF_4, [bmim]NTf_2 and [bmmim]BF_4 [16]. In the reaction the electronic effects of the acid chloride afforded better yields for the neutral or electron-donating groups, as depicted in Scheme **8.5**.

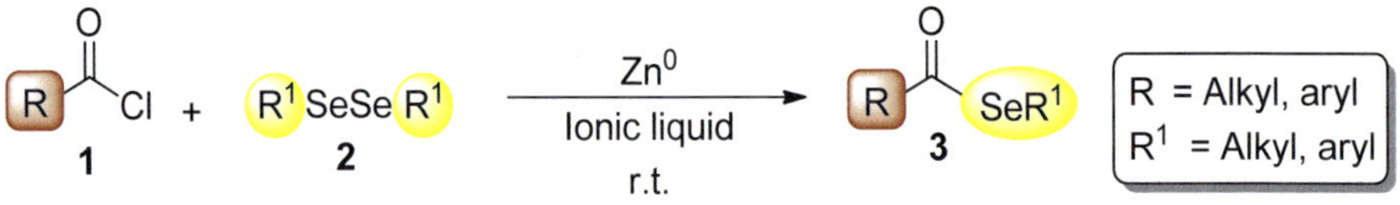

Scheme 8.5: Preparation of organic selenides and selenoesters.

The preparation of unsymmetrical selenides **15**, chiral *N*-Boc selenoamine **16**, and selenoesters **17** using a bimetallic $SnCl_2/CuBr_2$ system and diphenyl diselenide **14** in ionic liquids has been described [17]. The cationic and anionic part of the ionic liquids had a remarkable effect, where [bmim]BF_4 was more effective than [bmim]PF_6, [bmim]NTf_2, [bmmim]BF_4 or [bPy]BF_4 for the preparation of all the selenium derivatives (Scheme **8.6**).

The synthesis of diaryl or alkyl-aryl selenides **12** *via* palladium-catalyzed cross-coupling of aryl and alkyl halides **11** and $R^1SeSnBu_3$ **18** has also been performed in ionic liquids (Scheme **8.7**) [18]. It was reported that [bmim]PF_6 was the most effective solvent and $Pd(PPh_3)_4$ was the best catalyst. The reaction was performed at 80 °C and showed sensitivity in terms of substituents for the organic halides. Electron-donating groups attached to the respective halides afforded slightly less reactivity compared with electron-withdrawing groups. Interestingly, the reaction with organic halides bearing hydroxyl and amino groups allowed moderate yields, due to the strong coordination ability of these groups. The rate acceleration and increased yield compared with traditional solvents are some of the advantages of the use of an ionic liquid in this reaction. The reuse of an ionic liquid and the catalyst were evaluated and afforded the same yield after 3 runs.

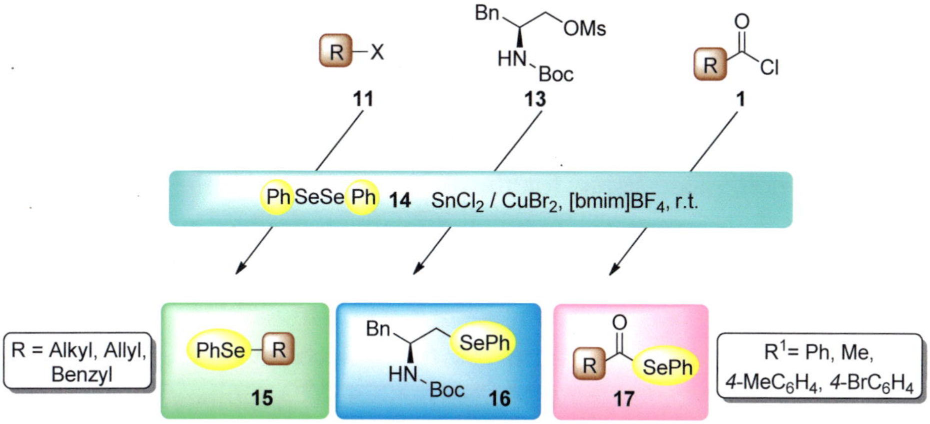

Scheme 8.6: Preparation of diorganyl selenides, chiral selenoamines and selenoesters.

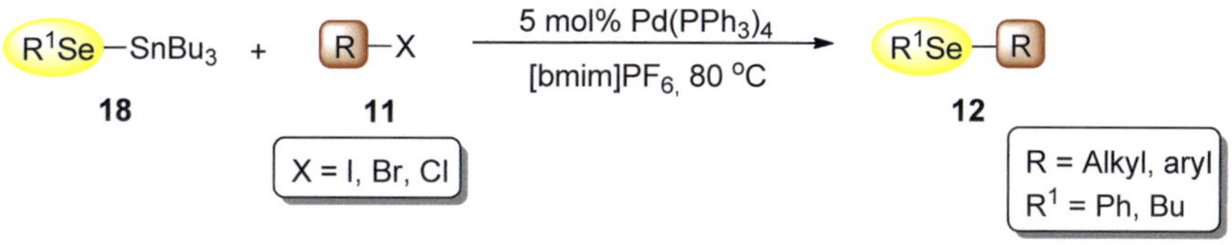

Scheme 8.7: Synthesis of diaryl or alkyl-aryl selenides *via* palladium-catalyzed cross-coupling reactions.

Recently, the metal-free cleavage of diphenyl diselenide **14** has been achieved in ionic liquids/triphenyl phosphine at 75 °C, affording unsymmetrical selenides **15** and selenoesters **17** [19]. The [pmim]Br was compared with conventional organic solvents, such as ethanol, toluene, dichloromethane and dioxane affording the desired compounds in higher yields and without side products. A variety of alkyl halides **11** and acid chlorides **1** were employed, affording the respective chalcogenides **15** and selenoesters **17** in good yields, as depicted in the Scheme **8.8**. The ionic liquid was highlighted as a fundamental compound for this reaction, probably acting in the stabilization of the intermediates.

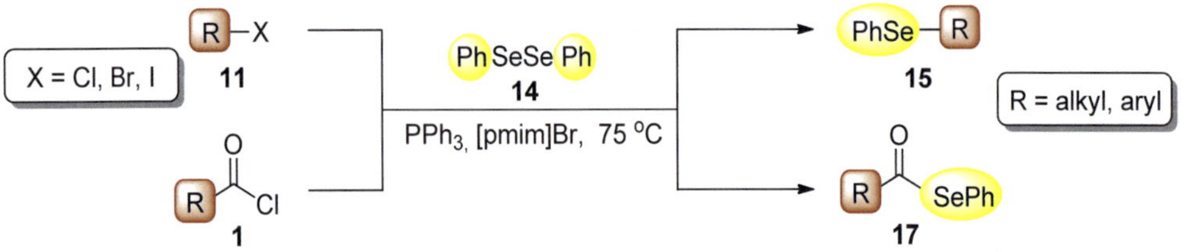

Scheme 8.8: Metal-free synthesis of diorganoyl selenides and selenoesters.

The hydroselenation of alkynes **19** has been carried out employing [bmim]BF_4 as the recyclable solvent, diphenyl diselenide **14** and $NaBH_4$ as the reducing agent [20]. The respective vinyl selenides **20-23** were obtained as a mixture of *Z* and gem, with the preference being for the *Z* compound, with the exception of the propargylic alcohol, where the anti-Markovnikov product was obtained in a higher amount than the Markovnikov product. The reaction takes place at 60 °C and only in the presence of the reducing agent. In addition, the reaction was sensitive to steric hindrance, where the regioselectivity was increased by the addition of selenolate for the most hindered alkynes. Additionally, the formation of the *(E)*-1,2-bis-phenylselenostyrene **22** was observed when the protocol was performed using phenylacetylene. Furthermore, the influence of the ionic liquid was highlighted due to the reduced reaction time of the hydroselenation in this medium, compared with conventional solvents (Scheme **8.9**).

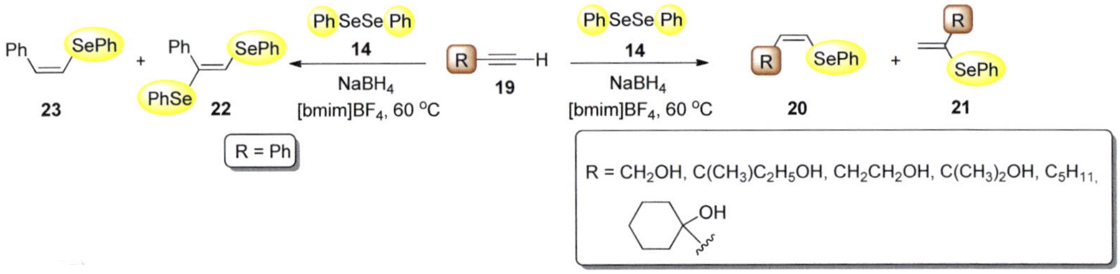

Scheme 8.9: Hydroselenation of alkynes.

Regioselective heterocycle ring opening has also been performed in the presence of ILs. For instance, a regioselective epoxide **24** ring opening was performed in ionic liquids using aryl selenol **25**, affording the respective hydroxyselenides **26** and **27** as depicted in (Scheme **8.10**, A) [21]. Three ILs were tested and [bmim]BF_4 afforded better yields than [bmim]Br and [bmim]Cl, showing the influence of the anion on this reaction. The influence of the temperature was investigated and mild heating to 50 °C was more effective compared with carrying out the reaction at room temperature. In terms of regiochemistry, in the reaction the nucleophilic attack occurred exclusively on the less hindered carbon of the epoxide, with the exception of the epoxide derived from styrene **24f**, from which a mixture of regioisomers has been obtained. The recyclability of the ILs was evaluated and similar yields were observed after 4 cycle runs. The influence of the IL can be attributed to this compound acting as a solvent and as a Lewis acid catalyst in this reaction, increasing the reactivity of the epoxide in the selenolate attack (Scheme **8.10**, B).

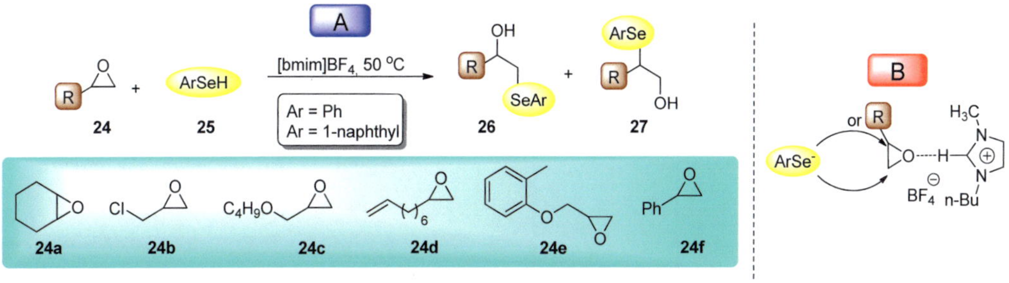

Scheme 8.10: A - Selenium epoxide ring-opening in the presence of IL. B - Mechanistic proposal for the selenium epoxide ring-opening.

Aziridine **28** ring opening has been carried out in ILs, affording the corresponding chiral selenoamines **29**. For instance, in one protocol, four different aziridines and two different protecting groups were employed in this reaction; diphenyl diselenide was used as a chalcogenium source and zinc as a reducing agent (Scheme **8.11**, **i**). The products **29** were obtained with yields ranging from 60 to 90 %, at 80 °C with a high regioselectivity. The IL was reused four times, affording the respective compound in good yields [22]. Similarly, (phenylselenenyl)zinc bromide (PhSeZnBr) in IL has been used for aziridine ring opening, affording the chiral selenoamines **29** [23]. The reaction was conducted in different organic solvents and in three ILs, observing the best efficiency for [bmim]BF_4 compared with the other solvents (Scheme **8.11**, **ii**). The reaction was temperature-sensitive, affording better yields at high temperatures, compared with room temperature. The recyclability of the [bmim]BF_4 was studied and the respective selenide was obtained with similar yields after 4 successive runs. The authors also explored the versatility of the methodology and alkylic selenides and selenoesters were similarly prepared [23]. In another protocol, chiral aziridine **28** ring opening was performed using CuO nanoparticles as the catalyst, different diselenides and KOH, furnishing the respective β-selenoamines **29** [24]. The optimized protocol involved the use of [bmim]BF_4 as the solvent at 80°C (Scheme **8.11**, **iii**). In terms of electronic effects in the diselenide moiety, the reaction was more effective for electron-withdrawing groups attached to the aromatic ring. Additionally, dialkyl diselenides were also employed affording the respective compounds in good yields. The recyclability of the IL was studied and after 4 successive runs the respective β-selenoamines **29** were obtained with similar yield.

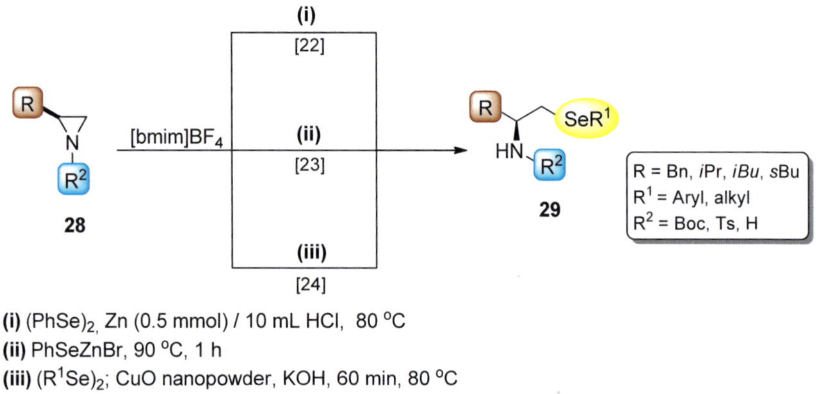

Scheme 8.11: Organoselenium reagents in the aziridine ring-opening reactions.

8.2.2. Water

Water represents one of the most benign solvents for reaction processes and thus various organic transformations mediated in this solvent are described in the literature. In organoselenium chemistry, some protocols employing this solvent have been reported, allowing the preparation of a variety of organoselenium compounds in the greenest possible fashion.

Zinc selenium salts, such as PhSeZnX, are one of the most important classes of nucleophilic selenium reagents for methodologies in water medium. The treatment of commercially available phenylselenenyl bromide PhSeCl and PhSeBr with Zn dust, for instance, allowed the preparation of the corresponding PhSeZnCl **30** and PhSeZnBr **31** selenolates [25]. These reagents were employed as the nucleophile in the epoxide **24** ring opening reactions in water affording the respective hydroxyselenides **32** and **33** in excellent yields and with regioselectivity through the attack on the less hindered carbon [25a]. For the epoxide derived from styrene, the regioselectivity was inverted, indicating that in this example the reaction probably proceeds *via* a partially stabilized carbocation. The reaction in water shows greater efficiency and was faster than that in THF (Scheme **8.12**).

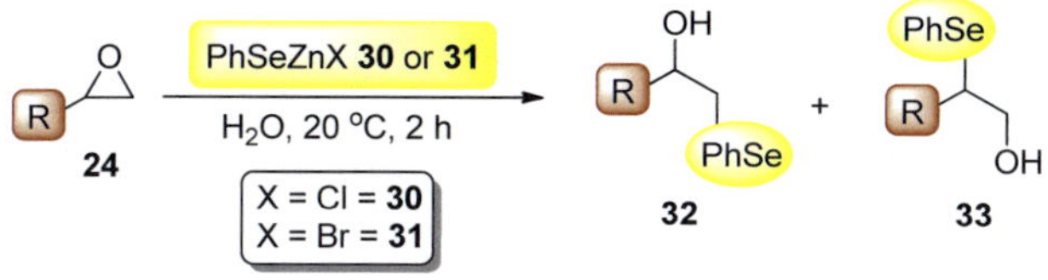

Scheme 8.12: PhSeZnX reagent in the epoxide ring-opening reactions.

In addition, the authors highlighted preliminary results for the application of the PhSeZnCl selenolate **30** in a variety of substrates in water affording a small library of selenides **34-39**, showing the versatility of this reagent, as depicted in Scheme **8.13**.

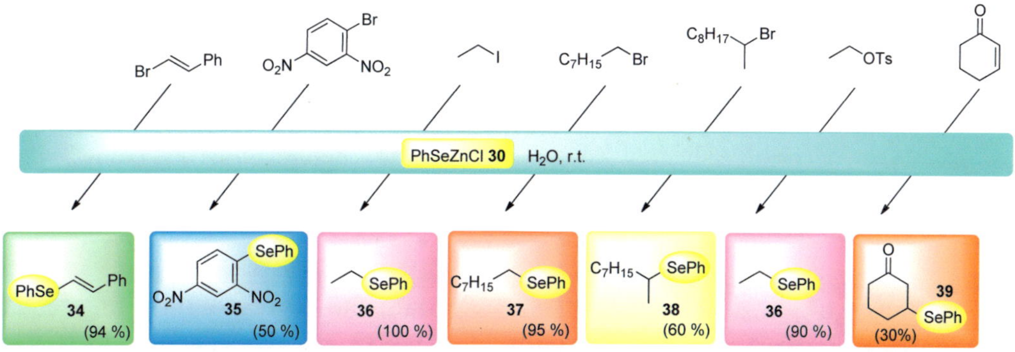

Scheme 8.13: Application of PhSeZnCl 30 in nucleophilic reactions with different substrates.

Recently, Santi and coworkers reported the Michael-addition of PhSeZnCl **30** to unsaturated ketones **40** and electron-deficient alkynes **41**, leading to synthetically useful β-seleno derivatives **42** and vinyl selenides **43**, carried out under "on water" conditions [26] (Scheme **8.14**). THF and water were employed as solvents in this reaction and it was observed that the organic solvent generally afforded better yields compared with water for the unsaturated ketones. Interesting, the yield increased when the reaction was performed in a light-protection system due the formation of unreactive PhSeSePh, probably through a photoactivated radical mechanism. The respective Michael-addition products were observed for a variety of conjugated ketones **40**, but for unsaturated esters, aldehydes, nitro derivatives, nitriles and selenones the methodology was not effective and the related substrates were completely unreactive under the described reaction conditions. In addition, PhSeZnBr **31** was not reactive for the addition in either solvent and in these cases the starting enone was recovered. In contrast with the unsaturated ketones, the PhSeZnCl **30** with electron-deficient alkynes was more efficient and the reaction faster in a water suspension than in THF, affording a near quantitative yield of the corresponding vinyl selenides **43** in 2 h at 23 °C. The reaction was highly stereoselective or stereospecific, following the behavior of benzeneselenol Michael-addition to activated alkynes leading to (*Z*)-vinylic selenides **43** as the major isomer, as depicted in Scheme **8.14**.

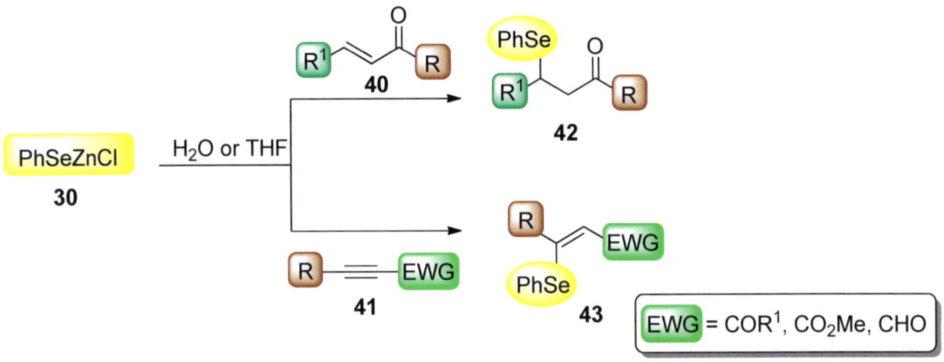

Scheme 8.14: Michael-type addition reactions promoted by PhSeZnCl.

The synthesis of vinyl selenides **45** in water and the corresponding theoretical study were performed starting from vinilic halides **44** and PhSeZnCl **30** [27]. The reaction proceeds at room temperature employing vinyl bromides and chlorides with good yields and selectivity, affording the products (with only one exception) with the geometry retention (Scheme **8.15**). In this regard, a DFT calculation was performed in order to investigate the reaction mechanism, the stereochemistry of the process and the role of the zinc atom.

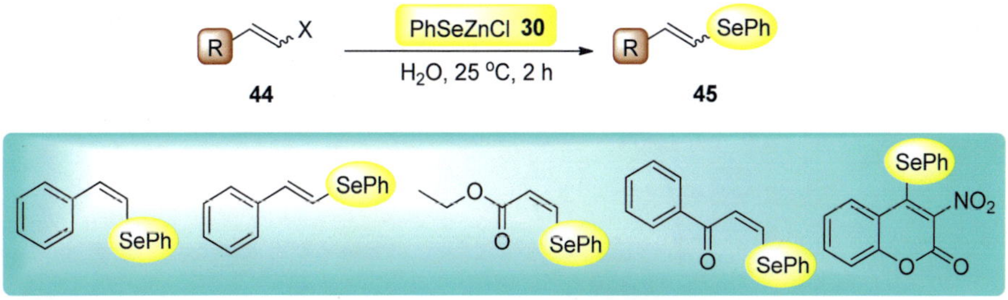

Scheme 8.15: Preparation of vinyl selenides using PhSeZnCl.

Aryl **47** -and the vinyl selenides **49** were prepared in water medium, from aryl iodides **46** or vinyl bromides **48**, diphenyl diselenide **14** and zinc dust in the presence of catalytic copper nanoparticles [28]. A variety of aryl halides with electron-donating and withdrawing-groups attached to the aromatic ring were employed. For the vinylic bromides, the reaction shows configuration retention for the isomers (*E*) and the loss of stereochemistry for the isomers (*Z*). This result was rationalized considering the steric interaction between the PhSe and the Ar group in the case of the (*Z*)-vinyl bromide. In the absence of zinc, the reaction afforded only marginal yield and without the presence of copper nanoparticles the reaction did not preceded at all. These results suggest the importance of the combination of Zn/Cu nanoparticles in this reaction, as depicted in the Scheme **8.16**.

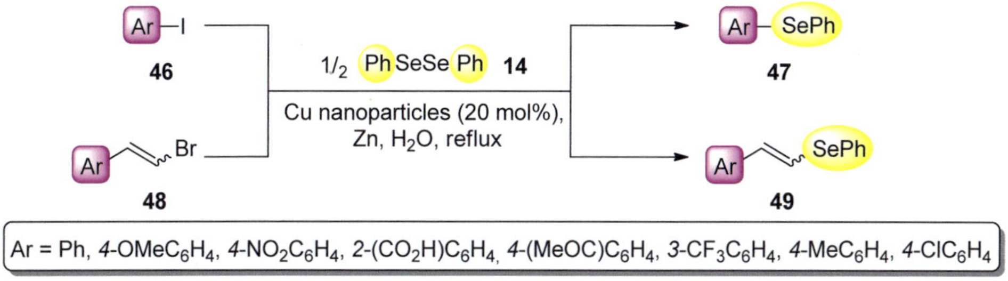

Scheme 8.16: Preparation of aryl -and vinyl selenides.

The authors suggest that the mechanism of this reaction involves the Cu(0) nanoparticles undergoing oxidative addition with diphenyl diselenide **14** to give the intermediate $(PhSe)_2Cu(II)$ **50** (Scheme **8.17**). Following reduction by Zn, this intermediate affords the intermediate PhSeCu(I) **51**, which reacts with aryl iodides **46** to give the product ArSePh **47** *via* a transient Cu(III) **52** intermediate.

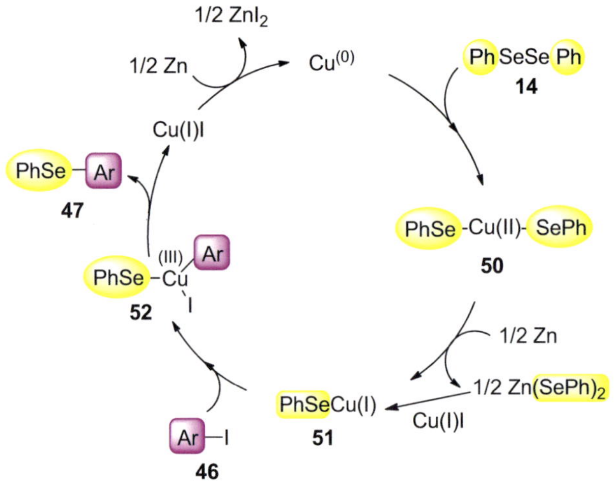

Scheme 8.17: Proposed copper mechanism in the synthesis of aryl selenides.

Symmetrical selenides **54** have been obtained from potassium selenocyanate **53** (KSeCN), base and aryl iodides **46** using water as the solvent, CuI as the catalyst and *trans* 1,2-diaminocyclohexane **55** as the ligand [29]. The reaction proceeds in a variety of substrates, affording the respective compounds in good yields (Scheme **8.18**). The presence of Cs_2CO_3 as a base and CuI was essential and the products were not observed in the absence of these components. The reaction provides slightly better yields with electron-donating groups attached to the aromatic ring, compared with electron-withdrawing groups. Halogen atom in the aryl moiety shows a significant influence: aryl bromides afford the respective selenides in moderate yields whereas aryl chlorides furnish only traces of the desired compound.

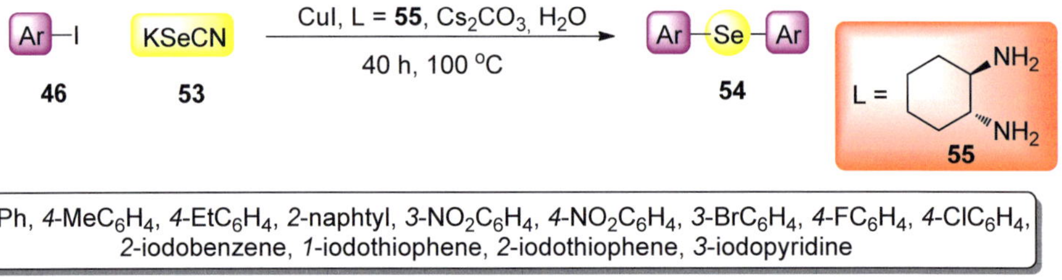

Scheme 8.18: Preparation of symmetrical substituted selenides using KSeCN.

The synthesis of selenoesters **17** from acid chlorides and PhSeZn-halides in water medium has been performed under mild reaction conditions and the reaction provided higher yields when water was used rather than THF or the ionic liquid [bmim]BF_4 (Scheme **8.19**) [30]. In terms of selenium reagent, the PhSeZnBr **31** was more effective than PhSeZnCl **30** and this was correlated with the softer Lewis acid character of the zinc atom and the lower polarization of the Zn-Se interaction. Studies on the reuse of water were carried out and the need to adjust the pH to 7 after each cycle has been reported as necessary in order to obtain good yields for up to ten cycles.

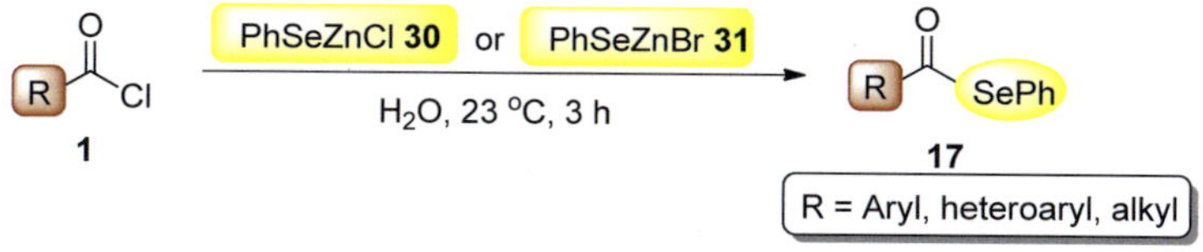

Scheme 8.19: Preparation of selenoesters using PhSeZn-halides.

The mechanism was investigated and the ability of PhSeZnBr to react faster with acyl chlorides **1** in water was evaluated. A concerted mechanism has been suggested involving simultaneously a "soft-soft" Lewis acid activation of the carbonyl group and the nucleophilic attack of the selenium atom (Scheme **8.20**) structure **56**. This affords the intermediate **57** that spontaneously evolves to the selenoester **17**.

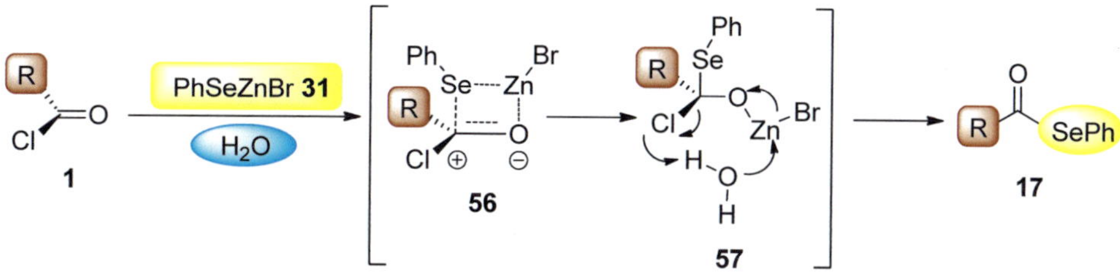

Scheme 8.20: Proposed mechanism for the synthesis of selenoesters.

β-hydroxyselenides **32** were obtained from epoxides **24** and PhSeH **58** in water using supramolecular catalysis. β-cyclodextrin (CD) was employed to promote the epoxide ring opening under mild reaction conditions and with good yields [31]. The process shows a high selectivity, derived from the attack of the selenium nucleophile on the less hindered carbon of the epoxide (Scheme **8.21**). The methodology was compatible with some functionalities, such as chloro, methyl, methoxy and methoxyethyl group. The preparation of the β-hydroxyselenides was also effective with α-cyclodextrin and the respective catalysts were recovered after the reaction work-up. The use of CD in the reaction is probably associated with the reversible formation of host-guest complexes *via* non-covalent bonding between the oxygen of the epoxide and the hydroxyl group of the cyclodextrin, and this interaction activates the oxirane driving the selenium nucleophilic attack. This complex was isolated and characterized by NMR and the non-reactivity of the epoxide without the presence of the catalyst has been demonstrated.

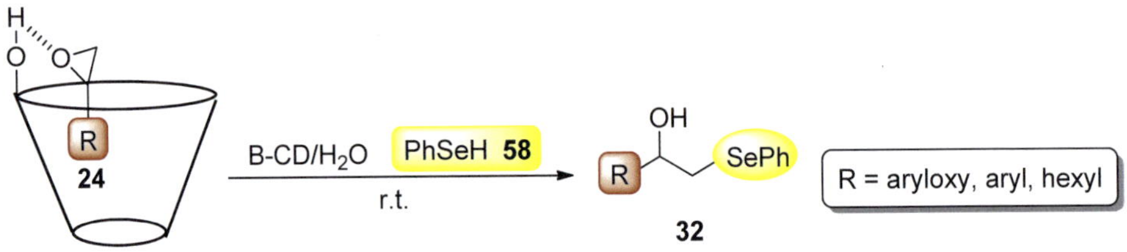

Scheme 8.21: Preparation of β-hydroxyselenides using β-cyclodextrin.

8.2.3. Glycerol, Ethanol and Polyethylene Glycol

Epoxide ring-opening reaction has been performed using selenium nucleophiles in ethanol as a green solvent, allowing the preparation of a variety of glycerol selenide thioethers. The preparation of substituted glycerol selenothioethers **60**, for instance, was carried out starting from thioepoxides **59**, diorganoyl diselenides **2**, using $NaBH_4$ as the reducing agent, with NaOH under microwave irradiation in ethanol under argon protection [32]. A variety of diselenides **2** were employed affording the respective glycerol selenide thioethers in good yields, without a notable variation in the reaction efficiency, regardless of the nature of the diselenide (alkyl or aryl substituted). The reaction was faster under MW irradiation (15 min) when compared with conventional heating (9-13 h) and the products were furnished in similar yields, as depicted in Scheme **8.22**.

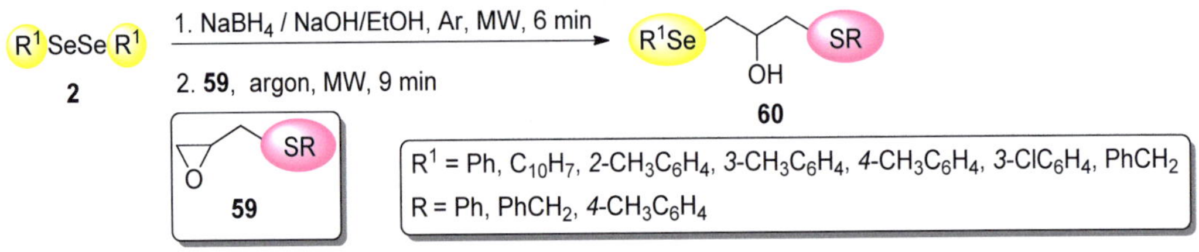

Scheme 8.22: Preparation of substituted glycerol selenide thioethers.

Employing the methodology used for the synthesis of glycerol selenide thioethers, 1-arylseleno-3-alkoxy-2-propanols **63** and **64** have also been prepared, using microwave irradiation, from epoxides **61** and **62**, respectively, and diselenides **2** [33]. The protocol allowed the efficient preparation of the respective compounds in good yields and in a short reaction time (Scheme **8.23**).

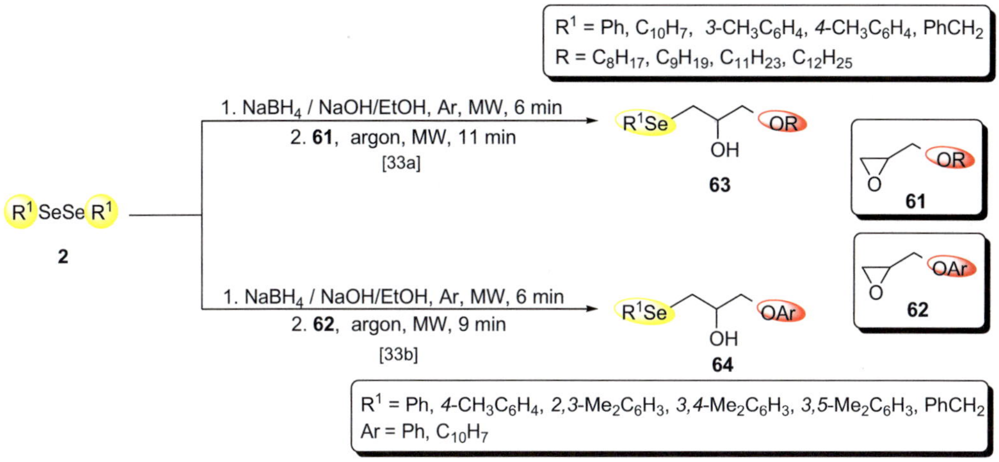

Scheme 8.23: Preparation of substituted glycerol selenide ethers.

Glycerol (obtained as a byproduct in biodiesel synthesis) can be used as a recyclable medium for some organic transformations. In the case of organoselenium chemistry, some publications have described using this compound, particularly in copper-coupling reactions. Glycerol was used, for example, as a recyclable solvent for the synthesis of diarylselenides **7** from arylboronic acids **5** and diaryl diselenides **9** in the presence of stoichiometric amount of DMSO as an additive [34]. A variety of copper catalysts were tested, including $CuCl_2$, $Cu(OAc)_2$, CuCl, CuO NPs and CuI, and the best yields were obtained with CuO NPs and CuI (78 and 91 %, respectively) under the standard conditions (Scheme **8.24**). Dimethyl sulfoxide was a more efficient additive for this coupling reaction compared with acetonitrile, zinc, magnesium and water, affording the corresponding products in higher yields. A possible explanation is that DMSO acts as oxidant activating required for the activity of the metal species in the cross-coupling reaction. The optimum amount of CuI was 5 mol%. The use of 1 and 3 mol% decreased the yield and with the use of 10 mol% the result was not significantly different to that obtained with 5 mol%. The reaction was not sensitive to electronic effects related to the boronic acid and diselenide moiety affording similar yields with electron-donating or electron-withdrawing groups attached to the aromatic ring.

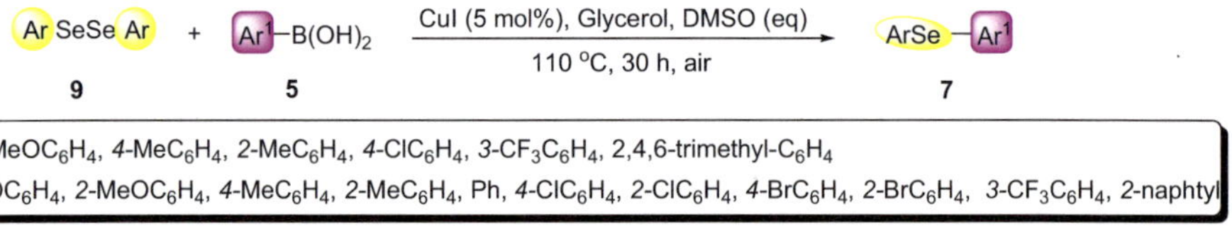

Scheme 8.24: Synthesis of diarylselenides in glycerol as solvent.

The cross coupling of vinyl bromides **48** and diaryl diselenides **9** in the presence of zinc, CuI as the catalyst and glycerol afforded the vinyl selenides **65** [35]. Copper salts were employed, such as CuI, CuCl, CuCN, CuO, $CuCl_2$ and $Cu(OAc)_2$, and 5 mol% of CuI was the most effective catalyst (Scheme **8.25**). The Zn played a crucial role in the reaction producing a significant variation in the yield from 43 to 95 % yield, respectively. In the absence of copper the reaction failed and no product was detected. The scope of the methodology allowed the preparation of a variety of vinyl selenides in good to excellent yield and with a good stereoselectivity, generally keeping the same *(E):(Z)* ratio of the starting halide for the (*E*)-isomer and with a slight decrease in the *(Z):(E)* ratio using the (*Z*)-isomer. The reuse of the glycerol was performed and the respective vinyl halides were obtained in good yields in the five cycles.

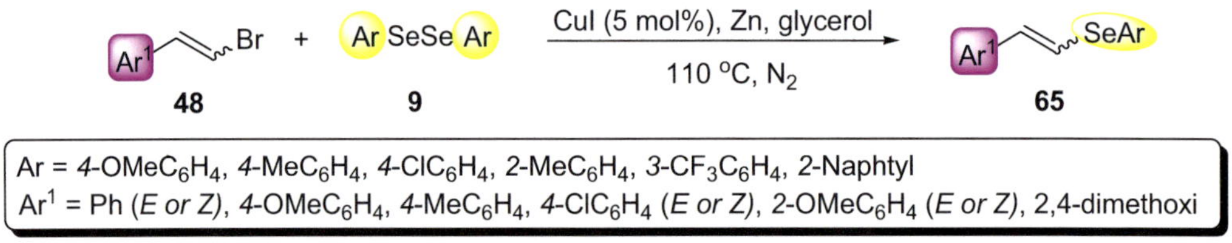

Scheme 8.25: Synthesis of vinyl selenides in glycerol.

Poly(ethylene)glycol (PEG) has also been employed as a solvent for organic transformations and shows interesting behavior, particularly its recyclability. In organoselenium chemistry, some protocols were optimized using this solvent. As an example, epoxide **24** was subjected to nucleophilic ring opening reaction using diphenyl diselenide **14** in the presence of sodium dithionite and PEG-400/H_2O as the solvent, under ultrasound irradiation. The β-hydroxy selenides **32** and **33** were obtained with high regioselectivity and in good to excellent yields [36]. Toluene, 2-Methyl THF and the ionic liquid [bmim]BF_4 were used as solvents, affording only traces of the desired products. On changing the solvents to DMF, DMSO, PEG-400 and PEG-400/H_2O the respective β-hydroxy selenides were obtained, with higher yield for the latter solvent. In terms of base, Et_3N, Cs_2CO_3, and KF·$2H_2O$ afforded 68 %, 80 %, 62 % yields, respectively, while K_2CO_3 yielded the β-hydroxyselenide more efficiently. The amount of sodium dithionite had an influence on the reaction, this being more effective in the presence of 3 equiv. and no reaction was observed in its absence, even for a long reaction time. In terms of regioselectivity, the hydroxyselenides have been, almost exclusively, derived from the attack of the nucleophilic selenium to the less hindered carbon of the epoxide. Additionally, when the chiral epoxides (*S*)-2-(chloromethyl)oxirane **24c** or (*R*)-2-(chloromethyl)oxirane **24h** were used under the optimized conditions, the optically pure epoxide was converted into (*R*)-1-chloro- 3-(phenylselanyl)propan-2-ol and (*S*)-1-chloro-3-(phenylselanyl) propan-2-ol in excellent yields without any racemization or inversion, as depicted in Scheme **8.26**.

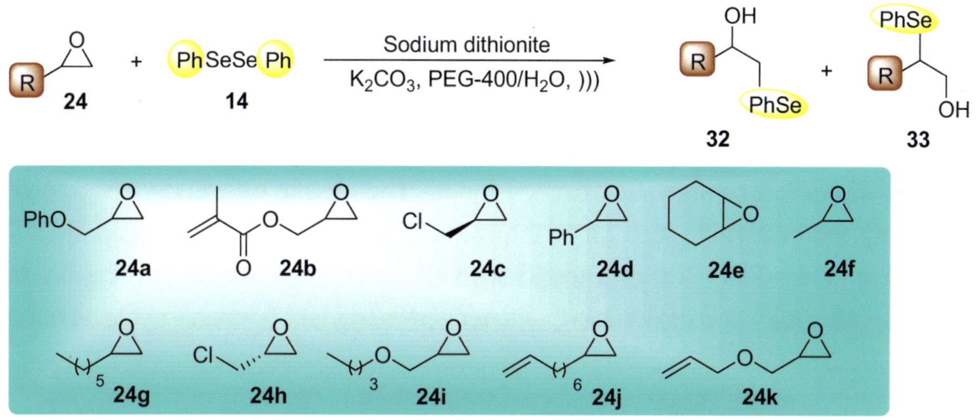

Scheme 8.26: Selenium epoxide ring-opening in PEG/H_2O.

The preparation of alkenyl chalcogenides **20** and **21** *via* addition of selenolate, generated *in situ* from the reaction of the respective diphenyl diselenide **14** with $NaBH_4$, *via* hydrochalcogenation of terminal alkynes **19**, has been reported [37]. PEG-400 and glycerin were used as recyclable solvents and the respective vinylic chalcogenides were obtained preferentially in the (Z)-configuration **20** (Scheme **8.27**). A series of propargylic alcohols and phenylacetylene were employed, affording the desired products in good yields. In terms of regiochemistry, the propargylic alcohols afforded a mixture of anti-Markovnikov and Markovnikov products, with the preference being for the adduct anti-Markovnikov(*Z*). Steric effects appear to be a relevant factor because both the anti-Markovnikov and Markovnikov products ratio, and the reaction time, increase with the size of the R group. The PEG-400 was reused up to 4 times without previous treatment with comparable yields and selectivity.

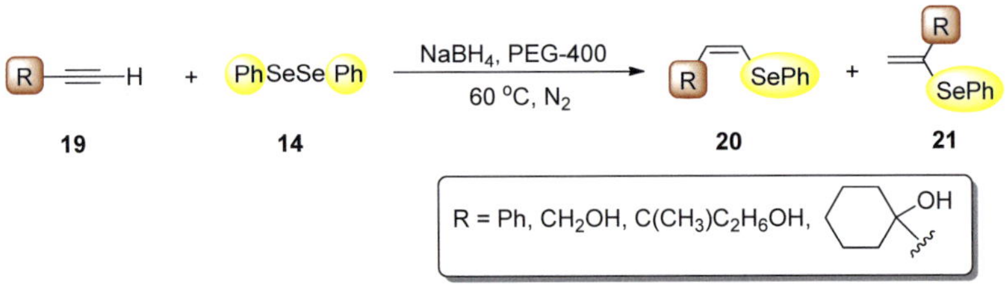

Scheme 8.27: Synthesis of alkenyl chalcogenides.

The α selenenylation reaction of carbonyl compounds **66** (aldehydes and ketones) was selectively performed using a solid-supported catalyst (KF/Al_2O_3) and PEG-400 as the solvent [38]. The protocol allowed the preparation of a variety of 2-phenylselenoaldehydes and ketones **67** in a clean, recyclable medium and in good to excellent yields (Scheme **8.28**). Under the standard conditions, it was observed that the temperature is an important factor, since the products were afforded effectively at 60 °C whereas at room temperature the reaction failed. The atomic efficiency was improved on reducing the diselenide amount to 0.5 mmol/1 mmol of carbonyl compound. The recyclability of the PEG-400 was tested without previous treatment and the products were observed after 5 successive runs, with a decreased yield for every cycle.

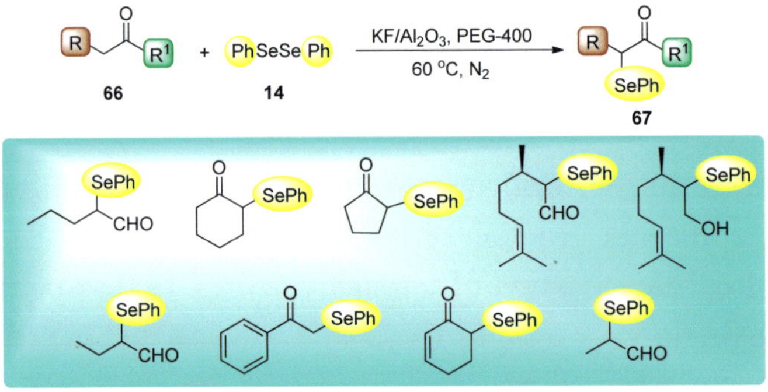

Scheme 8.28: Electrophilic addition of selenium species in carbonyl compounds.

8.2.4. Solvent-Free

Chemical solvent-free methodologies are one of the highlights in Green Chemistry. This approach is currently receiving much attention, particularly by the elimination of organic solvents, many of which have high toxicity, allowing the reactions to become more environmentally-friendly or, in other words, greener. The development of solvent-free methodologies in organoselenium chemistry is also a fast growing field and some methodologies for the synthesis of classes of organoselenium compounds are described in the literature.

β-hydroxyselenides **26** and **27** (Scheme **8.29**), for instance, have been efficiently prepared from epoxides **24**, using different nucleophilic selenium sources. ArSeH **25** in the presence of $Ti(O^iPr)_4$ as a Lewis acid under solvent-free conditions allowed the preparation of the respective compounds in good yields [39]. $Ni(NO_3)_2$, $SnCl_4.5H_2O$, $BF_3.Et_2O$, $AlCl_3$, $ZnCl_2$ and $Yb(OTf)_3$ were also used as catalysts (10 mol%). Titanium afforded the best result, furnishing the respective product with 96 % yield, although zinc, aluminum and ytterbium containing catalysts were also effective. In general, the reaction proceeds with a high regioselectivity, the major product being derived from the nucleophilic selenium attack on the less hindered carbon of epoxide, similarly to an SN_2 mechanism. The styrene epoxide afforded the other regioisomer, indicating that the reaction probably occurs *via* a partially stable carbocation intermediate.

Furthermore, β-hydroxyselenides **32** and **33** (Scheme **8.29**) were prepared using diphenyl diselenide **14** as a selenium source and Rongalite® as a reducing agent in an Al_2O_3 media [40]. The reaction was performed under grinding conditions and after 5 to 20 min the respective products were obtained in high yields. A series of neutral, basic and acidic Al_2O_3 based support were employed and the best yield was obtained using acidic Al_2O_3. Concerning the bases, Et_3N, Cs_2CO_3, $KF{\cdot}2H_2O$ and K_3PO_4 afforded 24 %, 80 %, 49 % and 66 % yield respectively, while K_2CO_3 provided the β-hydroxyselenide in 96 % yield after 5 min of grinding. The influence of Rongalite® has been investigated and the best yield was obtained with 3 equiv. whereas no reaction was observed even for a longer time in the absence of Rongalite®. For unsymmetrical epoxides, the ring opening reaction proceeds with high regioselectivity on the less hindered carbon to afford only the β-hydroxyselenide **32**. The use of dialkyl diselenides such as 1,2-dibenzyldiselenide and 1,2-dimethyldiselenide did not provided the expected compounds.

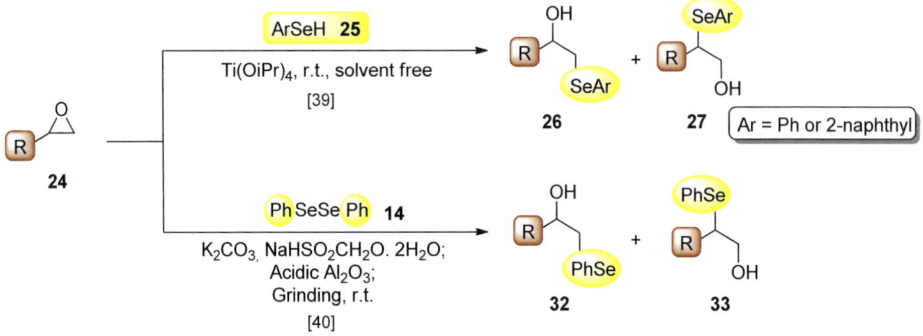

Scheme 8.29: Preparation of β-hydroxyselenides under solvent-free conditions.

The synthesis of selenoesters **3** from acid chlorides **1**, diorganoyldiselenide **2** and Zn under microwave irradiation in a solvent-free conditions was also performed [41]. The reactions were carried out in 2 min at 80 °C, using a variety of diselenides and acid chlorides. The temperature appears to be an important feature, the reaction affording better yields at 80°C than at 50 or 130 °C. The microwave power shows an interesting association with the reaction efficiency and, at the same temperature, 50 W and 150 W led to lower yields compared with 100 W. Electronic effects associated with the diselenide **2** moiety and the acid chloride molecules **1** showed a remarkable influence, furnishing higher yields for electron-withdrawing groups attached to the aromatic ring compared with electron-donating groups, for both substrates. Aliphatic diselenides have also been used, affording the respective seleno-aliphatic esters in good yields. The reaction was also sensitive to steric effects and more hindered alkyl acid chlorides afforded lower yields than less substituted substrates. In addition, selenoesters derived from 9-fluorenylmethyl, benzyl and ethyl chloroformate were synthesized, affording the selenium protecting groups, as depicted in Scheme **8.30**.

The mechanism was also investigated and the authors proposed the formation of a complex of di(organoylselenyl)zinc **68** from the cleavage of the diselenide **2** bridge *via* Zn dust, which reacts with acyl chloride **1** furnishing the selenoesters **3** and RSeZnCl. Subsequently, R^1SeZnCl **69** undergoes a nucleophilic attack on another equivalent of acyl chloride **1**, giving another equivalent of the desired selenoesters **3** (Scheme **8.31**).

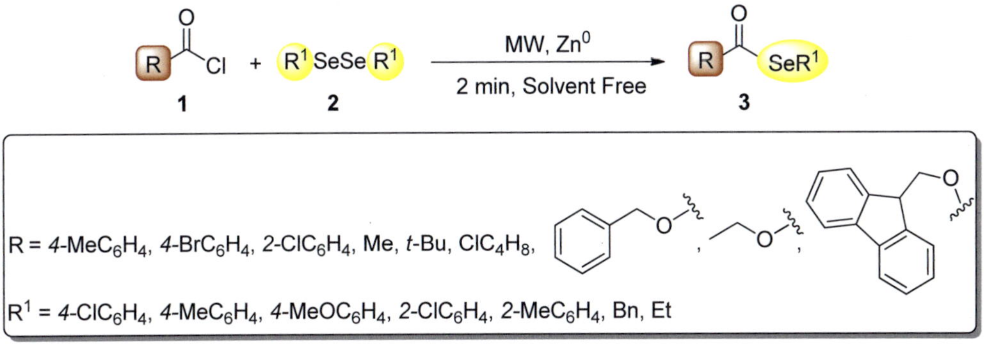

Scheme 8.30: Synthesis of selenoesters under solvent-free conditions.

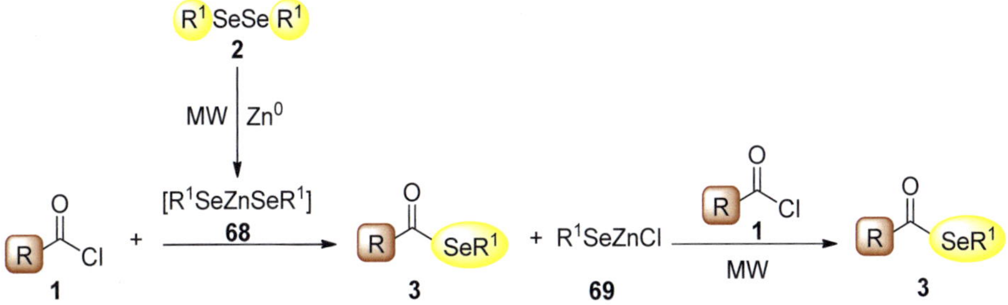

Scheme 8.31: Proposed mechanism for the synthesis of selenoesters.

The palladium-catalyzed addition of dichalcogenides **9** to acetylenes **19** has been carried out under solvent-free conditions, with 100 % atom economy, with no waste materials formed and with a high stereoselectivity for the major alkene *(Z)*-isomer **70** [42]. An important feature in the Pd-catalyzed solvent-free addition was the use of an excess of phosphine ligand. The authors explained that in the absence of an excess of PPh_3 rapid polymerization of the palladium catalyst was observed resulting in a dark brown precipitate. Without the phosphine no reaction was observed and the protocol showed the same efficiency under air or inert atmosphere. The amount of $Pd(PPh_3)_4$ was varied from 0.001 to 1 mol% and for all quantities the desired addition product was observed. Nevertheless, the best combination of high yield and lower catalyst charge was obtained with 0.1 mol% of $Pd(PPh_3)_4$. The difference in the reactivity of the chalcogenides was less notable under solvent-free conditions compared with the use of a solvent medium, as depicted in Scheme **8.32**. In addition to the ecological advantages, the solvent-free conditions showed an interesting rate acceleration due to the effects on the concentration and on the catalyst efficiency.

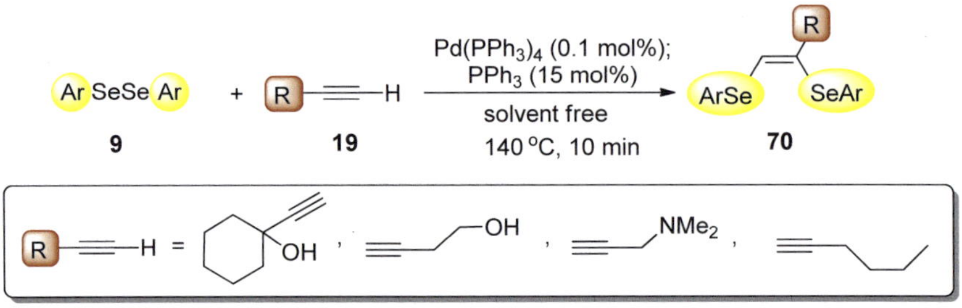

Scheme 8.32: Pd catalyzing addition of diselenide to acetylenes.

The preparation of vinyl selenides from acetylenes, using diselenides and $NaBH_4$ was carried out efficiently under microwave irradiation and solvent-free conditions. For the acetylenic ketones **71**, the standard protocol proceeded with the use of a solid support of Al_2O_3, 0.080g of $NaBH_4$ as the reducing agent, diselenide **2** (0.5 mmol) and 4-phenyl-3-butyn-2-one (1 equiv) (Scheme **8.33**) [43]. At room-temperature, the product was obtained in 1.5 h with 62 % yield and a **72***(Z)*:**73***(E)* ratio of 96:4. Under MW irradiation at 548 W, the vinylic selenides were obtained in a very short reaction time (1.5 min), with yields and selectivity similar to those obtained at room temperature. On changing the potency to 353 W the reaction was not complete, despite the longer reaction time (3 min) and the starting materials were recovered. Without the use of $NaBH_4$ the reaction does not proceed as well as it does without the presence of Al_2O_3, and the yield was lower with several side products. Performing the reaction under conventional heating, the yield was lower than that obtained with MW and it was observed a decrease in the *(Z)* selectivity. On comparing the different methodologies, it was found that the MW-assisted reactions are more efficient and convenient and also cleaner. A similar protocol was described for the preparation of β-phenylchalcogeno-α,β-unsaturated esters from methyl propiolate derivatives **74** [44]. The reaction allowed the synthesis of the desired compounds in good yields with stereoselective control, affording the predominantly *(Z)*-isomer **75** against the *(E)*-isomer **76** (Scheme **8.33**).

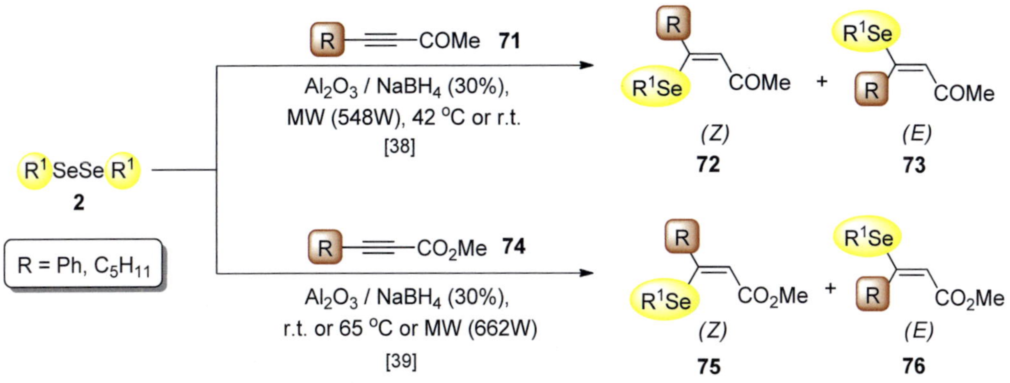

Scheme 8.33: Preparation of vinylic chalcogenides esters from acetylenes.

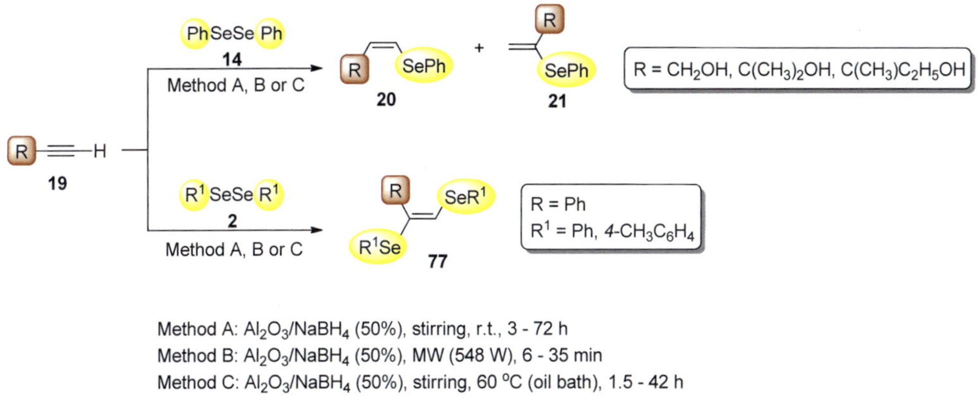

Scheme 8.34: Preparation of vinylic chalcogenides from acetylenes.

In the same way, a variety of acetylenes **19**, such as propargylic alcohols, sodium borohydride and alumina, have been used as the solid support under solvent-free conditions for the preparation of a series of vinyl selenides **20-21** starting from diaryl diselenides (Scheme **8.34**) [45]. The compounds were synthesized at room temperature, with conventional heating and under MW irradiation. The authors observed that under MW irradiation the reactions were faster compared with conventional heating (10 min *vs.* 46 h respectively) affording the respective vinyl selenides with yields and selectivity similar to those of the conventional reaction. Without the use of $NaBH_4$ the reaction did not proceed completely and the starting materials were recovered. In terms of regiochemistry, the anti-Markovnikov product was the major one and the regioselectivity was dependent on the steric effects, affording higher selectivities for most hindered R groups.

When the same protocol was applied to phenyl acetylene, the (*E*)-bis-organochalcogen alkenes **77** were obtained in good yields and high selectivity.

8.3. GREEN APPLICATION OF ORGANOSELENIUM REAGENTS AS CATALYSTS

Selenium-containing organic molecules have been applied successfully in several types of reactions. A further interesting aspect, which has emerged in recent years, concerns the use of catalytic amounts of an organoselenium compound. The development of efficient catalytic processes allows the application of organoselenium compounds as organocatalysts within a "greener perspective" and probably represents an important recent advance reported in this field. In this section we will describe the involvement of this type of compound in different green processes. The application of these compounds as ligands for metal-catalyzed processes will be not covered here, since this issue has been disclosed recently [3f], as well as their potential as efficient mimetics for selenoenzymes [5c-d].

8.3.1. Selenenylation-Deselenenylation Reaction

Selenenylation-deselenenylation is one of the reactions most commonly employed in organic chemistry for the functionalization of non-activated carbon-carbon double bonds with electrophilic selenium reagents [46]. The discovery that electrophilic selenium reagents can be added stereospecifically to alkenes was first described in the late 1950s, although the selenium reagents were already used as stoichiometric reagents [46a] and most of these reactions require several synthetic steps to prepare the selenium reagent [3a]. As a consequence of these drawbacks, special emphasis has been placed on the development of one-pot selenenylation-deselenenylation sequences, which can be performed with only catalytic amounts of the organoselenium compounds [4]. Another interesting aspect, which has emerged in recent years, concerns the catalytic chiral one-pot transformation promoted by optically-active electrophilic selenium reagents.

The use of organoselenium catalysts in oxidative selenenylation-deselenenylation is reported in Scheme **8.35**. The electrophilic species of selenium **78**, which is produced by *in situ* oxidation of the corresponding diselenide, reacts with an alkene **79** in a stereospecific *anti* addition to produce the seleniranium ion intermediate **80**, which is immediately opened in the presence of an external nucleophile Nu^- to produce the selenenylated compound **81**. The pathway followed by the deselenenylation reaction step depends on the solvent employed and on the structure of the starting unsaturated compound and can lead either to substitution or to elimination products. Thus, the oxidation of the organoselenium moiety in **82** allows the regeneration of the selenium reagent through β-elimination giving **83** or substitution to produce **84**. The great synthetic importance of this reaction lies in the fact that the adduct product **81**, which contains the selenium functionality, can be used as a precursor for a series of useful transformations in subsequent reactions [47].

Two main strategies have been explored to effect these catalytic transformations: an electrochemical approach and the use of a chemical oxidant. Thus several catalytic one-pot selenenylation-deselenenylation reactions have been described. The results obtained for these reactions will certainly stimulate further research in this area, especial concerning the green aspects and the asymmetric version.

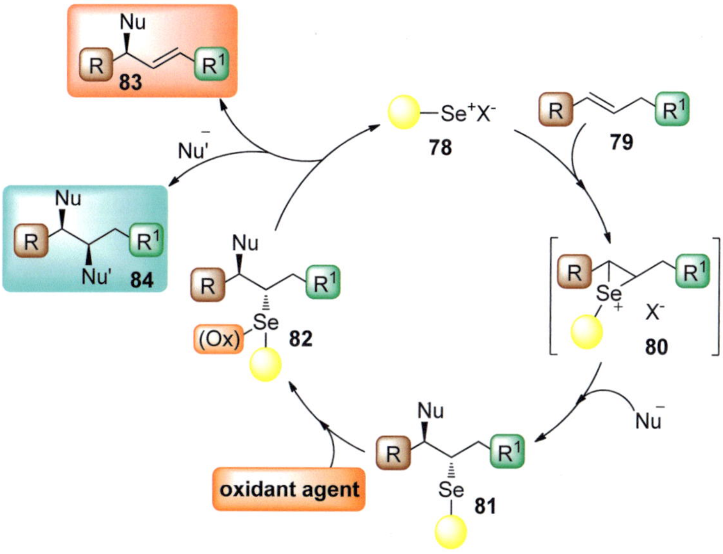

Scheme 8.35: Catalytic cycle of one-pot selenenylation-deselenenylation reaction.

The first use of catalytic amounts of selenium reagents in one-pot oxyselenenylation-deselenenylation was described by Torri and coworkers in 1981 [48]. They reported the preparation of allylic alcohols and ethers from alkenes using diphenyl diselenide as catalyst. The electrophilic species **85** was produced from diphenyl diselenide **14** by electrochemical oxidation in the presence of the olefin and magnesium sulfate in methanol or water. As exemplified in Scheme **8.36**, the reaction with alkene **86** occurs to form the β-alkoxy selenide **87** which is then electrochemically oxidized to afford the selenoxide **88** which undergoes *syn* elimination to give the allylic alcohols or ethers **89** and, at the same time, generate phenylselenenic acid **90**, which continues the cycle by re-addition to the alkene **86**.

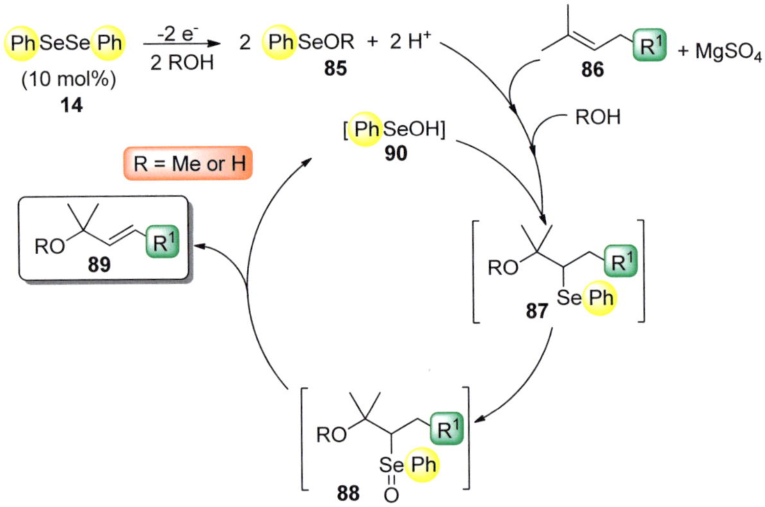

Scheme 8.36: Electrochemical catalytic oxyselenenylation-desenenylation of alkenes.

In the early 1990s, Tomoda described a chemical method to effect functionalizations through the selenenylation-deselenenylation sequence [49]. In this procedure sodium persulfate was used as the oxidizing agent with copper(II) nitrate as the co-oxidant in the presence of catalytic amounts of nitrogen substituted diselenides **92** or **93** (Scheme **8.37**). The reaction proceeds essentially in the same way, for the conversion of alkenes **79** into allylic ethers and esters **91**. Selenoxides were proposed as the reactive intermediates responsible for the elimination reaction, which affords the final products. It was suggested that the selenenic acid is stabilized by intramolecular coordination with the tertiary nitrogen atom thus preventing its oxidation to the seleninic acid. The use of molecular sieves increases the rate and the efficiency of the process.

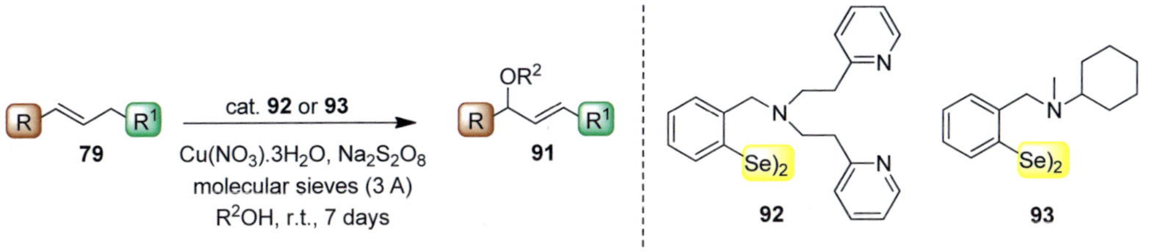

Scheme 8.37: Catalytic conversion of alkenes into allylic ethers and esters using nitrogen-containing diselenides and sodium persulfate as the oxidant.

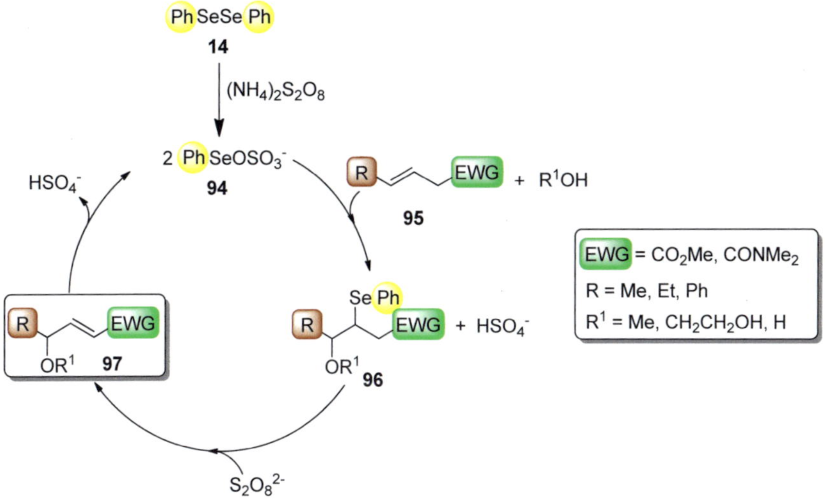

Scheme 8.38: One-pot selenenylation-deselenenylation of β,γ-unsaturated esters and amides into the γ-alkoxy or the γ-hydroxy-α,β-unsaturated derivatives, catalyzed by phenylselenenyl sulfate.

A new method which does not involve selenoxides during the elimination step was proposed by Tiecco and coworkers in 1993 [50]. In relation to this approach, the author described the preparation of a new electrophilic phenylselenyl sulfate by the reaction of diphenyl diselenide with ammonium persulfate [51]. Phenylselenyl sulfate is a very efficient electrophilic reagent and seems to have a potentially broader application in the synthesis of a range of different compounds, since this reagent leads to clean reactions. As exemplified in Scheme **8.38**, this sequence is carried out with the phenylselenenyl sulfate **94** as the electrophile, which is generated by the reaction of the diphenyl diselenide **14** and ammonium persulfate in methanol, ethylene glycol or in water. The reaction with the β,γ-unsaturated esters and amides **95** occurs to form addition products **96** that is then oxidized by the excess of ammonium persulfate, affording the deselenenylated γ-alkoxy or the γ-hydroxy-α,β-unsaturated derivatives **97** and, at the same time, regenerating the phenylselenenyl sulfate **94**. The presence of electron-withdrawing groups in the allylic position seems to be important for the success of the reaction.

After the work of Tomoda [49] and Tiecco [50], several other papers appeared in the literature employing the use of persulfate as the oxidant in one-pot selenenylation-deselenenylation reaction of alkenes (Scheme **8.39**).

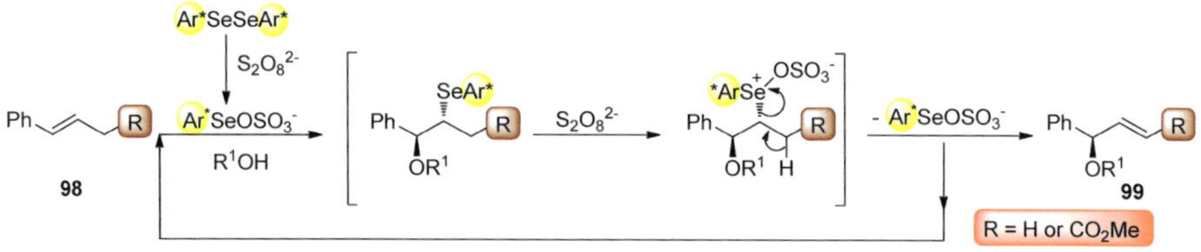

Scheme 8.39: Selenenylation-deselenenylation reaction catalyzed by phenylselenyl sulfate.

In 1994 Tomoda and co-workers described the first catalytic asymmetric oxyselenenylation-oxidative deselenenylation reaction promoted by diaryl diselenide **100** with chiral pyrrolidine rings in the *ortho*-position (Scheme **8.40**) [52]. The treatment of *trans*-β-methylstyrene **98** with ammonium persulfate and a catalytic amount of diselenide **100** gave rise to the formation of optically active allylic ether **99** with only 32% *ee* (Scheme **8.39**). Although the enantiomeric excess achieved was low, this result opened the way to a collective effort in the search for more efficient chiral diselenides which involved several research groups and resulted in a few years in enormous improvements in terms of yields and enantioselectivities (Scheme **8.40**).

A common feature of the overwhelming majority of these diselenides is the close proximity of an electron-rich heteroatom such oxygen, nitrogen or sulfur to the selenium atom (Scheme **8.40**). The efficiency and the selectivity of the catalyst as well as the rate of the reaction have a close relationship with the nature of the substituents attached to the phenyl framework. In this regard, different experiments have been performed to show the importance of intramolecular interaction between the lone pair of electrons of the heteroatom and the σ^* orbital of the selenium atom [3a]. This interaction will induce the chiral center to come close to the reaction center during the addition of selenenylating reagent to the alkene and this will result in asymmetric induction [53].

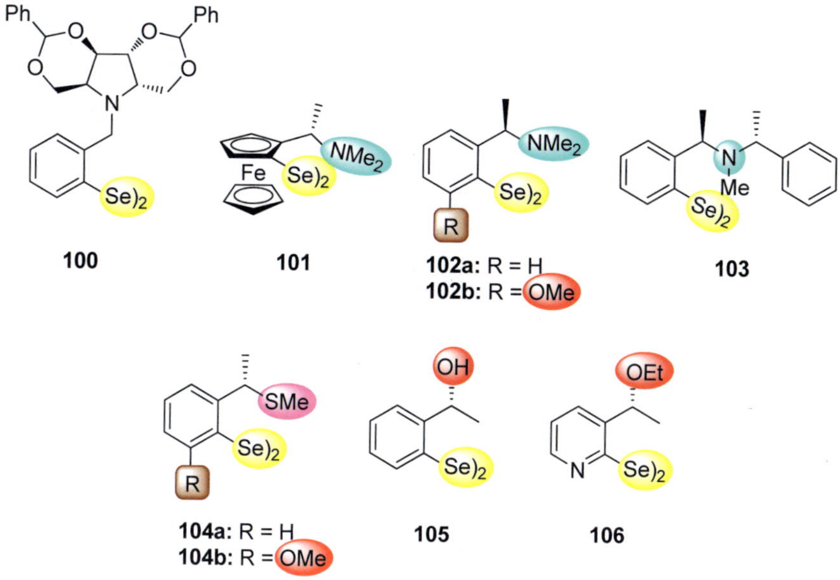

Scheme 8.40: Chiral diselenides employed in asymmetric one-pot selenenylation-deselenenylation.

Since the first report by Tomoda and co-workers in 1994 on the one pot asymmetric selenenylation-deselenenylation reaction of *trans*-β-methylstyrene **98** a huge variety of chiral diselenides have been used for this reaction. In 1997, Fukuzawa and co-workers [54] reported the use of chiral diferrocenyl diselenide **101**. However, the reaction with *trans*-β-methylstyrene **98** gave the essentially racemic allylic ether **99**. In 1998 Wirth and co-workers [55] reported the use of diselenides with nitrogen-containing substituents **102a-b** for the catalytic asymmetric oxyselenenylation-oxidative deselenenylation reaction. The chiral diselenide **102b** showed the best catalytic performance affording the allylic ether **99** in up to 75% *ee* (Scheme **8.39**). A further interesting example is the catalytic asymmetric version of the one-pot conversion of β,γ-unsaturated esters **98** into the allylic ethers and alcohols **99** employing diselenides containing-nitrogen **103** [56] and sulfur **104b** [57]. The good results obtained in the asymmetric selenomethoxylations and selenohydroxylations of β,γ-unsaturated esters in the presence of catalytic amounts of a new sulfur-containing chiral diselenide suggested that interaction of selenium with sulfur is probably more important than interaction with oxygen or nitrogen.

One-pot selenenylation-deselenenylation reactions have also gained popularity in the synthesis of heterocyclic compounds in recent years with regards to the preparation of complex molecules as well as natural products. For this purpose, the use of ammonium persulfate as an oxidant in the presence of a catalytic amount of organoselenium compound has been reported (Scheme **8.41**). This method is very versatile and allows the synthesis of a wide variety of heterocyclic structures. However, the heterocyclic compounds prepared using one-pot selenocyclization the heterocycles containing oxygen and/or nitrogen as an internal nucleophilic group have been the most explored due to their biological properties.

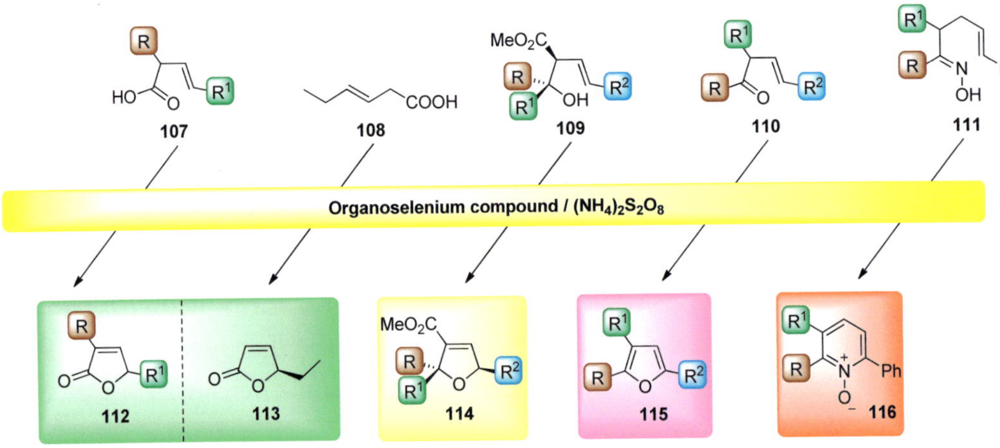

Scheme 8.41: Selenocyclization reaction with ammonium persulfate catalyzed by organoselenium compounds.

As reported in Scheme **8.41**, the butenolides **112** were obtained in good yields from the catalytic one-pot cyclization of β,γ-unsaturated acids **107** [58]. In this case, the carboxy group acts both as the internal nucleophile and as the electron withdrawing group, which facilitates the elimination. Good chemical yield and moderate enantiomeric excess were obtained in the cyclization of γ-alkenoic acid **108** into the butenolide **113** using the enantiomerically pure diselenide **104b** [57]. Two further examples are reported in Scheme **8.41**, concerning the preparation of the heterocycles containing oxygen. The first case refers to the direct conversion of 2-carbomethoxy-3-alkenols **109** into the 2,5-dihydrofurans **114** [59] and the second to that of α-vinyl β-dicarbonyl compounds **110** into the furans **115** [60]. However, less common nitrogen heterocycles have also been prepared in one-pot selenenylation-deselenenylation reactions (Scheme **8.41**). Under the usual conditions, the oximes **111** react with phenylselenyl sulfate to afford the pyridine *N*-oxides **116** [61].

The application of hypervalent iodine compounds as oxidants to form electrophiles from diselenides is known [62]. Recently, Wirth reported a new one-pot method for a catalytic addition-elimination reaction using selenium electrophiles [63]. A new method to cyclize a range of β,γ-butenoic acids **117** to the corresponding butenolides **118** (Scheme **8.42**) in the presence of catalytic amounts of diphenyl diselenide **14** and $PhI(OCOCF_3)_2$ has been described. When the reaction was performed in the presence of chiral diselenides **105** or **106** (Scheme **8.40**), the product **120** was obtained with enantiomeric ratios (er) of 57:43 (84 % yield) and 61:39 (46 % yield), respectively, as shown in Scheme **8.42**. The same catalytic approach has been used for cyclization of γ,δ-unsaturated carboxylic acids **121** to the corresponding 3,6-dihydro-2*H*-pyran-2-ones **122** in good to excellent yields [64].

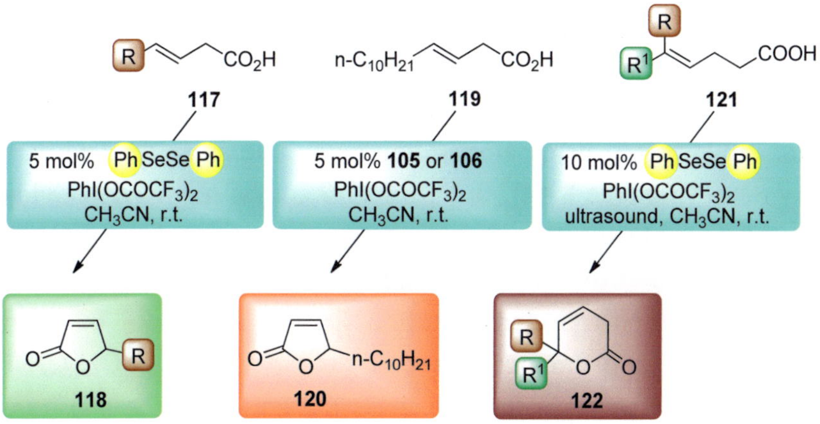

Scheme 8.42: Selenocyclization reaction with hypervalent iodine catalyzed by organoselenium compounds.

The catalytic use of diphenyl diselenide **14** in the electrochemical oxyselenenylation-deselenenylation sequence for the conversion of alkenes **86** into allylic compounds **89** has been reported [48] (Scheme **8.36**). In this context, Wirth and co-workers developed a new method using tetraethylammonium bromide and electrolysis to prepare the arylselenenyl bromide, which promotes the selenofunctionalization leading to products **124** or **125** depending on the substrate **123** used [65]. This new protocol using chiral diselenides afforded the enantiomerically pure **125** (Scheme **8.43**).

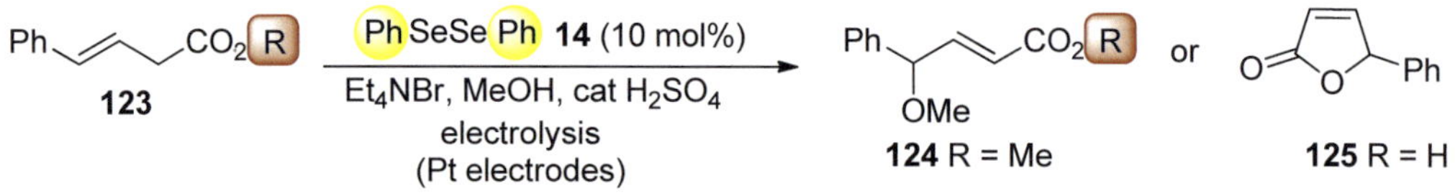

Scheme 8.43: Catalytic electrochemical oxyselenenylation-elimination sequence.

In all of the catalytic one-pot selenenylation-deselenenylation reactions described above the addition product reacted with the oxidant undergoing deselenenylation to afford the elimination product. The catalytic one-pot procedure through which the deselenenylation occurs with substitution is less common [66]. Most of the reactions in which the deselenenylation occurs with substitution require the use of stoichiometric diphenyl diselenide in order to prevent the large excess of oxidant causing the consumption of the starting material and leading to a complex mixture of products.

An interesting application of the sequential selenenylation-deselenenylation reaction followed by substitution in the dihydroxylation of olefins [67] and in the oxidation of alkynes [68] using ammonium persulfate as an oxidant was recently reported by Santi (Scheme **8.44**). In both reactions the oxidized selenium moiety is substituted by a molecule of water. Following this protocol 1,2-diols **127** or **128** were prepared from hydroxyselenenylation of unsaturated terpenes **126** promoted by the electrophilic phenylselenyl sulfate generated *in situ* by oxidation of diphenyl diselenides with ammonium persulfate in a mixture of CH_3CN-H_2O. The diastereoselectivity is strongly dependent on the nature of the substrate and the method proved to be efficient also when the sulfur containing chiral diselenide **104a** was used instead of diphenyl diselenide **14**. Alkynes **129** are easily oxidized to the corresponding dicarbonyl compounds **130** in moderate yields using a similar methodology. The excess of ammonium persulfate activates the phenylselenium moiety to carry out the nucleophilic substitution with a molecule of water, followed by a rapid oxidation of the hydroxy group to afford the corresponding 1,2-dicarbonyl compounds. Interestingly, when R^1 is replaced by hydrogen, the α-ketoaldehydes were converted *in-situ* to the corresponding hemiacetals **131** by treatment with silica gel.

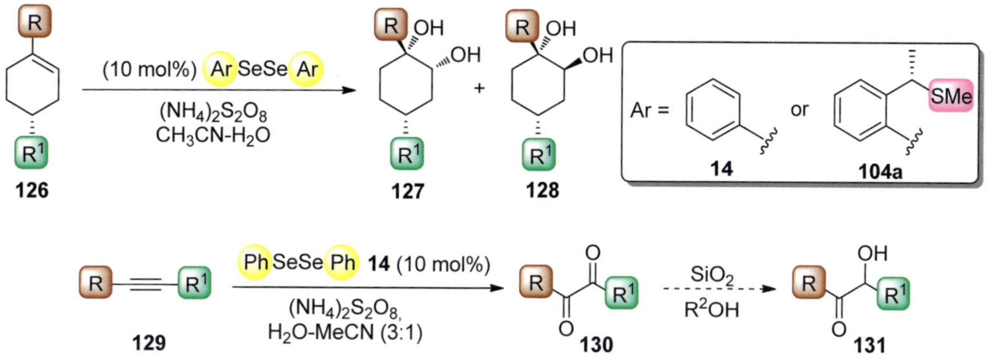

Scheme 8.44: Diselenide catalyzed one-pot conversions of unsaturated terpenes into 1,2-diols and alkynes into 1,2-dicarbonyl compounds in the presence of ammonium persulfate as the oxidant.

8.3.2. Baeyer-Villiger Reaction

Baeyer-Villiger oxidation has been used by organic chemists in the preparation of many different organic molecules starting from carbonyl compounds [69]. Since the discovery of this reaction, the Baeyer-Villiger reaction has become a valuable tool in organic chemistry and the search for new strategies to improve its efficiency has been the subject of renewed interest. Moreover, recently there has been considerable effort to make catalytic Baeyer-Villiger oxidation and at the same time preserve the high regio- and stereoselectivity of the reaction [70]. More than one century has passed and the standard protocol remains almost the same, that is, Baeyer-Villiger reactions are still performed with organic peracids in more than stoichiometric amounts in chlorinated solvents. In order to develop new green protocols, attention has been focused on the use of hydrogen peroxide in the presence of a catalyst. Hydrogen peroxide represents a clean oxidant, and offers many advantages: it is safe and cheap, the amount of active oxygen is high and it is clean since the byproduct is water. Two distinct categories of organoselenium compounds have emerged as powerful catalysts for oxidative reactions: substituted seleninic acids and selenoxides, which are usually prepared from diaryl diselenides and selenides, respectively.

The activation of H_2O_2 by seleninic acids for use in the Baeyer-Villiger oxidation reaction was first described in the late 1970s, although the seleninic acids were already being used as stoichiometric reagents [71]. The seleninic acid **133** is readily prepared by the reaction of the corresponding diselenide **132** with hydrogen peroxide. Further oxidation of **133** with H_2O_2 leads to perseleninic acid **134**, which is the active oxygen transfer agent [72]. Subsequently, **134** reacts with a carbonyl compound **135** to produce the adduct **136**, which after migration of R^1 or R^2 leads to the formation of the product **137** regenerating seleninic acid **133** for the catalytic cycle (Scheme **8.45**).

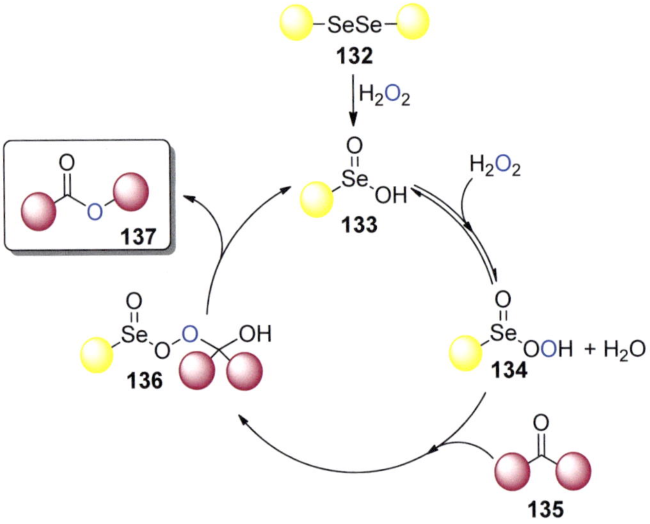

Scheme 8.45: Catalytic cycle for seleninic acid with H_2O_2 in the Baeyer-Villiger reaction.

Following this first report regarding the use of benzeneseleninic acid **138** [71], diselenides with different substituents in the phenyl ring were prepared in order to improve the efficiency and the selectivity of the catalyst (Scheme **8.46**). Thus, seleninic acids with electron-withdrawing groups (*e.g.* nitro **139** [72,73] or trifluoromethyl **140** [74] were prepared. Compared to SeO_2 and benzeneseleninic acid **138** they are better catalysts for Baeyer-Villiger oxidations. Among these, **141**, with two CF_3 groups in the *meta* position, is a superior catalysts, and one explanation for this lies in the fact that increasing the electron-poor nature of the catalyst leads to an increase in the activity and selectivity [74].

The nitro-substituted seleninic acids **139** also manifested remarkable catalytic activity in the Baeyer-Villiger reaction with hydrogen peroxide. However, although seleninic acids with nitro substituents **139** are demonstrated to be stronger acids than those carrying a CF_3 group **140**, the former precatalysts (diselenides) are poorer catalysts. This can be attributed to the sluggish oxidation of diselenide **142** with H_2O_2 to produce seleninic acid **139** (Scheme **8.47**) [74]. The utilization of preformed perseleninic acids could eliminate this drawback; however, the overoxidation of the seleninic acid to the perseleninic acid is also disfavored by strong electron-withdrawing nitro groups.

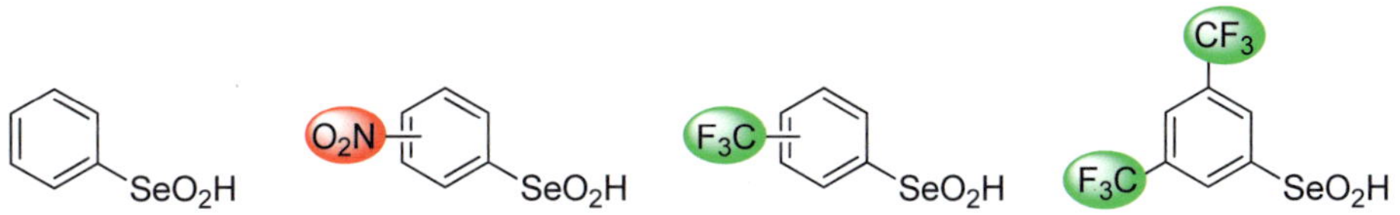

138 **139** **140** **141**

Scheme 8.46: Seleninic acids used as catalysts in Bayer-Villiger reaction.

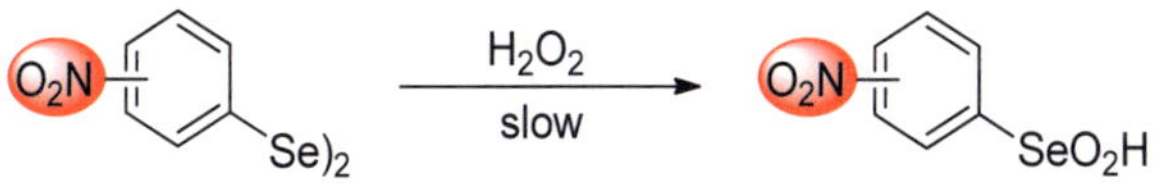

142 **139**

Scheme 8.47: Conversion of diselenides bearing nitro groups **142** to the corresponding seleninic acids **139**.

Novel catalysts designed to activate H_2O_2 for Baeyer-Villiger oxidation have been prepared with the aim to improve the efficiency and selectivity (Scheme **8.48**). Polystyrene has been used as a solid support to immobilize seleninic acid **143** [75]. In all cases, the recovered polymer could be used in subsequent reactions with no further treatment and no loss of reactivity. Unfortunately, the availability of structurally modified resins does not allow the preparation of a wide range of catalysts, especially concerning electronic effects. The catalyst **144** prepared by ten Brink and co-workers [76] proved to be more active than 3,5-bis(trifluoromethyl)benzeneseleninic acid **141**. Concerning the green aspects, the authors developed a triphasic system that allowed facile separation and recycling of the catalyst phase. Under this condition the catalyst **144** could be recycled, with a slight decrease in the formation of products after repeated reaction cycles. Catalysts **145** showed good results in the Baeyer-Villiger oxidation of ketones, when R = OTf.

On the other hand, when R = H or Me the product was not obtained [77]. Chiral diselenide **146** was the first catalyst to be used in the asymmetric Baeyer-Villiger oxidation of ketones; although the yields of lactones were good, the enantiomerically excess was very poor [78]. Selenoxides **147** has been used as efficient H_2O_2 activators for Baeyer-Villiger reactions. Their reactivity is also modulated by the electronic nature of the substituents, benzyl 3,5-bis(trifluoromethyl) phenyl selenoxide (R = 3,5-CF_3) being the most active catalyst [79].

143 **144** **145** **146** **147**

Scheme 8.48: Different catalysts employed as activators of hydrogen peroxide in the Bayer-Villiger reaction.

The control of the reaction conditions is of great importance in Baeyer-Villiger oxidation. The solvents generally used in these reactions are halogenated compounds like 1,1,1,3,3,3-hexafluoropropan-2-ol, 2,2,2-trifluoroethanol, 1,2-dicloroethane and dichloromethane. Dichloromethane is the solvent of choice when the products are sensitive toward hydrolysis. Overall, good conversion rates and selectivity are achieved on using these solvents [74]. Although halogenated solvents lead to better results, environmental and/ or economic issues make their use somewhat prohibitive. Therefore, the development of new alternatives, especially the use of aqueous solutions, would be of great interest. Another possibility is to replace the halogenated solvents with THF [71,80] or sulfolane [74]; however, in this latter case a reduced selectivity was observed.

Interestingly, the nature of the substrate **148**, the reaction conditions and the catalyst seem to have a considerable influence on the selectivity and in the formation of the final product. Depending on these features, product **149a** or **149b** may be isolated, or these products can undergo hydrolysis (Scheme **8.49**). The regiochemistry in the formation of product **149a** or **149b** is related to the ability of the group R^1 or R^2 to migrate. A number of examples are reported in scheme **8.49**.

Cyclic ketones **150** are easily oxidized under mild conditions to the corresponding lactones **151** in near quantitative yield and with selectivity higher than 90%. Increasing the ring strain as well as the electron density higher reaction rates have been observed. Linear ketones **152** react more slowly under the same conditions and, in general, yields are not very good, partly because of the lack of ring strain. In addition, the ester **153** is also sensitive under hydrolysis conditions and the formation of carboxylic acid **154** can be observed. In order to minimize the formation of this by-product a biphasic systems can be used. Thus, the contact of the ester with water will be minimized and the hydrolysis process will be reduced. The preparation of carboxylic acids from the corresponding aldehydes usually involves strong inorganic oxidants, such as $KMnO_4$, CrO_3, fuming HNO_3, Jones reagent and others. However, Baeyer-Villiger type reactions have been used to prepare these acids within a green perspective. In the case of aliphatic aldehydes **155**, these compounds are smoothly oxidized to carboxylic acids **156** when treated with seleninic acid. Interesting results are obtained for electron-poor **157** and electron-rich **159** aldehydes. When the aromatic ring is substituted with electron-withdrawing groups, the product of the oxidation is the carboxylic acid **158** and, in this case, the electron-withdrawing substituents favor hydrogen migration to yield acids. On the other hand electron-donating substituents **159** favor ring migration to afford the corresponding formiates, which are subsequently hydrolyzed to phenols **160**. When *tert*-butyl hydroperoxide (TBHP) was used as the oxidant in the presence of the catalyst Ebselen **161**, the corresponding carboxylic acids **162** are exclusively obtained from electron-withdrawing as well as electron-donating substituents [81].

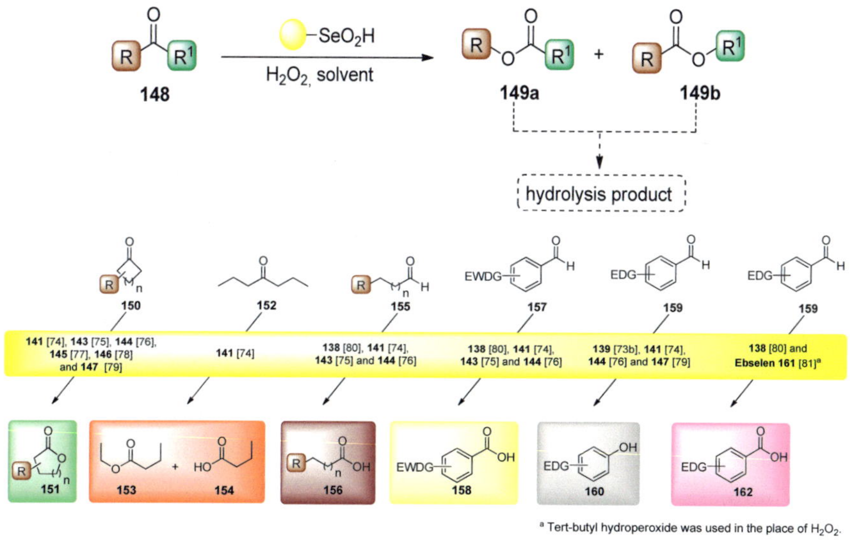

Scheme 8.49: Substrate scope and selectivity in the Baeyer-Villiger reaction catalyzed by organoselenium derivatives.

8.3.3. Oxidation of Carbon-Carbon Double Bonds

The synthesis of epoxides continues to attract interest, especially because of the possibility for the regio- and stereoselective ring opening of epoxides with a range of nucleophiles to afford more sophisticated compounds [82]. Therefore, the search for new methodologies for the epoxidation of carbon-carbon bonds is an area of current interest. Epoxidation reactions mediated by H_2O_2 in the presence of catalytic amounts of cheap, relatively nontoxic metals are a good way to obtain the epoxides [83].

Grieco and co-workers [84] first reported the use of organoselenium compounds in a stoichiometric amount for the epoxidation of olefins in the presence of H_2O_2 as the oxidant. The first example of a catalytic approach was reported one year later by Reich [85], who described the epoxidation of olefins using catalytic amounts of benzeneseleninic acid. Seleninic acid **133** is the precursor of the active oxidizing agent, perseleninic acid **134**, which is generated by the reaction of diselenide with H_2O_2. The electrophilic species perseleninic acid reacts with the olefin **163** producing the product **164** and regenerating **133** for the catalytic cycle. In general, the epoxide **164** is the major product in this reaction; however, depending on the reaction conditions, the substrate and the catalyst, the *vic*-diols **165** can also be isolated as by-products (Scheme **8.50**). Recently, epoxidation catalyzed by SeO_2 [86] and by benzeneseleninic acid **138** [87] combined with a subsequent ring-opening reaction leading to dihydroxylated products has been reported. The catalyst loading for an efficient transformation is usually in a range of 1-10 mol% (of the corresponding seleninic acid). Considerable efforts have been made recently in order to make these reactions "greener". In this context, it could be considered the use of glycerol-based solvents as reaction media [88].

As in Baeyer-Villiger oxidation, the structure of the catalyst is very important in epoxidation reactions promoted by hydrogen peroxide. Specifically, the electronic demand of selenium plays a crucial role in modulating the activity. The insertion of electron-withdrawing groups in the aryl moiety of aryseleninic acids significantly enhances the performance of the catalysts. For instance, nitro substituents, such as in seleninic acids **166a-b**, give better results, both in terms of yield and in the selectivity compared to **138** (Scheme **8.51**) [85]. Nevertheless

Reich [85] and Sharpless [89] showed that if the presence of nitro substituents at the *ortho*-position **166a** improves the catalysis, a second nitro-group placed at the *para-* position **166b** did not further increase the reactivity. Interesting results were obtained by Sheldon and co-workers [90] using catalysts with nitro groups **166a**, **167** and **168**. Catalysts **166a** and **168** react slowly leading to the *vic*-diols, common side products in epoxidation reactions in the presence of sensitive epoxides. This result is attributed to the strong electron-withdrawing properties combined with the resonance effect of the nitro substituent. These two electronic effects increase the acidity of the seleninic acid promoting the hydrolysis of the epoxides, yielding *vic*-diols [91]. On the other hand, the seleninic acid **167** substituted with a nitro-group at the *meta* position showed only an inductive effect and gave slightly better results than **166a** and **138**. Seleninic acid **141** containing two CF_3 groups in the *meta* position of the aryl moiety is the best catalyst. Compared to other compounds, **141** exhibits enhanced performance in terms of efficiency and selectivity [90].

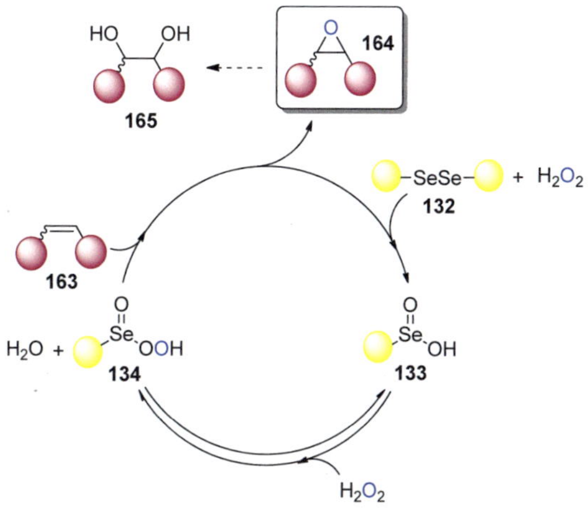

Scheme 8.50: Epoxidation promoted by H_2O_2 activated by seleninic acids.

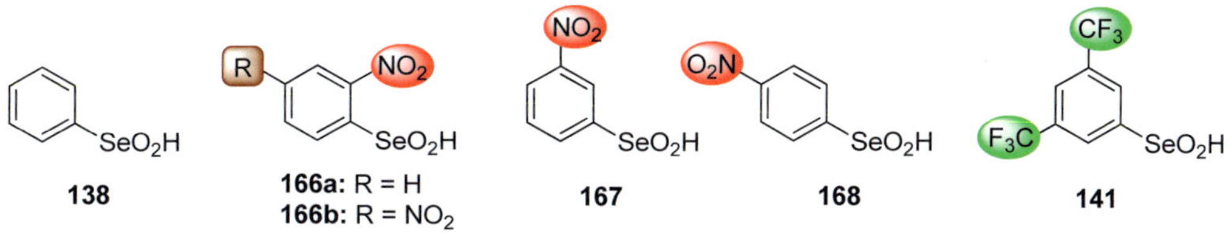

Scheme 8.51: Representative organoselenium catalyst for epoxidation reaction.

The search for different ways to reuse the catalyst in a new reaction cycle constitutes an area of great interest from the point of view of green chemistry. In this context, seleninic acid bearing perfluoroalkyl substituents **144**, when used together with perfluorinated solvents, allows the reuse of the reaction media (catalyst and solvent) at least ten times without a decrease in the yield [92] (Scheme **8.52**). Chiral seleninic acid **169** is also an efficient catalyst, however, it is not able to transfer chirality to the epoxide obtained from the oxidation of prochiral olefins [46]. Selenoxides **147a-d** can also be used as efficient catalysts for the epoxidation of olefins [36]. As in the case of seleninic acids, the ability of selenoxides is strongly dependent on the electronic effects of the substituents attached to the aryl moiety. The best catalyst was benzyl 3,5-bis(trifluoromethyl)phenyl-selenoxide **147d**, which is 13 times more active than selenoxide **147a**. Recently, Back and co-workers [93] developed a new catalyst for oxidative reactions using hydrogen peroxide. The seleninate esters **170a-c** were efficiently used in an oxidation of sulfides to sulfoxides, enamines to α-hydroxyketones and epoxidation of olefins. Interestingly, the catalyst **170a** gave better results, both in terms of yield and in the selectivity compared to benzeneseleninic acid **138**. Under the same reactions conditions and in the presence of magnesium sulfate, benzeneseleninic acid **138** yielded *vic*-diols whereas the seleninate ester **170a** exclusively the epoxide. Seleninate esters **170b** and **170c** were less selective than **170a**.

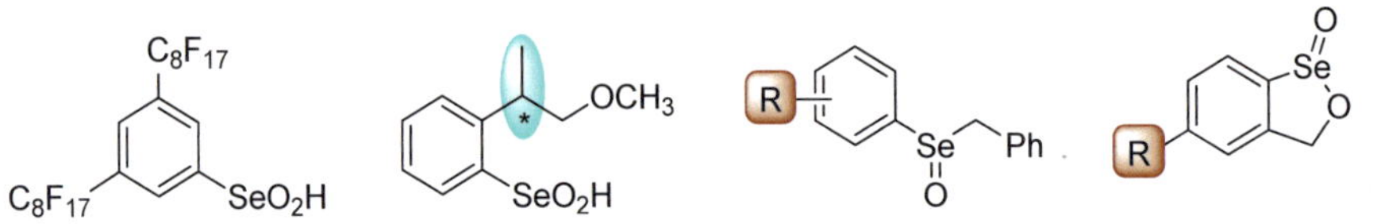

144 **169**

147a: R = H
147b: R = 4-OMe
147c: R = 4-NO2
147d: R = 3,5-CF3

170a: R = H
170b: R = OMe
170c: R = F

Scheme 8.52: Organoselenium catalyst used in the epoxidation reaction.

As mentioned above, the rate of the reaction as well as the selectivity in the formation of products is not only modulated by the nature of the catalyst, it is also affected by the reaction conditions. Among the common solvents used for epoxidation, 2,2,2-trifluoroethanol, which is miscible in H_2O_2 solution, is by far the best solvent. The success of this solvent has been attributed to its ability to make H_2O_2 a better nucleophile and hence a more active oxidant [94]. Very recently, the successful use of glycerol-based solvents for epoxidation reactions catalyzed by diselenides was reported [88]. Some of these glycerol-based solvents showed activity similar to 2,2,2-trifluoroethanol and better results were obtained when compared with dichloromethane. Furthermore, another important feature of glycerol-based solvents is the possibility to reuse the catalytically active phase after removal of the epoxide, which can be subsequently recharged with fresh H_2O_2 and the substrate. This solvent is a promising alternative to halogenated one due to the environmental advantages associated with its use, since they are renewable, non-toxic and non-volatile [95]. The use of a catalytic amount of base has a positive effect on the course of epoxidation [90]. The role of the base is to neutralize the acidic character of the H_2O_2 solution (pH 2.5-5.0) [96] and thus avoid the formation of *vic*-diols. The formation of these side products can also be conveniently reduced by the use of a drying agent such as $MgSO_4$ as a scavenger of the excess of water [89].

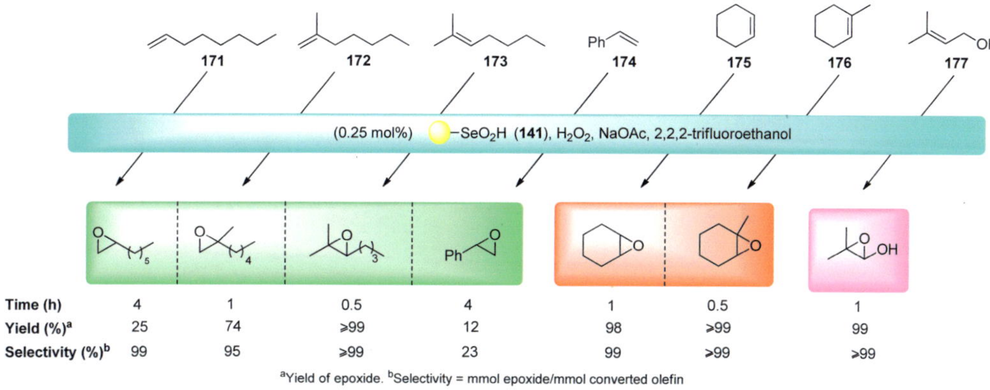

Scheme 8.53: Seleninic acid **141**-catalyzed epoxidation of olefins with hydrogen peroxide.

The results obtained in the reactions of different olefins with H_2O_2 catalyzed by seleninic acid **141** containing two CF_3 groups in the *meta* position of the aryl moiety are illustrated in Scheme **8.53** [90]. Primary alkyl **171** and aryl olefins **174** are more resistant toward epoxidation, and this is clearly evident since the reaction rates are much longer and the yields are low. On the other hand, secondary **172** and tertiary **173** olefins are much more reactive under the same conditions. This trend can also be noted in the cyclic olefins, for example, 1-

methylcyclohexene **176**, as expected, reacted faster than the less substituted cyclohexene **175**. Allylic alcohol **177** is selectively oxidized, albeit at a relatively low rate, due to decreased electron density at the double bond. This trend in the substrate reactivity demonstrates that electron rich olefins are better substrates in this valuable synthetic transformation.

Very recently, Santi and co-workers reported the first general eco-friendly protocol using diphenyl diselenide **14** as catalyst to prepare *vic*-diols [87]. These compounds were prepared through the epoxidation of alkene **179** by perseleninic acid **178** followed by the ring opening of epoxides **180** (Scheme **8.54**) in good yields and diastereoisomeric excess. The *syn/anti*-selectivity in the formation of products **181** and **182** is associated with the electronic and steric properties of the starting alkene [87]. An important feature of this methodology is the possibility to recover the catalyst as benzeneseleninic acid at the end of the reaction. The same research group demonstrated that sulfur-containing chiral diselenide can catalyze the enantioselective dihydroxylation of 1-phenyl-1-cyclohexene using hydrogen peroxide, leading to good enantiomeric excess (92% *ee*) when the reaction was performed at -10°C.

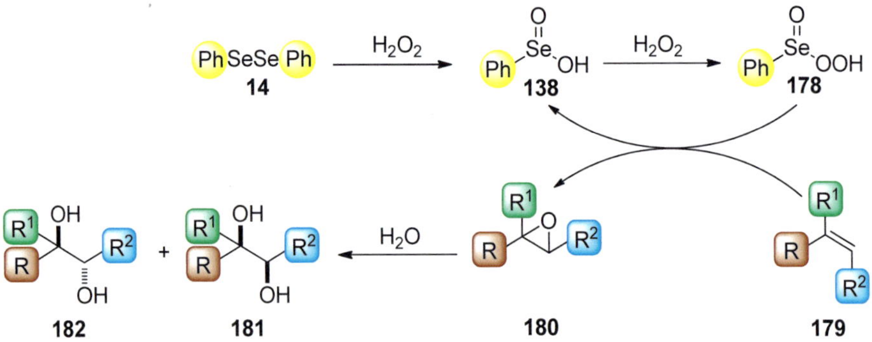

Scheme 8.54: Alkene dihydroxylation catalyzed by diphenyl diselenide.

Inspired in the GPx mechanism, Santi and coworkers reported the use of *L*-selenocystine **183** as a pre-catalyst for the synthesis of 1,2-diols (Scheme **8.55**) [97]. This new approach represents an "eco-friendly" protocol, since low-toxicity reagents, non-toxic solvents and mild conditions are used. Using water as a solvent and hydrogen peroxide as an oxidant, *L*-selenocystine **183** proved to be a better catalyst showing higher turnover numbers when compared with diphenyl diselenide **14**. This better performance in terms of catalytic activity can be attributed to the fact that the carboxylic moiety of *L*-selenocystine can act in a similar way to the epoxide hydrolases [98]. The presence of the catalyst perseleninic acid **185** was demonstrated by NMR spectroscopy analysis of ^{77}Se-NMR and ^{1}H-NMR. The results indicated that, once again, the stereocontrol in this reaction is dependent on the electronic and steric properties of the starting alkene **179**. Thus, alicyclic olefins afforded exclusively the *anti*-diol isomer **182** in good yields. However, if an aromatic substituent is present on the carbon-carbon double bond the stereoselectivity is affected and a mixture of *syn*-**181** and *anti*-1,2-diols **182** are obtained, respectively. These results could be explained considering that, in these epoxide intermediates the ring-opening reaction mainly occurs through the formation of a stabilized benzyl carbocation, leading to a mixture of *syn*- and *anti*-1,2-diols. Good enantiomeric ratios are obtained for 1,2-diols in some cases. One important feature of this protocol is the possibility to recover both the catalyst and the reaction medium, increasing the sustainability of the process. Thus, the system was reused for five runs and it was observed that the yields were almost constant for the first two runs, followed by a decrease for the following three runs [97]. When the reaction was performed using methanol as the co-solvent α-methoxyalcohol was obtained, however, the selectivity of the product is affected negatively by the non-catalyzed side oxidation reaction.

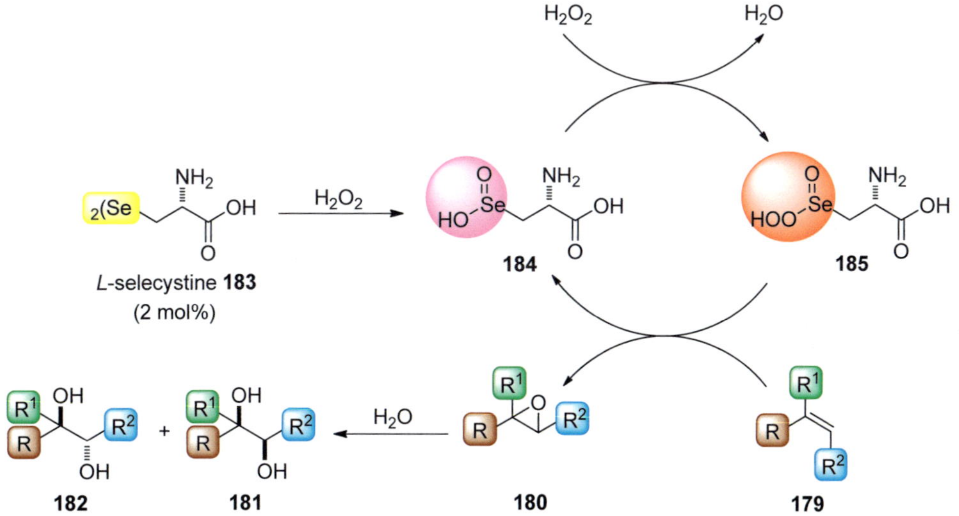

Scheme 8.55: *L*-Selenocystine-catalyzed stereoselective dihydroxylation of alkenes.

8.3.4. Oxidative Bromination

Halogenated compounds are important building blocks as they are precursors of a number of important molecules [99]. In this way brominated compounds have received considerable attention in recent years. In nature, biological systems use haloperoxidase enzymes to catalyze the oxidation of ion halides by hydrogen peroxide to form a halogenating reagent which is responsible for the halogenation of organic substrates [100]. However, the traditional method for the bromination of organic substrates in the laboratory is usually carried out with hazardous, toxic reagents in the presence of halogenated solvents. In this context, the wide applicability of brominated compounds in modern synthetic chemistry has driven the development of new methodologies for their preparation, not only in an efficient manner, but also using environmentally-sustainable protocols [101].

The ability of organoselenium compounds to activate peroxides for a number of oxidation reactions is already well known. Various substituted aryl seleninic acids and selenoxides were found to be appropriate catalysts for the oxidation of bromide salts employing H_2O_2 as the green oxidant, in other words, organoselenium compounds mimic enzymes like bromoperoxidazes that activate hydrogen peroxide [102].

The reaction sequence for the oxidation of bromide salts by H_2O_2 promoted by organoselenium catalysts is depicted in Scheme **8.56**. These reactions are normally carried out in phosphate buffer pH 6 or, depending on the organic substrate solubility, in a 1:1 (v:v) mixture of buffer and an organic solvent, most commonly dichloromethane, diethyl ether or 1,4-dioxane. Two possible pathways are illustrated for both seleninic acids **133**, cycle “A” [103], and selenoxides **186**, cycle “B” [104], and the exact mechanism associated with the oxidation of bromine salts to “Br^+” catalyzed by organoselenium compounds is not completely elucidated to date. Seleninic acids **133**, cycle “A”, or selenoxides **186** cycle “B”, when subjected to reaction with H_2O_2 are in equilibrium with the corresponding perseleninic acid **134** or hydroxy perhydroxy selenane **187**, respectively, and in both cases the bromide can react with **134** cycle “A” or **187** cycle “B” to produce the species **188** or **189**, respectively. This, in turn, may itself act as a brominating agent, reacting with the substrate or with another equivalent of bromide to deliver

Br_2. In both hypotheses, seleninic acid **133** cycle "A" or selenoxide **186** cycle "B" are regenerated to restart the catalytic cycle with H_2O_2. On the other hand, bromide would attack the terminal oxygen of perseleninic acid **134** cycle "A" or the terminal oxygen of perhydroxy selenane **187**, affording HOBr as the positive bromine species and seleninic acid **133** or selenoxides **186** to resume the catalytic cycle.

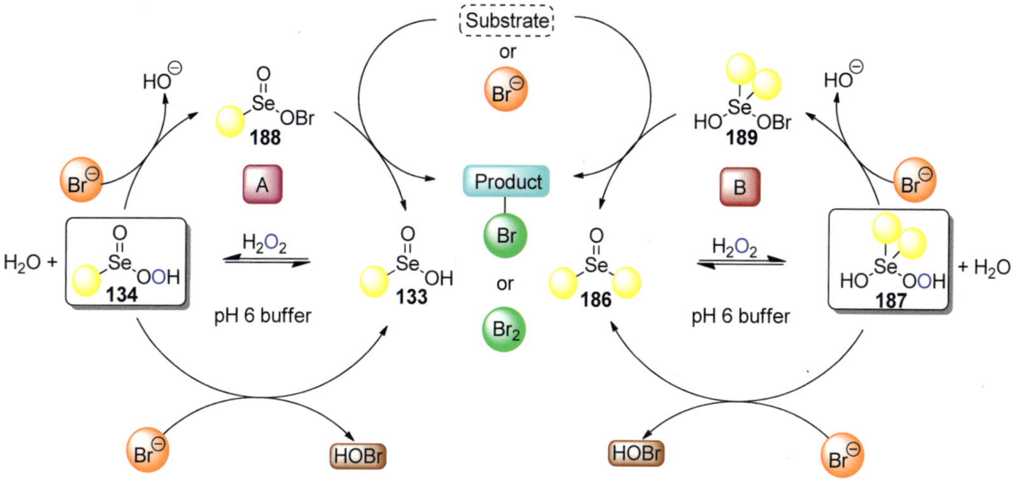

Scheme 8.56: Catalytic cycles for the bromination reaction of different organic substrates catalyzed by seleninic acids **133** (Cycle A) and selenoxides **186** (Cycle B) in the presence of hydrogen peroxide.

Similarly to epoxidation and Baeyer-Villiger reactions, the rate of oxidation of bromide to "Br^+" promoted by H_2O_2 activated by organoselenium catalysts is dependent on the electronic demand of the catalysts. However, in contrast to the Baeyer-Villiger and epoxidation reactions, which are accelerated by organoselenium catalysts bearing electron poor groups, the oxidation of bromide salts is slightly faster when electron-rich organoselenides are employed.

All seleninic acids shown in Scheme **8.57**, which are prepared *in situ* from the corresponding diselenide, are able to promote the bromolactonization of 4-pentenoic acid **190** with NaBr and H_2O_2 in two-phase systems of ether and phosphate buffer pH 6. Benzeneseleninic acid **138** and 4-methoxyphenylseleninic acid **192a** are more efficient when compared with the other seleninic acids. The catalyst 4-nitrophenylseleninic acid **168** is slightly slower than **192a**, however, the catalyst 3,5-*bis*(trifluoromethyl)phenylseleninic acid **141**, which is the best catalyst in other oxidation reactions promoted by H_2O_2, was the less efficient in this case. Sterically-hindered seleninic acid **192c** is also an efficient catalyst [103]. On the other hand, the 4-dimethylaminophenylseleninic acid **192b** does not follow the trend of the electronic demand in aryl seleninic acids in the oxidation of bromide salts, in spite of the fact that this seleninic acid has the most electron-donating substituent among the seleninic acids tested. Its lower efficiency may be reasoned as being due to the contribution of the equilibrium between the resonance species **192b** and **193** (Scheme **8.57**, B). This equilibrium might be responsible for a slower formation of the corresponding perseleninic acid from **192b**, affecting the reaction rate.

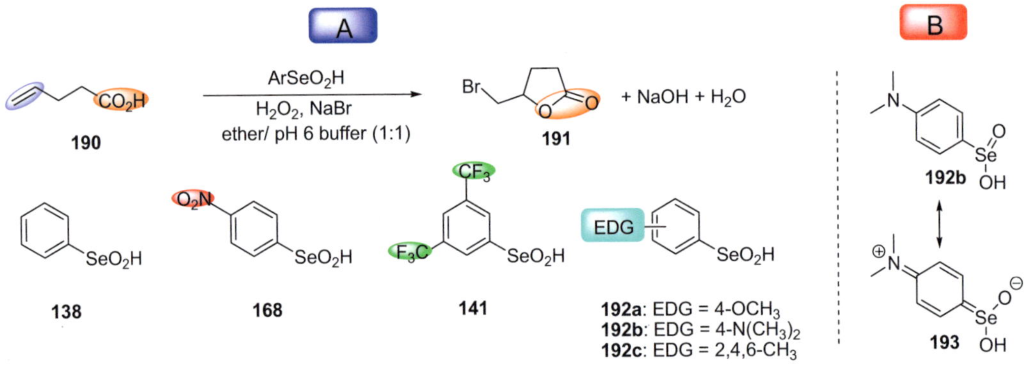

Scheme 8.57: A – Seleninic acids with different electronic demands for bromolactonization reaction of 4-pentenoic acid **190**. B – Resonance equilibrium of 4-dimethylaminophenylseleninic acid **192b**.

More recently, a water-soluble imidazolium-containing diselenide was reported by Detty and co-workers as an efficient catalyst for bromination reactions in the absence of organic solvent (Scheme **8.58**) [105]. An interesting aspect of these new organoselenium compounds is the water solubility, a characteristic uncommon for organic diselenides. This important feature is due to the presence of the *N*-methylimidazolium group which gives freely soluble compounds in water and improves the catalytic activity of these new compounds. Thus, in a similar way, compounds **194-196** were converted *in situ* into seleninic acids by the reaction with hydrogen peroxide for the bromination of 4-pentenoic acid **190** with NaBr and phosphate buffer pH 6. The results showed that the *ortho*-substituted catalysts **194** and **195** are approximately 3-fold more effective than

196 with selenium functionally in the *para*-position. It was demonstrated that the pH had a notable influence on the reaction course. When the reaction was carried out at pH 4.2 (NaH_2PO_4) it was approximately two times faster than the reaction carried out at pH 6.1. The addition of dioxane as a co-solvent with phosphate buffer pH 4.2 increased the yield. Under these conditions the reaction is homogeneous and the product is easily extracted from the 3/1 buffer/dioxane solvent. When diphenyl diselenide was used under the optimized conditions it was found to be a less effective catalyst compared to the diselenide **194**.

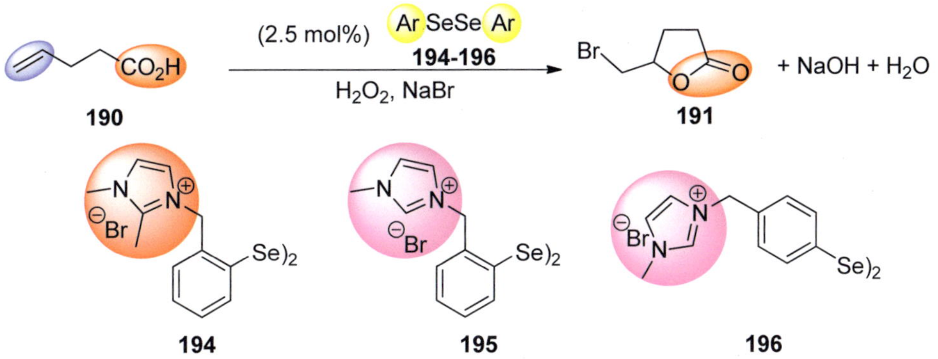

Scheme 8.58: New imidazolium-containing diselenides.

The oxidation of bromine salts by the selenoxides in a two-phase mixture of dichloromethane and phosphate buffer pH 6 follows the same trend observed for seleninic acids (Scheme **8.59**): the reaction rate is enhanced by selenoxides bearing electron donating groups. Assigning diphenyl selenoxide **197** as a standard catalyst for comparison purposes, its activity is almost 7-fold lower than the electron-richer dibenzyl selenoxide **198** and approximately 4-fold lower than benzyl phenyl selenoxide **147a**. The performances of 4-methoxyphenyl benzyl **147b** and 4-dimethylaminophenyl benzyl selenoxides **199**, rise up to 6 and 18-fold more efficient, respectively, if compared to **197**. Additionally, selenoxide **200**, bearing a chelating group, showed remarkable activity, catalyzing the oxidation of bromide and making it 28 times faster than the same reaction catalyzed by **197**. Furthermore, the activity of selenoxide **147d**, carrying a strong electron-withdrawing group CF_3, is much lower than that exhibited by electron-rich catalysts [104]. This effect could be explained on the basis of the fact that electron-donating and chelating groups should increase the basicity of the selenoxide oxygen and, consequently, the stability of the intermediate proposed by Detty and coworkers, promoting the faster formation of the hydroxy perhydroxy selenane **187** (Scheme **8.56**) and thus increasing the reaction rate. On the other hand, electron-withdrawing groups would decrease the basicity and the stability of this intermediate, resulting in a slower reaction rate [104].

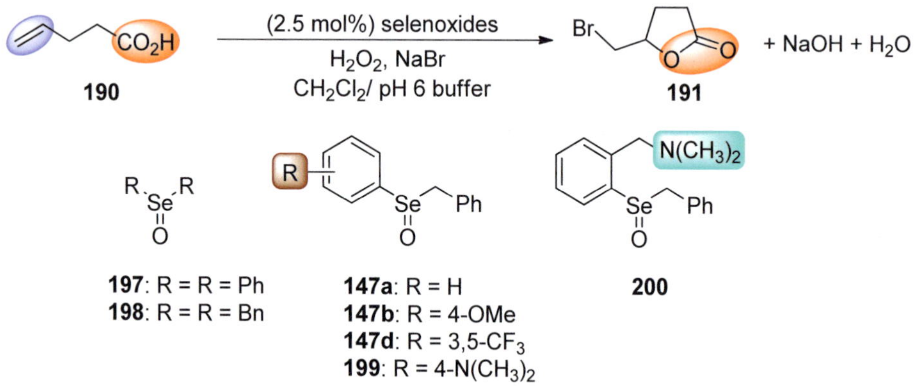

Scheme 8.59: Selenoxides with different electronic demands for bromolactonization reaction of 4-pentenoic acid.

Catalysts bound to dendrimers with phenylseleno groups have been described by Detty and co-workers [106]. These catalysts have been used as excellent activators of H_2O_2 for the oxidation of bromine salts (Scheme **8.60**). The performance of these catalysts **201-203** is closely associated with the number of selenium groups attached to the dendrimer core and the rate of catalysis for dendrimers increased as the number of phenylseleno groups in each molecule increased. Impressive turnover numbers (TON) of >60,000 moles of H_2O_2 consumed and TOF values of >2,000 moles of brominated product formed per mole of catalyst can be achieved using a dendrimer with 12 PhSe groups, as in dendrimer **203**. The enhanced catalytic activity was attributed to the autocatalysis, in which each PhSe group is converted to the selenoxide by the positive bromine species generated in the background reaction [107]. One important feature of the catalyst **203** is the possibility to recover it from the organic phase of the reaction and reusing it in further reactions without apparent loss of activity [106].

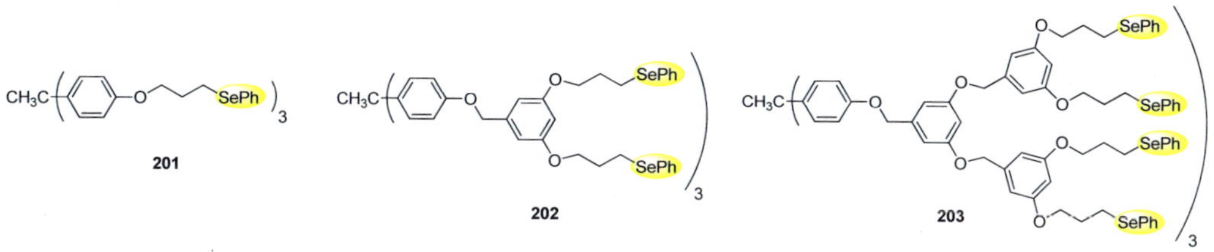

Scheme 8.60: Dendrimer catalysts for the activation of H_2O_2 for oxidation of bromine salts.

The same group recently developed an efficient catalyst selenoxide **204**, which is sequestered within halide-permeable xerogel, formed from 10/90 (mol/mol) of 3-aminopropyltriethosysilane/tetramethoxysilane **205** through bromination reactions (Scheme **8.61**) [108]. It was observed that the sequestration not only makes the catalysts 23-times faster compared to the xerogel-free catalyst in solution, but allows the easy separation of the reaction mixture by filtration which can then be used again without apparent loss of activity for at least four cycles. A decrease in its catalytic activity can be detected after the production of 80 mol of brominated product per mol of **205**.

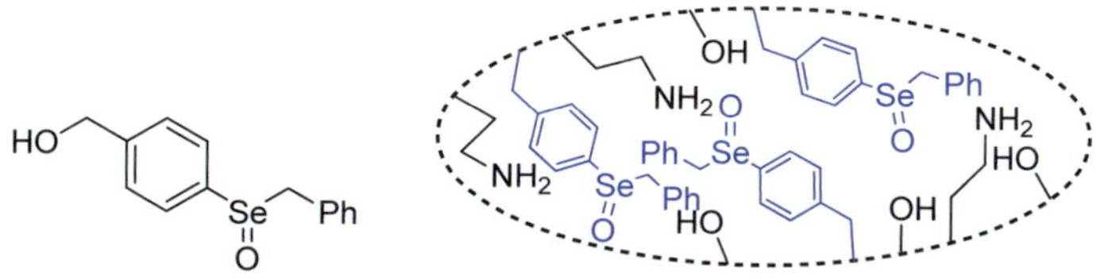

Scheme 8.61: Xerogel-sequestered selenoxide catalyst.

Synthetically, the oxidative halogenation of different organic substrates allows access to a series of halogenated organic compounds, such as those depicted in Scheme **8.62**. In this context alkenes and activated aryl compounds are suitable substrates to react with the positive bromine species formed in the oxidation of bromide salts. These reactions are usually conducted in a phosphate buffer solution (pH 4.2-6.1) in the presence of a solvent. Bromolactonization of alkenoic acids **206** in some cases produces a mixture of the dibromo carboxylic acid **207** and bromolactone **208**. However, if the mixture is stirred for an additional time, bromolactone **208** is isolated as the only product. The bromination of 2-(1-cyclohexenyl)acetic acid **209** gives a mixture of β- **210** and γ-bromolactones

211 (69:13) respectively. Unsaturated alcohols **212** also undergo reaction to afford cyclic bromoethers **213** and dibromoalcohol **214**. Under two-phase reaction conditions, CH_2Cl_2/aqueous phosphate buffer pH 6, cyclohexene **215** is converted into *trans*-1,2-dibromocyclohexane **216** and *trans*-2-bromocyclohexanol **217**. Despite the faster formation of brominated products in catalyzed reactions compared to the uncatalyzed ones, the selectivity is somehow independent of the action of the catalyst, affording mixtures in the range of 1:1.2-1:1.9 of **216** and **217**, respectively. Aromatic compounds bearing electron donating groups **218** are brominated, with a preferential regioselectivity toward the formation of *para*-aryl bromides **219**.

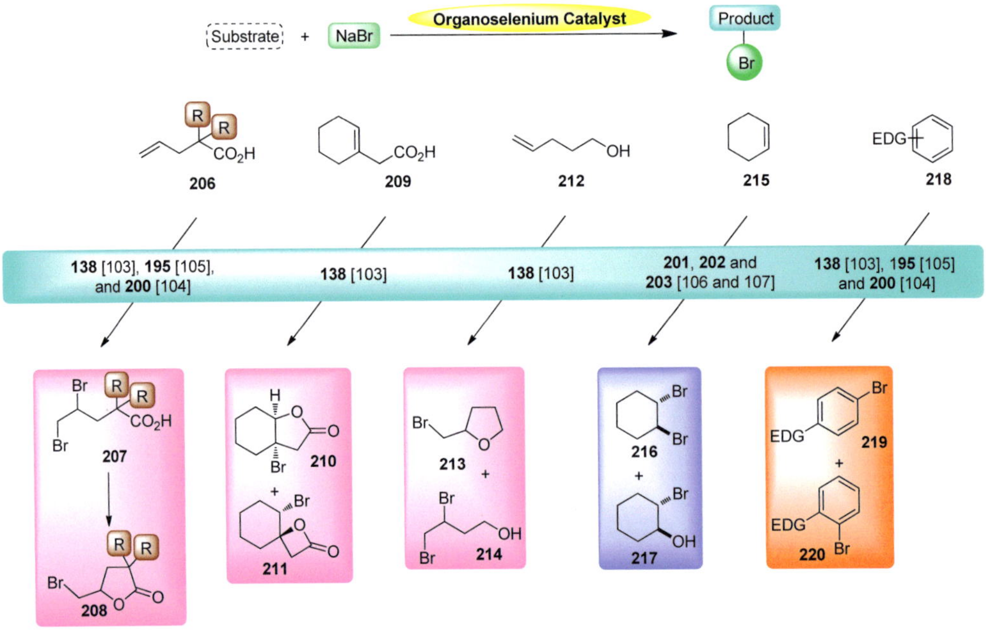

Scheme 8.62: Substrate scope and selectivity in oxidative halogenation reaction promoted by different organoselenium compounds.

8.3.5. Miscellaneous Reactions

Although the most extensively explored synthetic transformations mentioned in this chapter are those discussed in the first sections, the full potential of organoselenium compounds as eco-friendly catalysts for transformations involving different organic substrates has yet to be realized. In this context, the use of small organic selenium-containing molecules for different transformations has emerged in recent years and can be applied in the continuous search for "green reactions".

N-bromosuccinimide (NBS) and *N*-chlorosuccinimide (NCS) have been used by Tunge and co-workers as "preoxidized" halogen sources in organoselenium-catalyzed oxidative halogenation reactions [109], such as halolactonization [110], allylic halogenation [111] and α-halogenation of β-ketoesters and ketones [112]. The mechanisms of these reactions proceed *via* electrophilic halogenation in the presence of a selenium catalyst, which is responsible for enhancing the electrophilicity of the oxidized halogen compounds. In this regard, diphenyl diselenide **14** has been used as an efficient catalyst for the bromolactonization of carboxylic acids not only accelerating the reaction but, in some cases, allowing the regiocontrol in favor of γ-lactone formation. The mechanism suggested by the authors involves the activation of NBS **221** by nucleophilic attack of diphenyl diselenide **14** affording the cationic selenium halide with a succinimide counterion **222** (Scheme **8.63**). It was proposed that in the next step the succinimide anion deprotonated the carboxylic acid **223** and, through a nucleophilic displacement of the selenium-coordinated bromonium ion **224**, the lactone ring **225** was obtained regenerating the catalyst diphenyl diselenide **14**.

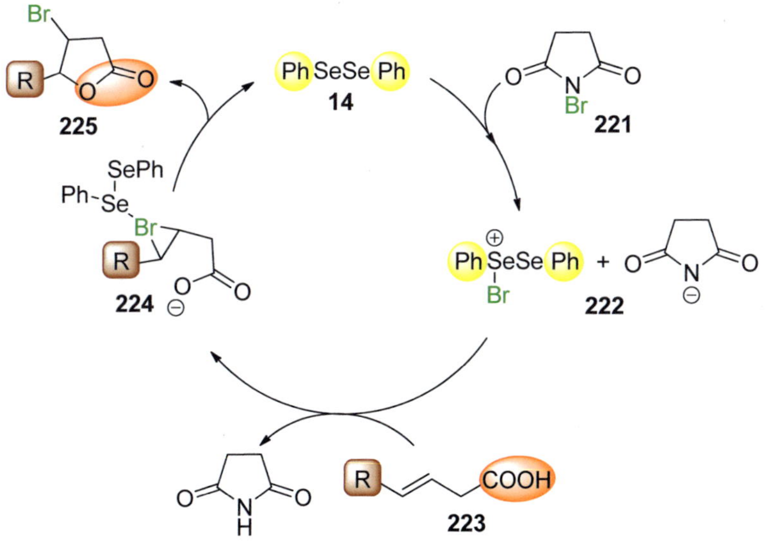

Scheme 8.63: Diphenyl diselenide catalyzed bromolactonization, using NBS as halogen source.

Phenylselenenyl chloride **226** has found application as a catalyst in the allylic chlorination of allylic acids, esters, arenes, nitriles and prenyl olefins in the presence of NCS as a halogen source (Scheme **8.64**) [111]. The selectivity of the reaction is strongly dependent on the electron-withdrawing groups present in the substrate **227** which are responsible for the formation of conjugate products **228**. The selectivity of the halogenation of prenyl olefins **229** can also be achieved by sterically differentiating the protons capable of elimination. Thus, the selenocatalytic chlorination leads to the allylic product **230** with high selectivity, since this product requires the adoption of a low-energy conformation during the selenium elimination. However, from a "green" chemistry standpoint NCS are less appealing when compared with other halogen sources such as NaBr and NaCl. Tunge showed the possibility to use NaCl as a halogen source in a biphasic system, H_2O/CH_2Cl_2, for the allylic chlorination of 2-methyl-2-hexene, however, the corresponding chlorinated product was obtained in low yield [109].

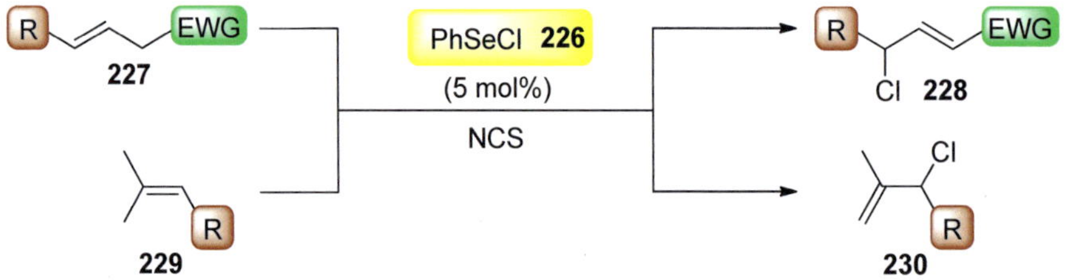

Scheme 8.64: Phenylselenenyl chloride catalyzed allylic chlorination.

Under similar conditions, commercially available polymer-supported selenenyl bromide **231** has been used as an efficient catalyst for the allylic chlorination of polyprenoids **232** in the presence of NCS [113]. The reaction showed high regioselectivity toward the functionalization of the terminal double bond (Scheme **8.65**). Moreover, the reaction was compatible with the presence of different functional groups.

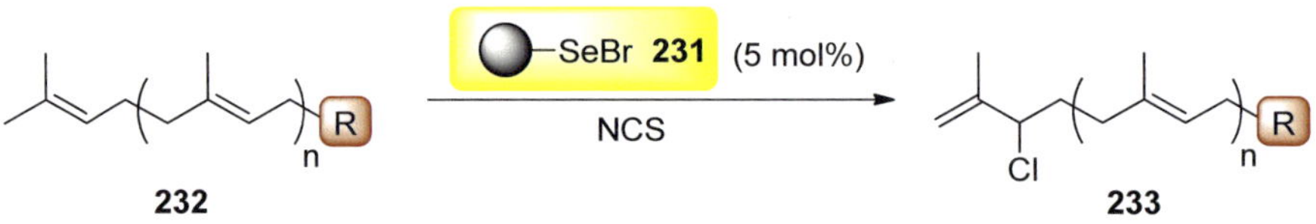

Scheme 8.65: Selenocatalytic allylic chlorination of polyprenoids.

The α-halogenation of the carbonyl compounds **234** is an important reaction in organic chemistry, due to the relevance of the products as synthetic precursors and pharmaceuticals. [114] To accomplish this transformation phenylselenenyl chloride **226** was used as a catalyst for the activation of the electrophilic chlorine source NCS, and proved to be more efficient than diphenyl diselenide, providing halogenated products **235** in good yields (Scheme **8.66**) [112]. Ebselen was also used as a catalyst for the α-halogenation of β-ketoester, however, the rate of catalysis was slow due to its low solubility.

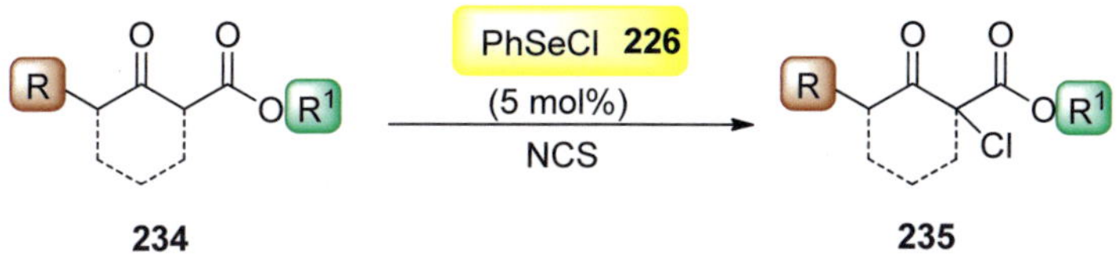

Scheme 8.66: α-Halogenation of β-ketoesteres.

The oxidation of allylic systems using selenium reagents is an important tool for the introduction of a hydroxyl or carbonyl group into various organic structures. Selenium dioxide is frequently used in a stoichiometric amount to accomplish this transformation. In 1985, Crich and Barton described the use of 2-pyridineseleninic anhydride **236**, prepared *in situ* by oxidation of the corresponding diselenide with iodoxybenzene, as a stoichiometric oxidant for the conversion of olefins to alkenones (Scheme **8.67**) [115]. Compared to benzeneseleninic anhydride **237**, this new organoselenium reagent is a better catalyst for the allylic oxidation of alkenes to alkenones. The enhanced performance of 2-pyridineseleninic anhydride **236** over benzeneseleninic anhydride **237** can be attributed to the influence of the electron-withdrawing properties of the pyridine ring which according to the authors renders the Se=O bond a better eneophile. An improvement in terms of efficiency and catalyst turnover was achieved by introducing stronger electron-withdrawing groups in the aromatic ring. In this context, 2-(*N*-oxide)pyridineseleninic anhydride **238** and pentafluorobenzeneseleninic acid **239** were designed as catalysts to promote the efficient allylic oxidation of alkenes in the presence of re-oxidant *t*-butyl hydroperoxide (TBHP) and (diacetoxyiodo) benzene respectively [116]. Unfortunately these were found to have a number of undesirable characteristics, not least of which was a tendency toward explosive decomposition. A further improvement came with the development of a new fluorous seleninic acid **240** (Scheme **8.67**) which was not associated with the same problems as **239** [117].

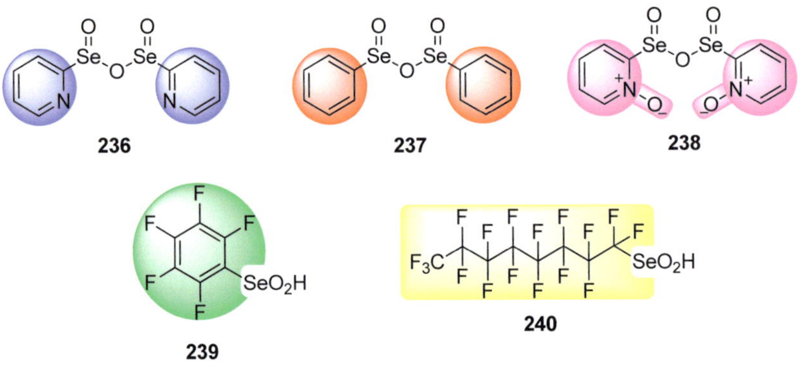

Scheme 8.67: Organoselenium compounds used as catalysts in the allylic oxidation of alkenes.

The catalyst **240** developed by Crich proved to be quite versatile and can be used for different reactions, such as in the allylic oxidation of alkenes **241** to enones **242** [117], in the oxidation of aryl alkyl ketones **243** to the corresponding ketoacids **244** [118] in the presence of stoichiometric amounts of iodoxybenzene as the oxidant (Scheme **8.68**). A notable advantage in using this catalytic system is that the catalyst can be recovered in the form of *bis*(perfluorooctyl) diselenide **245**, through a reductive workup with sodium metabisulfite, which itself serves as a convenient catalyst precursor. For both oxidations reactions, the results obtained for the recovery of *bis*(perfluorooctyl) diselenide **245** are comparable with those observed for perfluorooctylseleninic acid **240**.

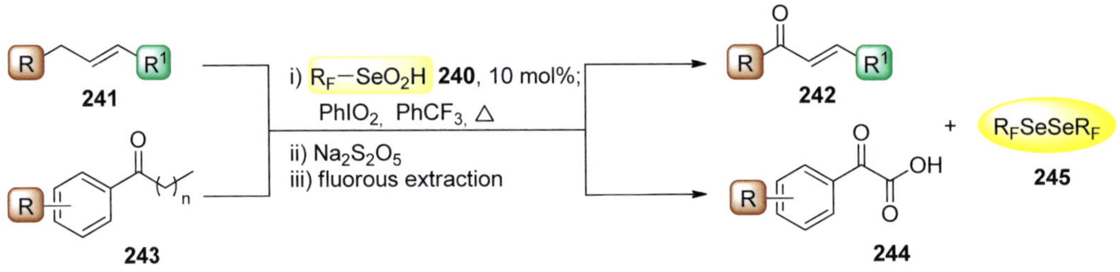

Scheme 8.68: Allylic oxidation, oxidation of ketones and of aryl alkyl ketones catalyzed by perfluorooctylseleninic acid with $PhIO_2$.

The conversion of alcohols to carbonyl compounds is one of the most important transformations in synthetic organic chemistry. This transformation has been the subject of renewed interest due to its widespread relevance [119]. Benzeneseleninic acid anhydride **237** was introduced as a stoichiometric reagent in the oxidation of alcohols in the late 1970s by Barton and co-workers [120]. With the aim of avoiding the use of stoichiometric amounts of seleninic anhydride, Kuwajima introduced the use of *t*-butyl hydroperoxide as the oxidizing agent, which is responsible for oxidizing aromatic diselenide to the corresponding anhydrides [121]. In the light of gaining a deeper investigation regarding how catalytic amounts of benzeneseleninic acid anhydride **237** can affect the oxidation of alcohols, a revision of its mechanism has been recently proposed using diphenyl diselenide together with TBHP as the oxidizing agent (Scheme **8.69**) [122]. The revised mechanism considers the initial reaction of diphenyl diselenide **14** with TBHP to produce the corresponding anhydride **237**. After this, **237** reacts with the alcohol **246** to afford the adduct **247** and seleninic acid **138**. Subsequently, **247** undergoes rearrangement to yield the desired product **248** and selenenic acid **90**. Both selenenic and seleninic acids **90** and **138**, respectively, react with another equivalent of TBHP to regenerate the active anhydride **237** and restart the catalytic cycle. The substrate scope is basically restricted to activated alcohols such as benzyl and allyl alcohols, while alkyl alcohols are generally inert. On the other hand, the levels of selectivity are high, furnishing aldehydes as major products, especially when the concentration of TBHP is kept low during the course of the reaction and the water formed during the reaction is removed with drying agents like $MgSO_4$ or molecular sieves 4 Å.

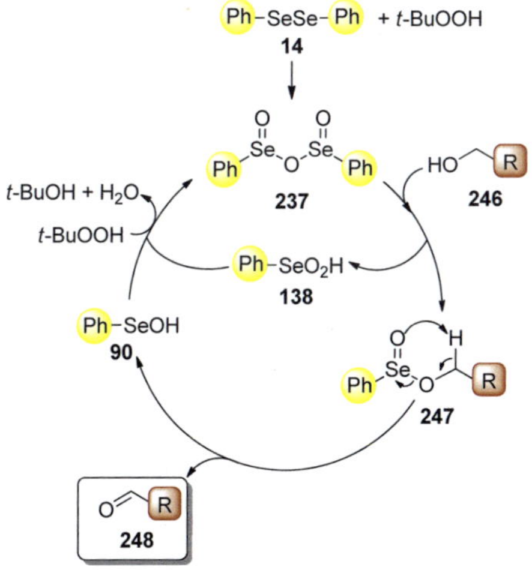

Scheme 8.69: Catalytic cycle for the oxidation of alcohols with TBHP and diselenide.

Another interesting reaction where organoselenium catalysts are also suitable activators of peroxides involves the oxidation of sulfides **249** to sulfoxides **250** or sulfones **251** (Scheme **8.70**) [85]. The selectivity in the formation of products is affected by the reaction conditions and a shorter reaction time together with a lower concentration of peroxide leads to sulfoxides **250**. On the other hand, with a longer reaction time and higher concentration of peroxide sulfones **251** are the preferred products. A broad array of catalysts has been successfully employed in this valuable transformation: 2-nitrobenzeneseleninic acid **166a** [85], 2-penhylselenobenzoic acid **252** [123], Ebselen **161** [124] and its silica imprinted analogue **253** [125]. From a "green" chemistry standpoint the silica-supported catalyst **253** is more attractive since it can be easily recovered from the reaction mixture and reused. A further versatile catalyst, which emerged very recently, is the cyclic seleninate ester **170a**. This catalyst has been used with success in the epoxidation of different olefins [93]. When **170a** was used in the oxidation of sulfide in the presence of trifluoroacetic acid it showed the best overall performance, as it minimized the overoxidation to sulfones **251**.

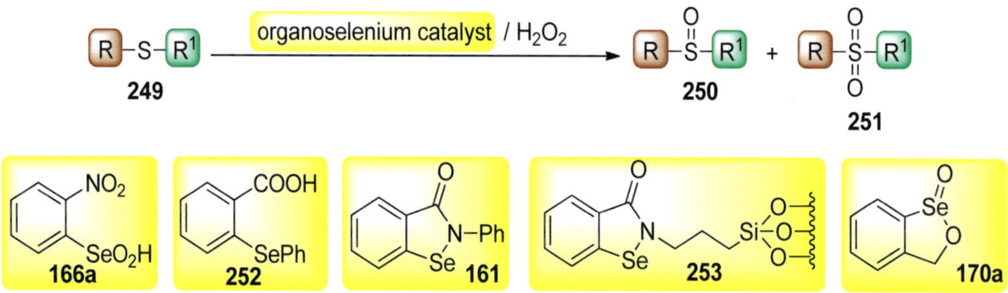

Scheme 8.70: Oxidation of sulfides to sulfoxides or sulfones catalyzed by organoselenium compounds in the presence of hydrogen peroxide as the oxidant.

Seleninate ester **170a** has also been applied as a catalyst in the oxidation of different enamines to α-hydroxyketones in the presence of hydrogen peroxide [93]. The oxidation of enamine **254** under optimized conditions afforded the dimeric product, which can be isolated or can undergo hydrolysis *in situ* with HCl, leading to the desired α-hydroxyketones **255** (Scheme **8.71**). The substrate scope is basically restricted to morpholine-based enamines of cyclic and acyclic ketones, while a more hindered enamine failed under these conditions.

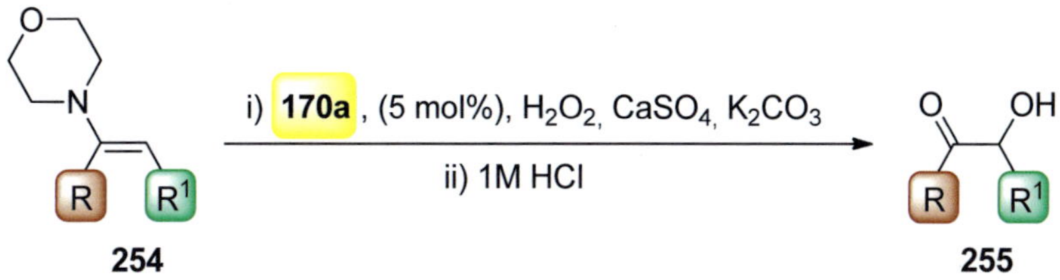

Scheme 8.71: Oxidation of enamines to α-hydroxyketones catalyzed by seleninate ester 170a.

Nitroso compounds are valuable building blocks for a variety of organic transformations and different methods have been used to prepare these compounds. Organoselenium catalysts have been used to obtain these compounds [126]. As an example, perseleninic acid generated from diphenyl diselenide **14** in the presence of H_2O_2 has been used successfully for the oxidation of aromatic amines **256** to aromatic nitroso compounds **257** [127]. This synthetic transformation allows the preparation *in situ* of **257**, which is subsequently reacted with a diene **258** in a hetero-Diels-Alder reaction to afford a variety of products **259** (Scheme **8.72**). Successful hetero-Diels-Alder can be affected with cyclic and acyclic dienes and with different aniline substrates. The method is very versatile and allows a wide variety of heterocyclic structures to be constructed with good selectivity.

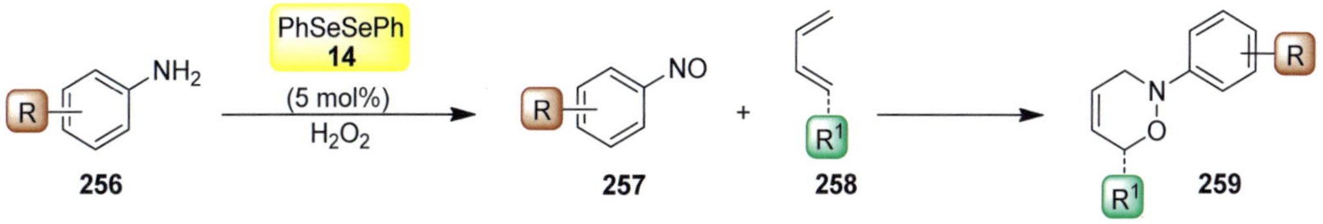

Scheme 8.72: Synthesis of nitroso derivatives **257** by oxidation of aryl amines catalyzed by perseleninic acid.

It is known that the treatment of cycloalkanones in the presence of selenium dioxide and hydrogen peroxide results in a skeletal rearrangement *via* a cyclopropanone intermediate to give carboxylic acids and this reaction is known as the Favorskii rearrangement [128]. Unfortunately, the yields obtained using selenium dioxide are low since the reaction leads to a mixture of products. The use of hydrogen peroxide with an organoselenium catalyst has emerged as a more efficient and selective reaction. In this context, Mlochowski described the use of an easily accessible poly(bis-1,2-antracenylene) diselenide **260** as the catalyst in the oxidative conversion of cycloalkanones **261** into cycloalkanecarboxylic acids **262** in moderate yields using hydrogen peroxide as the oxidant [129]. This methodology is suitable for the synthesis of acids, particularly those possessing five, six and seven-membered rings (Scheme **8.73**). Moreover, the yield obtained with diselenide **260** using hydrogen peroxide is better than that obtained with selenium dioxide.

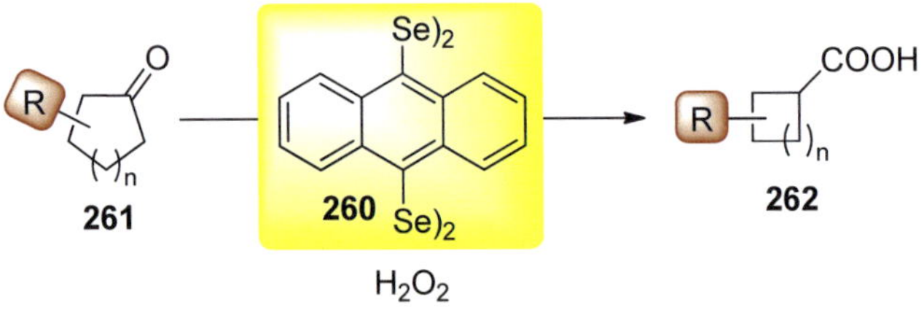

Scheme 8.73: Oxidative conversion of cycloalkanones into cycloalkanecarboxylic acids with H_2O_2 and poly(bis-1,2-antracenylene) diselenide.

Organoselenium catalysts are also used in the oxidation of 1,4-hydroquinones **263** to 1,4-benzoquinones **264** by H_2O_2 even on a kilogram scale (Scheme **8.74**). Catalytic loadings of diphenyl diselenide **14** can effectively promote this transformation in the presence of a phase transfer catalyst, which enhances the rate of reaction in the biphasic system of aqueous H_2O_2 solution and substrate. Substituents such as alkyl, halogen or alkoxy in the substrate framework do not affect the course of the reaction [130].

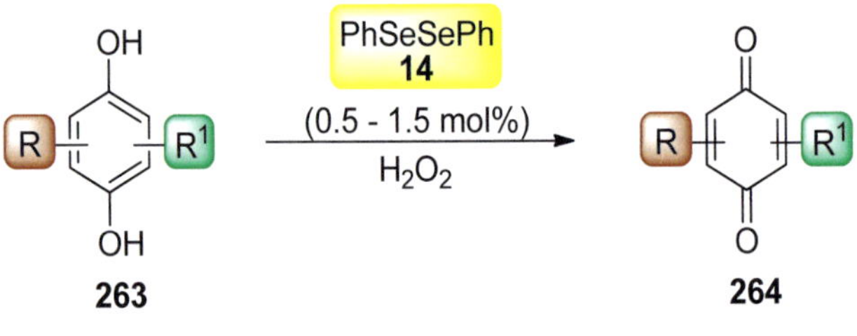

Scheme 8.74: Oxidation of 1,4-hydroquinones with H_2O_2 and diphenyl diselenide.

Nitrogen-containing compounds can also be oxidized employing catalytic amounts of organoselenium compounds like 2-nitro-phenylseleninic acid **166a** or Ebselen **161** in the presence of green oxidants (hydrogen peroxide) (Scheme **8.75**). Through this approach, aliphatic and aromatic aldoximes **265** have been oxidized in the presence of different alcohols to the corresponding esters **266** [131]. Esterification products are only obtained when primary and secondary alcohols are used, as tertiary alcohols do not lead to the formation of products **266**. In the same way, *N,N*-dimethyl aryl hydrazones **267** are converted into aryl nitriles **268** and aryl ketazines **269** are oxidized to the corresponding carbonyl compounds **270** in the presence of Ebselen **161** [132].

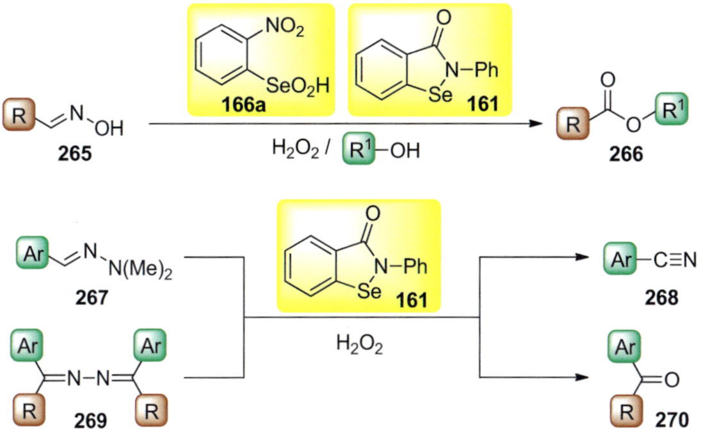

Scheme 8.75: Oxidation of nitrogen compounds using catalytic amounts of organoselenium compounds and H_2O_2.

The search for new organoselenium catalysts that can be used in more than one transformation has been an area of intense research. In this regard, very recently, a new class of isoselenazoles was synthesized and applied by Kumar and co-workers [133] as a catalyst for the activation of bromine in different reactions. Among these new catalysts, difluoro derivative **271** was used with success in the bromolactonization of a series of alkenoic acids **272** to afford a series of bromo lactone rings **273**, in the bromoesterification of different alkenes **274** with different carboxylic acids **275** and in the oxidation of secondary alcohols **277** to ketones **278**. All these reactions were performed in the presence of bromine or NBS as the stoichiometric oxidant (Scheme **8.76**).

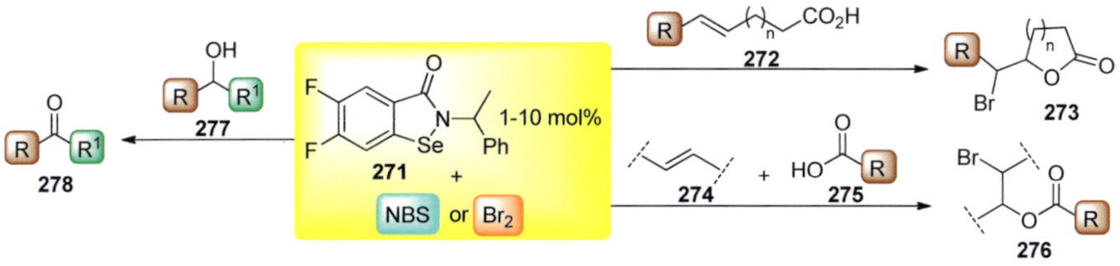

Scheme 8.76: A new class of isoselenazoles as a catalyst for oxidation reactions.

ACKNOWLEDGEMENT

The authors are grateful to the CNPq, CAPES, FAPESP (2013/06558-3), FAPESC and INCT-Catálise for financial support.

CONFLICT OF INTEREST

The authors confirm that this chapter contents have no conflict of interest.

REFERENCES

[1] Berzelius, J. J.; Eggertz, H. P.; Arfvedson, A. *Afhandlingar I Fysik, Kemi Och Mineralogi, Part 6*, **1818,** 42-144, particularly 68-74.

[2] (a) Rotruck, J. T.; Pope, A. L.; Ganther, H. E.; Swanson, A. B.; Hafeman, D. G.; Hoekstra, W. G. Selenium-Biochemical Role as a Component of Glutathione Peroxidase. *Science*, **1973**, *179*, 588-590. (b) Flohe, L.; Günzler, E. A.; Schock, H. H. Glutathione Peroxidase-Selenoenzyme. *FEBS Lett.*, **1973**, *32*, 132-134. (c) Flohe, L.; Andreesen, J. R.; Brigelius-Flohe, R.; Maiorino, M.; Ursini, F. Selenium, the element of the moon, in life on earth. *IUBMB Life*, **2000**, *49*, 411. (d) Klayman, D. L.; Günter, H. H. *Organoselenium Compounds: Their Chemistry and Biology*, Wiley, New York, **1973**. (e) Shamberger, R. J., *Biochemistry of Selenium*, Plenum Press, New York, **1983**.

[3] (a) Wirth, T. Chiral Selenium Compounds in Organic Synthesis. *Tetrahedron,* **1999**, *55*, 1-28. (b) Wirth, T. Organoselenium Chemistry in Stereoselective Reactions. *Angew. Chem. Int. Ed.,* **2000**, *39*, 3740-3749. (c) Nishibayashi, Y. Uemura, S. In: *Selenium Compounds as Ligands and Catalysts*; Wirth, T. Ed.; Springer-Verlag: Berlin, **2000**; Vol. *208*, pp. 235-255. (d) Braga, A. L.; Lüdtke, D. S.; Vargas, F.; Braga, R. C. Catalytic Applications of Chiral Organoselenium Compounds in Asymmetric Synthesis. *Synlett,* **2006**, 1453-1466. (e) Braga, A. L.; Lüdtke, D. S.; Vargas, F. Enantioselective Synthesis Mediated by Catalytic Chiral Organoselenium Compounds. *Curr. Org. Chem.*, **2006**, *10*, 1921-1938. (f) Godoi, M.; Paixão, M. W.; Braga, A. L. Chiral organoselenium-transition-metal catalysts in asymmetric transformations. *Dalton Trans.*, **2011**, *40*, 11347-11355.

[4] Freudendahl, D. M.; Santoro, S.; Shahzad, S. A.; Santi, C.; Wirth, T. Green Chemistry with Selenium Reagents: Development of Efficient Catalytic Reactions. *Angew. Chem. Int. Ed.*, **2009**, *48*, 8409-8411.

[5] (a) Mugesh, G.; du Mont, W.-W.; Sies, H. Chemistry of Biologically Important Synthetic Organoselenium Compounds. *Chem. Rev.*, **2001**, *101*, 2125-2180. (b) Nogueira, C. W.; Zeni, G.; Rocha, J. B. T. Organoselenium and Organotellurium Compounds:□ Toxicology and Pharmacology. *Chem. Rev.*, **2004**, *104*, 6255-6286. (c) Bhabak, K. P.; Mugesh, G. Functional Mimics of Glutathione Peroxidase: Bioinspired Synthetic Antioxidants. *Acc. Chem. Res.*, **2010**, *43*, 1408-1419. (d) Alberto, E., Braga, A. L. In: *Activation of Peroxides by Organoselenium Catalysts: A Synthetic and Biological Perspective*; Woollins, J. D.; Laitinen. R. Ed.; Springer-Verlag: Berlin, **2011**; pp. 251-283.

[6] United Nations Environment Programme: Environment for development. http://www.unep.org/Documents.Multilingual/Default.asp?documentid=97&articleid=1503 (accessed October 15, **2012**).

[7] Anastas, P. T.; Warner, J. C.; In: *Green Chemistry: Theory and Practice*, University Press, USA, **2000**.

[8] Welton, T.; Room-Temperature Ionic Liquids. Solvents for Synthesis and Catalysis. *Chem. Rev.,* **1999**, *99,* 2071-2083.

[9] Wasserscheid, P.; Welton T. In: *Ionic Liquids in Synthesis*, 2nd ed Wiley-VHC, **2007**.

[10] Tabarelli, G.; Alberto, E. E.; Deobald, A. M.; Marin, G.; Rodrigues, O. E. D.; Dornelles, L.; Braga, A. L. Ionic liquid: an efficient and reusable media for seleno- and thioester synthesis promoted by indium. *Tetrahedron Lett.,* **2010**, *51*, 5728-5731.

[11] Singh, D.; Narayanaperumal, S.; Gul, K.; Godoi, M.; Rodrigues, O. E. D.; Braga, A. L. Efficient synthesis of selenoesters from acyl chlorides mediated by CuO nanopowder in ionic liquid. *Green Chem.,* **2010**, *12*, 957-960.

[12] Freitas, C. S.; Barcellos, A. M.; Ricordi, V. G.; Pena, J. M.; Perin, G.; Jacob, R. G.; Lenardão, E. J.; Alves, D. Synthesis of diaryl selenides using electrophilic selenium species and nucleophilic boron reagents in ionic liquids. *Green Chem.,* **2011**, *13*, 2931-2938.

[13] Narayanaperumal, S.; Gul, K.; Kawasoko, C. Y.; Singh, D.; Dornelles, L.; Rodrigues, O. E. D.; Braga, A. L. Transition Metal Oxide Nanopowder and Ionic Liquid: an Efficient System for the Synthesis of Diorganyl Selenides, Selenocysteine and Derivatives. *J. Braz. Chem. Soc.,* **2010**, *21*, 2079-2087.

[14] Narayanaperumal, S.; Alberto, E. E.; Gul, K.; Rodrigues, O. E. D.; Braga, A. L. Synthesis of Diorganyl Selenides Mediated by Zinc in Ionic Liquid. *J. Org. Chem.,* **2010**, *75*, 3886-3889.

[15] Narayanaperumal, S.; Alberto, E. E.; de Andrade, F. M.; Lenardão, E. J.; Taube, P. S.; Braga, A. L. Ionic liquid: an efficient and recyclable medium for synthesis of unsymmetrical diorganyl selenides promoted by InI. *Org. Biomol. Chem.*, **2009**, *7*, 4647-4650.

[16] Narayanaperumal, S.; Alberto, E. E.; Gul, K.; Kawasoko, C. Y.; Dornelles, L.; Rodrigues, O. E. D.; Braga, A. L. Zn in ionic liquid: an efficient reaction media for the synthesis of diorganyl chalcogenides and chalcogenoesters. *Tetrahedron*, **2011**, 67, 4723-4730.

[17] Gul, K.; Narayanaperumal, S.; Dornelles, L.; Rodrigues, O. E. D.; Braga, A. L. Bimetallic system for the synthesis of diorganyl selenides and sulfides, chiral β-seleno amines, and seleno- and thioesters. *Tetrahedron Lett.,* **2011**, *52*, 3592-3596.

[18] Zhao, H.; Hao, W.; Xi, Z.; Cai, M. Palladium-catalyzed cross-coupling of $PhSeSnBu_3$ with aryl and alkyl halides in ionic liquids: a practical synthetic method of diorganyl selenides. *New J. Chem.*, **2011**, *35*, 2661-2665.

[19] Banerjee, S.; Adak, L.; Ranu, B. C. Ionic liquid/PPh_3 promoted cleavage of diphenyl disulfide and diselenide: a straight-forward metal-free one-pot route to the synthesis of unsymmetrical sulfides and selenides. *Tetrahedron Lett.,* **2012**, *53*, 2149-2152.

[20] Lenardão, E. J.; Dutra, L. G.; Saraiva, M. T.; Jacob, R. G.; Perin, G. Hydroselenation of alkynes using $NaBH_4$/$BMIMBF_4$: easy access to vinyl selenides. *Tetrahedron Lett.,* **2007**, *48*, 8011-8013.

[21] Yang, M-H; Yan, G-B; Zheng, Y-F. Regioselective ring-opening reactions of 1,2-epoxides with thiols and arylselenols directly promoted by [Bmim]BF_4. *Tetrahedron Lett.,* **2008**, *49*, 6471-6474.

[22] Vargas, J.; Narayanaperumal, S.; Gul, K.; Ravanello, B.; Dornelles, L.; Soares, L. C.; Alves, C. F. S.; Schneider, T.; Vaucher, R. de A.; Santos, R. C. V.; Rodrigues, O. E. D.

Synthesis of chiral β-chalcogen amine derivatives and Gram-positive bacteria activity. *Tetrahedron,* **2012**, *68*, 10444-10448.

[23] Salman, S. M.; Schwab, R. S.; Alberto, E. E.; Vargas, J.; Dornelles, L.; Rodrigues, O. E. D.; Braga, A. L. Efficient Ring Opening of Protected and unprotected Aziridines Promoted by Stable Zinc Selenolate in Ionic Liquid. *Synlett*, **2011**, 69-72.

[24] Salman, S. M.; Narayanaperumal, S.; Schwab, R. S.; Bender, C. R.; Rodrigues, O. E. D.; Dornelles, L. CuO nano particles and [bmim]BF_4: an application towards the synthesis of chiral β-seleno amino derivatives *via* ring opening reaction of aziridines with diorganyl diselenides. *RSC Advances*, **2012**, *2*, 8478-8482.

[25] (a) Santi, C.; Santoro, S.; Batistelli, B.; Testaferri, L.; Tiecco, M. Preparation of the First Bench-Stable Phenyl Selenolate: an Interesting "On Water" Nucleophilic Reagent. *Eur. J. Org. Chem.* **2008**, 5387-5390. (b) Tidei, C. Phenylselenyl Zinc Halides. *Synlett*, **2012**, *23*, 2865-2866. (c) Santi, C. *Phenylselenenylzinc Halides*, In: *Encyclopedia of Reagents for Organic Synthesis*; Paquette, L. A., Ed.; John Wiley & Sons: New York, **2011**.

[26] Batistelli, B.; Lorenzo, T.; Tiecco, M.; Santi, C. "On-Water" Michael-Type Addition Reactions Promoted by PhSeZnCl. *Eur. J. Org. Chem.,* **2011**, 1848-1851.

[27] Santoro, S.; Batistelli, B.; Testaferri, L.; Tiecco, M.; Santi, C. Vinylic Substitutions Promoted by PhSeZnCl: Synthetic and Theoretical Investigations. *Eur. J. Org. Chem.,* **2009**, 4921-4925.

[28] Saha, A.; Saha, D.; Ranu, B. C. Copper nano-catalyst: sustainable phenyl-selenylation of aryl iodides and vinyl bromides in water under ligand free conditions. *Org. Biomol. Chem.*, **2009**, *7*, 1652-1657.

[29] Kumar, A. V.; Reddy, V. P.; Reddy, C. S.; Rao, K. R. Potassium selenocyanate as an efficient selenium source in C-Se cross-coupling catalyzed by copper iodide in water. *Tetrahedron Lett.,* **2011**, *52*, 3978-3981.

[30] Santi, C.; Batistelli, B.; Testaferri, L.; Tiecco, M. On water preparation of phenylselenoesters. *Green Chem.,* **2012**, *14*, 1277-1280.

[31] Sridhar, R.; Srinivas, B.; Surendra, K.; Krishnaveni, N. S.; Rao, K. R. Synthesis of β-hydroxy selenides using benzeneselenol and oxiranes under supramolecular catalysis in the presence of β-cyclodextrin in water. *Tetrahedron Lett.,* **2005**, *46*, 8837-8839.

[32] Wang, J-X; Wu, X.; Hu, Y.; Zhao, K.; Liu, Z. Synthesis of Substituted Glycerol Selenide Ethers under Microwave Irradiation. *J. Chem. Research (S),* **1999**, 688-689.

[33] (a) Wang, J-X; Wu, X.; Huang, D.; Hu, Y. The synthesis of 1-arylseleno- 3-alkoxy-2-propanols under microwave irradiation conditions. *Synth. Commun.* **2002**, *32*, 1903-1908. (b) Wang, J-X; Xi, Y.; Hu, Y. Microwave assisted regioselective opening reaction of epoxides with diphenyl diselenides. *J. Chem. Research (S),* **2000**, 588-560.

[34] Ricordi, V. G.; Freitas, C. S.; Perin, G.; Lenardão, E. L.; Jacob, R. G.; Savegnago, L.; Alves, D. Glycerol as a recyclable solvent for copper-catalyzed cross-coupling reactions of diaryl diselenides with aryl boronic acids. *Green Chem.*, **2012**, *14*, 1030-1034.

[35] Gonçalves, L. C.; Fiss, G. F.; Perin, G.; Alves, D.; Jacob, R. G.; Lenardão, E. J. Glycerol as a promoting medium for cross-coupling reactions of diaryl diselenides with vinyl bromides. *Tetrahedron Lett.,* **2010**, *51*, 6772-6775.

[36] Cheng, T.; Zheng, X.; Ke, Q. Ultrasound assisted ring-opening reaction of epoxides with 1,2-diphenyldiselenide. *J. Chem. Research (S),* **2011**, 522-524.

[37] Lenardão, E. J.; Silva, M. S.; Sachini, M.; Lara, R. G.; Jacob, R. G.; Perin, G. Synthesis of alkenyl selenides and tellurides using PEG-400. *Arkivoc*, **2009**, (xi), 221-227.

[38] Victoria, F. N.; Radatz, C. S.; Sachini, M.; Jacob, R. G.; Perin, G.; da Silva, W. P.; Lenardão, E. J. KF/Al_2O_3 and PEG-400 as a recyclable medium for the selective α-selenation of aldehydes and ketones. Preparation of potential antimicrobial agents. *Tetrahedron Lett.*, **2009**, *50*, 6761-6763.

[39] Yang, M-H; Yuan, C-Y; Pan, Y.; Zhu, C-J; Regioselective Ring-opening Reaction of 1,2-Epoxides with Arylselenol under Solvent-free Conditions and Application to the Synthesis of 1,3-Oxazolidin-2-ones. *Chin. J. Chem.*, **2006**, 669-673.

[40] Lv, G.; Li, T.; Hu, R.; Chen, J.; Ding, J.; Wu, H. Rongalite®-promoted odourless and highly regioselective synthesis of β-hydroxyselenides under solvent-free conditions. *J. Chem. Research,* **2010**, 549-552.

[41] Godoi, M.; Ricardo, E. W.; Botteselle, G. V.; Galetto, F. Z.; Azeredo, J. B.; Braga, A. L. Synthesis of selenol esters from diorganyl diselenides and acyl chlorides under solvent-free conditions and microwave irradiation. *Green Chem.,* **2012**, *14*, 456-460.

[42] (a) Ananikov, V. P.; Beletskaya, I. P. Palladium-catalyzed addition of disulfides and diselenides to alkynes under solvent free conditions. *Org. Biomol. Chem.,* **2004**, 284-287. (b) Ananikov, V. P.; Beletskaya, I. P. Solvent free palladium catalyzed addition of diaryl dichalcogenides to alkynes. *Russ. Chem. Bull.*, **2004**, 561-565.

[43] Perin, G.; Jacob, R. G.; de Azambuja, F.; Mendes, S. R.; Silva, M. S.; Lenardão, E. J.; dos Santos, P. C. Synthesis of β-Phenylchalcogeno-α,β-unsaturated Ketones *via* Hydrochalcogenation of Acetylenes using Microwave and Solvent-Free Conditions. *Synth. Commun.,* **2006**, *36*, 2587-2595.

[44] Perin, G.; Jacob, R. G.; Dutra, L. G.; de Azambuja, F.; Botteselle, G. V.; Siqueira, G. M.; Freitag, R. A.; Lenardão, E. J. The first synthesis of β-phenylchalcogeno-α,β-unsaturated esters *via* hydrochalcogenation of acetylenes using microwave and solvent-free conditions. *Tetrahedron Lett.*, **2005**, *46*, 1679-1682.

[45] Perin, G.; Jacob, R. G.; Dutra, L. G.; de Azambuja, F.; dos Santos, G. F. F.; Lenardão, E. J. Addition of chalcogenolate anions to terminal alkynes using microwave and solvent-free conditions: easy access to bis-organochalcogen alkenes. *Tetrahedron Lett.*, **2006**, *47*, 935-938.

[46] (a) Hözle, G.; Jenny, W. Zur Kenntnis der Sulfen- und Selenensäuren und ihrer Derivate. 9. Mitteilung Additions- und Substitutionsreaktionen organischer Selenverbindungen mit unpolaren und polaren Äthylenen. *Helv. Chim. Acta.*, **1958**, *41*, 593-603. (b) Back, T. G. In: *Organoselenium Chemistry: A Practical Approach (Practical Approach in Chemistry Series)*, 1st ed.; Oxford University Press: USA, **1999**. (c) Tiecco, M. In: *Electrophilic Selenium, Selenocyclizations*; Wirth, T. Ed.; Springer-Verlag: Berlin, **2000**; Vol. *208*, pp. 7-54.

[47] Browne, D. M.; Wirth, T. New Developments with Chiral Electrophilic Selenium Reagents. *Curr. Org. Chem.*, **2006**, *10*, 1893-1903.

[48] Torri, S.; Uneyama, K.; Ono, M.; Bannou, T. Generation and Recycle Use of Selenenylating Reagents in Electrochemical Oxyselenenylation-Deselenenylation of Olefins. *J. Am. Chem. Soc.*, **1981**, *103*, 4606-4608.

[49] Iwaoka, M.; Tomoda, S. Catalytic Conversion of Alkenes into Allylic Ethers and Esters using Diselenides having Internal Tertiary Amines. *J. Chem. Soc., Chem. Commun.*, **1992**, 1165-1167.

[50] Tiecco, M.; Testaferri, L.; Tingoli, M.; Bagnoli, L. Santi, C. Catalytic Conversion of β,γ-Unsaturated Esters, Amides and Nitriles into γ-Alkoxy or γ-Hydroxy α,β-Unsaturated

Derivatives induced by Persulfate Anion Oxidation of Diphenyl Diselenide. *J. Chem. Soc., Chem. Commun.*, **1993**, 637-639.

[51] (a) Tiecco, M.; Testaferri, L.; Tingoli, M. Chianelli, D.; Bartoli, D. The reaction of diphenyl diselenide with peroxydisulphate ions in methanol a convenient procedure to effect the methoxyselenenylation of alkenes. *Tetrahedron Lett.,* **1989**, *30*, 1417-1420. (b) Tiecco, M.; Testaferri, L.; Tingoli, M. Bagnoli L. Marini, F. Ring-closure reactions of alkenyl oximes induced by persulfate anion oxidation of diphenyl diselenide. Formation of 1,2-oxazines and cyclic nitrones. *J. Chem. Soc. Perkin Trans 1*, **1989**, 1989-1903.

[52] Fujita, K.; Iwaoka, M.; Tomoda, S. Synthesis of Diaryl Diselenides Having Chiral Pyrrolidine Rings with C_2 Symmetry. Their Application to the Asymmetric Methoxyselenenylation of *trans*-β-Methylstyrenes. *Chem. Lett.*, **1994**, *23*, 923-926.

[53] (a) Wirth, T.; Fragale, G. Asymmetric Addition Reactions with Optimized Selenium Electrophiles. *Chem. Eur. J.*, **1997**, *3*, 1894-1902. (b) Zhao, L.; Li, Z.; Wirth, T.; Preparation of Novel Chiral Non-Racemic Diselenides and Applications in Asymmetric Synthesis. *Eur. J. Org. Chem.*, **2011**, 176-182. (c) Tiecco, M.; Testaferri, L.; Santi, C.; Tomassini, C.; Marini, F.; Bagnoli, L.; Temperini, A. Asymmetric Azidoselenenylation of Alkenes: A Key Step for the Synthesis of Enantiomerically Enriched Nitrogen-Containing Compounds. *Angew. Chem. Int. Ed.,* **2003**, *42*, 3131-3133. (d) Santi, C.; Santoro, S.; Tomassini, C.; Pascolini, F.; Testaferri, L.; Tiecco, M. Enantioselective Methoxyselenenylation of α,β-Unsaturated Aldehydes. *Synlett,* **2009**, 743-746. (e) Wirth, T.; Fragale, G.; Spichty, M. Mechanistic Course of the Asymmetric Methoxyselenenylation Reaction. *J. Am. Chem. Soc.*, **1998**, *120*, 3376-3381.

[54] Fukuzawa, S.; Takahashi, K.; Kato, H.; Yamazaki, H. Asymmetric Methoxyselenenylation of Alkenes with Chiral Ferrocenylselenium Reagents. *J. Org. Chem.*, **1997**, *62,* 7711-7716.

[55] Wirth, T.; Häuptli, S.; Leuenberger, M. Catalytic asymmetric oxyselenenylation-elimination reactions using chiral selenium compounds. *Tetrahedron: Asymmetry*, **1998**, *9*, 547-550.

[56] Tiecco, M.; Testaferri, L.; Santi, C.; Tomassini, C.; Marini, F.; Bagnoli, L.; Temperini, A. New nitrogen containing chiral diselenides: synthesis and asymmetric addition reactions to olefins. *Tetrahedron: Asymmetry*, **2000**, *11*, 4645-4650.

[57] Tiecco, M.; Testaferri, L.; Santi, C.; Tomassini, C.; Marini, F.; Bagnoli, L.; Temperini A. Preparation of a New Chiral Non-Racemic Sulfur-Containing Diselenide and Applications in Asymmetric Synthesis. *Chem. Eur. J.*, **2002**, *8*, 1118-1124.

[58] Tiecco, M.; Testaferri, L.; Tingoli, M.; Bagnoli, L.; Santi, C. Selenium Catalysed Conversion of β,γ-Unsaturated Acids into Butenolides. *Synlett*, **1993**, 798-800.

[59] Tiecco, M.; Testaferri, L.; Santi, C. Catalytic Oxyselenenylation-Deselenenylation Reactions of Alkenes -Stereoselective One-Pot Conversion of 3-Alkenols into 2,5-Dihydrofurans. *Eur. J. Org. Chem.*, **1999**, 797-803.

[60] Tiecco, M.; Testaferri, L.; Tingoli, M.; Marini, F. Selenium Promoted Conversion of α-Substituted β,γ-Unsaturated Ketones into 2,3,5-Trisubstituted Furans. *Synlett*, **1994**, 373-374.

[61] Tiecco, M.; Testaferri, L.; Bagnoli, L.; Marini, F.; Santi, C.; Temperini, A. Selenium Catalyzed Conversion of δ-Phenyl-γ-Alkenyl Oximes into 2-Phenylpyridines. **1996**, *Heterocycles*, *43*, 2679-2686.

[62] (a) Tingoli, M.; Tiecco, M.; Chianelle, D.; Balducci, R.; Temperini, A. Novel azido-phenylselenenylation of double bonds. Evidence for a free-radical process. *J. Org. Chem.*,

1991, *56*, 6809-6813. (b) Tingoli, M.; Tiecco, M.; Testaferri, L.; Balducci, R. Reactions of Terminal Alkynes with Iodobenzene Diacetate and Diphenyl Diselenide: Synthesis of Phenyl Alkynyl Selenides. *Synlett*, **1993**, 211-212.

[63] Browne, D. M.; Niyomuraa, O.; Wirth, T. Catalytic Use of Selenium Electrophiles in Cyclizations. *Org. Lett.*, **2007**, *9*, 3169-3171.

[64] Singh, F. T.; Wirth, T. Selenium-Catalyzed Regioselective Cyclization of Unsaturated Carboxylic Acids Using Hypervalent Iodine Oxidants. *Org. Lett.*, **2011**, *13*, 6504-6507.

[65] Niyomura, O.; Cox, M.; Wirth, T. Electrochemical Generation and Catalytic Use of Selenium Electrophiles. *Synlett*, **2006**, 251-254.

[66] (a) Tiecco, M.; Testaferri, L.; Tingoli, M.; Chianelli, D.; Bartoli, D. Selenium Catalyzed Conversion of Vinyl Halides into α-Alkoxy Acetals. *Tetrahedron*, **1988**, *44*, 2273-2282. (b) Tiecco, M.; Testaferri, L.; Tingoli, M.; Bartoli, D. Selenium-Catalyzed Conversion of Methyl Ketones into α-Keto Acetals. *J. Org. Chem.*, **1990**, *55*, 4523-4528.

[67] Santi, C.; Tiecco, M.; Testaferri, L.; Tomassini, C.; Santoro, S.; Bizzoca, G. Diastereo and Enantioselective Synthesis of 1,2-Diols Promoted by Electrophilic Selenium Reagents. *Phosphorus Sulfur*, **2008**, *183*, 956-960.

[68] Santoro, S.; Battistelli, B.; Gjoka, B.; Si, S. C.-w.; Testaferri, L.; Tiecco, M.; Santi C. Oxidation of Alkynes in Aqueous Media Catalyzed by Diphenyl Diselenide. *Synlett*, **2010**, 1402-1406.

[69] (a) Baeyer, A.; Villiger, V. Einwirkung des Caro'schen Reagens auf Ketone. *Ber. Dtsch. Chem. Ges.*, **1899,** *32*, 3625-3633. (b) Krow, G. R. The Baeyer-Villiger Oxidation of Ketones and Aldehydes. Org. *React.*, **1993,** *43*, 251-798. (c) Bolm, C.; Beckmann, O.; Luong, O. In: *Metal-catalyzed Baeyer-Villiger reactions*. Wiley-VCH, New York **1998**. (d) Renz, M.; Meunier, B. 100 Years of Baeyer-Villiger Oxidations. *Eur. J. Org. Chem.,* **1999**, 737-750. (e) Mihovilovic, M. D.; Rudroff, F.; Grötzl, B. Enantioselective Baeyer-Villiger Oxidations. *Curr. Org. Chem.,* **2004**, *8*, 1057-1069.

[70] ten Brink, G. J.; Arends, I. W. C. E.; Sheldon, R. A. The Baeyer-Villiger Reaction: New Developments toward Greener Procedures. *Chem. Rev.,* **2004**, *104*, 4105-4123.

[71] Grieco, P. A.; Yokoyama, Y.; Gilman, S.; Ohfune, Y. Conversion of Ketones into Lactones with Benzeneseleninic Acid and Hydrogen Peroxide (Benzeneperoxyseleninic Acid): a New Reagent for the Baeyer-Villiger Reaction. *J. Chem. Soc. Chem. Commun.,* **1977**, 870-871.

[72] Syper, L.; Mlochowski, J. Benzeneperoxyseleninic Acids - Synthesis and Properties. *Tetrahedron*, **1987**, *43*, 207-213.

[73] (a) Syper, L. Reaction of α,β-Unsaturated Aldehydes with Hydrogen Peroxide Catalyzed by Benzeneseleninic Acids and their Precursors. *Tetrahedron*, **1987**, *43*, 2853-2871. (b) Syper, L. The Baeyer-Villiger Oxidation of Aromatic Aldehydes and Ketones with Hydrogen Peroxide Catalyzed by Selenium Compounds. A Convenient Method for the Preparation of Phenols. *Synthesis*, **1989**, 167-172.

[74] ten Brink, G.-J.; Vis, J.-M.; Arends, I. W. E.; Sheldon, R. A. Selenium-Catalyzed Oxidations with Aqueous Hydrogen Peroxide. 2. Baeyer-Villiger Reactions in Homogeneous Solution. *J. Org. Chem.*, **2001**, 66, 2429-2433.

[75] (a) Taylor, R. T.; Flood, L. A. Polystyrene-Bound Phenylseleninic Acid: Catalytic Oxidations of Olefins, Ketones, and Aromatic Systems. *J. Org. Chem.*, **1983**, 48, 5160-5164. (b) Qian, H.; Shao, L.X.; Huang, X. Polystyrene-bound phenylseleninic acid:

catalytic oxidation of aldehydes to carboxylic acids with hydrogen peroxide. *J. Chem. Research (S)*, **2002**, 514-515.

[76] ten Brink, G.-J.; Vis, J. M.; Arends, I. W. C. E.; Sheldon, R. A. Selenium catalyzed oxidations with aqueous hydrogen peroxide. Part 3: Oxidation of carbonyl compounds under mono/di/triphasic conditions. *Tetrahedron*, **2002**, *58*, 3977-3983.

[77] Ichikawa, H.; Usami, Y.; Arimoto, M. Synthesis of novel organoselenium as catalyst for Baeyer-Villiger oxidation with 30 % H_2O_2. *Tetrahedron Lett.*, **2005**, *46*, 8665-8668.

[78] Miyake, Y.; Nishibayashi, Y.; Uemura, S. Asymmetric Baeyer-Villiger Oxidation of Cyclic Ketones Using Chiral Organoselenium Catalysts. *Bull. Chem. Soc. Jpn.,* **2002**, *75*, 2233-2237.

[79] Goodman, M. A.; Detty, M. R. Selenoxides as Catalysts for Epoxidation and Bayer-Villiger Oxidation with Hydrogen Peroxide. *Synlett*, **2006**, 1100-1104.

[80] Choi, J.-K.; Chang, Y.-K.; Hong, S. Y. Catalytic Oxidation of Aldehydes to Carboxylic Acids with Hydrogen Peroxide as Oxidant. *Tetrahedron Lett.*, **1988**, *29*, 1967-1970.

[81] Wóijtowicz, H.; Brzaszcz, M.; Kloc, K.; Mlochowski, J. Selective oxidation of aromatic aldehydes to arenecarboxylic acids using ebselen-tert-butyl hydroperoxide catalytic system. *Tetrahedron*, **2001**, *57*, 9743-9748.

[82] (a) Smith, G. J, Synthetically Useful Reactions of Epoxides. *Synthesis*, **1984**, 629-656. (b) Marco-Contelles, J.; Molina, M. T.; Anjum, S. Naturally Occurring Cyclohexane Epoxides: Sources, Biological Activities, and Synthesis. *Chem. Rev.,* **2004**, *104*, 2857-2899.

[83] Lane, B. S.; Burgess, K. Metal-Catalyzed Epoxidations of Alkenes with Hydrogen Peroxide. *Chem. Rev.*, **2003**, *7*, 2457-2474.

[84] Griecco, P. A.; Yokoyama, Y.; Gilman, S.; Nishizawa, M. Organoselenium Chemistry, Epoxidation of Olefins with Benzeneseleninic Acid and Hydrogen Peroxide ("Benzeneperoxyseleninic Acid"). *J. Org. Chem.*, **1977**, *42*, 2034-2036.

[85] Reich, H. J.; Chow, F.; Peake, S. L. Seleninic Acids as Catalysts for Oxidation of Olefins and Sulfides Using Hydrogen Peroxide. *Synthesis*, **1978**, 299-301.

[86] Gogoi, P.; Sharma, S. D.; Konwar, D. $SeO_2/H_2O_2/H_2O$-Dioxane: A New Catalytic System for *Trans* Dihydroxylation of Olefins. *Lett. Org. Chem.*, **2007**, *4*, 249-252.

[87] Santoro, S.; Santi, C.; Sabatini, M.; Testaferri, L.; Tiecco, M. Eco-Friendly Olefin Dihydroxylation by Diphenyl Diselenide. *Adv. Synth. Catal.*, **2008**, *350*, 2881-2884.

[88] (a) García-Marín, H.; van der Toorn, J. C.; Mayoral, J. A.; García, J. I.; Arends, I. W. C. E. Glycerol-based solvents as green reaction media in epoxidations with hydrogen peroxide catalyzed by bis[3,5-bis(trifluoromethyl)-diphenyl] diselenide. *Green. Chem.*, **2009**, *11*, 1605-1609. (b) García-Marín, H.; van der Toorn, J. C.; Mayoral, J. A.; García, J. I.; Arends, I. W. C. E. Epoxidation of cyclooctene and cyclohexene with hydrogen peroxide catalyzed by bis[3,5-bis(trifluoromethyl)-diphenyl] diselenide: Recyclable catalyst-containing phases through the use of glycerol-derived solvents. *J. Mol. Cat. A: Chem.*, **2011**, *334*, 83-88.

[89] Hori, T.; Sharpless, K. B. Synthetic Applications of Arylselenenic and Arylseleninic Acids. Conversion of Olefins to Allylic Alcohols and Epoxides. *J. Org. Chem.*, **1978**, *43*, 1689-1697.

[90] ten Brink, G.-J.; Fernandes , C. M.; van Vliet, M. C. A.; Arends, I. W. C. E.; Sheldon, R. A. Selenium catalysed oxidations with aqueous hydrogen peroxide. Part I: epoxidation reactions in homogeneous solution. *J. Chem. Soc., Perkin Trans.,* **2001**, 224-228.

[91] Al-Ajlouni, A. M.; Espenson, J. H. Epoxidation of Styrenes by Hydrogen Peroxide As Catalyzed by Methylrhenium Trioxide. *J. Am. Chem. Soc.*, **1995**, *117*, 9243-9250.

[92] Betzemeier, B.; Lhermitte, F.; Knochel, P. A Selenium Catalyzed Epoxidation in Perfluorinated Solvents with Hydrogen Peroxide. *Synlett*, **1999**, 489-491.

[93] Mercier, E. A.; Smith, C. D.; Parvez, M.; Back, T. G. Cyclic Seleninate Esters as Catalysts for the Oxidation of Sulfides to Sulfoxides, Epoxidation of Alkenes, and Conversion of Enamines to α-Hydroxyketones. *J. Org. Chem.*, **2012**, *77*, 3508-3517.

[94] (a) de Visser, S. P.; Kaneti, J.; Neumann, R.; Shaik, S. Fluorinated Alcohols Enable Olefin Epoxidation by H_2O_2: Template Catalysis. *J. Org. Chem.*, **2003**, *68*, 2903-2912. (b) Berkessel, A.; Adrio, J. A.; Hüttenhain, D.; Neudörfl, J. M. Unveiling the "Booster Effect" of Fluorinated Alcohol Solvents: Aggregation-Induced Conformational Changes and Cooperatively Enhanced H-Bonding. *J. Am. Chem. Soc.,* **2006**, *128*, 8421-8426. (c) Berkessel, A.; Adrio, J. A. Dramatic Acceleration of Olefin Epoxidation in Fluorinated Alcohols: Activation of Hydrogen Peroxide by Multiple H-Bond Networks. *J. Am. Chem. Soc.,* **2006**, *128*, 13412-13420.

[95] Anastas, P.; Eghbali, N. Green Chemistry: Principles and Practice. *Chem. Soc. Rev.*, **2010**, *39*, 301-312.

[96] Anelli, P. L.; Banfi, S.; Montanari, F.; Quici, S. Synergistic Effect of Lipophilic Carboxylic Acids and Heterocyclic Axial Ligands in Alkene Epoxidation by Hydrogen catalysed by Manganese(III) Tetra-aryl Porphiyrins. *J. Chem. Soc., Chem. Commun.,* **1989**, 779-780.

[97] Santi, C.; Di Lorenzo, R.; Tidei, C.; Bagnoli, L.; Wirth, T. Stereoselective Selenium catalyzed Dihydroxylation and Hydroxymethoxylation of Alkenes. *Tetrahedron*, **2012**, *68*, 10530-10535.

[98] Moussou, P.; Archelas, A.; Baratti, J.; Furstoss, R. Microbiological Transformations. 38. Clues to the Involvement of a General Acid Activation during Hydrolysis of Para-Substituted Styrene Oxides by a Soluble Epoxide Hydrolase from *Syncephalastrum racemosum. J. Org. Chem.*, **1998**, *63*, 3532-3537.

[99] Dagani, M. J.; Barda, H. J.; Benya, T. J.; Sanders, D. C. In: *Ulmann's Encyclopedia of Industrial Chemistry: Bromine Compounds, Wiley-VHC, Weinheim,* **2002**.

[100] (a) Neidleman, S. L.; Geigert, J. *Biohalogenation*; Wiley: New York, NY, **1986**. (b) A.; Walker, J. V. Marine Haloperoxidases. *Chem. Rev.*, **1993**, *93*, 1937-1944. (c) Vaillancourt, F. H.; Yeh, E.; Vosburg, D. A.; Garneau-Tsodikova, S.; Walsh, C. T. Nature's Inventory of Halogenation Catalysts: Oxidative Strategies Predominate. *Chem. Rev.*, **2006**, *106*, 3364-3378.

[101] Podgoršek, A.; Zupan, M.; Iskra, J. Oxidative Halogenation with "Green" Oxidants: Oxygen and Hydrogen Peroxide. *Angew. Chem. Int. Ed.*, **2009**, *48*, 8424-8450.

[102] Wever, R.; Kreen, M. B. E. In: *Vanadium in Biological Systems*; Chasteen, N. D., Ed.; Kluwer Academic Publishers: Dordrecht, The Netherlands, **1990**, 81-97.

[103] Drake, M. D.; Bateman, M. A.; Detty, M. R. Substituent Effects in Arylseleninic Acid-Catalyzed Bromination of Organic Substrates with Sodium Bromide and Hydrogen Peroxide. *Organometallics*, **2003**, *22*, 4158-4162.

[104] Goodman, M. A.; Detty, M. R. Substituent Effects in Arylseleninic Acid-Catalyzed Bromination of Organic Substrates with Sodium Bromide and Hydrogen Peroxide. *Organometallics*, **2004**, *23*, 3016-3020.

[105] Alberto, E. E.; Braga, A. L.; Detty, M. R. Imidazolium-containing diselenides for catalytic oxidations with hydrogen peroxide and sodium bromide in aqueous solutions. *Tetrahedron*, **2012**, *68*, 10476-10481.

[106] Francavilla, C.; Bright, F. V.; Detty, M. R. Dendrimeric Catalysts for the Activation of Hydrogen Peroxide. Increasing Activity per Catalytic Phenylseleno Group in Successive Generations. *Org. Lett.*, **1999**, *1*, 1043-1046.

[107] (a) Francavilla, C.; Drake, M. D.; Bright, F. V.; Detty, M. R. Dendrimeric Organochalcogen Catalysts for the Activation of Hydrogen Peroxide: Improved Catalytic Activity through Statistical Effects and Cooperativity in Successive Generations. *J. Am. Chem. Soc.*, **2001**, *123*, 57-67. (b) Drake, M. D.; Bright, F. V.; Detty, M. R. Dendrimeric Organochalcogen Catalysts for the Activation of Hydrogen Peroxide: Origins of the "Dendrimer Effect" with Catalysts Terminating in Phenylseleno Groups. *J. Am. Chem. Soc.*, **2003**, *125*, 12558-12566.

[108] Bennett, S. M.; Tang, Y.; McMaster, D.; Bright, F. V.; Detty, M. R. A Xerogel-Sequestered Selenoxide Catalyst for Brominations with Hydrogen Peroxide and Sodium Bromide in an Aqueous Environment. *J. Org. Chem.*, **2008**, *73*, 6849-6852.

[109] Mellegaard-Waetzig, S. R.; Wang, C.; Tunge, J. A. Selenium-catalyzed oxidative halogenation. *Tetrahedron*, **2006**, *62*, 7191-7198.

[110] Mellegaard, S. R.; Tunge, J. A. Selenium-Catalyzed Halolactonization: Nucleophilic Activation of Electrophilic Halogenating Reagents. *J. Org. Chem.*, **2004**, *69*, 8979-8981.

[111] Tunge, J. A.; Mellegaard, S. R. Selective Selenocatalytic Allylic Chlorination. *Org. Lett.*, **2004**, *6*, 1205-1207.

[112] Wang, C.; Tunge, J. Selenocatalytic α-halogenation. *Chem. Commun.*, **2004**, 2694-2695.

[113] Barrero, A. F.; del Moralm J. F. Q.; Herrador, M. M.; Cortés, M.; Arteaga, P.; Catalán, J. V.; Sánchez, E. M.; Arteaga, J. F. Solid-Phase, Selenium-Catalyzed Selective Allylic Chlorination of Polyprenoids: Facile Syntheses of Biologically Active Terpenoids. *J. Org. Chem.*, **2006**, *71*, 5811-5814.

[114] Kimpe, N. de, Verhe, R. In; *The Chemistry of α-Haloketones, α-Haloaldehydes, and α-Haloimines*; John Wiley & Sons, New York, **1990**.

[115] Barton, D. H. R.; Crich, D. Oxidation of olefins with 2-Pyridineseleninic Anhydride. *Tetrahedron*, **1985**, *41*, 4359-4364.

[116] Barton, D. H. R.; Wang, T.-L. Novel Allylic Oxidation Reagents. *Tetrahedron Lett.*, **1994,** *35*, 5149-5152.

[117] Crich, D.; Zou, Y. Catalytic Allylic Oxidation with a Recyclable, Fluorous Seleninic Acid. *Org. Lett.*, **2004**, *6*, 775-777.

[118] Crich, D.; Zou, Y. Catalytic Oxidation Adjacent to Carbonyl Groups and at Benzylic Positions with a Fluorous Seleninic Acid in the Presence of Iodoxybenzene. *J. Org. Chem.*, **2005**, *70*, 3309-3311.

[119] Tojo, G.; Fernandez, M. In: *Oxidation of alcohols to aldehydes and ketones*. Springer, Berlin, **2006**.

[120] Barton, D. H. R.; Brewster, A. G.; Hui, R. A. H. F.; Lester, D. J.; Ley, S. V. Oxidation of Alcohols using Benzeneseleninic Anhydride. *J. Chem. Soc. Chem. Commun.*, **1978**, 952-954.

[121] (a) Shimizu, M.; Kuwajima, I. Highly Efficient Method for Oxidation of Alcohol with tert-butyl Hydroperoxide Catalyzed by Diaryl Diselenide. *Tetrahedron Lett.*, **1979**, *20*, 2801-2804. (b) Kuwajima, I.; Shimizu, M.; Urabe, H. Oxidation of Alcohols with *tert*-Butyl Hydroperoxide and Diaryl Diselenide. *J. Org. Chem.*, **1982**, *47*, 837-842.

[122] van der Toorn, J. C.; Kemperman, G.; Sheldon, R. A.; Arends, I. W. C. E. Diphenyldiselenide-Catalyzed Selective Oxidation of Activated Alcohols with *tert*-Butyl Hydroperoxide: New Mechanistic Insights. *J. Org. Chem.*, **2009**, *74*, 3085-3089.

[123] Drabowicz, J.; Lyzwa, P.; Luczak, J.; Mikolajczyk, M.; Laur, P. New Procedures for the Oxidation of Sulfides to Sulfoxides and Sulfones. *Phosphorus Sulfur Silicon*, **1997**, *120 & 121,* 425-426.

[124] Mlochowski, J.; Gurg, M.; Kubicz, E.; Said, S. B. Benzisoselenazol-3(2H)-ones and *bis*(2-Carbamoyl)phenyl Diselenides as New Catalysts for Hydrogen Peroxide Oxidation of Organic Compounds. *Synth Commun.*, **1996**, *26*, 291-300.

[125] Wójtowicz, H.; Soroko, G.; Mlochowski, J. New Recoverable Organoselenium Catalyst for Hydroperoxide Oxidation of Organic Substrates. *Synth Commun.*, **2008**, *38*, 2000-2010.

[126] Gebhardt, C.; Priewisch, B.; Irran, E.; Rück-Braun, K. Oxidation of Anilines with Hydrogen Peroxide and Selenium Dioxide as Catalyst. *Synthesis*, **2008**, 1889-1894.

[127] Zhao, D.; Johansson, M.; Bäckvall, J.-E. *In Situ* Generation of Nitroso Compounds from Catalytic Hydrogen Peroxide Oxidation of Primary Aromatic Amines and Their One-Pot Use in Hetero-Diels-Alder Reactions. *Eur. J. Org. Chem.*, **2007**, 4431-4436.

[128] (a) Payne, G. B.; Smith, C. W. Reactions of Hydrogen Peroxide. II. A Novel Use of Selenium Dioxide as Catalyst for the Ring Contraction of Cycloalkanones to Cycloalkanecarboxylic Acids. *J. Org. Chem.*, **1957**, *22*, 1680-1682. (b) Sonoda, N.; Tsutsumi, S. Hydrogen Peroxide Oxidation. I. A New Selenium Dioxide-Catalyzed Synthesis of Carboxylic Acids from Aliphatic Ketones Accompanied by Rearrangement of Alkyl Groups. *Bull. Chem. Soc. Jpn.*, **1959**, *32*, 505-510.

[129] Giurg, M.; Mlochowski, J. Oxidative Ring Contraction of Cycloalkanones: A Facile Method for Synthesis of Medium Ring Cycloalkanecarboxylic Acids. *Synth. Commun.*, **1999**, *29*, 2281-2291.

[130] Pratt, D. V.; Ruan, F.; Hopkins, P. B. Practical Large-Scale Oxidation of 1,4-Hydroquinones to 1,4-Benzoquinones Using Hydrogen Peroxide/Catalytic Diphenyl Diselenide. *J. Org. Chem.*, **1987**, *52*, 5053-5055.

[131] (a) Said, S. B.; Skarzewski, J.; Mlochowski, J.; Oxidative Conversion of Aldoximes into Carboxylic Acid Esters. *Synth. Commun.*, **1992**, *22*, 1851-1862. (b) Giurg, M.; Wóijowicz, H. Mlochowski, J.; Hydroperoxide Oxidation of Azomethines and Alkylarenes Catalyzed by Ebselen. *Polish J. Chem.*, **2002**, *76*, 537-542.

[132] Mlochowski, J.; Giurg, M.; Kubicz, E.; Said, S. B.; Benzisoselenazol-3(2H)-ones and *bis*(2-Carbamoyl)phenyl Diselenides as New Catalysts for Hydrogen Peroxide Oxidation of Organic Compounds. *Synth. Commun.*, **1996**, *26*, 291-300.

[133] Balkrishna, S. J.; Prasad, C. D.; Panini, P.; Detty, M. R.; Chopra, D.; Kumar, S. Isoselenazolones as Catalysts for the Activation of Bromine: Bromolactonization of Alkenoic Acids and Oxidation of Alcohols. *J. Org. Chem.*, **2012**, *77*, 9541-9552.

268 *Organoselenium Chemistry Between Synthesis and Biochemistry*, 2014, 268-327

CHAPTER 9

Biochemistry and Nutrition of Selenium: From Inorganic Forms to Endogenous Proteins

Bartolini Desirée, Ciffolilli Silvia, Piroddi Marta, Murdolo Giuseppe, Tortoioli Cristina and Francesco Galli[*]

Department of Pharmaceutical Sciences, Nutrition and Clinical Biochemistry Lab, University of Perugia, Istituti biologici di Via del Giochetto, 06126 Perugia, Italy

Abstract: Selenium (Se) is an essential micronutrient with proposed role in the protection of DNA, proteins and lipids from oxidative damage, and in the homeostasis of the immune system, thyroid gland and spermatogenesis. Environmental (soil and water) availability influences Se food content and thus the Se status of the different populations. Recommended daily intake in USA and Europe is 55 and 60 μg/day, with tolerable upper intake levels of 400 and 300 μg, respectively. Biological activities are dependent on the organization of the elemental form in the structure of the amino acids selenomethionine and selenocysteine, with the latter as functional component of endogenous proteins with role in redox catalysis (glutathione peroxidases, thioredoxin reductases, and iodothyronine deiodinases), signaling, storage, transport and structural activities. A series of epidemiology and intervention studies have suggested that populations exposed to high levels of intake may benefit of health promoting effects and this has made Se supplements quite popular. However, such positive effects are not always supported by sufficient evidence and a number of studies, on the other hand, have suggested that higher intakes may cause adverse effects increasing the risk of cancer and diabetes. Chronic toxicity of Se remains in fact matter of concern representing a valid reason for further clinical investigation on the safety of current supplementation and food fortification protocols. This chapter provides a systematic analysis of the literature available on these aspects particularly focusing on nutritional and biochemistry features of Se.

Keywords: Selenoproteins, Selenocysteine, Selenomethionine, Se intake, Selenosis, Selenium deficiency, Thyroid hormones, Diabetes, Glutathione peroxidase, Redox catalysis.

9.1. INTRODUCTION

Discovered in 1817 by the Swedish chemist Jöns Jacob Berzelius, the non-metal

***Address correspondence to Francesco Galli:** Department of Pharmaceutical Sciences, Nutrition and Clinical Biochemistry Lab, University of Perugia, Istituti biologici di *Via* del Giochetto, 06126 Perugia, Italy; Tel/Fax: +39 075 585 7490; E-mail: francesco.galli@unipg.it

Claudio Santi (Ed)

element Selenium (Se) was defined as an essential micronutrient with significant health benefits only in the mid-1950's; in the early 1970's its biological activity was first attributed to the newly discovered amino acid selenocysteine (Sec) [1]. This step opened the way to the protein biochemistry of this element. Organization of Se into biological structures depends on the environmental (soil and water) availability and on the fixation in living structures of this element into the amino acid structure of methionine (Met) and cysteine (Cys) to form selenomethionine (Se-Met or Sem) and Sec, respectively. Only this latter amino acid enters the structure of human proteins in a significant amount while Sem is of minor importance in this context. Se-amino acids thus represent a sort of Se trap for the human organism with liver, kidney, muscle and blood as main sites of distribution of this element in the protein form (Fig. **9.1**). Distribution of Se into protein structures is a highly specialized and finely regulated process. These aspects demonstrate the strategic importance of Se for living organisms. Specific biosynthetic machinery has been developed in either unicellular organisms or mammalian cells to insert Sec in the primary structure of proteins and twenty-five selenoprotein genes have been identified in the human genome [2]. Selenoproteins have a number of functions, comprising various catalytic roles (glutathione peroxidases, thioredoxin reductases, and iodothyronine deiodinases), structural roles, and storage and transport activities (see section 9.9).

Redox chemistry and biological roles of Se are intimately connected with those of sulphur that is present in proteins as Cys residues. Although Se and S belong to the same family, Sec and Cys possess distinct chemical properties that are responsible of different roles in protein structure and functioning [1, 3]. While Cys is one of the most abundant amino acid residues in proteins, Sec is a rare component with strategic distribution and role in redox-active proteins. Its formation and incorporation in protein structure is thus a critical step in redox biology of living organisms. As an essential micronutrient Se reaches human tissues through the diet. Se is found in soil and water and consequently it enters the food chain through the root ways of plants and aquatic organisms. As a consequence, plant organisms and food items contain this element to an extent that varies on a geographical base. Food processing and technology aspects may also influence the dietary availability of Se.

In the last few decades, there has been increasing interest in several nutritional Se compounds particularly for disease-preventing activities, but environmental and toxicological properties are also of main importance. Among the essential human micronutrients, selenium is peculiar due to its beneficial physiological activity and toxicity that are close to each other in a narrow window of intake [4]. These are associated with three levels of biological activity: (1) trace concentrations are required for normal growth and development; (2) moderate concentrations can be stored, and homeostatic functions maintained and (3) elevated concentrations can result in toxic effects.

This chapter describes the route that brings the essential micronutrient Se from inorganic to organic form thus leading to incorporation into proteins of living organisms for physiological functions and disease prevention effects. Toxicological aspects are also reviewed.

9.2. THE FOOD CHAIN OF SELENIUM

Under an evolutionary point of view, it is possible to hypothesize that Se similarly to Iodine became available to living organisms thanks to primordial blue-green algae that near to 3 billions of years ago, developing primitive photosynthetic structures to produce oxygen, first used these redox-active elements abundantly present in sea waters [5]. The food chain of Se thus developed from these marine organisms through fish species and then to reptiles and other animal species that entered the chain spreading Se in other environments (freshwater basins and land) to make it available for all other evolutionary steps in plant and animal kingdoms up to our days.

In the contemporary food chain of Se, plant organisms are former dietary sources of Se for the animal kingdom. Se in the form of selenate or selenite is taken up by plant roots and mainly transformed into Se-Met in plant tissues, as observed for instance in cereal grains [6]. Se content of plants is thus directly affected by the levels of this element in the soil in which these plant organisms are grown varying geographically both within and between different regions throughout the globe. Therefore, Se content of animal products reflects the levels of Se in their consumed diet [7], with herbivores as first ring of this level of the food chain.

Plant absorption of Se mainly depends on the concentration and physicochemical forms obtainable in the soil. Factors such as the type of rocks, pH and redox potential in the soil, the existence of some organic and inorganic compounds, soil moisture and salinity, soil sulfate concentration, plant species, soil-management practices, oxidation state of the element (the absorption of Se 6+ is higher than that of Se 4+), nature of draining waters, and climatic conditions, all influence the distribution, status and availability of this element [8].

In acid soils Se is generally present as selenite, which has very low solubility and plant availability. In alkaline soils, Se is oxidized to selenate, which is more soluble and more available for uptake in the crops.

In many regions of the world Se soil levels generally reflect the Se status in human populations [9] as well as in animals that provide an important dietary of this element. Actually, Se also can be found in some meats and seafood, and animals that eat grains and plants that were grown in selenium-rich soil have higher levels of selenium in their muscle. For example, soils in the high plants of northern Nebraska and the Dakotas have very high levels of Se. People living in those regions generally have the highest Se intakes in the United States [10]. In regions of high Se soil concentrations (>5 ppm), it is possible to have a net excess of this element as observed in Canada, Ireland, some regions of the western USA, and some zones of China, France, Germany, *etc.* [8a,11].

There are some zones where Se levels in soil are very low (< 0.05 ppm), such as parts of China, Russia, Finland and New Zealand. A deficiency of this essential micronutrient is often reported in those regions because most food in those areas is grown and eaten locally. In these areas, the appearance of an associated heart disease and bone disorders are described as heath effects caused by Se deficiency in livestock that can be corrected with dietary selenium. In this respect, efforts can be made to increase the Se content in plants by adding Se to the soil [12]. Food distribution patterns across a region or between regions help prevent people living in low-selenium geographic areas from having low dietary Se intakes.

9.3. METABOLISM AND BIOAVAILABILITY

Available data indicate that selenium-containing amino acids and probably other selenium forms, such as selenite and selenate, can be converted to selenide in mammals [13]. Selenide is a central metabolic form of selenium [14], which is utilised for the formation of selenocysteine, incorporated into specific selenoproteins, and in case of high exposure, into excretory products such as dimethyl selenide (which is exhaled) and trimethylselenonium ions (which are excreted into urine).

Selenomethionine and selenocysteine formed by transsulfuration of selenomethionine can be nonspecifically incorporated into protein as analogues to methionine and cysteine. Other forms of protein bound selenium may also occur [15].

Relevant forms of dietary Se are described in Table **9.1** and include selenium salts and organic forms, as selenomethionine and selenocysteine.

Table 9.1: Forms of selenium in foods and their characteristics [5,16]

Selenomethionine	This is the selenium derivative of the amino acid, methionine; found in plant sources (notably cereals), selenium yeast, and other selenium supplements. It is incorporated non-specifically into body proteins in place of methionine (*e.g.*, selenomethionine in albumin contributes to selenium measured in plasma); supplements containing selenomethionine therefore seem to have more bioavailable selenium.
Selenocysteine	It is selenium analogue of the amino acid, cysteine; found in animal foods (from their selenoproteins).
Selenoneine (2-selenyl-Nα,Nα,Nα-trimethyl-L-histidine)	Discovered as the major selenium compound in fish such as tuna and mackerel; lower concentrations in squid, tilapia, pig, and chicken. It has strong radical-scavenging activity.
Se-methylselenocysteine and γ-glutamyl-Se-methylselenocysteine	Found in plant sources such as selenium-enriched yeast, garlic, onions, and broccoli. These are metabolized to methyl selenol, which is reported to have anticancer effects.
Sodium selenite and selenate	Components of dietary supplements; selenate occasionally appears in water supplies. Some selenate is found in fish and plant sources (*e.g.* cabbage).

Once introduced with the diet, Se is absorbed efficiently by the intestine over a wide range of concentrations and across a variety of different molecular forms [17]. Bioavailability studies [18] describe that Se in blood or serum is most effectively raised by selenium-rich wheat or yeast selenium (the latter may vary in quality as described in 9.6), probably because of non-specific incorporation of selenomethionine into proteins. Kinetic studies indicated that blood plasma contains at least four components with half-lives between 1 and 250 hours [19]. Inorganic selenium as selenate and selenite can be incorporated specifically into selenium proteins *via* selenide as selenocysteine and increase seleno-enzyme activity until saturation [20]. A few studies have also compared selenium bioavailability from different foods. Van der Torre and coworkers [21] found that supplementation with selenium-rich forms of bread and meat gave similar increases in circulating selenium levels. Christensen and coworkers [22], using a triple stable-isotope method, found that the absorption of selenium from selenite was 36% and that from intrinsically labelled poultry meat was 71%. Selenium consumed from fish had no apparent effect on the amount of selenium incorporated into functional selenoproteins and a low effect on levels of selenium in plasma [23]. Given different bioavailability and differences in non-specific incorporation of selenium compounds from different sources such as cereals, meat, fish and organic and inorganic supplements, the selenium concentration in whole blood will relate differently to the total intake of selenium [15b,24].

The content of Se of normal adult humans can vary widely (5-15 mg). Values from 3 mg in New Zealanders to 14 mg in some Americans reflect the profound influence of the natural environment on the selenium contents of soils, crops, and human tissues. One of the earlier studies in US reported that in the presence of a daily intake of 62 μg with a standard diet, human body contained on average 14.6 mg of Se (range 13-20 mg) [25].

Few organs contain the majority of body's Se (Fig. **9.1**). Approximately 30 percent of tissue selenium is contained in the liver, 15 percent in kidney, 30 percent in muscle, and 10 percent in blood plasma. Much of tissue selenium is found in proteins as seleno-analogues of sulphur amino acids; other metabolically active forms include selenotrisulphides and other acid-labile selenium compounds.

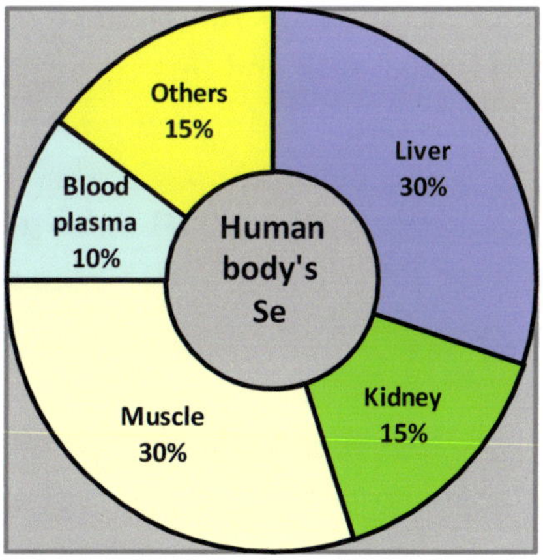

Figure 9.1: Relative distribution of Se in human organs. Sections of the ring chart are as % of total body's content.

9.4. DAILY REQUIREMENTS AND INTAKE

9.4.1. Daily Requirements

US recommendations for selenium are provided in the Dietary Reference Intakes (DRIs) developed by the Institute of Medicine [26]. DRIs are a set of reference values used for planning and assessing nutrient intake for healthy people and include: Recommended Dietary Allowances (RDA), Adequate Intakes (AI), and Tolerable Upper Intake Levels (UL). The RDA recommends the average daily dietary intake level that is sufficient to meet the nutrient requirements of nearly all (97–98%) healthy individuals in each age and gender group. An AI is set when there is insufficient scientific data available to establish a RDA. AIs meet or exceed the amount needed to maintain a nutritional state of adequacy in nearly all members of a specific age and gender group. The UL, on the other hand, is the maximum daily intake unlikely to result in adverse health effects. The amount of dietary selenium (as DL-selenomethionine) required to saturate the selenium need of extracellular GSH-Px was used as one of the approaches to define a Dietary Reference Intake for Selenium in the USA in 2000, and a RDA value of 55 μg/day was established for adult men and women [26]. This value increases to 60 and 70 μg/day in adult women during pregnancy and lactation, respectively, and decreases to 40 μg/day in children 9 to 13 yrs old, and to 30 and 20 μg/day in 4-8 yr and 1-3 yr old children, respectively.

Being the available information to establish an RDA for infants insufficient, an AI value has been defined that is based on the amount of selenium consumed by healthy infants who are fed breast milk and this corresponds to 15 and 20 μg/day in 0–6 months and 7–12 months (male and female) infants, respectively.

In Europe the recommended daily intake (RDI) is 65 μg with a Tolerable Upper Intake of 300 μg/day. A joint FAO/IAEA/WHO Expert Consultation [27] gave several modes for the calculation of requirements of the individual and populations. For a 65 kg reference man the average normative requirement of individuals for selenium was estimated to be 26 μg/day, and from this value the lower limit of the need of population mean intakes was estimated to be 40 μg/day. The corresponding values for a 55 kg reference woman were 21 and 30 μg selenium/day, respectively. The latter value was estimated to increase to 39 μg/day throughout pregnancy and to attain the values of 42, 46 and 52 μg selenium/day at 0-3, 3-6 and 6-12 months of lactation, respectively. The Nordic Nutrition Recommendations [28] have set a recommended intake of 50 μg/day for men, an average requirement of 35 μg/day and a lower limit of needed intake of 20 μg/day, the corresponding values for women being 40, 30 and 20 μg/day, respectively.

9.4.2. Actual Intake as Measured by Surrogate Biomarkers of Se Status

Surrogate biomarkers of selenium status (this should reflect both the Se content in the body, or in the main organs, and its biological function) are commonly used to verify the biological compliance to Se supplements or dietary intake. These markers include serum or plasma Se levels, glutathione peroxidase (GPx) activity in plasma and blood cells as well as thioredoxin reductase (TR) activity and peroxide-induced DNA damage (DNA stability test) [29]. Other and large population studies on Se status have assessed toenail Se concentrations [10,30]. Yet the approach to define a biochemical index for the saturation of the functional selenium requirement using a limited number of selenoproteins has given variable results [31]. The estimations are also complicated by the fact that different forms of dietary selenium (organic *vs*. inorganic) give variable responses in different measures of selenium status [20c] and the physiological relevance of the 'saturation of selenium dependent enzymes approach' can be questioned [15c]. Further studies are thus required to optimize protocols of investigation of the Se status.

Notwithstanding, the aforementioned surrogate biomarkers, have been used to assess actual intakes of selenium in several regions. Unlike to many other micronutrients, Se intake shows extreme variability among different areas of a region [2b,32]; an example of this is reported in Table **9.2** in which dietary Se intakes of some countries are reported. Intakes are high in the USA, Venezuela, Canada and Japan, and lower in Europe, principally in eastern Europe. China has region of both selenium deficiency and excess. Intakes in New Zealand, which were previously low, have enhanced after the introduction of high-selenium Australian wheat [32a].

Table 9.2: Se intake of selected countries

Country	Se (μg/day)	References
Belgium	28-61	[36]
Canada	98-224	[37]
China	3-11	[38]
Finland		[8a, 39]
Before using Se fertilizer	25	
After using Se fertilizer	67-110	
Greece	39	[40]
Libya	13-44	[41]
Lithuania	100	[42]
México	61-73	[43]
The Netherlands	67	[44]
New Guinea	20	[45]
Norway	80	[46]
Scotland, United Kingdom	30-60	[47]
Sweden	38	[38]
Switzerland	70	[38]
Spain (south-eastern)	73	[48]
Turkey	30	[38]
United Kingdom	34	[47b]
USA	60-160	[10]

A mean values of dietary selenium intake (supplements that contain selenium are included in the calculation) of 40 μg per day is observed in Europe. More in detail, report of the European Food Safety Agency [33] described in European countries mean intakes for non-vegetarian adults of: Belgium 28-61 μg/day, Denmark 41-57 μg/day, Finland 100-110 μg/day, France 29-43 μg/day, United Kingdom 63 μg/day, The Netherlands 40-54 μg/day, Norway 28-89 μg/day, Spain 79 μg/day, and Sweden 24-35 μg/day.

In the USA, mean intakes of 93 and 134 μg per day are observed in women and men, respectively, and 50% of the population takes dietary supplements [32b].

Results of the National Health and Nutrition Examination Survey (NHANES III—1988–94) indicated that diets of most Americans provide recommended amounts of selenium [34]. The INTERMAP study examined nutrient intakes of almost 5,000 middle-aged men and women in four countries in the late 1990s, including the United States. The primary aim of the study was to evaluate the effect of dietary micronutrients on blood pressure. Each study participant completed four, 24-hour dietary recalls, during which they were asked to record everything consumed (food, beverages, and dietary supplements) over the previous 24 hours. Selenium intake was lowest among residents of China, the country with the highest known rate of selenium deficiency. Mean dietary intake of selenium of U.S. participants was 153 μg for men and 109 μg for women. Both values exceed the recommended selenium intake for adults that further supports the evidence of an adequate selenium intake in the United States [35].

9.5. SELENIUM TOXICITY

Selenium is also toxic and this toxicity depends not only on the Se compound and dose applied, but also on the method of administration, exposure time, idiosyncrasy, physiological status, and interaction with other metals, nutrients, *etc.* [9c]. The mechanism of Se toxicity has not been clarified, but it is mostly attributed to its ability to induce oxidative stress [49]. In fact, at high doses, Se has the potential to induce toxic side effects such as induction of DNA damage [50]. It is likely that several molecular mechanisms operate in different biological contexts to cause oxidative stress and these are expected to vary among different selenium compounds. These mechanisms include: redox cycling of auto-oxidisable selenium metabolites, glutathione depletion, protein synthesis inhibition, depletion of S-adenosyl-methionine (cofactor for selenide methylation), general replacement of sulphur and reactions with critical sulphydryl groups of proteins and cofactors.

As far as acute toxicity concerns, inorganic Se (selenate, selenite) is more toxic and less bioavailable than organic forms [4a], and one potential reason for differences in genotoxicity observed among selenocompounds is their distinct metabolism [51]. Intake of 250 mg selenium as a single dose or multiple doses of 27-31 mg resulted in acute toxicity with nausea, vomiting, nail changes, dryness of hair, hair loss, tenderness and swelling of fingertips, fatigue, irritability and garlicky breath [52]. In Sweden, several cases of toxicity in children have been reported due to accidental overconsumption of selenium tablets. Acute symptoms such as vomiting have been observed, but so far no serious cases of toxicity have been recorded [15c].

Chronic exposure in humans or animals results in selenosis [53]. Animals show growth reduction, liver changes, anaemia, pancreatic enlargement and some domestic animals also exhibit neurotoxicity following selenium exposure above 0.03-0.4 mg/kg bw [24]. Growth reduction in experimental animals is apparently caused by selective selenium accumulation and toxicity in growth hormone producing cells of the anterior pituitary gland [54]. In humans, symptoms may include bad (garlic) breath, diarrhea, hair and nail loss, disorders of the nervous system and skin, poor dental health, and paralysis [55]. Other related toxic effects are a disruption of endocrine function, that beside growth hormones may influence synthesis of thyroid hormones and an insulin-like growth factor metabolism [56].

9.5.1. No Observed Adverse Effect Level (NOAEL) and Tolerable Upper Intake Level (UL)

The Scientific Committee on Food of EFSA calculated for adults a tolerable upper intake level (UL, that is the maximum level of total chronic daily intake of a nutrient from all sources) judged to be unlikely to pose a risk of adverse health effects to humans, of 300 μg Se/day [33]. This value covers selenium intake from all sources of food, including supplements, and should be considered to apply also to pregnant and lactating women, as well as to children after extrapolation on a body weight basis.

This UL was calculated using a "no observed adverse effect level" (NOAEL) of 850 μg/day for clinical selenosis in the study on 349 subjects of [57], and an uncertainty factor of 3 (this parameter is used to correct the NOAEL for scientific uncertainties associated with extrapolation/analysis of data used in deriving the upper level). Higher NOAEL values of dietary Se are however reported in other studies (1,540–1,600 μg/day) [58] that may lead to consider this UL level of 300 μg Se/day largely reassuring.

The Food and Nutrition Board of the Institute of Medicine set a UL for selenium in adults at 400 μg/day and this value decreases progressively from 280 μg/day in children 9-13 years to 45 μg/day in infants 0-6 months [26].

9.6. DIETARY SOURCES

Diet is the most important Se source, and intake of this essential element depends on its concentration in food and amount of food consumed [56]. Different forms of Se are present in food (Table **9.1**). These show diverse concentration and bioavailability, which is higher for organic Se forms [56,59]. According with the detailed information provided below in this section and in Table **9.2**, Se content in food is influenced by geographical location, seasonal changes, protein levels, and food processing method. As a result, food levels of Se can vary many folds not only between countries, but also between regions in a country. A food may have more than tenfold difference in Se content, depending on where it was produced. Protein content is an important factor influencing the presence of Se in food. In fact, Se can replace sulphur in the amino acids thus appearing in protein structures as Se-Met, Sec and selenocystathionine. Inorganic Se-compounds are used in the synthesis of these Se-amino acids and finally incorporated in vegetable proteins. Thus, the Se forms included in the vegetable proteins of animal feed would ultimately be employed in the synthesis of animal's own proteins, facilitating their accumulation in livestock. Most plants do not have the ability to accumulate large amounts of Se (concentrations rarely exceed 100 μg/g, dry weight). However, various plant species such as garlic (*Allium sativum*), Indian mustard (*Brassica juncea*), canola (*Brassica napus*), and some mushrooms have been recognized as Se accumulators. They have the ability to take up large amounts of Se without exhibiting any negative effect. This is mainly due to the organization of the intracellular Se concentration into the structures of Sec and Se-Met, which are normally incorporated into proteins. When consumed in appropriate amounts, these foods can be a significant dietary source of Se [38].

Se intake is mainly in the form of organic compounds ingested in grains, meat, yeast, and vegetables [60]. In the United States, meats and bread are common sources of dietary Se [61]. Diaz-Alarcon and coworkers [48] showed that the same food items are the main contributors to daily Se intake in healthy individuals from South-eastern Spain. Here, a total of 55% of daily Se intake came from these two food groups due to their high Se concentrations and/or consumption. These findings are in agreement with those obtained by other researchers [45,62] and suggest that some dietary regimens and particularly vegetarians may experience a decreased intake, and thus a suboptimal Se status.

Although Se content of foods can vary widely, estimates have been done and tables of composition for common food items are available for assessing the dietary intake in the general population. For example, Food Composition Database of the U.S. Department of Agriculture lists the selenium content of several types of food items [63]. A detailed analysis of food content of Se is also available by the EU organism EFSA through a call for data on Se and chromium levels in food and beverages (the EXPOCHI project on long-term dietary exposure to Se in young children living in 12 different European countries) [64]. Data covered food testing carried out in the period from 2000 to 2008 and were reported as lower bound (LB) and upper bound (UB) mean selenium concentration of 45 communal food groups (Table **9.3**). Beside this, a detailed description of food items recognized to represent an important source of dietary Se is reported in Table **9.4** as evaluated in selected countries (adapted from [65]). This latter chart highlights the interplay between geographical origin of food and food choice in determining Se intake of individuals or populations of subjects.

Table 9.3: Mean Se concentration per food group as assessed in the EFSA project EXPOCHI on long-term dietary exposure to selenium in young children living in 12 different European countries [66]

Food Group		**Mean Selenium Concentration (µg/kg)** [a]	
Nr	**Name**	**LB**	**UB**
1	Composed foods-cereal based mixed dishes and cereal-based desserts	51.05	62.25
2	Vegetables excl. dried vegetables	47.26	61.04
3	Nuts/seeds	379.13	390.61
4	Coffee/tea in concentrated and in powdered form	240.67	288.39
5	Chocolate/chocolate products	259.50	375.76
6	Fruit excl. dried fruit	3.62	17.42
7	Dried fruit	0.00	0.00
8	Fresh and dried herbs, spices, seasonings and condiments	12.01	22.72
9	Food supplements	32308.72	32379.06
10	Waters	1.32	2.97

Table 9.3: contd...

11	Sugar, sweeteners and sugar products (*e.g.* sugar based confectionery, chewing gum and decorations)	27.59	51.27
12	Fats, oils and fat emulsions (also *e.g.* rice milk, no soy milk)	65.09	87.54
13	Composed foods: meat based mixed dishes	40.10	49.37
14	Composed foods: fish based mixed dishes	255.44	258.29
15	Dried vegetables	0.00	0.00
16	Pulses/legumes	176.13	187.17
17	Soy milk/soy based dessert	76.00	91.71
18	Milk/dairy drinks	12.48	23.45
19	Cheese	58.05	80.87
20	Dairy based products	8.68	43.09
21	Salt	428.14	499.57
22	Fish	353.15	355.13
23	Molluscs	488.51	493.14
24	Cephalopods	475.10	475.10
25	Crustaceans	360.32	362.10
26	Other seafood (echinoderms)	478.11	478.11
27	Beer/malt beverages	1.58	13.10
28	Wine/substitutes	0.39	8.38
29	Other alcoholic beverages	111.25	122.77
30	Fruit juices/nectars(a)	0.48	11.44
31	Vegetable juices/nectars	2.99	12.00
32	Soft drinks/edible ices	6.61	20.65
33	Cereals/cereal products (no cereal based desserts or cereal based mixed dishes)	53.21	81.01
34	Other food for special dietary uses	7073.16	1098.18
35	Infant formulae, follow up formulae, food for young children and infant formulae and follow up formulae for medical purposes	40.30	65.45
37	Miscellaneous foods	45.35	127.69
38	Liver/kidney	490.93	492.18
39	Offal except liver/kidney	166.71	169.61
40	Types of vegetarian substitutes for meat/fish	0.00	0.00
41	Fresh meat	114.11	237.60
42	Processed meat	0.00	0.00
43	Eggs	233.63	246.80

[a] Two mean selenium concentrations were calculated per food group: lower bound (LB) and upper bound (UB). The LB mean concentrations were calculated, where appropriate, by assigning zero to the samples with a level below limit of

reporting (LOR), determination (LOD) or quantification (LOQ). The UB mean concentrations were calculated by assigning LOR, LOD or LOQ to the samples with a level below the respective limits. Fluctuations in daily concentrations are assumed to average out in the long run.

Table 9.4: Se content (μg/kg, range or mean) in main food from different countries (adapted from [65])

	Beef	Bread	Cereals	Cheese	Chicken	Eggs	Fish	Fruits	Milk	Pork	Rice	Vegetables
Australia	72–121	92.6–125	62.9	70–78.9	116–280	190–414	20–632	4.5–76	2.5–25.9	94–205	25	0.5–32
Canada	30–310[a]	–	10–1,350	60	150	60	46–1,570	1–23	10	–	–	10–119
Finland	10–70[a]	–	5–115	10–40	50–100	110–180	180–980	2–30	2–20	–	–	1–2
Germany	130–280[a]	–	30–880	100	150	180	240–530	10–41	10	–	–	4–98
Greece	33.5–63.1	37.9–150.2	19.1–20.2	14.3–127.9	76.3–82.4	56.4–181.1	28.7–519.9	1.1–7.9	10.7–22.2	90–98.2	17.7–20.5	1.2–15.9
Ireland	61–105	15–158	–	9.5–11.5	86–147	56–282	268–298	–	14–22	82–129	10–17	10–38
Italy	15–446[a]	12	0–43	30–140	15–416	29–89	118–293	1–13	10–35	–	20.1	1–25
New Zealand	22.3–83	31.6–59.4	–	23	137–145	157–161	195–512	–	1–14	19.3–150	0	0–2.5
Thailand	72–226	–	–	–	156–271	145–420	196–1,137	–	19–36	142–250	29–65	1–127
UK	30–76	43–92	20–530	7.4–12	60–70	90–120	200–500	5	10–15	140	4–13	3–22
USA	134–190	282–366	300–560	13.9	190–276	225–308	126–502	1–13	20–21	144–450	75	1–1,180

[a] Se content in red muscle meats.

The contribution by different food items to dietary intake of Se reported in these tables is supported by the available literature examined in the next paragraphs of this section. However, variations of selected food items can hardly interfere with nutritional evaluation (dietary assessment) of Se intake in individuals or population groups, which needs analytical validation by proper biomarkers. For example, the U.S. Department of Agriculture Food Composition Database lists the selenium content of Brazil nuts as 544 mcg of Se per ounce, but values from others analyses markedly differ from these numbers [67]. Cooked and processed foods contain considerably less selenium than raw foods.

Selected food sources of selenium are described as main food groups in the following paragraphs and include beside Brazil nuts and grains (wheat germ, barley, brown rice, oats), sunflower seeds, mushrooms, onions, eggs and a variety of meats (beef, liver, lamb, pork, and poultry as chicken and turkey) and fish such as tuna, halibut, sardines, flounder, salmon, and shellfish such as oysters, mussels, shrimp, clams, scallops.

9.6.1. Meat, Chicken, Fish and Eggs

Meat, chicken, fish and eggs are protein-rich foods containing high levels of Se [68]. In these food groups, Se levels range from 87.6 to 737 ng/g [69]. Fish and eggs show the highest Se concentrations [40,70]. Meat, fish and eggs contribute the major part of dietary Se in several countries such as Greece, Portugal and Japan [40, 69-70]. In Japan, fish is the greatest Se contributor (up to 60% of daily total intake) rather than staple foods (rice and vegetables) [70]. Literature on Se content in fish from different locations ranged between 62.7 and 506.7 ng/g in Greece [40], 120.0–632.0 ng/g in Australia [71], 126 to 502 ng/g in the USA [72], and 195 to 512 ng/g in New Zealand [73]. Se content in eggs from Australia have a mean concentration of 90 ng/g in white and 260 ng/g in yolk (boiled eggs) [74].

Mean concentrations of Se in meat from Greece ranged from 48.8 to 94.1 ng/g, with pork measuring significantly higher than beef [40,71]. In sausages from Spain, Se levels range from 89.0 to 739 ng/g [48].

9.6.2. Fruits and Vegetables

Selenium occurs in staple foods such as corn, wheat, and soybean as selenomethionine, the organic selenium analogue of the amino acid methionine [75]. Fruit contains low concentrations of Se. This fact could be explained by the low protein fraction (and therefore, the high water content) of these products. Fresh vegetables were also shown to be poor sources of Se [68c].

However, it is known that vegetables such as *B. juncea* and the better known species of the *Brassica genus* (broccoli, Brussel sprouts, cabbage, cauliflower, collards, kohlrabi, mustards and kale), garlic, chives and onions tend to have higher Se concentrations and the extent to which they are consumed is reflected in the Se content of human tissue and body fluids [38,76]. These plants have a greater fraction of sulphur containing amino acids and their derivatives, but they also contain other sulphur compounds like glycosinolates or sulfoxides. Adequate analogues of these can be formed by substitution of sulphur with Se, resulting in higher Se levels [76c].

Garlic and onions seem to be a good dietary source of Se, and both have valuable anti-carcinogenic activities. Additionally, their intake does not result in excess accumulation of Se in tissues; nor could any perturbation in the action of Se enzymes be observed, even at high Se intakes [38]. Similarly, Manjusha and coworkers [77] found high Se content in mushrooms (1.34 μg/g). Some, but not all mushrooms tend to accumulate Se because they are another vegetable species with a high content of sulphur containing compounds. *Agaricus bisporus* is one of the most commonly studied mushrooms for Se speciation purposes and is also the most commonly consumed mushroom in Europe and the USA. Other mushrooms that accumulate Se are *Boletus edulis* and *B. macrolepiota* [38]. Plants that accumulate Se may be used as a natural source of mineral supplements for both animals and human beings, especially in areas that are Se deficient.

9.6.3. Nuts, Legumes and Cereals

Pappa and coworkers [40] reported that the Se content in legumes from Greece ranged from 24.4 to 443.9 ng/g, with a mean value of 165.2 ng/g, lentils presenting the highest concentration. They encountered Se concentrations between 7.0 and 32.27 ng/g in nuts. Some nuts are also sources of selenium [67a]. It is wise to eat Brazil nuts only occasionally because of their very high selenium content. Pistachios proved to be the richest, whereas almonds were the poorest Se source. Protein-rich nuts (pistachios, walnuts) present higher Se concentration than other products [38,76c]. Manjusha and coworkers [77] encountered a mean content of Se in Brazil nuts of 3800 ng/g. Brazil nuts are known for their high Se concentration and one single Brazil nut could exceed the RDA for Se [38]. The proteins found in Brazil nuts are very high in Se-containing amino acids, mainly Se-Met.

Levels of Se in cereals of 10.0–550.0 ng/g (referred to fresh weight) have been reported [38]. Marro [78] described Se levels in white bread of 80.0–109.0 ng/g (mean of 92.6 ng/g) and in whole meal bread of 100.0–152.0 ng/g (mean of 125.0 ng/g). Tinggi and coworkers [79] reported that the major source of Se in Australia comes from wheat products such as bread (60.0–150 ng/g) and pasta (10.0–100 ng/g) that is in agreement with the information reported by Diaz-Alarcon and coworkers [48]. Mean Se concentrations of bread ranging from 70.0 to 131.8 ng/g [40]. The differences of Se content between brown, whole-wheat and white bread were not statistically significant ($p<0.01$), although brown bread seemed to be much richer in Se than the others.

9.6.4. Milk and Dairy Products

It has been found that Se concentrations in milk from different animal species decreases in the following order: human, sheep, goat and cow. It is known that Se concentrations in milk are negatively correlated with its fat content [40]. A similar trend was observed by Barclay and coworkers [80] in cheeses. A survey of Se content of Australian cow milk showed a wide variation with higher levels in summer (23.8±4.6 ng/l) than in winter milk (20.7±4.2 ng/l). Cabrera and coworkers [81] determined Se content in dairy products and observed a wide variability among the data due to the different concentrations of Se present in milk, eggs, cereal, fruit and other foodstuffs used in their manufacture. All authors consider that milk and dairy products contribute a considerable fraction of the total dietary intake of Se, particularly for infants [71,82].

9.6.5. Miscellaneous

Water Se content is usually trivial compared to the content of this element in food [26]. Drinks and potable water are generally poor sources of Se [83].

Se content in fast food from the UK encounter the highest levels in products based on certain mushrooms, spinach and fish [47b], which however varied their Se content from sample to sample, even in cases of the same product. In baby foods from Spain such as sole with vegetables or with potatoes, angler fish with vegetables and hake with rice, Viñas and coworkers [84] detected Se levels that ranged from 21.5±1.3 to 72.8±5.3 ng/g. Se levels in virgin olive oil, olive oil and marc oil produced in Southern Spain ranging from undetectable to 178.51 ng/g [85]. No statistically significant differences were found between the three types of oil. Se content in Indian spices and condiments ranging from 12 to 670 ng/g. The highest levels were detected in turmeric (500 ng/g) and sweet neem (670 ng/g) [86].

9.6.6. Dietary Supplements

Selenomethionine, the former of the organic molecules of Se in staples and meats, is generally considered to be the best absorbed and utilized form of selenium, serving as a vehicle for selenium storage in organs and tissues. As such it is widely used in dietary supplements to be readily incorporated into body proteins in place of methionine. Selenium supplements may also contain sodium selenite and sodium selenate, two inorganic forms of selenium (Table **9.1**).

Selenium is also available in 'high selenium yeasts', which may contain as much as 1,000 to 2,000 micrograms of selenium per gram [75a]. Most of the selenium in these yeasts is in the form of selenomethionine. This form of selenium was used in a large-scale cancer prevention trial in 1983, which demonstrated that taking a daily supplement containing 200 micrograms of selenium per day could lower the risk of developing prostate, lung, and colorectal cancer [87]. However, some yeasts may also contain inorganic forms of selenium, which are not readily available for absorption and incorporation into human proteins.

Early studies suggested that the organic forms of selenium increase blood selenium concentration to a greater extent than inorganic forms [29b]. However, this effect does not significantly improve in all treated subjects the activity of the selenium-dependent enzyme glutathione peroxidase, a biomarker of Se status commonly used to verify the biological compliance to Se treatments [29c]. As a general recommendation, human Se supplementation should be accompanied by this type of post-treatment verification as well as by the characterization of Se status at baseline. These steps should also help at excluding effects of accumulation and damage and should be operated using appropriate laboratory biomarkers (see section 9.4.2).

The database of the European Food Safety Agency (EFSA) Panel on Dietetic Products, Nutrition and Allergies (consultable at http://www.efsa.europa.eu/) contains several requests of authorization for health claims of Se containing supplements submitted in the last years, such as Se-Met and Se-containing yeast products which have been confirmed to represent the best choice in Se supplementation. Some of these requests have received positive opinions, and in 2009 this agency accepted health claims linking the dietary intake of selenium to: "protection of DNA, proteins and lipids from oxidative damage, normal function of the immune system, normal thyroid function and normal spermatogenesis" [88].

In the majority of developed countries, dietary sources are expected to provide enough Se to obtain an optimal Se status. Therefore, supplementation is not required in these cases and body's Se and the functioning of Se-proteins can be maintained with an optimal diet. It is likely that Se supplementation benefits certain populations only, such as Se-deficient or genetically predisposed groups, or that Se supplementation is particularly effective at lowering the risk of some health disorders and not others. These aspects need to be further investigated and appropriate evaluations of the Se status have to be made in order to personalize requirements thus maximizing health benefits of an optimal Se intake.

9.7. SELENIUM DEFICIENCY

A deficiency in Se in human is rare. However, in some cases, a lack of adequate sources of this trace mineral is proposed to expose the general population lead to significant health risks. There are few major signs that an individual may be suffering from a selenium deficiency, which include general fatigue, muscle and joint pain, unhealthy hair and white spots on the fingernails, hypothyroidism (mainly due to autoimmune thyroiditis) and reproductive disorders (changes in menstruation, miscarriage).

Chronic deficiency in selenium in human is considered a risk factor for developing Hashimoto's disease, an auto-immune disease of thyroid gland.

Non-endemic deficiencies of selenium include forms caused by severe malnutrition. Indeed, biochemical evidence of selenium depletion is not uncommon in subjects maintained on parenteral or enteral feeding for long periods using preparations not fortified with this micronutrient [89]. Inappropriate selenium contents of commercial infant formulas may also result in a fall of selenium intake [90]. Clinical manifestations of deficiency arising from such situations are however uncommon and poorly defined [91].

9.7.1. Endemic Forms

The Keshan disease and the Kashin-Beck disease are considered typical endemic forms of Se deficiency. The former is a cardiomyopathy endemic in certain Se-deficient areas of China that is completely preventable with Se supplementation

[35]. This disease was first described in Chinese medical literature more than one century ago, but not until 40 years after its widespread occurrence in 1935 was it discovered that selenium deficiency was an important factor in its etiology [92]. Endemic in children aged 2-10 years and in women of childbearing age, this disease has a geographic distribution covering localities from northeast to southwest China and has marked seasonal fluctuation in incidence. Typical manifestations are fatigue after even mild exercise, loss of appetite, and severe cardiac comorbidity associated with multifocal myocardial necrosis and fibrosis, in the absence of damage to coronary arteries. Ultrastructural studies show that membranous organelles, such as mitochondria or sarcolemma, are affected earliest. Once the disease is established, selenium is of little or no therapeutic value. Prophylaxis consisting of oral administration of selenium 3 months before the periods of highest anticipated risk is highly effective.

Geographic similarities in the distribution of Keshan disease and the selenium and vitamin E-responsive white muscle disease in animals prompted successful investigation of the relevance of a low selenium status. However, the disease appears to have a multifactorial origin. A viral myocarditis probably attributable to enhancement of the virulence of a coxsackievirus during its passage through selenium-deficient host tissues has been postulated as causal mechanism [93] and other nutritional variables, such as a marginal vitamin E status, may also be involved. Notwithstanding, the finding of extremely low selenium contents in staple crops of affected areas and convincing demonstrations of the prophylactic effectiveness of selenium administration, leave no hesitation that selenium deficiency is among the primary pathogenic factors [94].

The Kashin-Beck disease is a selenium-responsive bone and joint disease (osteoarthropathy) detected in children aged 5-13 years in the same pandemic areas of Keshan disease where the availability of soil selenium for crop growth is low, *i.e.* China and less extensively south-east Siberia. The disease is characterized by joint necrosis - epiphyseal degeneration of the arm and leg joints resulting in structural shortening of the fingers and long bones with consequent growth retardation and stunting [92]. The selenium contents of hair and whole blood are abnormally low and blood levels of GPx are reduced. Although it is ameliorated by selenium therapy, other factors such as the frequent presence of mycotoxins in cereal grains grown in the area may be involved. A spontaneous decrease in incidence from 1970s (44 %) to 1986 (1 %) has been observed and is attributed to general improvements in the nutritional status of Chinese rural communities.

The observation that cardiac lesions of Keshan disease may involve a coxsackie virus infection, beside selenium deficiency, suggests that selenium status may influence the susceptibility to, and the virulence of, a viral infection.

Animal studies have confirmed that selenium-deficient mice are particularly susceptible to infection by this virus and the passage of the virus through selenium-deficient subjects enhanced its virulence [93] with acquisition of myocarditic virulence even in viral strains which normally were not myopathogenic; this virulence was maintained even during passages in animals with normal selenium status [95].

9.7.2. Se and Thyroid Hormones

The thyroid gland has the highest selenium concentration of all tissues [96]. Selenium modulates different thyroid functions. Thyroid hormone formation requires selenoproteins, which also regulate thyroid hormone activity by inactivating it. These selenoproteins are known as deiodinases enzymes, D1, D2 and D3. The conversion of the thyroid hormone precursor, T4, to the active T3 hormone is mediated by D1 or D2, while both D1 and D3 can inactivate T3 [96,97]. Mice genetically lacking the D2 selenoprotein exhibit reduced thyroid stimulating hormone regulation and reduced capacity to metabolize fat to keep warm (brown fat thermogenesis) [98]. Interestingly the thyroid expresses as many as 11 different selenoproteins that appears importantly implicated in the protection from the oxidative stress induced by hydrogen peroxide, a byproduct generated during thyroid hormone synthesis.

The importance of selenium for thyroid hormone metabolism appears evident from changes in the T3-T4 ratio which develops after relatively mild selenium depletion in infants and elderly (65+ years) subjects [99]. Decreases in the T3-T4 ratio, a hallmark of decreased thyroid hormone balance, have been detected when serum selenium falls below 0.9 μmol/l [100]. In a Scottish study this variation correlated with a decline in dietary and plasma selenium after the replacement of selenium-rich wheat from Canada and the United States with selenium-deficient wheat from European sources [101]. However, no evidence of an effect on thyroid function or on the ratio of free or total T4 to T3 was shown in a randomized controlled trial of selenium supplementation in apparently euthyroid, elderly adults with low-to-moderate selenium status [102]. Communities with increased prevalence of myxedematous cretinism, in addition to iodine deficiency, display low plasma selenium status and GPx activity. However, restoration of iodine supply, particularly if excessive, may induce a high peroxidative stress through the action of iodide peroxidase, the first step in iodine utilization by the thyroid. It has been postulated that necrosis and thyroid fibrosis leading to irreversible hypothyroidism result if a concurrent deficiency of selenium limits peroxide destruction by the protective action of the selenium-dependent enzymes, GPx and, more probably, thioredoxin reductase [103]. In areas where myxedematous cretinism is endemic and characterized by persistent hypothyroidism, dwarfism, and stunting, that attempts to introduce iodine therapy for mildly affected individuals should be preceded by an assessment of Se status and appropriate supplementation in case of Se deficiency [104]. Although this suggestion appears compatible with experimental observations on hypothyroid rats differing in Se status, its validity in humans has yet to be validated [103,105].

On the other hand, selenium, in the form of GPx3, protects thyroid cells from the hydrogen peroxide generated during the biosynthesis of T3 and T4 [96]. This activity may partly explain the inverse association between Se status and thyroid volume, as well as the relationship between the incidence of thyroid cancer and low prediagnostic serum-Se concentration [106]. Moreover, several studies demonstrated that Se supplementation (80 μg or 200 μg per day as sodium selenite or selenomethionine) positively modulates the natural history of Hashimoto's thyroiditis, the most common form of autoimmune thyroid disease characterized by the presence of complement-fixing autoantibodies to thyroid peroxidase [107]. Indeed, Se supplementation significantly lowered thyroid peroxidase autoantibody titre at 3 months [107a]. Of note, in pregnant women with autoimmune thyroiditis, Se supplementation (*i.e.*, 200 μg per day selenomethionine) was associated with a reduction of either thyroid inflammatory activity or occurrence of post-partum thyroid disease and permanent hypothyroidism [108].

Se is also effective in autoimmune hyperthyroidism- *i.e.*, Graves' disease. In a randomized controlled trial of 100 μg sodium selenite twice daily, or pentoxifylline (600 mg), or placebo for 6 months, Se treatment alone was significantly associated with slower progression of Graves' orbitopathy, less eye involvement and improved quality of life [109].

9.8. HEALTH PROMOTING AND ADVERSE EFFECTS

While Se deficiency in humans is rare (discussed in the previous section), there is evidence that less clear changes in Se status may affect key aspects of human health such as the immune response, cardiovascular disease, diabetes, cancer and neuronal disorders [29a,110]. On the other hand, long-term adverse effects of the exposure to high levels of Se have been repeatedly reported with age and sex dependent differences, and this has stimulated an intense debate on safety of fortified food and supplements containing Se.

9.8.1. Se and Cancer

The possibility that increased intakes of Se might protect against the development of cancer in humans has generated great interest in the scientific community [110a,111]. An increased intake of Se appears to stimulate tumorigenesis in some animal models of pancreatic and skin cancer. In contrast, the protective effect of higher exposures to Se observed in several animal studies, together with small but statistically significant differences in blood levels of Se detected in some retrospective-prospective studies of subgroups of people developing cancer, explains the continuing interest in the anti-carcinogenic potential of selenium.

Supplementation of Se become of public interest when a remarkable decrease in cancer mortality in persons consuming 200 μg of Se daily (approximately 4 times the daily requirement; see the NPC trial described below) was reported [87]. As a general rule, Se may have anticarcinogenic effects at low concentrations, whereas at concentrations higher than those recommended for nutrition, it can be genotoxic and carcinogenic [112], and its toxic level is relatively close to that required for normal health (see above for detailed description of Se toxicity and nutritional levels). Moreover, a number of earlier epidemiologic studies showed no relationship between selenium and cancer risk [113]. Nevertheless, the mechanisms of action of Se compounds, either *via* a pro-oxidant pathway, as seen in cytotoxicity and apoptosis, or *via* an antioxidant pathway, as proposed in cancer chemoprevention, are still unclear [114].

As a result, the question of whether selenium protects against cancer was considered "wide open" by Birt end of 1980s [115] and remains as such nowadays. The benefits of Se supplementation pertaining to cancer have been a focus of recent clinical studies, which have yielded varying results. One study conducted by the Nutritional Prevention of Cancer (NPC) study group indicated that Se supplementation significantly decreased the risk of lung, colorectal, and prostate cancers [87,116]. This led to the largest, most comprehensive study on the effects of Se intake, alone and with vitamin E, on cancer: the Selenium and Vitamin E Cancer Prevention Trial (SELECT). This randomized, placebo-controlled trial included 35.533 men from 427 participating sites in the United States, Canada, and Puerto Rico and was designed as a 12-year study. However, trial supplements were discontinued early (in year 7) predominantly based on a lack of evidence of benefit against cancer for either study agent [117].

Different factors could explain such apparently divergent findings between the trials. First, selenium might be more important in prevention of prostate cancer progression, hence its stronger effect could be obtained against advanced than primary disease [2b]. Second, selenium might show an effect on risk only over a particular range of status, neither too low nor too high – *e.g.*, when the selenium concentrations in the group under investigation range from below to above the concentrations needed to optimize the activity of selenoenzymes such as GPx and SEPP1. Indeed for selenium, as for many nutrients, several human studies have provided evidence of a U-shaped relationship between intake or status and protection from cancer [32a,118]. Third, the interaction between selenium intake or status and genetic background could be important. Single nucleotide polymorphisms (SNPs) in selenoprotein genes can affect the efficiency of selenoprotein synthesis as well as, its activity and concentration in plasma, and thus its effects on the risk of disease [119]. Different SNPs in GPx and SEPS have been linked with the risk of prostate, lung, breast, colorectal, bladder and thyroid cancers and survival. Moreover, these selenoprotein SNPs (and SNPs in related pathways) can interact with selenium status such that individuals with specific genetic variants might benefit more from additional selenium than others [93].

Overall, the association between selenium intake and cancer risk, although clearly of some interest, is still in need of further investigation before a conclusion can be reached.

The relationships between Se supplementation and cancer yielded discrepant and conflicting results. Prospective studies have provided some evidence for a beneficial effect of selenium on the risk of lung [120], bladder [121], colorectal [118a], liver [122], oesophageal or gastric-cardia [123], thyroid [106b] and prostate cancers [118a,119,124] (See Table **9.5**).

9.8.2. Se and Diabetes

A preventive role of selenium on the risk of diabetes has been reported and ascribed to the "insulin-like" activity of selenium and to antioxidant properties of the selenoenzymes [125].

Animal model and epidemiologic investigations reported correlations between Se deficiency and abnormal glucose or lipid metabolism [126], as well as low plasma Se levels or selenoperoxidase activity in diabetic subjects [127]. Thus, supplementing Se was perceived as an effective strategy to prevent and treat diabetes.

However, mice overexpressing the Se-dependent cellular glutathione peroxidase-1 (GPX1; involved in reduction of peroxides), the most abundant selenoprotein, developed type 2 diabetes-like phenotypes [128].

Investigating health effects of Se at the clinical level, researchers have often obtained conflicting results. As an example of this, observational studies and a few randomized trials have showed both lower and higher risk of type-2 diabetes linked to the level of selenium intake and status [129].

Table 9.5: Meta-analyses of prospective studies of selenium and cancer risk, by tissue type (modified from [110a])

	RR/OR (95% CI)	Study Design	Comparison	Author(s)
Lung				
16 studies	0·74 (0·57 to 0·97)*	Cohort	Higher *vs* lower selenium exposure (serum, toenails, intake)	[120]
Bladder				
7 studies	0·61 (0·42 to 0·87)*	Cohort, nested	Highest *vs* lowest selenium status (serum, toenails)	[121]
Prostate				
5 studies	0·74 (0·61 to 1·39)	Case-control	Any *vs* lowest selenium exposure†	[124c]
5 studies	0·74 (0·39 to 1·39)	Case-control	Moderate *vs* lowest selenium exposure†	[124c]
11 studies	0·72 (0·61 to 0·84)*	Cohort	Any *vs* lowest selenium exposure†	[124c]
11 studies	0·74 (0·61 to 0·90)*	Cohort	Moderate *vs* lowest selenium exposure†	[124c]
20 studies	-0·23 (-0·40 to -0·05)*	Cohort, nested, case-control	Overall pooled standardised mean difference (serum, plasma, toenail selenium)	[124b]

RR=relative risk. OR=odds ratio. *Significant finding. †Any intake: average between the first and fourth quintile (plasma, serum, toenail concentrations, or intake) or the first and third quartile. Moderate intake: average between the second and fourth quintile or the second and third quartile. The lowest level (first quartile) was used as the reference group.

A strong positive correlation between erythrocyte GPX1 activity and insulin resistance in nondiabetic pregnant women was reported in 2003 [130]. Moreover, posthoc analysis of the Nutritional Prevention of Cancer Trial (NPC) revealed a >2-fold increase in type 2 diabetes incidence in the Se *versus* the placebo group [131], resulting in a hazard ratio of 2.7 in subjects in the highest tertile of plasma Se concentrations. In this study supplementation was carried out with 200 μg/d selenium as high-selenium yeast for 7.7 years. Regression analysis of the ORDET cohort study [132] comprising 71women [133] from northern Italy for a median follow-up of 16 years indicated that the odds ratio for diabetes comparing the highest to the lowest quintile of Se intake was 2.39 (P for linear trend = 0.005);

the odds ratio for diabetes associated with an increase in Se intake of only 10 μg/day was 1.29.

Cross-sectional analyses within the U.S. Third National Health and Nutritional Examination also revealed a strong positive correlation between serum Se concentrations and the prevalence of type 2 diabetes higher fasting plasma glucose and glycosylated hemoglobin levels [134]. Notably, high body Se status also was associated with adverse plasma lipid profiles in adults in the United States, United Kingdom, and Taiwan.

In contrast, mixed or positive effects of Se on decreasing diabetic risk have been shown in several studies in the United States and Europe [30,125b,135].

A recent intervention study by Rayman and coworkers [136] investigating adiponectin levels in elderly individuals of relatively low selenium status, showed baseline adverse association between Se status and adiponectin levels, but did not show a diabetogenic effect (in terms of adiponectin and other metabolic parameters) of a six-month supplementation with selenium.

Two studies found lower serum selenium concentrations in diabetic patients than in control subjects [137] while in the Health Professionals Follow-up Study, toenail concentrations were lower in diabetic men than in non-diabetic controls [30]. Randomized trials again showed discordant findings. In the SUVIMAX trial, despite positive correlations between plasma selenium and blood glucose levels both at baseline and at the end of the follow-up, no effect of supplementation with a mixture of antioxidants, including 100 μg/day selenium, on plasma glucose levels after 7.5 years of follow-up, was found [138]. Finally, in SELECT, supplementation of 35.533 American men with 200 μg selenium per day as selenomethionine had no effect on risk of T2D after a median follow-up of 5.5 years [117].

The lower selenium status measured in men with T2D in case-control studies might be an effect of the disease, or its associated inflammation, on the selenium status. It has been shown that a systemic inflammatory response, may inhibit the expression of SEPP1 and reduce plasma selenium [139]. Accordingly, insulin resistance can be triggered by oxidative stress and ameliorated by antioxidant treatments [140]. It is thus plausible that plasma selenium levels below the concentrations needed for achieving an optimal activity of GPx may fail to prevent the insulin resistance effect of inflammatory cytokines and oxidative stress. On the contrary, the increased risk of diabetes with high selenium intake or supplementation might be explained by the effects of high selenium on insulin signalling. Binding of insulin to its receptor initiates an intracellular signalling cascade, which is accompanied by a burst of hydrogen peroxide that acts as a second messenger [129]. High activity of GPx1, which removes hydrogen peroxide, might thus interfere with insulin signalling. Experimental animal models and data in pregnant women, partly confirmed that GPx1 overexpression is consistently associated with insulin resistance [128,141]. However, besides GPx1 other selenoproteins are also emerging as potentially relevant for diabetes risk. SEPP1, a selenoprotein that requires a higher selenium intake than GPx1 to achieve maximum plasma concentrations, has been suggested to act as a negative insulin modulator either by inhibiting the insulin-induced burst of reactive oxygen species *in vitro* or by inactivating AMP-activated protein kinase, a positive regulator of insulin synthesis and secretion in pancreatic islet β cells [142]. Clinical studies have shown that serum SEPP1 concentrations are significantly correlated with glycated haemoglobin A1c and fasting plasma glucose, and raised in a stepwise manner in patients with prediabetes/T2D as compared with individuals with normal glucose tolerance [142,143]. Different reasons may thus underlie the apparent paradox of increased risk of T2D with selenium supplementation in the NPC, but not in SELECT trial. First, the small number of participants included in NPC trial was at much greater risk of "chance findings" than the larger SELECT. Alternatively, the higher baseline selenium status of men in SELECT (median serum selenium = 136 μg/L *vs.* mean plasma selenium= 114 μg/L) had already raised selenoprotein expression or activity to plateau or pass the "threshold of risk" before supplementation. In support of this assertion, the number of cases in the placebo group per 1000 person-years in SELECT was higher than in the NPC trial (14.1 *vs.* 8.4). Finally, in animal models, low levels of expression of GPx1 and other stress-related selenoproteins (regulated by the amount of selenium in the diet) are as damaging as high levels of expression with respect to insulin resistance and hyperglycaemia [144]. Hence a U-shaped association between selenoproteins and T2D risk might explain some of the apparently contradictory findings. In conclusion, more basic and clinical research is needed to better elucidate the role of Se in glucose homeostasis and metabolism as well as to clarify the optimal level of Se intake for preventing diabetes [127].

9.8.3. Se and Cardiovascular Disease

A significant increase in cardiovascular comorbidity and death in individuals with a lower Se status has reported in some epidemiological and observational studies. Large cohort studies that used a cutoff point of serum selenium of 45 μg/liter observed increased scores for illness and death from cardiovascular disease, and a significant difference in deaths from stroke ([145] and references therein). Another study in middle-aged and elderly men found an increased risk of cardiovascular disease in men with serum selenium levels below 79 μg/liter [146]. In the EURAMIC Study, a multi-center trial in Europe, toenail selenium levels and risk of myocardial infarction (heart attack) were only associated in the centers in which selenium levels were the lowest [147]. These observations have found the support of many *in vitro* and *in vivo* investigations that suggested how optimizing selenoenzyme activity could decrease the risk of cardiovascular disease by decreasing lipid peroxidation and influencing the metabolism and signaling of inflammatory mediators such as prostaglandins.

However, several other studies found no clear inverse association between selenium nutritional status and cardiovascular disease risk [148]. Prospective studies in humans have not demonstrated strong support for the cardioprotective effects of selenium and randomized clinical trials have not performed so far to confirm the causal effect of selenium on cardiovascular disease and risk factors. Cross-sectional evaluations carried out in the Young Finns study and the Olivetti Heart study observed positive associations of selenium status with serum lipids that represent an important cardiometabolic concern, but no evidence to support the causality of this link has been obtained in longitudinal analyses of these cohorts [149]. This type of evidence has been reported in other studies that have been recently reviewed by Stranges and coworkers [150], and these Authors conclude that "at the present time the widespread use of selenium supplements or other strategies that artificially increase selenium status above the level required for optimal selenoprotein activity is not justified and should not be encouraged".

9.8.4. Immune Function and Susceptibility to Viral Infections

An enhanced virulence of coxsackievirus with a (sub-clinical or manifest) selenium deficiency has been described as a possible causal factor of Keshan disease (see section 9.7.1). The fact that this Se-dependent effect could be common to other viral diseases cannot be ruled out. The early pre-clinical stages of development of human immunodeficiency virus (HIV) infection are accompanied by a marked decline in plasma selenium and the virulence of other RNA viruses such as hepatitis B and those associated with development of haemolytic anaemias, are similarly enhanced by a poor selenium status. Sub-clinical malnutrition play major clinical roles during the development of these infections and there are strong indications that only the extent of the decline in selenium status has predictive value with respect to both the rate of development of AIDS and its mortality [151]. Only a few trials of selenium supplementation in HIV-infected individuals have been published. Encouraging positive results have obtained particularly in a recent randomized, double-blind, placebo-controlled trial in 174 HIV-1-positive individuals in which 200 μg/day of Se-enriched yeast were administered for nine months. This supplementation protocol resulted in increased CD4 T cell counts and no progression of the HIV-1 viral load [152].

The mechanisms underlying these effects are not yet resolved. There are indications that the loss of protective antioxidant functions dependent on selenium and vitamin E are both involved and that the resulting structural changes in viral nucleotide sequences may have a role; these changes cause additional selenoprotein synthesis [153], which may further deplete previously diminished pools of physiologically available selenium accelerating pathologic responses [154]. Animal models of selenium deficiency have shown that the microbicidal activity of blood neutrophils is severely impaired even though phagocytic activity remains intact [155].

9.8.5. Se and Mood Disorders

The presence of selenium in the central nervous system is believed to be essential for the normal brain functionality. In fact, in case of insufficient intake of this micronutrient, its concentration significantly decreases in blood and organs, but not in brain, in which the levels of Se remain is high [156]. Moreover, a number of disorders ranking from the neurodegenerative ones, as Alzheimer's and Parkinson's disease [157], to tumors [158], multiple sclerosis [159] and retardation [160] have been associated to changes of selenium concentration in different cerebral regions and in serum. The mechanisms of these physiopatological implications are not completely understood but the neuroprotective role of Se can be ascribe to three main features: first of all, oxidative stress is associated with the development of neuronal and neuromuscular disorders [161], therefore the antioxidant activity of selenium, mainly under the form of the enzyme glutathione peroxidase is expected to contribute a role in the prevention of ROS-induced cell damages of the nervous system. Secondly, a catalytic component of deiodinase proteins, Se is required for the normal synthesis and metabolism of thyroid hormones and thus, for the homeostatic control of brain and of the overall body. The third aspect is connected with the toxicity of heavy metals, which is reduced in the presence of selenium. In fact, the organic Se is known to possess metal-binding functions, especially in the form of the enzyme SelP [162].

Recently, this element has been correlated to mood disorders. Low levels of dietary selenium have been proposed in a 2012 study to increase the risk of developing major depressive disorder (MDD) [163]. Throughout a nested case-control study focused on a representative sample of Australian women, dietary selenium intake was measured using detailed semi-quantitative food frequency questionnaire, together with lifestyle measures. The data support the hypothesis suggesting that selenium supplementation could be important for the treatment and prevention of depression.

9.9. BIOCHEMISTRY AND MOLECULAR BIOLOGY OF SEC AND SELENOPROTEINS (Table 9.6)

9.9.1. Why do Selected Proteins have Substituted Cys with Sec?

Biological roles of Se are intimately connected with those of sulphur. Although Se and S belong to the same family, Sec and Cys possess distinct biochemical properties that are responsible of different roles in protein structure and functioning (Fig. **9.2**).

While Cys is one of the most abundant amino acid residues in proteins, Sec is a rare component with strategic distribution and role in redox-active proteins. Its formation and incorporation in protein structure is thus a critical step in redox biology of living organisms.

At the certain point in the evolution of living organisms, Se and S have combined to provide fine regulatory mechanisms for biological reactions. The substitution of specific Cys residues with Sec in selected proteins is expected to introduce some biological advantage thus having strategic function along the evolution. Chemistry of selenols and thiols provides basic rationalization for such an evolutionary step. A selenol is three orders of magnitude more acidic [pKa (Se-H) ≈ 5.3] than a thiol [pKa (S–H) ≈ 8.4] [164] also showing higher reduction potential [165]. Reactions of simple selenols at pH 7 are up to four orders of magnitude faster than their sulfur analogues (depending on reaction type). This explains why as free amino acid, Sec is less stable than Cys. Indeed, upon storage of Sec in acidic aqueous solution, precipitates of red elemental selenium are observed, while Cys is stable. In controlled environments, Sec shows stronger nucleophilicity than Cys, arising from the higher stability of Sec selenolate compared to Cys thiolate [1,3]. These basic properties suggest that selenoproteins have higher activity than sulfur proteins. Indeed, certain selenoproteins that have Cys in place of Sec at their active sites, exhibit up to 100-fold lower in catalytic efficiency [166]. This may lead to assume that evolutionary selection could have operated on the basis of an increased efficiency of Se-proteins in redox catalysis that has been counterbalanced by availability of Se. Such a straightforward interpretation may help to explain the production of a complex distribution of Sec- and Cys-containing homologues throughout nature [167].

However, native selenoproteins do not necessarily have higher activity than native sulfur proteins, actually, human peroxiredoxin 2 is a sulfur protein that has a similar rate of reaction with H_2O_2 as the selenoprotein GPx [168]. Mammals TrxR feature a selenosulfide bond (GCUG motif) in the active site, but sulfur analogs of these proteins in other organisms often have very similar or even higher activities [169]. Site-specific mutation experiments in *Drosophila melanogaster* showed that the GCUG active site of TrxR has approximately the same activity as the SCCS wild-type active site [170].

S-Proteins	Se-Proteins
1. Proteins contain S as Cys and Met	1. Proteins contain Se essentially as Sec
2. Cys is a semi-essential AA abundant in plant and animal organisms	2. Sec is much less abundant than Cys in plant and animal organisms
3. the free form is abundant in the extracellular milieu - within the cells it is incorporated in proteins or in the tripeptide GSH (1-5 mM)	3. No free pool of Sel exists in the cell. Instead, cells store Se in the less reactive selenide form (H2Se).
4. It is synthesized from Ser with Met as S donor	4. Sel biosynthesis and encoding occur on a specialized tRNA and mRNA structures
5. Cys thiol group is nucleophilic and easily oxidized. The reactivity is enhanced when the thiol is ionized to the thiolate form	5. Sec reacts as an electrofilic agent under condition close to physiological
6. Cys has both structural and functional roles in proteins such as folding, protein-protein interaction, catalytic activity, turnover, signaling, etc.	6. Sec has the functional role of catalytic center of redox proteins that influence the redox of protein and free Cys residues

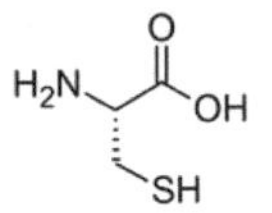

Cys

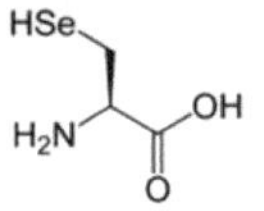

Sec

Figure 9.2: Biochemical characteristics and roles of Se and S as components of amino acid and protein structures.

Moreover, evolutionary biologists have demonstrated the interchangeability of Sec and Cys residues in proteins. Castellano [171] observed evolutionary changes through different species in which Sec was replaced by Cys in most selenoprotein families thus providing alternative explanation to an advantage based on the superior activity of Sec.

Therefore other factors besides kinetics are likely to be relevant in the control of Sec and Cys function in proteins, which could contribute to understand evolutionary aspects and biochemistry characteristics of selenoproteins.

A suggestive hypothesis by Nauser and coworkers [172], for instance, points to a protective effect of Sec residues that have substituted selected Cys residues. The one-electron oxidized product of Sec, the selanyl radical, is not oxidizing enough to modify or destroy proteins, whereas the cysteine-thiyl radical can do this very rapidly. In fact, selenols and selanyl radicals are less oxidizing than thiyl radicals and therefore these are more easily produced and controlled within the structure of proteins with role in redox catalysis or other ROS-dependent processes [173].

9.9.2. Selenoproteins: General Characteristics

While small molecular weight selenocompounds influence human health, a crucial step in understanding the biological effects of dietary Se is determining the functions of the proteins into which Se is incorporated. This section provides a general description of biochemical characteristics and function of selenoproteins, which are summarized in Table **9.6**. For an extensive and systematic description see [2a,2d] and references therein.

Table 9.6: Selenoproteins: biological function and distribution

Se-Proteins	Functions	Tissue Distribution and Subcellular Localization	Catalytic Role(s)	Refs.
Glutathione Peroxidase 1 (GPx1)	**Antioxidant** - Protects human cells from UV-induced DNA damage. - Protects mice from colon cancer. - Protects rodents from chemical toxicity. - Protects mice from viral associated cardiomyopathy. - Protects stressed neurons in - Parkinson's Disease.	ubiquitous (cytosolic localization)	GPx1 uses GSH to reduce ROS, producing GSSG in the process, which is converted back to GSH by the enzyme glutathione reductase.	[174]
Glutathione Peroxidase 2 (GPx2)	**Antioxidant** - Reduces metastatic properties of human cancer cells. - Protects mice from intestinal cancer.	Gastro intestinal tract and in particular liver (cytoplasmic localization)	Protects intestinal epithelium from oxidative stress.	[175]
Glutathione Peroxidase 3 (GPx3)	**Antioxidant** - Protects from cerebral clots and stroke. - Protects from atherosclerosis and cardiovascular disease.	Kidney and blood. Present in plasma or extracellular compartments (secreted)	Gpx3 may use GSH to reduce hydrogen peroxide or *tert*-butylhydroxide.	[176]

Table 9.6: contd....

Glutathione Peroxidase 4 (GPx4)	**Antioxidant - redox signaling** - Role as oxidative stress sensor and transducer of cell death signals. - Supports normal brain development in mice. - Protects mice from Parkinson's-like neurodegeneration. - Involved in sperm production, motility and fertility in men.	Phospholipid hydroperoxide in membrane or intracellular space. Turns from enzyme to structural protein in the mitochondrial sheath of spermatozoa.	Its main substrate are phospholipid hydroperoxides in membranes. Reversing oxidation of lipid peroxides is a critical role for GPx4.	[97, 151b, 177]
Glutathione Peroxidase 6 (GPx6)	**Unknown** - It has not been characterized in terms of function or disease.	Developing embryo and olfactory epithelium in adults.	It has not been characterized in terms of catalytic role	[2a]
Thioredoxin reductases (3 enzymes)	**Antioxidant - redox signaling** - Protein thiol redox regulation - Vitamin C recycling	All including the thyroid gland (cytoplasmic, nuclear, mitochondrial)	NADPH-oxidoreductases that catalyze the reduction of oxidized thioredoxin	[178]
Iodothyronine deiodinases (3 enzymes: D1, D2 and D3)	**Thyroid Function** (see section 9.7.2): - T4 activation, T3 inactivation. - D2 maintains healthy metabolic rate in mice	Liver, kidneys, thyroid, Central Nervous System (CNS), pituitary, Brown adipose tissue and placenta	D1 or D2 carried out activation of T4 prohormone to T3: which catalyze an outer ring monodeiodination reaction. T4 and T3 are inactivated *via* inner ring monodeiodination catalyzed by D3.	[116, 179]
Selenoprotein H (Sel H)	**Antioxidant – redox signaling** - in response to redox status, it regulates the expression of genes involved in *de novo* synthesis of GSH and phase II detoxification.	All tissue and relatively high in early stages of embryonic development. Located to the nucleus	-	[180]
Selenoprotein I (Sel I)	**Unknown** - Possibly involved in phospholipid biosynthesis	Ubiquitously expressed in multiple tissues.	-	[181]

Table 9.6: contd....

Selenoprotein K (Sel K)	**Unknown** - The function of Sel K is unclear - over expression in cardiomyocytes was shown to decrease ROS (mechanism of action is unknown).	High in human heart (Localized to ER, plasma membrane)	-	[182]
Selenoprotein M (Sel M)	**Antioxidant – redox signaling** - Thiol-disulfide oxidoreductase - most probably involved in antioxidation, neuroprotection, and intracellular calcium regulation (proposed role in preventing the onset and progression of Alzheimer's disease)	Human brain Localized to ER	-	[183]
Selenoprotein 15 (Sep15)	**Antioxidant and redox regulation** - Thiol-disulfide oxidoreductase (involved in glycoprotein folding).	Located in the ER	-	[184]
Selenoprotein N (Sel N)	**Antioxidant** - Supports normal muscle development in mice - Regulates calcium mobilisation required for early muscle development; mutations cause myopathies including - multiminicore disease	Skeletal muscle, brain, lung and placenta. Located in the ER		[87, 184]
Selenoprotein O (Sel O)	**Unknown** - Contains a Cys-X-X-Sec motif suggestive of redox function	No information regarding its tissue distribution, subcellular location	Unknown physiological role	[2a]
Selenoprotein P (Sel P)	**Antioxidant** - Prevention of neurodegeneration and infertility in mice - Possible role in cancers prevention in humans - protects humans from AD. - Protection against diquat-induced liver necrosis in selenium-deficient rats	Plasma	-	[143, 174a, 185]

Table 9.6: contd....

Selenoprotein R (Sel R)	**Antioxidant and redox regulation** - Functions as a methionine sulfoxide reductase - Sel R knockouts show mild damage to oxidative insult.	All tissues (high levels in liver and kidney) Cytoplasmic localization	It is a part of the enzyme methionine sulfoxide reductase. ROS can oxidize mathionine residues in proteins to produce methionine sulfoxide, which is reduced by MsrA and MsrB enzymes.	[186]
Selenoprotein S (Sel S)	**Anti-inflammatory** - Protects cells from ER stress-induced apoptosis - linked to glucose metabolism and insulin sensitivity	All tissues Located in plasma and the ER	-	[184, 187]
Selenoprotein T (Sel T)	**Unknown** - Supports neurite/nerve formation in rodent neuronal cells	Located in the ER	Unknown	[188]
Selenoprotein V (Sel V)	**Unknown** - Expression of this protein appears to be restricted to testes (function unknown). - Conserved Cys-X-X-Sec motif suggesting redox functions.	Testes	Unknown	[2a, 189]
Selenoprotein W (Sel W)	**Antioxidant** - Putative antioxidant role - perhaps important in muscle growth	Muscle (cytoplasmic localization)	It interacts with GSH and it plays an antioxidant role in cells.	[190]
Selenophosphate-synthetase 2(SPS2)	**Synthesis of all other selenoproteins.** - Involved in Se metabolism and Se-protein biosynthesis - metabolic control in humans	All tissues (cytoplasmic localization)	It is involved in the biosynthesis of selenoproteins.	[191]

The fact that Se is an essential oligoelement for humans, animals, and some bacteria depends on its presence as component of several selenoproteins with essential biological functions [4a] that in human cells include oxidoreductions, redox signaling, antioxidant defense, thyroid hormone metabolism, and immune responses. Some of these selenoproteins are selenoenzymes, such as thioredoxin reductase and glutathione peroxidase, which represent fundamental antioxidative systems for the preservation of cellular redox homeostasis [29a]. These proteins are thus involved in key physiological processes and are believed to influence human diseases [29a,192]. Indeed, selenoproteins are: a) essential antioxidant enzymes that fight cancer and other chemical toxicities, b) regulators of thyroid hormone and thyroid function (described in 9.7.2), c) structural proteins in sperm required for fertility, and d) immunomodulators that can reduce the virulence associated with certain viral infections including HIV-1, also playing roles in the control of inflammatory reactions.

There are at least 25 human selenoproteins and 24 in the mouse, each characterized by the incorporation of selenium into the primary sequence as the amino acid Sec [2a,193].

The members of Se-protein family exhibit diverse patterns of tissue distribution (Table **9.6**), ranging from ubiquitous to tissue-specific expression. Subcellular localization also differs in the various members with some selenoproteins exclusively expressed in certain organelles or as transmembrane proteins, while others are secreted in the extracellular spaces or plasma. These and other properties have provided important information regarding physiological roles for selenoproteins. The process of Sec incorporation into selenoproteins exists in different forms in the bacteria, archaea, and eukarya kingdoms. However, not all species within these kingdoms have the capacity for selenoprotein synthesis. The mechanism by which selenoproteins are synthesized is characterized by a series of peculiarities that have been thoroughly described elsewhere as regards bacteria archaea and eukarya [1,2d,194]. The most important feature is represented by the fact that Sec residues in the active sites of Se-proteins is encoded by TGA in DNA and a unique mechanism is used to decode the UGA codon in mRNA to co-translationally incorporate Sec into the growing polypeptide because there is no free pool of Sec. UGA Sec codon is reprogrammed to mean Sec instead of stop. This effect on ribosomes results from the action of a complex machinery comprising RNA and protein factors. The discovery of selenocysteine synthase has represented a key step in clarification of Sec biosynthesis. In bacteria, monoselenophosphate is the activated form of Se used as the substrate by selenocysteine synthase. In eukaryotes, selenophosphate synthetase 1 (SPS1) and selenophosphate synthetase 2 (SPS2) were described to catalyze formation of this compound, from ATP and presumably selenide.

Selenoproteins are collectively essential for life, as demonstrated by the transgenic mouse model involving deletion of the gene encoding the tRNA dedicated to their synthesis, which results in embryonic lethality [195]. Certain model systems have shown that reduced selenoprotein activity is counterbalanced by a cytoprotective response mediated by the transcription factor Nrf2, which may represent a parallel, essential system involved in maintaining cellular redox homeostasis and viability [196].

Several selenoproteins have been characterized as antioxidant enzymes, serving to mitigate the damage caused by ROS. However, the emerging concept of ROS as secondary messengers of cell signaling has required closer examination of potential roles for selenoproteins as redox modulators signal transduction. Not all selenoproteins are enzymes involved in regulating cellular ROS, and the list of selenoprotein functions is expanding at an exciting pace [2d].

Selenoproteins constitute one of the most important classes of antioxidant enzymes the glutathione peroxidases (GPx). In the human body there are eight forms of GPxs (GPx1–GPx8) and of these, five (GPx1–GPx4, and GPx6) are selenoproteins [2a]. In addition to these well studies antioxidant enzyme systems, at least seven additional selenoproteins (TR1, TR2, TR3, Sel P, Sel R and Sel W) have been suggested to play a role in humans in protecting against oxidative stress. Se availability and the expression of GPx1, are key factors in preventing the stimulation of an excess of cellular ROS and the pro-tumoral activity of xenobiotics as UV light and the genotoxic carcinogen azoxymethane [197]. GPx2 is primarily express in the gut epithelium and mice engineered to lack GPx2 (GPx2 knockouts) have an increase incidence of intestinal cancer [175a]. Further, *in vitro* cultures of human cancer cells in which production of GPx2 was blocked demonstrated an increase in invasive and migratory properties associated with metastatic potential [175b]. In human studies, increased risks for prostate cancer and lung cancer (Non Small Cell Lung Carcinoma) have been associated with low Sel P production [185c,d]. These studies support the observation that selenium deficient diets may increase the risk of developing cancer and that increasing selenium intake may boost the antioxidant capabilities of selenoproteins in cancer therapy and prevention. In addition to being more susceptible to DNA damage and cancer, GPX1 knockout mice demonstrate increase rates of sickness and death when exposed to ROS-inducing chemicals such as paraquat [198] and also to coxsackievirus-associated cardiac disease [176a].

Further, GPx1 may protect the central nervous system from ROS induced neurodegeneration. Microglia and astrocytes protect neurons in the brain from oxidative stress (ROS induced damage). Studies in rodents have shown that microglia and astrocytes express high levels of GPx1 in rodents [199] and humans with Parkinson's and dementia show microglia with elevated GPx1 activity associated with distressed neurons [200]. GPx4 depletion in the hippocampus of mice has been shown to cause oxidative-damage in brain neurons leading to cell death like that seen in Parkinson's Disease and β-amyloid production like that seen in Alzheimer's Disease [201]. In addition to preventing neurodegenerative diseases in adults, GPx4 antioxidant activity also supports normal brain development in mice [174a, 202]. Knockout of the selenoprotein, Sel P, in mice also leads to neurodegeneration [174b]. Moreover, elevated Sel P is found in post-mortem analysis of β-amyloid plaques of Alzheimer's Disease patients [203] further suggesting a role for oxidative stress in neurodegenerative diseases and the potential for selenium and selenoproteins to prevent their onset.

Selenoproteins are also important for cardiovascular health. Human clinical studies show that people with low Gpx3 activity have reduced capacity to metabolize ROS toxins and thus present with increase risk for both arterial and cerebral clots and stroke [174c,d,f, 204]. Further, in a mouse model of atherosclerosis, over-expression of GPx4 leads to reduced lipid peroxidation and associated atherosclerotic plaque formation [174e]. People with specific genetic variants of the selenoprotein Sel S, have also been shown to be at a higher risk for cardiovascular disease and stroke although this increased risk has not been associated with ROS metabolism [205].

Another important class of selenoproteins is associated with thyroid gland function (see section 9.7.2). The thyroid gland produces thyroid hormones which in turn regulates a variety of metabolic events in humans. Thyroid hormone formation requires selenoproteins and selenoproteins can also regulate thyroid hormone activity by inactivating it. These selenoproteins are the deiodinases enzymes, D1, D2 and D3. The conversion of the thyroid hormone precursor, T4, to the active T3 hormone is mediated by D1 or D2 while both D1 and D3 can inactivate T3 [97]. Mice genetically lacking the D2 selenoprotein exhibit reduced thyroid stimulating hormone regulation and reduced capacity to metabolize fat to keep warm (brown fat thermogenesis) [98].

ACKNOWLEDGEMENT

This research is part of the scientific activity of the multidisciplinary group "SeS redox and catalysis" at the Department of Pharmaceutical Sciences, University of Perugia, Italy.

CONFLICT OF INTEREST

The authors confirm that this chapter contents have no conflict of interest.

REFERENCES

[1] Allmang, C.; Krol, A., Selenoprotein synthesis: UGA does not end the story. *Biochimie* **2006,** *88* (11), 1561-71.

[2] (a) Kryukov, G. V.; Castellano, S.; Novoselov, S. V.; Lobanov, A. V.; Zehtab, O.; Guigo, R.; Gladyshev, V. N., Characterization of mammalian selenoproteomes. *Science* **2003,** *300* (5624), 1439-43; (b) Rayman, M. P., Selenium in cancer prevention: a review of the evidence and mechanism of action. *Proc Nutr Soc* **2005,** *64* (4), 527-42; (c) Stoytcheva, Z. R.; Berry, M. J., Transcriptional regulation of mammalian selenoprotein expression. *Biochim Biophys Acta* **2009,** *1790* (11), 1429-40; (d) Lu, J.; Holmgren, A., Selenoproteins. *J Biol Chem* **2009,** *284* (2), 723-7.

[3] Muttenthaler, M.; Alewood, P. F., Selenopeptide chemistry. *J Pept Sci* **2008,** *14* (12), 1223-39.

[4] (a) Letavayova, L.; Vlasakova, D.; Spallholz, J. E.; Brozmanova, J.; Chovanec, M., Toxicity and mutagenicity of selenium compounds in Saccharomyces cerevisiae. *Mutat Res* **2008,** *638* (1-2), 1-10; (b) Hamilton, S. J.; Holley, K. M.; Buhl, K. J.; Bullard, F. A.; Weston, L. K.; McDonald, S. F., Evaluation of flushing of a high-selenium backwater channel in the Colorado River. *Environ Toxicol* **2004,** *19* (1), 51-81.

[5] Rayman, M. P.; Infante, H. G.; Sargent, M., Food-chain selenium and human health: spotlight on speciation. *Br J Nutr* **2008,** *100* (2), 238-53.

[6] Sathe SK, M. A., Rodibaugh R, Weaver CM., Chemical form of selenium in soybean (Glycine max L.) lectin.. *J Agric Food Chem* **1992,** *40*, 2084–91.

[7] Barclay, L. R.; Artz, J. D.; Mowat, J. J., Partitioning and antioxidant action of the water-soluble antioxidant, Trolox, between the aqueous and lipid phases of phosphatidylcholine membranes: 14C tracer and product studies. *Biochim Biophys Acta* **1995,** *1237* (1), 77-85.

[8] (a) Aro, A.; Alfthan, G.; Varo, P., Effects of supplementation of fertilizers on human selenium status in Finland. *Analyst* **1995,** *120* (3), 841-3; (b) Combs, G. F., Jr., Selenium in global food systems. *Br J Nutr* **2001,** *85* (5), 517-47.

[9] (a) Goyer, R. A., T.M. Clarkson, Toxic effects of metals. *Klaassen, C.D., ed Casarett & Doull's toxicology. New York: McGraw-Hill* **2001,** *Chapter 23*, pp. 811-868; (b) Goyer, R. A.; Bachmann, J.; Clarkson, T. W.; Ferris, B. G., Jr.; Graham, J.; Mushak, P.; Perl, D. P.; Rall, D. P.; Schlesinger, R.; Sharpe, W.; *et al.*, Potential human health effects of acid rain: report of a workshop. *Environ Health Perspect* **1985,** *60*, 355-68; (c) Burk, R. F., Selenium, an antioxidant nutrient. *Nutr Clin Care* **2002,** *5* (2), 75-9.

[10] Longnecker, M. P.; Taylor, P. R.; Levander, O. A.; Howe, M.; Veillon, C.; McAdam, P. A.; Patterson, K. Y.; Holden, J. M.; Stampfer, M. J.; Morris, J. S.; *et al.*, Selenium in diet, blood, and toenails in relation to human health in a seleniferous area. *Am J Clin Nutr* **1991,** *53* (5), 1288-94.

[11] Simonoff, M.; Sergeant, C.; Garnier, N.; Moretto, P.; Llabador, Y.; Simonoff, G.; Conri, C., Antioxidant status (selenium, vitamins A and E) and aging. *EXS* **1992,** *62*, 368-97.

[12] McNaughton, S. A.; Marks, G. C., Development of a food composition database for the estimation of dietary intakes of glucosinolates, the biologically active constituents of cruciferous vegetables. *Br J Nutr* **2003,** *90* (3), 687-97.

[13] Young, V. R.; Nahapetian, A.; Janghorbani, M., Selenium bioavailability with reference to human nutrition. *Am J Clin Nutr* **1982,** *35* (5), 1076-88.

[14] Zeng, H.; Combs, G. F., Jr., Selenium as an anticancer nutrient: roles in cell proliferation and tumor cell invasion. *J Nutr Biochem* **2008,** *19* (1), 1-7.

[15] (a) Sunde, R. A., Molecular biology of selenoproteins. *Annu Rev Nutr* **1990,** *10*, 451-74; (b) Meltzer, H. M.; Haug, E., Oral intake of selenium has no effect on the serum concentrations of growth hormone, insulin-like growth factor-1, insulin-like growth factor-binding proteins 1 and 3 in healthy women. *Eur J Clin Chem Clin Biochem* **1995,** *33* (7), 411-5; (c) Johansson L., A. B., Alexander J., Avaibility of selenium from soils in relation to human nutritional requirements in Sweden-Is there a need for supplementation. *Swedish Environmental Protection Agency, Stockholm* **1997**, 104.

[16] Fairweather-Tait, S. J.; Collings, R.; Hurst, R., Selenium bioavailability: current knowledge and future research requirements. *Am J Clin Nutr* **2010,** *91* (5), 1484S-1491S.

[17] Bates, C. J.; Prentice, A.; Birch, M. C.; Delves, H. T.; Sinclair, K. A., Blood indices of selenium and mercury, and their correlations with fish intake, in young people living in Britain. *Br J Nutr* **2006,** *96* (3), 523-31.

[18] (a) Mutanen, M., Bioavailability of selenium. *Ann Clin Res* **1986,** *18* (1), 48-54; (b) Neve, J., Methods in determination of selenium states. *J Trace Elem Electrolytes Health Dis* **1991,** *5* (1), 1-17; (c) Néve, Assessing the biological activity of selenium supplements: Interest of blood selenium and glutathione peroxidase. *Proceedings of the Fifth*

International Symposium on the Uses of Selenium and Tellerium. Grimbergen: Selenium-Tellerium Development Association **1994**, pp 123-130.

[19] Patterson, B. H.; Levander, O. A.; Helzlsouer, K.; McAdam, P. A.; Lewis, S. A.; Taylor, P. R.; Veillon, C.; Zech, L. A., Human selenite metabolism: a kinetic model. *Am J Physiol* **1989,** *257* (3 Pt 2), R556-67.

[20] (a) Levander, O. A., Considerations in the design of selenium bioavailability studies. *Fed Proc* **1983,** *42* (6), 1721-5; (b) Levander, O. A.; Alfthan, G.; Arvilommi, H.; Gref, C. G.; Huttunen, J. K.; Kataja, M.; Koivistoinen, P.; Pikkarainen, J., Bioavailability of selenium to Finnish men as assessed by platelet glutathione peroxidase activity and other blood parameters. *Am J Clin Nutr* **1983,** *37* (6), 887-97; (c) Alfthan, G.; Aro, A.; Arvilommi, H.; Huttunen, J. K., Selenium metabolism and platelet glutathione peroxidase activity in healthy Finnish men: effects of selenium yeast, selenite, and selenate. *Am J Clin Nutr* **1991,** *53* (1), 120-5.

[21] van der Torre, H. W.; Van Dokkum, W.; Schaafsma, G.; Wedel, M.; Ockhuizen, T., Effect of various levels of selenium in wheat and meat on blood Se status indices and on Se balance in Dutch men. *Br J Nutr* **1991,** *65* (1), 69-80.

[22] Christensen, M. J.; Janghorbani, M.; Steinke, F. H.; Istfan, N.; Young, V. R., Simultaneous determination of absorption of selenium from poultry meat and selenite in young men: application of a triple stable-isotope method. *Br J Nutr* **1983,** *50* (1), 43-50.

[23] (a) Bibow, K.; Meltzer, H. M.; Mundal, H. H.; Paulsen, I. T.; Holm, H., Platelet selenium as indicator of wheat selenium intake. *J Trace Elem Electrolytes Health Dis* **1993,** *7* (3), 171-6; (b) Srikumar, T. S.; Kallgard, A.; Lindeberg, S.; Ockerman, P. A.; Akesson, B., Trace element concentrations in hair of subjects from two South Pacific Islands, Atafu (Tokelau) and Kitava (Papua New Guinea). *J Trace Elem Electrolytes Health Dis* **1994,** *8* (1), 21-6; (c) Akesson, B.; Bellew, T.; Burk, R. F., Purification of selenoprotein P from human plasma. *Biochim Biophys Acta* **1994,** *1204* (2), 243-9; (d) Svensson, B. G.; Schutz, A.; Nilsson, A.; Akesson, I.; Akesson, B.; Skerfving, S., Fish as a source of exposure to mercury and selenium. *Sci Total Environ* **1992,** *126* (1-2), 61-74; (e) Huang, W.; Akesson, B.; Svensson, B. G.; Schutz, A.; Burk, R. F.; Skerfving, S., Selenoprotein P and glutathione peroxidase (EC 1.11.1.9) in plasma as indices of selenium status in relation to the intake of fish. *Br J Nutr* **1995,** *73* (3), 455-61; (f) Akesson B., S. T., Occurrence of low- molecular-weight and high molecular-weight selenium compounds in fish *Food Chem* **1994,** *51*, 45-49.

[24] Alexander J., M. H. M., Risk evaluation of essential trace elements- essential *versus* toxic level of intake. *Selenium. In Oskarsson A Ed.* **1995**, 15-65.

[25] Schroeder, H. A.; Frost, D. V.; Balassa, J. J., Essential trace metals in man: selenium. *J Chronic Dis* **1970,** *23* (4), 227-43.

[26] Institute of Medicine, F. a. N. B. *Dietary Reference Intakes: Vitamin C, Vitamin E, Selenium, and Carotenoids*; 2000.

[27] Food and Agriculture Organization of the United Nations.; World Health Organization., *Biotechnology and food safety : report of a joint FAO/WHO consultation, Rome, Italy, 30*

September-4 October 1996.c. Food and Agriculture Organization of the United Nations: Rome, 1996; p 27 p.

[28] (SNO), S. N. R. O., Basis for general advice on food consumption for healthy adults by H Enghardt Barbieri and C Lindvall. **1996**.

[29] (a) Rayman, M. P., The importance of selenium to human health. *Lancet* **2000,** *356* (9225), 233-41; (b) Neve, J., Human selenium supplementation as assessed by changes in blood selenium concentration and glutathione peroxidase activity. *J Trace Elem Med Biol* **1995,** *9* (2), 65-73; (c) Karunasinghe, N.; Han, D. Y.; Zhu, S.; Duan, H.; Ko, Y. J.; Yu, J. F.; Triggs, C. M.; Ferguson, L. R., Effects of supplementation with selenium, as selenized yeast, in a healthy male population from new zealand. *Nutr Cancer* **2013,** *65* (3), 355-66.

[30] Rajpathak, S.; Rimm, E.; Morris, J. S.; Hu, F., Toenail selenium and cardiovascular disease in men with diabetes. *J Am Coll Nutr* **2005,** *24* (4), 250-6.

[31] Neve, J., Physiological and nutritional importance of selenium. *Experientia* **1991,** *47* (2), 187-93.

[32] (a) Rayman, M. P., Food-chain selenium and human health: emphasis on intake. *Br J Nutr* **2008,** *100* (2), 254-68; (b) Fairweather-Tait, S. J.; Bao, Y.; Broadley, M. R.; Collings, R.; Ford, D.; Hesketh, J. E.; Hurst, R., Selenium in human health and disease. *Antioxid Redox Signal* **2011,** *14* (7), 1337-83.

[33] EFSA - Scientific Panel on Dietetic Products, N. a. A. *Tolerable Upper Intake Levels for Vitamins and Minerals*; European Food Safety Authority: 2006.

[34] Bialostosky, K.; Wright, J. D.; Kennedy-Stephenson, J.; McDowell, M.; Johnson, C. L., Dietary intake of macronutrients, micronutrients, and other dietary constituents: United States 1988-94. *Vital Health Stat 11* **2002,** (245), 1-158.

[35] Zhou, B. F.; Stamler, J.; Dennis, B.; Moag-Stahlberg, A.; Okuda, N.; Robertson, C.; Zhao, L.; Chan, Q.; Elliott, P., Nutrient intakes of middle-aged men and women in China, Japan, United Kingdom, and United States in the late 1990s: the INTERMAP study. *J Hum Hypertens* **2003,** *17* (9), 623-30.

[36] Robberecht, H. J.; Hendrix, P.; Van Cauwenbergh, R.; Deelstra, H. A., Actual daily dietary intake of selenium in Belgium, using duplicate portion sampling. *Z Lebensm Unters Forsch* **1994,** *199* (4), 251-4.

[37] (a) Kuhnlein, H. V.; Receveur, O.; Soueida, R.; Berti, P. R., Unique patterns of dietary adequacy in three cultures of Canadian Arctic indigenous peoples. *Public Health Nutr* **2008,** *11* (4), 349-60; (b) Gissel-Nielsen, G., Effects of selenium supplementation of field crops. *Environmental chemistry of selenium* **1998,** (6).

[38] Dumont, E.; Vanhaecke, F.; Cornelis, R., Selenium speciation from food source to metabolites: a critical review. *Anal Bioanal Chem* **2006,** *385* (7), 1304-23.

[39] Anttolainen, M.; Valsta, L. M.; Alfthan, G.; Kleemola, P.; Salminen, I.; Tamminen, M., Effect of extreme fish consumption on dietary and plasma antioxidant levels and fatty acid composition. *Eur J Clin Nutr* **1996,** *50* (11), 741-6.

[40] Pappa, E. C.; Pappas, A. C.; Surai, P. F., Selenium content in selected foods from the Greek market and estimation of the daily intake. *Sci Total Environ* **2006,** *372* (1), 100-8.

[41] El-Ghawi, U. M.; Al-Sadeq, A. A.; Bejey, M. M.; Alamin, M. B., Determination of selenium in Libyan food items using pseudocyclic instrumental neutron activation analysis. *Biol Trace Elem Res* **2005,** *107* (1), 61-72.

[42] Golubkina, N. A.; Shagova, M. V.; Spirichev, V. B.; Alftan, J.; Laaksonen, P.; Kumpulainen, I.; Puura, L., [Selenium intake by the population of Lithuania]. *Vopr Pitan* **1992,** (1), 35-7.

[43] Valentine, J. L.; Cebrian, M. E.; Garcia-vargas, G. G.; Faraji, B.; Kuo, J.; Gibb, H. J.; Lachenbruch, P. A., Daily selenium intake estimates for residents of arsenic-endemic areas. *Environ Res* **1994,** *64* (1), 1-9.

[44] Foster, L. H.; Sumar, S., Selenium in health and disease: a review. *Crit Rev Food Sci Nutr* **1997,** *37* (3), 211-28.

[45] Donovan, U. M.; Gibson, R. S.; Ferguson, E. L.; Ounpuu, S.; Heywood, P., Selenium intakes of children from Malawi and Papua New Guinea consuming plant-based diets. *J Trace Elem Electrolytes Health Dis* **1992,** *6* (1), 39-43.

[46] Meltzer, H. M.; Norheim, G.; Loken, E. B.; Holm, H., Supplementation with wheat selenium induces a dose-dependent response in serum and urine of a Se-replete population. *Br J Nutr* **1992,** *67* (2), 287-94.

[47] (a) Barclay, M. N.; MacPherson, A., Selenium content of wheat for bread making in Scotland and the relationship between glutathione peroxidase (EC 1.11.1.9) levels in whole blood and bread consumption. *Br J Nutr* **1992,** *68* (1), 261-70; (b) Molnar, J.; MacPherson, A.; Barclay, I.; Molnar, P., Selenium content of convenience and fast foods in Ayrshire, Scotland. *Int J Food Sci Nutr* **1995,** *46* (4), 343-52.

[48] Diaz-Alarcon, J. P.; Navarro-Alarcon, M.; Lopez-Garcia de la Serrana, H.; Lopez-Martinez, M. C., Determination of selenium in cereals, legumes and dry fruits from southeastern Spain for calculation of daily dietary intake. *Sci Total Environ* **1996,** *184* (3), 183-9.

[49] (a) Kitahara, J.; Seko, Y.; Imura, N., Possible involvement of active oxygen species in selenite toxicity in isolated rat hepatocytes. *Arch Toxicol* **1993,** *67* (7), 497-501; (b) Yan, L.; Spallholz, J. E., Generation of reactive oxygen species from the reaction of selenium compounds with thiols and mammary tumor cells. *Biochem Pharmacol* **1993,** *45* (2), 429-37.

[50] (a) Reid, J.; Svejstrup, J. Q., DNA damage-induced Def1-RNA polymerase II interaction and Def1 requirement for polymerase ubiquitylation *in vitro*. *J Biol Chem* **2004,** *279* (29), 29875-8; (b) Wycherly, B. J.; Moak, M. A.; Christensen, M. J., High dietary intake of sodium selenite induces oxidative DNA damage in rat liver. *Nutr Cancer* **2004,** *48* (1), 78-83.

[51] (a) Suzuki, K. T.; Ohta, Y.; Suzuki, N., Availability and metabolism of 77Se-methylseleninic acid compared simultaneously with those of three related selenocompounds. *Toxicol Appl Pharmacol* **2006,** *217* (1), 51-62; (b) Suzuki, K. T.; Somekawa, L.; Kurasaki, K.; Suzuki, N., Simultaneous tracing of 76Se-selenite and 77Se-selenomethionine by absolute labeling and speciation. *Toxicol Appl Pharmacol* **2006,** *217* (1), 43-50.

[52] (a) Jensen, G. E.; Clausen, J., Glutathione peroxidase and reductase, glucose-6-phosphate dehydrogenase and catalase activities in multiple sclerosis. *J Neurol Sci* **1984,** *63* (1), 45-53; (b) World Health Organization.; DANIDA., *Strengthening ministries of health for primary health care*. World Health Organization ; Obtained from WHO Publications Centre USA: Geneva Albany, N.Y., 1984; p 58 p.

[53] Goldhaber, S. B., Trace element risk assessment: essentiality *vs.* toxicity. *Regul Toxicol Pharmacol* **2003,** *38* (2), 232-42.

[54] Thorlacius-Ussing, O., Selenium-induced growth retardation. Histochemical and endocrinological studies on the anterior pituitaries of selenium treated rats. *Dan Med Bull* **1990,** *37* (4), 347-58.

[55] Johnson, C. C.; Fordyce, F. M.; Rayman, M. P., Symposium on 'Geographical and geological influences on nutrition': Factors controlling the distribution of selenium in the environment and their impact on health and nutrition. *Proc Nutr Soc* **2010,** *69* (1), 119-32.

[56] Navarro-Alarcon, M.; Cabrera-Vique, C., Selenium in food and the human body: a review. *Sci Total Environ* **2008,** *400* (1-3), 115-41.

[57] (a) Yang, G.; Yin, S.; Zhou, R.; Gu, L.; Yan, B.; Liu, Y., Studies of safe maximal daily dietary Se-intake in a seleniferous area in China. Part II: Relation between Se-intake and the manifestation of clinical signs and certain biochemical alterations in blood and urine. *J Trace Elem Electrolytes Health Dis* **1989,** *3* (3), 123-30; (b) Yang, G.; Zhou, R., Further observations on the human maximum safe dietary selenium intake in a seleniferous area of China. *J Trace Elem Electrolytes Health Dis* **1994,** *8* (3-4), 159-65.

[58] Whanger, P. D., Selenium and its relationship to cancer: an update. *Br J Nutr* **2004,** *91* (1), 11-28.

[59] (a) Simonoff M, S. G., Le selenium *et al.* Vie.. *Paris: Masson* **1991**; (b) Navarro-Alarcon, M.; Lopez-Martinez, M. C., Essentiality of selenium in the human body: relationship with different diseases. *Sci Total Environ* **2000,** *249* (1-3), 347-71.

[60] Cao, J. J.; Qin, W.; Zhu, S. F.; Cao, Y. Y., [Identification for genetically modified maize T14/T25 with real time fluorescent PCR method]. *Yi Chuan* **2004,** *26* (5), 689-94.

[61] Pennington, J. A.; Young, B. E., Total diet study nutritional elements, 1982-1989. *J Am Diet Assoc* **1991,** *91* (2), 179-83.

[62] Srikumar, T. S.; Kallgard, B.; Ockerman, P. A.; Akesson, B., The effects of a 2-year switch from a mixed to a lactovegetarian diet on trace element status in hypertensive subjects. *Eur J Clin Nutr* **1992,** *46* (9), 661-9.

[63] U.S. Department of Agriculture, A. R. S. USDA National Nutrient Database for Standard Reference. http://www.ars.usda.gov/ba/bhnrc/ndl.

[64] (EFSA), E. F. S. A. *Long-term dietary exposure to selenium in young children living in different European countries*; European Comunity: 2010; pp 1-38.

[65] Valdiglesias, V.; Pasaro, E.; Mendez, J.; Laffon, B., *In vitro* evaluation of selenium genotoxic, cytotoxic, and protective effects: a review. *Arch Toxicol* **2010,** *84* (5), 337-51.

[66] EFSA *Long-term dietary exposure to selenium in young children living in different European countries*; European Food Safety Agency: 2010; pp 1-38.

[67] (a) Thomson, C. D.; Chisholm, A.; McLachlan, S. K.; Campbell, J. M., Brazil nuts: an effective way to improve selenium status. *Am J Clin Nutr* **2008,** *87* (2), 379-84; (b) Bodo, E. T.; Stefanka, Z.; Ipolyi, I.; Soros, C.; Dernovics, M.; Fodor, P., Preparation, homogeneity and stability studies of a candidate LRM for Se speciation. *Anal Bioanal Chem* **2003,** *377* (1), 32-8.

[68] (a) Klapec, T.; Mandic, M. L.; Grgic, J.; Primorac, L.; Ikic, M.; Lovric, T.; Grgic, Z.; Herceg, Z., Daily dietary intake of selenium in eastern Croatia. *Sci Total Environ* **1998,** *217* (1-2), 127-36; (b) Klapec T, M. M., Grgic J, Primorac LG, Perl A, Krstanovic V., Selenium in selected foods grown or purchased in eastern Croatia. *Food Chem* **2004,** *85*, 445–52; (c) Sirichakwal PP, P. P., Polngam J, Kongkachuichai R., Selenium content of Thai foods. *J Food Compos* **2005,** *18*, 47–59.

[69] Ventura MG, F. M., Pacheco A, Van Meerten T, Wolterbeek HT., Selenium content in selected Portuguese foodstuffs. *Eur Food Res Technol* **2007,** *224*, 395–401.

[70] Haratake M, T. J., Ono M, Nakayama M., An assessment of Niboshi (a processed Japanese anchovy) as an effective food source of selenium. *J Health Sci* **2007,** *53*, 457–63.

[71] McNaughton SA, M. G., Selenium content of Australian foods: a review of literature values. *J Food Compos* **2002,** *15*, 169–82.

[72] USDA Food Recovery & Gleaning Initiative., *A citizen's guide to food recovery*. Rev. ed.; USDA: Washington, D.C., 1999; p v, 57 p.

[73] NZ-ICFRL,.(New Zealand Institute for Crop and Food ResearchLimited).. *New Zealand Food Composition Database* **2000,** *Christchurch: Ministry of Health.*

[74] Tinggi, U., Determination of selenium in meat products by hydride generation atomic absorption spectrophotometry. *J AOAC Int* **1999,** *82* (2), 364-7.

[75] (a) Schrauzer, G. N., Nutritional selenium supplements: product types, quality, and safety. *J Am Coll Nutr* **2001,** *20* (1), 1-4; (b) Schrauzer, G. N., The nutritional significance, metabolism and toxicology of selenomethionine. *Adv Food Nutr Res* **2003,** *47*, 73-112.

[76] (a) Ip, C.; Thompson, H.; Ganther, H., Activity of triphenylselenonium chloride in mammary cancer prevention. *Carcinogenesis* **1994,** *15* (12), 2879-82; (b) Ip, C.; el-Bayoumy, K.; Upadhyaya, P.; Ganther, H.; Vadhanavikit, S.; Thompson, H., Comparative effect of inorganic and organic selenocyanate derivatives in mammary cancer chemoprevention. *Carcinogenesis* **1994,** *15* (2), 187-92; (c) Ip C, G. H., Novel strategies in selenium cancer chemoprevention research. *Seleniuminbiology and human health. In: BurkRF, editor.* **1994,** *New York: Springer-Verlag*, p. 169-80; (d) Kapolna, E.; Fodor, P., Bioavailability of selenium from selenium-enriched green onions (Allium fistulosum) and chives (Allium schoenoprasum) after '*in vitro*' gastrointestinal digestion. *Int J Food Sci Nutr* **2007,** *58* (4), 282-96.

[77] Manjusha R, D. K., Karunasagar D., UV-photolysis assisted digestion of food samples for the determination of selenium by electrothermal atomic absorption spectrometry (ETAAS). *Food Chem* **2007,** *105*, 26–265.

[78] Marro, The 1994 Australian Market Basket Survey. *Camberra: Australian Government Publishing Service* **1996.**

[79] Tinggi, U.; Reilly, C.; Hahn, S.; Capra, M. J., Comparison of wet digestion procedures for the determination of cadmium and lead in marine biological tissues by Zeeman graphite furnace atomic absorption spectrophotometry. *Sci Total Environ* **1992,** *125*, 15-23.

[80] Barclay MNI, M. A., Dixon J., Selenium content of a range of UK foods. *J Food Compos* **1995,** *8*, 307-18.

[81] Cabrera, C.; Lorenzo, M. L.; De Mena, C.; Lopez, M. C., Chromium, copper, iron, manganese, selenium and zinc levels in dairy products: *in vitro* study of absorbable fractions. *Int J Food Sci Nutr* **1996,** *47* (4), 331-9.

[82] Papp, L. V.; Lu, J.; Striebel, F.; Kennedy, D.; Holmgren, A.; Khanna, K. K., The redox state of SECIS binding protein 2 controls its localization and selenocysteine incorporation function. *Mol Cell Biol* **2006,** *26* (13), 4895-910.

[83] Diaz, J. P.; Navarro, M.; Lopez, H.; Lopez, M. C., Determination of selenium levels in dairy products and drinks by hydride generation atomic absorption spectrometry: correlation with daily dietary intake. *Food Addit Contam* **1997,** *14* (2), 109-14.

[84] Vinas, P.; Pardo-Martinez, M.; Hernandez-Cordoba, M., Determination of copper, cobalt, nickel, and manganese in baby food slurries using electrothermal atomic absorption spectrometry. *J Agric Food Chem* **2000,** *48* (12), 5789-94.

[85] Roca A, C. C., Lorenzo ML, López MC., Levels of calcium, magnesium, manganese, zinc, selenium and chromium in olive oils produced in Andalusia.. *Int J Fats Oils* **2000,** *51*, 393-9.

[86] Singh V, G. A., Availability of essential trace elements in Indian cereals, vegetables and spices using INAA and the contribution of spices to daily dietary intake. *Food Chem* **2006,** *94*, 81-9.

[87] Clark, L. C.; Combs, G. F., Jr.; Turnbull, B. W.; Slate, E. H.; Chalker, D. K.; Chow, J.; Davis, L. S.; Glover, R. A.; Graham, G. F.; Gross, E. G.; Krongrad, A.; Lesher, J. L., Jr.; Park, H. K.; Sanders, B. B., Jr.; Smith, C. L.; Taylor, J. R., Effects of selenium supplementation for cancer prevention in patients with carcinoma of the skin. A randomized controlled trial. Nutritional Prevention of Cancer Study Group. *JAMA* **1996,** *276* (24), 1957-63.

[88] EFSA Panel on Dietetic Products, N. a. A. N., Scientific Opinion on the substantiation of health claims related to selenium and maintenance of normal hair (ID 281), maintenance of normal nails (ID 281), protection against heavy metals (ID 383), maintenance of normal joints (ID 409), maintenance of normal thyroid function (ID 410, 1292), protection of DNA, proteins and lipids from oxidative damage (ID 1292), and maintenance of the normal function of the immune system (ID 1750) pursuant to Article 13(1) of Regulation (EC) No 1924/2006. *EFSA Journal* **2010,** *8* (10), 1727 [18 pp.].

[89] Brennan, M. F.; Horowitz, G. D., Total parenteral nutrition in surgical patients. *Adv Surg* **1984,** *17*, 1-36.

[90] (a) Rossipal, E.; Tiran, B., Selenium and glutathione peroxidase levels in healthy infants and children in Austria and the influence of nutrition regimens on these levels. *Nutrition* **1995,** *11* (5 Suppl), 573-5; (b) Lombeck, I.; Kasperek, K.; Feinendegen, L. E.; Bremer, H.

J., Serum-selenium concentrations in patients with maple-syrup-urine disease and phenylketonuria under dieto-therapy. *Clin Chim Acta* **1975,** *64* (1), 57-61.

[91] Lombeck, I.; Ebert, K. H.; Kasperek, K.; Feinendegen, L. E.; Bremer, H. J., Selenium intake of infants and young children, healthy children and dietetically treated patients with phenylketonuria. *Eur J Pediatr* **1984,** *143* (2), 99-102.

[92] Ge, K.; Yang, G., The epidemiology of selenium deficiency in the etiological study of endemic diseases in China. *Am J Clin Nutr* **1993,** *57* (2 Suppl), 259S-263S.

[93] Levander, O. A.; Beck, M. A., Interacting nutritional and infectious etiologies of Keshan disease. Insights from coxsackie virus B-induced myocarditis in mice deficient in selenium or vitamin E. *Biol Trace Elem Res* **1997,** *56* (1), 5-21.

[94] Yang, G. Q.; Chen, J. S.; Wen, Z. M.; Ge, K. Y.; Zhu, L. Z.; Chen, X. C.; Chen, X. S., The role of selenium in Keshan disease. *Adv Nutr Res* **1984,** *6*, 203-31.

[95] Beck, M. A., The influence of antioxidant nutrients on viral infection. *Nutr Rev* **1998,** *56* (1 Pt 2), S140-6.

[96] Schomburg, L.; Kohrle, J., On the importance of selenium and iodine metabolism for thyroid hormone biosynthesis and human health. *Mol Nutr Food Res* **2008,** *52* (11), 1235-46.

[97] Bianco, A. C.; Salvatore, D.; Gereben, B.; Berry, M. J.; Larsen, P. R., Biochemistry, cellular and molecular biology, and physiological roles of the iodothyronine selenodeiodinases. *Endocr Rev* **2002,** *23* (1), 38-89.

[98] (a) Schneider, M. J.; Fiering, S. N.; Pallud, S. E.; Parlow, A. F.; St Germain, D. L.; Galton, V. A., Targeted disruption of the type 2 selenodeiodinase gene (DIO2) results in a phenotype of pituitary resistance to T4. *Mol Endocrinol* **2001,** *15* (12), 2137-48; (b) de Jesus, L. A.; Carvalho, S. D.; Ribeiro, M. O.; Schneider, M.; Kim, S. W.; Harney, J. W.; Larsen, P. R.; Bianco, A. C., The type 2 iodothyronine deiodinase is essential for adaptive thermogenesis in brown adipose tissue. *J Clin Invest* **2001,** *108* (9), 1379-85.

[99] (a) Arthur, J. R.; Nicol, F.; Beckett, G. J., Selenium deficiency, thyroid hormone metabolism, and thyroid hormone deiodinases. *Am J Clin Nutr* **1993,** *57* (2 Suppl), 236S-239S; (b) Corvilain, B.; Contempre, B.; Longombe, A. O.; Goyens, P.; Gervy-Decoster, C.; Lamy, F.; Vanderpas, J. B.; Dumont, J. E., Selenium and the thyroid: how the relationship was established. *Am J Clin Nutr* **1993,** *57* (2 Suppl), 244S-248S.

[100] Olivieri, O.; Girelli, D.; Stanzial, A. M.; Rossi, L.; Bassi, A.; Corrocher, R., Selenium, zinc, and thyroid hormones in healthy subjects: low T3/T4 ratio in the elderly is related to impaired selenium status. *Biol Trace Elem Res* **1996,** *51* (1), 31-41.

[101] MacPherson, A.; Milne, E. M.; Macpherson, A. J., Copper poisoning in ewes. *Vet Rec* **1997,** *141* (24), 631.

[102] Rayman, M. P.; Thompson, A. J.; Bekaert, B.; Catterick, J.; Galassini, R.; Hall, E.; Warren-Perry, M.; Beckett, G. J., Randomized controlled trial of the effect of selenium supplementation on thyroid function in the elderly in the United Kingdom. *Am J Clin Nutr* **2008,** *87* (2), 370-8.

[103] Contempre, B.; Le Moine, O.; Dumont, J. E.; Denef, J. F.; Many, M. C., Selenium deficiency and thyroid fibrosis. A key role for macrophages and transforming growth factor beta (TGF-beta). *Mol Cell Endocrinol* **1996,** *124* (1-2), 7-15.

[104] Vanderpas, J. B.; Contempre, B.; Duale, N. L.; Deckx, H.; Bebe, N.; Longombe, A. O.; Thilly, C. H.; Diplock, A. T.; Dumont, J. E., Selenium deficiency mitigates hypothyroxinemia in iodine-deficient subjects. *Am J Clin Nutr* **1993,** *57* (2 Suppl), 271S-275S.

[105] Ma, T.; Guo, J.; Wang, F., The epidemiology of iodine-deficiency diseases in China. *Am J Clin Nutr* **1993,** *57* (2 Suppl), 264S-266S.

[106] (a) Derumeaux, H.; Valeix, P.; Castetbon, K.; Bensimon, M.; Boutron-Ruault, M. C.; Arnaud, J.; Hercberg, S., Association of selenium with thyroid volume and echostructure in 35- to 60-year-old French adults. *Eur J Endocrinol* **2003,** *148* (3), 309-15; (b) Glattre, E.; Thomassen, Y.; Thoresen, S. O.; Haldorsen, T.; Lund-Larsen, P. G.; Theodorsen, L.; Aaseth, J., Prediagnostic serum selenium in a case-control study of thyroid cancer. *Int J Epidemiol* **1989,** *18* (1), 45-9.

[107] (a) Toulis, K. A.; Anastasilakis, A. D.; Tzellos, T. G.; Goulis, D. G.; Kouvelas, D., Selenium supplementation in the treatment of Hashimoto's thyroiditis: a systematic review and a meta-analysis. *Thyroid* **2010,** *20* (10), 1163-73; (b) Nacamulli, D.; Mian, C.; Petricca, D.; Lazzarotto, F.; Barollo, S.; Pozza, D.; Masiero, S.; Faggian, D.; Plebani, M.; Girelli, M. E.; Mantero, F.; Betterle, C., Influence of physiological dietary selenium supplementation on the natural course of autoimmune thyroiditis. *Clin Endocrinol (Oxf)* **2010,** *73* (4), 535-9.

[108] Negro, R.; Greco, G.; Mangieri, T.; Pezzarossa, A.; Dazzi, D.; Hassan, H., The influence of selenium supplementation on postpartum thyroid status in pregnant women with thyroid peroxidase autoantibodies. *J Clin Endocrinol Metab* **2007,** *92* (4), 1263-8.

[109] Marcocci, C.; Kahaly, G. J.; Krassas, G. E.; Bartalena, L.; Prummel, M.; Stahl, M.; Altea, M. A.; Nardi, M.; Pitz, S.; Boboridis, K.; Sivelli, P.; von Arx, G.; Mourits, M. P.; Baldeschi, L.; Bencivelli, W.; Wiersinga, W., Selenium and the course of mild Graves' orbitopathy. *N Engl J Med* **2011,** *364* (20), 1920-31.

[110] (a) Rayman, M. P., Selenium and human health. *Lancet* **2012,** *379* (9822), 1256-68; (b) Hoffmann, P. R.; Berry, M. J., The influence of selenium on immune responses. *Mol Nutr Food Res* **2008,** *52* (11), 1273-80; (c) Gromadzinska, J.; Reszka, E.; Bruzelius, K.; Wasowicz, W.; Akesson, B., Selenium and cancer: biomarkers of selenium status and molecular action of selenium supplements. *Eur J Nutr* **2008,** *47 Suppl 2*, 29-50; (d) Taylor, P. R.; Albanes, D., Selenium, vitamin E, and prostate cancer--ready for prime time? *J Natl Cancer Inst* **1998,** *90* (16), 1184-5.

[111] Ip, C.; Sinha, D. K., Enhancement of mammary tumorigenesis by dietary selenium deficiency in rats with a high polyunsaturated fat intake. *Cancer Res* **1981,** *41* (1), 31-4.

[112] Bronzetti, G.; Cini, M.; Andreoli, E.; Caltavuturo, L.; Panunzio, M.; Croce, C. D., Protective effects of vitamins and selenium compounds in yeast. *Mutat Res* **2001,** *496* (1-2), 105-15.

[113] Levander, O. A., A global view of human selenium nutrition. *Annu Rev Nutr* **1987,** *7*, 227-50.

[114] (a) Shen, C. L.; Song, W.; Pence, B. C., Interactions of selenium compounds with other antioxidants in DNA damage and apoptosis in human normal keratinocytes. *Cancer Epidemiol Biomarkers Prev* **2001,** *10* (4), 385-90; (b) Hurst, R.; Elliott, R. M.; Goldson, A.

J.; Fairweather-Tait, S. J., Se-methylselenocysteine alters collagen gene and protein expression in human prostate cells. *Cancer Lett* **2008,** *269* (1), 117-26.

[115] Birt, D. F., Effects of the intake of selected vitamins and minerals on cancer prevention. *Magnesium* **1989,** *8* (1), 17-30.

[116] Duffield-Lillico, A. J.; Dalkin, B. L.; Reid, M. E.; Turnbull, B. W.; Slate, E. H.; Jacobs, E. T.; Marshall, J. R.; Clark, L. C., Selenium supplementation, baseline plasma selenium status and incidence of prostate cancer: an analysis of the complete treatment period of the Nutritional Prevention of Cancer Trial. *BJU Int* **2003,** *91* (7), 608-12.

[117] Lippman, S. M.; Klein, E. A.; Goodman, P. J.; Lucia, M. S.; Thompson, I. M.; Ford, L. G.; Parnes, H. L.; Minasian, L. M.; Gaziano, J. M.; Hartline, J. A.; Parsons, J. K.; Bearden, J. D., 3rd; Crawford, E. D.; Goodman, G. E.; Claudio, J.; Winquist, E.; Cook, E. D.; Karp, D. D.; Walther, P.; Lieber, M. M.; Kristal, A. R.; Darke, A. K.; Arnold, K. B.; Ganz, P. A.; Santella, R. M.; Albanes, D.; Taylor, P. R.; Probstfield, J. L.; Jagpal, T. J.; Crowley, J. J.; Meyskens, F. L., Jr.; Baker, L. H.; Coltman, C. A., Jr., Effect of selenium and vitamin E on risk of prostate cancer and other cancers: the Selenium and Vitamin E Cancer Prevention Trial (SELECT). *JAMA* **2009,** *301* (1), 39-51.

[118] (a) Peters, U.; Takata, Y., Selenium and the prevention of prostate and colorectal cancer. *Mol Nutr Food Res* **2008,** *52* (11), 1261-72; (b) Duffield-Lillico, A. J.; Reid, M. E.; Turnbull, B. W.; Combs, G. F., Jr.; Slate, E. H.; Fischbach, L. A.; Marshall, J. R.; Clark, L. C., Baseline characteristics and the effect of selenium supplementation on cancer incidence in a randomized clinical trial: a summary report of the Nutritional Prevention of Cancer Trial. *Cancer Epidemiol Biomarkers Prev* **2002,** *11* (7), 630-9.

[119] Rayman, M. P., Selenoproteins and human health: insights from epidemiological data. *Biochim Biophys Acta* **2009,** *1790* (11), 1533-40.

[120] Zhuo, H.; Smith, A. H.; Steinmaus, C., Selenium and lung cancer: a quantitative analysis of heterogeneity in the current epidemiological literature. *Cancer Epidemiol Biomarkers Prev* **2004,** *13* (5), 771-8.

[121] Amaral, A. F.; Cantor, K. P.; Silverman, D. T.; Malats, N., Selenium and bladder cancer risk: a meta-analysis. *Cancer Epidemiol Biomarkers Prev* **2010,** *19* (9), 2407-15.

[122] Yu, M. W.; Horng, I. S.; Hsu, K. H.; Chiang, Y. C.; Liaw, Y. F.; Chen, C. J., Plasma selenium levels and risk of hepatocellular carcinoma among men with chronic hepatitis virus infection. *Am J Epidemiol* **1999,** *150* (4), 367-74.

[123] Wei, W. Q.; Abnet, C. C.; Qiao, Y. L.; Dawsey, S. M.; Dong, Z. W.; Sun, X. D.; Fan, J. H.; Gunter, E. W.; Taylor, P. R.; Mark, S. D., Prospective study of serum selenium concentrations and esophageal and gastric cardia cancer, heart disease, stroke, and total death. *Am J Clin Nutr* **2004,** *79* (1), 80-5.

[124] (a) Rayman, Bioactive compounds and cancer. *Selenium. In: Milner JA, Romagnolo DF, eds.. New York: Humana Press/Springer* **2010**, 411-48; (b) Brinkman, M.; Reulen, R. C.; Kellen, E.; Buntinx, F.; Zeegers, M. P., Are men with low selenium levels at increased risk of prostate cancer? *Eur J Cancer* **2006,** *42* (15), 2463-71; (c) Etminan, M.; FitzGerald, J. M.; Gleave, M.; Chambers, K., Intake of selenium in the prevention of prostate cancer: a systematic review and meta-analysis. *Cancer Causes Control* **2005,** *16* (9), 1125-31.

[125] (a) Mueller, A. S.; Pallauf, J., Compendium of the antidiabetic effects of supranutritional selenate doses. *In vivo* and *in vitro* investigations with type II diabetic db/db mice. *J Nutr Biochem* **2006,** *17* (8), 548-60; (b) Akbaraly, T. N.; Arnaud, J.; Rayman, M. P.; Hininger-Favier, I.; Roussel, A. M.; Berr, C.; Fontbonne, A., Plasma selenium and risk of dysglycemia in an elderly French population: results from the prospective Epidemiology of Vascular Ageing Study. *Nutr Metab (Lond)* **2010,** *7*, 21.

[126] Mueller, A. S.; Bosse, A. C.; Most, E.; Klomann, S. D.; Schneider, S.; Pallauf, J., Regulation of the insulin antagonistic protein tyrosine phosphatase 1B by dietary Se studied in growing rats. *J Nutr Biochem* **2009,** *20* (4), 235-47.

[127] Lei, X. G.; Vatamaniuk, M. Z., Two tales of antioxidant enzymes on beta cells and diabetes. *Antioxid Redox Signal* **2011,** *14* (3), 489-503.

[128] McClung, J. P.; Roneker, C. A.; Mu, W.; Lisk, D. J.; Langlais, P.; Liu, F.; Lei, X. G., Development of insulin resistance and obesity in mice overexpressing cellular glutathione peroxidase. *Proc Natl Acad Sci U S A* **2004,** *101* (24), 8852-7.

[129] Steinbrenner, H.; Speckmann, B.; Pinto, A.; Sies, H., High selenium intake and increased diabetes risk: experimental evidence for interplay between selenium and carbohydrate metabolism. *J Clin Biochem Nutr* **2011,** *48* (1), 40-5.

[130] Chen, C. J.; Huang, H. S.; Chang, W. C., Depletion of phospholipid hydroperoxide glutathione peroxidase up-regulates arachidonate metabolism by 12S-lipoxygenase and cyclooxygenase 1 in human epidermoid carcinoma A431 cells. *FASEB J* **2003,** *17* (12), 1694-6.

[131] Stranges, S.; Marshall, J. R.; Natarajan, R.; Donahue, R. P.; Trevisan, M.; Combs, G. F.; Cappuccio, F. P.; Ceriello, A.; Reid, M. E., Effects of long-term selenium supplementation on the incidence of type 2 diabetes: a randomized trial. *Ann Intern Med* **2007,** *147* (4), 217-23.

[132] Stranges, S.; Sieri, S.; Vinceti, M.; Grioni, S.; Guallar, E.; Laclaustra, M.; Muti, P.; Berrino, F.; Krogh, V., A prospective study of dietary selenium intake and risk of type 2 diabetes. *BMC Public Health* **2010,** *10*, 564.

[133] Kelly, D. P.; Scarpulla, R. C., Transcriptional regulatory circuits controlling mitochondrial biogenesis and function. *Genes Dev* **2004,** *18* (4), 357-68.

[134] (a) Laclaustra, M.; Navas-Acien, A.; Stranges, S.; Ordovas, J. M.; Guallar, E., Serum selenium concentrations and hypertension in the US Population. *Circ Cardiovasc Qual Outcomes* **2009,** *2* (4), 369-76; (b) Bleys, J.; Navas-Acien, A.; Guallar, E., Serum selenium and diabetes in U.S. adults. *Diabetes Care* **2007,** *30* (4), 829-34.

[135] Akbaraly, T. N.; Singh-Manoux, A.; Tabak, A. G.; Jokela, M.; Virtanen, M.; Ferrie, J. E.; Marmot, M. G.; Shipley, M. J.; Kivimaki, M., Overall diet history and reversibility of the metabolic syndrome over 5 years: the Whitehall II prospective cohort study. *Diabetes Care* **2010,** *33* (11), 2339-41.

[136] Rayman, M. P.; Blundell-Pound, G.; Pastor-Barriuso, R.; Guallar, E.; Steinbrenner, H.; Stranges, S., A randomized trial of selenium supplementation and risk of type-2 diabetes, as assessed by plasma adiponectin. *PLoS One* **2012,** *7* (9), e45269.

[137] (a) Kljai, K.; Runje, R., Selenium and glycogen levels in diabetic patients. *Biol Trace Elem Res* **2001,** *83* (3), 223-9; (b) Navarro-Alarcon, M.; Lopez, G. d. l. S. H.; Perez-Valero, V.; Lopez-Martinez, C., Serum and urine selenium concentrations as indicators of body status in patients with diabetes mellitus. *Sci Total Environ* **1999,** *228* (1), 79-85.

[138] Czernichow, S.; Bertrais, S.; Blacher, J.; Galan, P.; Briancon, S.; Favier, A.; Safar, M.; Hercberg, S., [Antioxidant supplements and risk of hypertension in the SU.VI.MAX trial: relationship to plasma antioxidants]. *Arch Mal Coeur Vaiss* **2006,** *99* (7-8), 665-8.

[139] (a) Renko, K.; Hofmann, P. J.; Stoedter, M.; Hollenbach, B.; Behrends, T.; Kohrle, J.; Schweizer, U.; Schomburg, L., Down-regulation of the hepatic selenoprotein biosynthesis machinery impairs selenium metabolism during the acute phase response in mice. *FASEB J* **2009,** *23* (6), 1758-65; (b) Hesse-Bahr, K.; Dreher, I.; Kohrle, J., The influence of the cytokines Il-1beta and INFgamma on the expression of selenoproteins in the human hepatocarcinoma cell line HepG2. *Biofactors* **2000,** *11* (1-2), 83-5; (c) Nichol, C.; Herdman, J.; Sattar, N.; O'Dwyer, P. J.; St, J. O. R. D.; Littlejohn, D.; Fell, G., Changes in the concentrations of plasma selenium and selenoproteins after minor elective surgery: further evidence for a negative acute phase response? *Clin Chem* **1998,** *44* (8 Pt 1), 1764-6.

[140] Houstis, N.; Rosen, E. D.; Lander, E. S., Reactive oxygen species have a causal role in multiple forms of insulin resistance. *Nature* **2006,** *440* (7086), 944-8.

[141] Chen, X.; Scholl, T. O.; Leskiw, M. J.; Donaldson, M. R.; Stein, T. P., Association of glutathione peroxidase activity with insulin resistance and dietary fat intake during normal pregnancy. *J Clin Endocrinol Metab* **2003,** *88* (12), 5963-8.

[142] Misu, H.; Takamura, T.; Takayama, H.; Hayashi, H.; Matsuzawa-Nagata, N.; Kurita, S.; Ishikura, K.; Ando, H.; Takeshita, Y.; Ota, T.; Sakurai, M.; Yamashita, T.; Mizukoshi, E.; Honda, M.; Miyamoto, K.; Kubota, T.; Kubota, N.; Kadowaki, T.; Kim, H. J.; Lee, I. K.; Minokoshi, Y.; Saito, Y.; Takahashi, K.; Yamada, Y.; Takakura, N.; Kaneko, S., A liver-derived secretory protein, selenoprotein P, causes insulin resistance. *Cell Metab* **2010,** *12* (5), 483-95.

[143] Yang, S. J.; Hwang, S. Y.; Choi, H. Y.; Yoo, H. J.; Seo, J. A.; Kim, S. G.; Kim, N. H.; Baik, S. H.; Choi, D. S.; Choi, K. M., Serum selenoprotein P levels in patients with type 2 diabetes and prediabetes: implications for insulin resistance, inflammation, and atherosclerosis. *J Clin Endocrinol Metab* **2011,** *96* (8), E1325-9.

[144] Labunskyy, V. M.; Lee, B. C.; Handy, D. E.; Loscalzo, J.; Hatfield, D. L.; Gladyshev, V. N., Both maximal expression of selenoproteins and selenoprotein deficiency can promote development of type 2 diabetes-like phenotype in mice. *Antioxid Redox Signal* **2011,** *14* (12), 2327-36.

[145] Virtamo, J.; Valkeila, E.; Alfthan, G.; Punsar, S.; Huttunen, J. K.; Karvonen, M. J., Serum selenium and the risk of coronary heart disease and stroke. *Am J Epidemiol* **1985,** *122* (2), 276-82.

[146] Suadicani, P.; Hein, H. O.; Gyntelberg, F., Serum selenium concentration and risk of ischaemic heart disease in a prospective cohort study of 3000 males. *Atherosclerosis* **1992,** *96* (1), 33-42.

[147] Kardinaal, A. F.; Kok, F. J.; Kohlmeier, L.; Martin-Moreno, J. M.; Ringstad, J.; Gomez-Aracena, J.; Mazaev, V. P.; Thamm, M.; Martin, B. C.; Aro, A.; Kark, J. D.; Delgado-Rodriguez, M.; Riemersma, R. A.; van 't Veer, P.; Huttunen, J. K., Association between toenail selenium and risk of acute myocardial infarction in European men. The EURAMIC Study. European Antioxidant Myocardial Infarction and Breast Cancer. *Am J Epidemiol* **1997,** *145* (4), 373-9.

[148] Salvini, S.; Hennekens, C. H.; Morris, J. S.; Willett, W. C.; Stampfer, M. J., Plasma levels of the antioxidant selenium and risk of myocardial infarction among U.S. physicians. *Am J Cardiol* **1995,** *76* (17), 1218-21.

[149] (a) Stranges, S.; Tabak, A. G.; Guallar, E.; Rayman, M. P.; Akbaraly, T. N.; Laclaustra, M.; Alfthan, G.; Mussalo-Rauhamaa, H.; Viikari, J. S.; Raitakari, O. T.; Kivimaki, M., Selenium status and blood lipids: the cardiovascular risk in Young Finns study. *J Intern*

Med **2011,** *270* (5), 469-77; (b) Stranges, S.; Galletti, F.; Farinaro, E.; D'Elia, L.; Russo, O.; Iacone, R.; Capasso, C.; Carginale, V.; De Luca, V.; Della Valle, E.; Cappuccio, F. P.; Strazzullo, P., Associations of selenium status with cardiometabolic risk factors: an 8-year follow-up analysis of the Olivetti Heart study. *Atherosclerosis* **2011,** *217* (1), 274-8.

[150] Stranges, S.; Navas-Acien, A.; Rayman, M. P.; Guallar, E., Selenium status and cardiometabolic health: state of the evidence. *Nutr Metab Cardiovasc Dis* **2010,** *20* (10), 754-60.

[151] (a) Baum, M. K.; Shor-Posner, G., Micronutrient status in relationship to mortality in HIV-1 disease. *Nutr Rev* **1998,** *56* (1 Pt 2), S135-9; (b) Baum, M. K.; Shor-Posner, G.; Lai, S.; Zhang, G.; Lai, H.; Fletcher, M. A.; Sauberlich, H.; Page, J. B., High risk of HIV-related mortality is associated with selenium deficiency. *J Acquir Immune Defic Syndr Hum Retrovirol* **1997,** *15* (5), 370-4; (c) Cirelli, A.; Ciardi, M.; de Simone, C.; Sorice, F.; Giordano, R.; Ciaralli, L.; Costantini, S., Serum selenium concentration and disease progress in patients with HIV infection. *Clin Biochem* **1991,** *24* (2), 211-4; (d) Dworkin, B. M., Selenium deficiency in HIV infection and the acquired immunodeficiency syndrome (AIDS). *Chem Biol Interact* **1994,** *91* (2-3), 181-6.

[152] Hurwitz, B. E.; Klaus, J. R.; Llabre, M. M.; Gonzalez, A.; Lawrence, P. J.; Maher, K. J.; Greeson, J. M.; Baum, M. K.; Shor-Posner, G.; Skyler, J. S.; Schneiderman, N., Suppression of human immunodeficiency virus type 1 viral load with selenium supplementation: a randomized controlled trial. *Arch Intern Med* **2007,** *167* (2), 148-54.

[153] Taylor, E. W.; Nadimpalli, R. G.; Ramanathan, C. S., Genomic structures of viral agents in relation to the biosynthesis of selenoproteins. *Biol Trace Elem Res* **1997,** *56* (1), 63-91.

[154] (a) Zazzo, J. F.; Chalas, J.; Lafont, A.; Camus, F.; Chappuis, P., Is nonobstructive cardiomyopathy in AIDS a selenium deficiency-related disease? *JPEN J Parenter Enteral Nutr* **1988,** *12* (5), 537-8; (b) Kavanaugh-McHugh, A. L.; Ruff, A.; Perlman, E.; Hutton, N.; Modlin, J.; Rowe, S., Selenium deficiency and cardiomyopathy in acquired immunodeficiency syndrome. *JPEN J Parenter Enteral Nutr* **1991,** *15* (3), 347-9; (c) Ramanathan, C. S.; Taylor, E. W., Computational genomic analysis of hemorrhagic fever viruses. Viral selenoproteins as a potential factor in pathogenesis. *Biol Trace Elem Res* **1997,** *56* (1), 93-106.

[155] (a) Serfass, R. E.; Ganther, H. E., Defective microbicidal activity in glutathione peroxidase-deficient neutrophils of selenium-deficient rats. *Nature* **1975,** *255* (5510), 640-1; (b) Boyne, R.; Arthur, J. R., The response of selenium-deficient mice to Candida albicans infection. *J Nutr* **1986,** *116* (5), 816-22.

[156] (a) Sun, Y.; Ha, P.-C.; Butler, J.A.; Ou, B.-R.; Yeh, J.-Y.; Whanger, P.D., Effect of dietary selenium on selenoprotein W and glutathione peroxidase in 28 tissues of rat. *J Nutr Biochem* **1998**, *9*, 23–27. (b) Yeh, J.-Y.; Beilstein, M.A.; Andrews, J.S.; Whanger, P.D., Tissue distribution and influence of selenium status on levels of selenoprotein W. *FASEB J* **1995,** *9,* 392–396.

[157] (a) Cardoso, B.R.; Ong, T. P.; Jacob-Filho, W.; Jaluul, O.; Freitas, M.I.; Cozzolino, S. M., Nutritional status of selenium in Alzheimer's disease patients. *Br J Nutr* **2010**, *103*, 803-6. (b) Khan Shoeb, Z.; Almas, S.; Iqbal, S.; Muzamil, A.; Sofian, S.; Fakhrul, I., Dose-

dependent protective effect of selenium in rat model of Parkinson's disease: neurobehavioral and neurochemical evidences. *J Neurochem* **2003**, *84*, 438–446.

[158] (a) Rayman, M. P., Selenium in cancer prevention: a review of the evidence and mechanism of action. *Proc Nutr Soc* **2005**, *64*, 527-42. (b) Zhu, Z.; Kimura, M.; Itokawa, Y.; Nakatsu, S.; Oda, Y.; Kikuchi, H., Effect of selenium on malignant tumor cells of brain. *Biol Trace Elem Res* **1995**, *49*, 1-7.

[159] Wikström, J.; Westermarcik, T.; Palo, J., Selenium, vitamin e and copper in multiple sclerosis. *Acta Neurol Scandinav* **1976**, *54*, 287-90.

[160] Ramaekers, V. Th.; Calomme, M.; Vanden Berghe, D.; Makropoulos W., Selenium deficiency triggering intractable seizures. *Neuropediatrics* **1994**, *25*(4), 217-223.

[161] Facchinetti, F.; Dawson, V. L.; Dawson, T. M., Free Radicals as Mediators of Neuronal Injury. *Cell Mol Neurobiol* **1998**, *18*, 667-682.

[162] Fujii, M.; Saijoh, K.; Sumino, K., Regulation of selenoprotein P mRNA expression in comparison with metallothionein and osteonectin mRNAs following cadmium and dexamethasone administration. *Kobe J Med Sci.* **1997**, *43* (1), 13-23.

[163] Pasco, J. A.; Jacka, F. N.; Williams, L. J.; Evans-Cleverdond, M.; Brennan, S. L.; Kotowicz, M. A.; Nicholson, G.C.; Ball, M. J.; Berk, M. Dietary selenium and major depression: a nested case-control study. *Complement Ther Med* **2012**, *20*, 119-23.

[164] Huber, R. E.; Criddle, R. S., Comparison of the chemical properties of selenocysteine and selenocystine with their sulfur analogs. *Arch Biochem Biophys* **1967,** *122* (1), 164-73.

[165] Besse, D.; Moroder, L., Synthesis of selenocysteine peptides and their oxidation to diselenide-bridged compounds. *J Pept Sci* **1997,** *3* (6), 442-53.

[166] Kim, H. Y.; Gladyshev, V. N., Different catalytic mechanisms in mammalian selenocysteine- and cysteine-containing methionine-R-sulfoxide reductases. *PLoS Biol* **2005,** *3* (12), e375.

[167] Lobanov, A. V.; Hatfield, D. L.; Gladyshev, V. N., Reduced reliance on the trace element selenium during evolution of mammals. *Genome Biol* **2008,** *9* (3), R62.

[168] Peskin, A. V.; Low, F. M.; Paton, L. N.; Maghzal, G. J.; Hampton, M. B.; Winterbourn, C. C., The high reactivity of peroxiredoxin 2 with H(2)O(2) is not reflected in its reaction with other oxidants and thiol reagents. *J Biol Chem* **2007,** *282* (16), 11885-92.

[169] Kanzok, S. M.; Fechner, A.; Bauer, H.; Ulschmid, J. K.; Muller, H. M.; Botella-Munoz, J.; Schneuwly, S.; Schirmer, R.; Becker, K., Substitution of the thioredoxin system for glutathione reductase in Drosophila melanogaster. *Science* **2001,** *291* (5504), 643-6.

[170] Gromer, S.; Johansson, L.; Bauer, H.; Arscott, L. D.; Rauch, S.; Ballou, D. P.; Williams, C. H., Jr.; Schirmer, R. H.; Arner, E. S., Active sites of thioredoxin reductases: why selenoproteins? *Proc Natl Acad Sci U S A* **2003,** *100* (22), 12618-23.

[171] Castellano, S., On the unique function of selenocysteine - insights from the evolution of selenoproteins. *Biochim Biophys Acta* **2009,** *1790* (11), 1463-70.

[172] Nauser, T.; Steinmann, D.; Koppenol, W. H., Why do proteins use selenocysteine instead of cysteine? *Amino Acids* **2012,** *42* (1), 39-44.

[173] Nauser, T.; Dockheer, S.; Kissner, R.; Koppenol, W. H., Catalysis of electron transfer by selenocysteine. *Biochemistry* **2006,** *45* (19), 6038-43.

[174] (a) Ufer, C.; Wang, C. C.; Fahling, M.; Schiebel, H.; Thiele, B. J.; Billett, E. E.; Kuhn, H.; Borchert, A., Translational regulation of glutathione peroxidase 4 expression through guanine-rich sequence-binding factor 1 is essential for embryonic brain development. *Genes Dev* **2008,** *22* (13), 1838-50; (b) Valentine, W. M.; Abel, T. W.; Hill, K. E.; Austin,

L. M.; Burk, R. F., Neurodegeneration in mice resulting from loss of functional selenoprotein P or its receptor apolipoprotein E receptor 2. *J Neuropathol Exp Neurol* **2008,** *67* (1), 68-77; (c) Voetsch, B.; Jin, R. C.; Bierl, C.; Benke, K. S.; Kenet, G.; Simioni, P.; Ottaviano, F.; Damasceno, B. P.; Annichino-Bizacchi, J. M.; Handy, D. E.; Loscalzo, J., Promoter polymorphisms in the plasma glutathione peroxidase (GPx-3) gene: a novel risk factor for arterial ischemic stroke among young adults and children. *Stroke* **2007,** *38* (1), 41-9; (d) Voetsch, B.; Jin, R. C.; Bierl, C.; Deus-Silva, L.; Camargo, E. C.; Annichino-Bizacchi, J. M.; Handy, D. E.; Loscalzo, J., Role of promoter polymorphisms in the plasma glutathione peroxidase (GPx-3) gene as a risk factor for cerebral venous thrombosis. *Stroke* **2008,** *39* (2), 303-7; (e) Guo, Z.; Ran, Q.; Roberts, L. J., 2nd; Zhou, L.; Richardson, A.; Sharan, C.; Wu, D.; Yang, H., Suppression of atherogenesis by overexpression of glutathione peroxidase-4 in apolipoprotein E-deficient mice. *Free Radic Biol Med* **2008,** *44* (3), 343-52; (f) Kenet, G.; Freedman, J.; Shenkman, B.; Regina, E.; Brok-Simoni, F.; Holzman, F.; Vavva, F.; Brand, N.; Michelson, A.; Trolliet, M.; Loscalzo, J.; Inbal, A., Plasma glutathione peroxidase deficiency and platelet insensitivity to nitric oxide in children with familial stroke. *Arterioscler Thromb Vasc Biol* **1999,** *19* (8), 2017-23.

[175] (a) Chu, F. F.; Esworthy, R. S.; Doroshow, J. H., Role of Se-dependent glutathione peroxidases in gastrointestinal inflammation and cancer. *Free Radic Biol Med* **2004,** *36* (12), 1481-95; (b) Banning, A.; Kipp, A.; Schmitmeier, S.; Lowinger, M.; Florian, S.; Krehl, S.; Thalmann, S.; Thierbach, R.; Steinberg, P.; Brigelius-Flohe, R., Glutathione Peroxidase 2 Inhibits Cyclooxygenase-2-Mediated Migration and Invasion of HT-29 Adenocarcinoma Cells but Supports Their Growth as Tumors in Nude Mice. *Cancer Res* **2008,** *68* (23), 9746-53.

[176] (a) Beck, M. A.; Esworthy, R. S.; Ho, Y. S.; Chu, F. F., Glutathione peroxidase protects mice from viral-induced myocarditis. *FASEB J* **1998,** *12* (12), 1143-9; (b) Beck, M. A.; Handy, J.; Levander, O. A., Host nutritional status: the neglected virulence factor. *Trends Microbiol* **2004,** *12* (9), 417-23; (c) Gladyshev, V. N.; Stadtman, T. C.; Hatfield, D. L.; Jeang, K. T., Levels of major selenoproteins in T cells decrease during HIV infection and low molecular mass selenium compounds increase. *Proc Natl Acad Sci U S A* **1999,** *96* (3), 835-9; (d) Beck, M. A.; Shi, Q.; Morris, V. C.; Levander, O. A., Rapid genomic evolution of a non-virulent coxsackievirus B3 in selenium-deficient mice results in selection of identical virulent isolates. *Nat Med* **1995,** *1* (5), 433-6.

[177] (a) Campa, A.; Shor-Posner, G.; Indacochea, F.; Zhang, G.; Lai, H.; Asthana, D.; Scott, G. B.; Baum, M. K., Mortality risk in selenium-deficient HIV-positive children. *J Acquir Immune Defic Syndr Hum Retrovirol* **1999,** *20* (5), 508-13; (b) Burbano, X.; Miguez-Burbano, M. J.; McCollister, K.; Zhang, G.; Rodriguez, A.; Ruiz, P.; Lecusay, R.; Shor-Posner, G., Impact of a selenium chemoprevention clinical trial on hospital admissions of HIV-infected participants. *HIV Clin Trials* **2002,** *3* (6), 483-91; (c) Kalantari, P.; Narayan, V.; Natarajan, S. K.; Muralidhar, K.; Gandhi, U. H.; Vunta, H.; Henderson, A. J.; Prabhu, K. S., Thioredoxin reductase-1 negatively regulates HIV-1 transactivating protein Tat-dependent transcription in human macrophages. *J Biol Chem* **2008,** *283* (48), 33183-90; (d) Ursini, F.; Heim, S.; Kiess, M.; Maiorino, M.; Roveri, A.; Wissing, J.; Flohe, L., Dual function of the selenoprotein PHGPx during sperm maturation. *Science* **1999,** *285* (5432), 1393-6.

[178] (a) Muecke, R.; Schomburg, L.; Glatzel, M.; Berndt-Skorka, R.; Baaske, D.; Reichl, B.; Buentzel, J.; Kundt, G.; Prott, F. J.; Devries, A.; Stoll, G.; Kisters, K.; Bruns, F.; Schaefer,

U.; Willich, N.; Micke, O., Multicenter, phase 3 trial comparing selenium supplementation with observation in gynecologic radiation oncology. *Int J Radiat Oncol Biol Phys* **2010,** *78* (3), 828-35; (b) Gilbert, H. F., Molecular and cellular aspects of thiol-disulfide exchange. *Adv Enzymol Relat Areas Mol Biol* **1990,** *63*, 69-172; (c) Arner, E. S.; Holmgren, A., Physiological functions of thioredoxin and thioredoxin reductase. *Eur J Biochem* **2000,** *267* (20), 6102-9.

[179] Duffield-Lillico, A. J.; Slate, E. H.; Reid, M. E.; Turnbull, B. W.; Wilkins, P. A.; Combs, G. F., Jr.; Park, H. K.; Gross, E. G.; Graham, G. F.; Stratton, M. S.; Marshall, J. R.; Clark, L. C., Selenium supplementation and secondary prevention of nonmelanoma skin cancer in a randomized trial. *J Natl Cancer Inst* **2003,** *95* (19), 1477-81.

[180] Panee, J.; Liu, W.; Nakamura, K.; Berry, M. J., The responses of HT22 cells to the blockade of mitochondrial complexes and potential protective effect of selenium supplementation. *Int J Biol Sci* **2007,** *3* (5), 335-41.

[181] Horibata, Y.; Nagatsuka, Y.; Greimel, P.; Ito, Y.; Hirabayashi, Y., Sensitivity of phosphatidylglucoside against phospholipases. *Anal Biochem* **2007,** *365* (1), 149-51.

[182] Lu, C.; Qiu, F.; Zhou, H.; Peng, Y.; Hao, W.; Xu, J.; Yuan, J.; Wang, S.; Qiang, B.; Xu, C.; Peng, X., Identification and characterization of selenoprotein K: an antioxidant in cardiomyocytes. *FEBS Lett* **2006,** *580* (22), 5189-97.

[183] Chen, P.; Wang, R. R.; Ma, X. J.; Liu, Q.; Ni, J. Z., Different Forms of Selenoprotein M Differentially Affect Abeta Aggregation and ROS Generation. *Int J Mol Sci* **2013,** *14* (3), 4385-99.

[184] Reeves, M. A.; Hoffmann, P. R., The human selenoproteome: recent insights into functions and regulation. *Cell Mol Life Sci* **2009,** *66* (15), 2457-78.

[185] (a) Foresta, C.; Flohe, L.; Garolla, A.; Roveri, A.; Ursini, F.; Maiorino, M., Male fertility is linked to the selenoprotein phospholipid hydroperoxide glutathione peroxidase. *Biol Reprod* **2002,** *67* (3), 967-71; (b) Vidlar, A.; Vostalova, J.; Ulrichova, J.; Student, V.; Krajicek, M.; Vrbkova, J.; Simanek, V., The safety and efficacy of a silymarin and selenium combination in men after radical prostatectomy - a six month placebo-controlled double-blind clinical trial. *Biomed Pap Med Fac Univ Palacky Olomouc Czech Repub* **2010,** *154* (3), 239-44; (c) Cooper, M. L.; Adami, H. O.; Gronberg, H.; Wiklund, F.; Green, F. R.; Rayman, M. P., Interaction between single nucleotide polymorphisms in selenoprotein P and mitochondrial superoxide dismutase determines prostate cancer risk. *Cancer Res* **2008,** *68* (24), 10171-7; (d) Gresner, P.; Gromadzinska, J.; Jablonska, E.; Kaczmarski, J.; Wasowicz, W., Expression of selenoprotein-coding genes SEPP1, SEP15 and hGPX1 in non-small cell lung cancer. *Lung Cancer* **2009,** *65* (1), 34-40; (e) Atkinson, J. B.; Hill, K. E.; Burk, R. F., Centrilobular endothelial cell injury by diquat in the selenium-deficient rat liver. *Lab Invest* **2001,** *81* (2), 193-200.

[186] Kim, H. Y.; Gladyshev, V. N., Methionine sulfoxide reductases: selenoprotein forms and roles in antioxidant protein repair in mammals. *Biochem J* **2007,** *407* (3), 321-9.

[187] Gao, Y.; Walder, K.; Sunderland, T.; Kantham, L.; Feng, H. C.; Quick, M.; Bishara, N.; de Silva, A.; Augert, G.; Tenne-Brown, J.; Collier, G. R., Elevation in Tanis expression alters glucose metabolism and insulin sensitivity in H4IIE cells. *Diabetes* **2003,** *52* (4), 929-34.

[188] Combs, G. F., Jr.; Clark, L. C.; Turnbull, B. W., Reduction of cancer mortality and incidence by selenium supplementation. *Med Klin (Munich)* **1997,** *92 Suppl 3*, 42-5.

[189] Dikiy, A.; Novoselov, S. V.; Fomenko, D. E.; Sengupta, A.; Carlson, B. A.; Cerny, R. L.; Ginalski, K.; Grishin, N. V.; Hatfield, D. L.; Gladyshev, V. N., SelT, SelW, SelH, and

Rdx12: genomics and molecular insights into the functions of selenoproteins of a novel thioredoxin-like family. *Biochemistry* **2007,** *46* (23), 6871-82.

[190] (a) Loflin, J.; Lopez, N.; Whanger, P. D.; Kioussi, C., Selenoprotein W during development and oxidative stress. *J Inorg Biochem* **2006,** *100* (10), 1679-84; (b) Aachmann, F. L.; Fomenko, D. E.; Soragni, A.; Gladyshev, V. N.; Dikiy, A., Solution structure of selenoprotein W and NMR analysis of its interaction with 14-3-3 proteins. *J Biol Chem* **2007,** *282* (51), 37036-44.

[191] Dumitrescu, A. M.; Liao, X. H.; Abdullah, M. S.; Lado-Abeal, J.; Majed, F. A.; Moeller, L. C.; Boran, G.; Schomburg, L.; Weiss, R. E.; Refetoff, S., Mutations in SECISBP2 result in abnormal thyroid hormone metabolism. *Nat Genet* **2005,** *37* (11), 1247-52.

[192] Combs, G. F., Jr.; Midthune, D. N.; Patterson, K. Y.; Canfield, W. K.; Hill, A. D.; Levander, O. A.; Taylor, P. R.; Moler, J. E.; Patterson, B. H., Effects of selenomethionine supplementation on selenium status and thyroid hormone concentrations in healthy adults. *Am J Clin Nutr* **2009,** *89* (6), 1808-14.

[193] Foster, C. B.; Aswath, K.; Chanock, S. J.; McKay, H. F.; Peters, U., Polymorphism analysis of six selenoprotein genes: support for a selective sweep at the glutathione peroxidase 1 locus (3p21) in Asian populations. *BMC Genet* **2006,** *7*, 56.

[194] (a) Squires, J. E.; Berry, M. J., Eukaryotic selenoprotein synthesis: mechanistic insight incorporating new factors and new functions for old factors. *IUBMB Life* **2008,** *60* (4), 232-5; (b) Hoffmann, P. R.; Berry, M. J., Selenoprotein synthesis: a unique translational mechanism used by a diverse family of proteins. *Thyroid* **2005,** *15* (8), 769-75; (c) Papp, L. V.; Lu, J.; Holmgren, A.; Khanna, K. K., From selenium to selenoproteins: synthesis, identity, and their role in human health. *Antioxid Redox Signal* **2007,** *9* (7), 775-806; (d) Bock, A., Molecular biology. Invading the genetic code. *Science* **2001,** *292* (5516), 453-4.

[195] Bosl, M. R.; Takaku, K.; Oshima, M.; Nishimura, S.; Taketo, M. M., Early embryonic lethality caused by targeted disruption of the mouse selenocysteine tRNA gene (Trsp). *Proc Natl Acad Sci U S A* **1997,** *94* (11), 5531-4.

[196] Suzuki, T.; Kelly, V. P.; Motohashi, H.; Nakajima, O.; Takahashi, S.; Nishimura, S.; Yamamoto, M., Deletion of the selenocysteine tRNA gene in macrophages and liver results in compensatory gene induction of cytoprotective enzymes by Nrf2. *J Biol Chem* **2008,** *283* (4), 2021-30.

[197] (a) Baliga, M. S.; Wang, H.; Zhuo, P.; Schwartz, J. L.; Diamond, A. M., Selenium and GPx-1 overexpression protect mammalian cells against UV-induced DNA damage. *Biol Trace Elem Res* **2007,** *115* (3), 227-42; (b) Carlson, B. A.; Xu, X. M.; Gladyshev, V. N.; Hatfield, D. L., Selective rescue of selenoprotein expression in mice lacking a highly specialized methyl group in selenocysteine tRNA. *J Biol Chem* **2005,** *280* (7), 5542-8.

[198] (a) Cheng, W. H.; Combs, G. F., Jr.; Lei, X. G., Knockout of cellular glutathione peroxidase affects selenium-dependent parameters similarly in mice fed adequate and excessive dietary selenium. *Biofactors* **1998,** *7* (4), 311-21; (b) de Haan, J. B.; Bladier, C.; Griffiths, P.; Kelner, M.; O'Shea, R. D.; Cheung, N. S.; Bronson, R. T.; Silvestro, M. J.; Wild, S.; Zheng, S. S.; Beart, P. M.; Hertzog, P. J.; Kola, I., Mice with a homozygous null mutation for the most abundant glutathione peroxidase, Gpx1, show increased susceptibility to the oxidative stress-inducing agents paraquat and hydrogen peroxide. *J Biol Chem* **1998,** *273* (35), 22528-36.

[199] (a) Trepanier, G.; Furling, D.; Puymirat, J.; Mirault, M. E., Immunocytochemical localization of seleno-glutathione peroxidase in the adult mouse brain. *Neuroscience* **1996,**

75 (1), 231-43; (b) Lindenau, J.; Noack, H.; Asayama, K.; Wolf, G., Enhanced cellular glutathione peroxidase immunoreactivity in activated astrocytes and in microglia during excitotoxin induced neurodegeneration. *Glia* **1998,** *24* (2), 252-6.

[200] Power, J. H.; Blumbergs, P. C., Cellular glutathione peroxidase in human brain: cellular distribution, and its potential role in the degradation of Lewy bodies in Parkinson's disease and dementia with Lewy bodies. *Acta Neuropathol* **2009,** *117* (1), 63-73.

[201] (a) Blackinton, J.; Lakshminarasimhan, M.; Thomas, K. J.; Ahmad, R.; Greggio, E.; Raza, A. S.; Cookson, M. R.; Wilson, M. A., Formation of a stabilized cysteine sulfinic acid is critical for the mitochondrial function of the parkinsonism protein DJ-1. *J Biol Chem* **2009,** *284* (10), 6476-85; (b) Seiler, A.; Schneider, M.; Forster, H.; Roth, S.; Wirth, E. K.; Culmsee, C.; Plesnila, N.; Kremmer, E.; Radmark, O.; Wurst, W.; Bornkamm, G. W.; Schweizer, U.; Conrad, M., Glutathione peroxidase 4 senses and translates oxidative stress into 12/15-lipoxygenase dependent- and AIF-mediated cell death. *Cell Metab* **2008,** *8* (3), 237-48; (c) Chen, L.; Na, R.; Gu, M.; Richardson, A.; Ran, Q., Lipid peroxidation up-regulates BACE1 expression *in vivo*: a possible early event of amyloidogenesis in Alzheimer's disease. *J Neurochem* **2008,** *107* (1), 197-207.

[202] Yant, L. J.; Ran, Q.; Rao, L.; Van Remmen, H.; Shibatani, T.; Belter, J. G.; Motta, L.; Richardson, A.; Prolla, T. A., The selenoprotein GPX4 is essential for mouse development and protects from radiation and oxidative damage insults. *Free Radic Biol Med* **2003,** *34* (4), 496-502.

[203] Bellinger, F. P.; He, Q. P.; Bellinger, M. T.; Lin, Y.; Raman, A. V.; White, L. R.; Berry, M. J., Association of selenoprotein p with Alzheimer's pathology in human cortex. *J Alzheimers Dis* **2008,** *15* (3), 465-72.

[204] Freedman, J. E.; Loscalzo, J.; Benoit, S. E.; Valeri, C. R.; Barnard, M. R.; Michelson, A. D., Decreased platelet inhibition by nitric oxide in two brothers with a history of arterial thrombosis. *J Clin Invest* **1996,** *97* (4), 979-87.

[205] (a) Alanne, M.; Kristiansson, K.; Auro, K.; Silander, K.; Kuulasmaa, K.; Peltonen, L.; Salomaa, V.; Perola, M., Variation in the selenoprotein S gene locus is associated with coronary heart disease and ischemic stroke in two independent Finnish cohorts. *Hum Genet* **2007,** *122* (3-4), 355-65; (b) Silander, K.; Alanne, M.; Kristiansson, K.; Saarela, O.; Ripatti, S.; Auro, K.; Karvanen, J.; Kulathinal, S.; Niemela, M.; Ellonen, P.; Vartiainen, E.; Jousilahti, P.; Saarela, J.; Kuulasmaa, K.; Evans, A.; Perola, M.; Salomaa, V.; Peltonen, L., Gender differences in genetic risk profiles for cardiovascular disease. *PLoS One* **2008,** *3* (10), e3615.

CHAPTER 10

Antimicrobial Activity of Organoselenium Compounds

Donatella Pietrella*

Department of Pharmaceutical Sciences, University of Perugia, Building B, Via del Giochetto, 06122 Perugia, Italy

Abstract: Selenium (Se) is a key component of a number of selenoproteins required for the normal human metabolism. Se is considered a scavenger of free radicals; it has been studied for its ability to improve the immune system functions and to prevent cancer. The initial studies on anti-inflammatory and antimicrobial activities of ebselen, an organoselenium compound, have furthered the research to identify new molecules with antimicrobial activities. The antimicrobial potential of many organoselenium compounds was studied in *in vitro* and *in vivo* assays against a wide range of microorganisms and suggested a potential use for them in different medical applications such as treatment of drug resistant microorganisms, and as an adjuvant in healing wounds. The interesting results observed for the Se-containing organic compounds also suggest their potential use to enhance the safety and the shelf life of food products. Another interesting application of organoselenium compounds is their use as a herbicide because of their activity against parasites and plant pathogens, although the residual toxicity of the compounds in food has to be investigated.

Keywords: Antibacterial activity, Antifungal activity, Antiviral activity, Antiparasitic activity, Food preservative, Wound dressing, Drug resistant bacteria, Herbicide, Se-enriched probiotics.

10.1. INTRODUCTION

Selenium (Se) is an essential micro-mineral element involved in the regulation of the cellular metabolism, it is essential in small amounts for humans but it is toxic at high levels. Humans and animals require Se for the function of enzymes known as selenoproteins. During selenoprotein synthesis, selenocysteine is incorporated in the amino acidic sequence of the active site of these enzymes. Unlike animals,

***Address correspondence to Donatella Pietrella:** Department of Pharmaceutical Sciences, University of Perugia, Building B, *Via* del Giochetto, 06122 Perugia, Italy; Tel: +39-0755857208;
E-mail: donatella.pietrella@unipg.it

plants do not appear to require selenium for survival. However, when selenium is present in the soil, plants incorporate it non-specifically into compounds that usually contain sulfur [1]. Se is involved in oxygen metabolism and it is considered a scavenger of free radicals [2,3]. It has been demonstrated that Se activity is involved in improving the immune system functions [4,5], in preventing cancer [6], in enhancing reproductive performance and in postponing aging because of its anti-oxidant activity [7]. Se is relatively scarce in most areas of the world and Se supplementation in humans and animal diets is a routine step toward good health [8,9]. Supplementation with sodium selenite as an inorganic source of Se has been limited because of its toxicity; organic Se is considered more efficient in improving the selenoenzyme activity of blood cells [10].

Under physiological conditions Se of selenocysteine, in its ionized form, is an efficient biological catalyst [11]. The organoselenium compounds are attractive synthetic targets due to their chemical characteristics such as chemo-, regio- and stereo-selective reactions [12]. The organic species of Se have greater bioavailability than inorganic selenium and they are also less toxic [13].

Infectious diseases are becoming a new challenge for human health due to increased resistance to antimicrobial drugs currently used against Gram-positive and Gram-negative bacteria, and to some intractable pathogens. The alarming rate of emerging microbial threats are major challenges to the public health and the synthesis of new antimicrobial drugs is becoming a must for the scientific community [14]. In this context, the antimicrobial potential of many organoselenium compounds against a wide range of microorganisms was studied in *in vitro* and *in vivo* assays.

10.2. ANTIMICROBIAL ACTIVITY

In the past two decades a variety of new compounds (Table **10.1**) containing Se have been synthesized and tested against bacteria, fungi, algae and viruses. Most of them have shown a good performance compared to the antimicrobial drugs in current use.

10.2.1. Ebselen

Ebselen (2-phenyl-1,2-benzisoselenazol-3(2H)-one) is a compound containing selenium that has antioxidant, antinflammatory and cytoprotective properties in

Table 10.1: Antimicrobial activity of Se-compounds against different microorganisms

Organoselenium Compounds	Microorganisms	References
Ebselen	*S. aureus, S. aureus* (MRSA), *M. luteus, B. subtilis, E. coli,* *P. aeruginosa, S. enteritidis* *S. simulans, K. pneumonia, Bacillus spp,* *S. cerevisiae, C. neoformans,* *C. albicans,* *A. niger* HSV-1, EMCV, VSV	[15-20]
Selenoquinoline	*C. albicans, Aspergillus flavus,* *A. niger, Fusarium oxysporum* *S. aureus, B. subtilis, S. pyogenes*	[24,25]
Diselenide	*S. aureus* *Leishmania infantum*	[26,27]
Selenomorpholine	*S. aureus, E. coli*	[28-30]
Piperidine	*Botrytis cinerea,* *Fusarium culmorum,* *Phytophthora cactorum* *Rhizoctonia solani*	[31]
Pyridazine	*C. albicans, Aspergillus flavus,* *A. niger, Fusarium oxysporum*	[32]
Selenoazoles	*Botrytis cinerea, Phonopsis spp,* *Drechslera graminea, Rhizoctonia solani* *S. aureus, B. subtilis, E. coli, P. aeruginosa* *C. albicans* *M. tuberculosis*	[33-35]
Gold(I)phosphine	*Plasmodium falciparum*	[37]
Amino acid derivatives containing Se	*S. aureus, S. typhimurium, E. coli B. subtilis*	[38,40]
Dideoxynucleosides analogues	Coxsackievirus, HIV, HBV	[52]
Imidazolium ionic liquids	*Prototheca zopfii*	[53]

mammalian cells and is cytotoxic against bacteria and fungi (Table **10.2**). Ebselen is able to inhibit the growth of Gram-positive bacteria and yeast (IC_{50} 2-5 μM) but it is less active against Gram-negative bacteria (IC_{50} >80 μM). The effect is time- and dose-dependent [14]. Its activity against a variety of bacteria and against drug resistant bacteria such as methicillin resistant *Staphylococcus aureus* (MRSA) has also been demonstrated [15]. The antimicrobial mechanism of action is still not clear although it has been proposed that the target of the ebselen is the plasma membrane H^+-ATPase (Pma1p). Pma1p is a transmembrane protein that works as an electrogenic proton pump that couples the ATP hydrolysis to proton (H^+) transport regulating the cellular pH and the proton gradient during nutrient uptake. Ebselen is able to block the activity of Pma1p not only in opportunistic pathogens such as *Cryptococcus neoformans* and *Candida albicans* but also in non pathogenic yeasts such as *Saccharomyces cerevisiae*. Moreover it is able to inhibit the growth of fluconazole-resistant *C. albicans* [16]. The fungicidal activity is due to the interference with the proton-translocating function and the ATPase activity of the Pma1p [17]. Comparing the activity of ebselen with other currently used antifungal agents, ebselen is ten-fold more potent than fluconazole but less active than other azoles and polyenes [16].

Table 10.2: Antimicrobial activity of Ebselen and its analogs

	Microorganisms	**References**
Ebselen (2-phenyl-1,2-benzoisoselenazol-3(2H)-one	*S. cerevisiae*	[17]
	Gram-positive bacteria (*S. aureus, M. luteus, B. subtilis*)	[17]
	Gram-negative bacteria (*E. coli, P. aeruginosa, S. enteritidis*)	[17]
	S. aureus (MRSA)	[15]
	Fluconazole resistant *C. albicans*	[16]
benzisoselenazol-3(2H)-one,2-pyridylbenzisoselenazol-3(2H)-ones and 7-azabenzisoselenazol-3(2H)-ones	Gram-positive bacteria (*S. aureus, Bacillus spp*)	[18]
	HSV-1, EMCV	[18]
2-(4-chlorophenyl)-1,2-benzisoselenazol-3(2H)-one	*S. cerevisiae* and *C. albicans*	[19]
	S. aureus	[19]
	E. coli	[19]
N-substituted benzisoselenazol-3(2H)-ones	*E coli, S. aureus, S. simulans, P. aeruginosa, K. pneumoniae*	[16,20]
	C. albicans, A. niger	[20]
	HSV-1, EMCV, *VSV*	[20]

Different analogues of ebselen have been tested for their antimicrobial activity. The ebselen-unsubstituted benzisoselenazol-3(2H)-one, 2-pyridylbenzisoselenazol-3(2H)-one and 7-azabenzisoselenazol-3(2H)-one have shown a strong antiviral activity against Herpes virus 1 (HSV-1) and encephalomyocarditis virus (EMCV) and a good antibacterial activity against Gram-positive *S. aureus* and *Bacillus spp.* Some of them were able to inhibit the growth of the yeast *C. albicans* and other filamentous fungi [18].

Bien and coworkers [19] have demonstrated that the selenorganic compounds benzoisoselenazolones inhibit the growth of the yeasts *S. cerevisiae* and *C. albicans*, the Gram-negative bacterium *E. coli* and Gram positive *S. aureus*.

Different N-substituted benzoisoselenazol-3(2H)-ones have been synthesized and tested for their antimicrobial activity against pathogenic microorganisms: *S. aureus, Staphylococcus simulans, E. coli, Pseudomonas aeruginosa, Klebsiella pneumoniae, C. albicans, Aspergillus niger*, HSV-1, EMCV and vesicular stomatitis virus (VSV). Most of these compounds have shown high antiviral activity against HSV-1 and EMCV, antibacterial activity against *S. aureus* and *S. simulans*, and antifungal activity against *C. albicans* and *A. niger* [20]. The structural modifications of ebselen replacing phenyl substituent in 2-position of heterocycling ring by a small non polar alkyl group leads to a strong enhancement of antimicrobial and antiviral activity.

10.2.2. Selenoquinoline

Quinoline drugs (quinine, quinidine, chloroquine, mefloquine, halofantrine) have been widely used as antimalarial agents. Quinoline derivatives posses a wide range of antifungal [21], antibacterial [22], antileishmanial [23] activities. New 4-methylquinoline-2(1H) selenone derivatives have shown a mild antifungal activity against *C. albicans, Aspergillus flavus, A. niger* and *Fusarium oxysporum* [24]. Quinolines can chelate Fe(II) and Fe(III) and prevent free radical production, moreover these compounds can be intercalated by DNA molecules. Wound healing is a complex multifactorial process that occurs as a sequence of steps: inflammation, contraction and closure of the wound by migration and proliferation of different cells types. During the inflammatory process, cells recruited in the wound produce the radical oxygen species (ROS) through the respiratory burst. ROS are deleterious in the wound healing because of their oxidative stress on cells and tissues. Organoselenium compounds are known for their anti-oxidant properties and they have been evaluated for their ability to protect tissues from radical damage and for their antimicrobial activity. The organoselenium compound 2-selenobenzo[h]quinoline-3-carbaldehyde has been tested for its antibacterial activity and its action in wound healing. The compound showed significant antimicrobial activity against Gram-positive bacteria (*S. aureus, B. subtilis* and *S. pyogenes*), an interesting wound healing activity and ability to protect against oxidative stress [25].

10.2.3. Diselenides

Other interesting organoselenium compounds with antimicrobial activity are diselenides. Xu and coworkers studied the antibacterial activity of novel organoselenium compounds, the diselenide-bridged bis(porphyrin)s, against *Staphylococcus aureus*. The analysis of the power-time curve of microbial growth in the presence of this selenium compound showed an inhibition of the growth probably due to the ability of Se to catalyze the production of reactive oxygen radical resulting in the oxidative damage [26].

The antimicrobial activity of diselenides against *Leishmania* infections has also been studied [27]. *Leishmania* is a parasite with two different microorganism forms in its life cycle: the extracellular flagellated promastigotes in the vector (female sandfly) and intracellular non-flagellated amastigotes (mammalian macrophages). A group of diselenides were synthesized and tested against both forms of *Leishmania infantum*, promastigotes and amastigotes, showing good cell growth inhibition (IC_{50}) with better selectivity index (SI) values than that observed for edelsofine and miltefosine already used in the treatment of leishmaniasis. *Bis*(4-aminophenyl)-diselenide exhibited the best activity against infected macrophages with high SI values. This new lead compound is still under investigation to establish its mechanism of action [27].

10.2.4. Selenomorpholines

The antibacterial effect of selenomorpholine derivatives on the growth of *S. aureus* and *E. coli* has been tested by microcalorimetry [28-30]. Three different selenomorpholine derivatives showed the capability to inhibit the metabolism of *S. aureus*. The effect was dose-dependent and the chemical groups of the compound influence the uptake into the cells [28]. The antimicrobial activity of certain selenomorpholine compounds is better that that observed for sodium selenite [30].

10.2.5. Piperidines

Organoselenium compounds have recently been investigated for their pesticidal properties. Zakrewski and Krawczyk (2011) demonstrated the antifungal activity of selenium derivatives of 2,2,6,6-tetramethylpiperidine: 4-isoselenocyanato- and N-formyl-, N-thioformyl-, N-selenoformyl of 2,2,6,6-tetramethylpiperidine. These compounds resulted active against four fungi pathogens for plants *Botrytis cinerea, Fusarium culmorum, Phytophthora cactorum* and *Rhizoctonia solani* [31].

10.2.6. Pyridazines

Heterocyclic pyridazines are attractive compounds because of their pharmacological activities, and they have been found to have herbicidal, insecticidal and fungicidal properties. A new series of selenolo[2,3-c]pyridazyne and pyrimido[4',5':4,5] selenolo[2,3-c]pyridazyne derivatives has been synthesized and investigated for their anti-inflammatory, analgesic and antimicrobial activities. The compounds showing the best anti-inflammatory and analgesic activities displayed a consistent antifungal effect on *Aspergillus flavus, A. niger, Fusarium oxysporum* and *C. albicans* [32].

10.2.7. Selenoazoles

Azoles are currently used as antimicrobial drugs against yeasts or filamentous fungi. In the attempt to improve the antifungal activity of azoles, Deidda and coworkers tested a series of heterocyclic compounds, thio- and seleno-azoles for their antimicrobial activity. The Se-containing compounds were more toxic than the corresponding sulfur ones, but they exerted an interesting inhibitory activity against plant pathogenic fungi such as *Botrytis cinerea,* which is the cause of grey mould in vineyards all over the world, and *Phonopsis spp*, *Drechslera graminea* and *Rhizoctonia solani,* common pathogens of leguminosae, cereal and potatoes respectively [33].

Selenium containing heterocyclic systems have been used as chemotherapeutic agents [34]. In 2008 Karnik and coworkers demonstrated that, among selenium containing heterocyclic compounds, 1,2,3-selenadiazoles had a consistent antimicrobial activity [35]. Benzopyran and naphthopyran compounds have various biological activities and they have been incorporated in 1,2,3-selenadiazole derivatives. The 4H-naphthol[1',2'-5,6]pyrano [3,4-d](1,2,3) selenadiazole and its derivative have shown moderate anti-bacterial activity against Gram-positive bacteria *Staphylococcus aureus* and *Bacillus subtilis*, and Gram-negative bacteria such as *Escherichia coli* and *Pseudomonas aeruginosa* [35]. More recently Chitra and coworkers synthesized a series of carbocycle mono and *bis*-1,2,3-selenadiazoles that resulted active against *E. coli*, *S. aureus* and *C. albicans* and showed good activity against *Mycobacterium tuberculosis* [34].

10.2.8. Gold(I)Phosphine Complexes Containing Se Ligands

The chemistry of Se containing gold compounds has not been widely studied. In the nineties, Scovill and Klayman tested the antimicrobial activity of 2-acetylpyridyl functionalized selenosemicarbazones against *Plasmodium falciparum* [36]. Recently, Molter and coworkers combined selenium ligands with known antimalarial activity with a gold phosphine unit [37]. The authors synthesized a series of mono- and dinuclear gold(I)phosphine complexes containing monoanionic seleno- and thiosemicarbazones as ligands. Some of the Se derivatives in this study showed only moderate antimalarial activity *in vitro*.

10.2.9. Amino Acid Derivatives Containing Se

Novel little amino acid derivatives containing Se have been prepared as a source of supplemental Se (*N*-succinyl L-selenomethonine, L-selenomethionine isopropylester hydrochloride, *N*-Carbonyl L-selenomethionine, L-selenomethionine hydration) [38]. Novel *bis*-alpha-amino acids containing sulfur and the selenium atom have been synthesized from hydroxyproline, wherein L-cysteine or L-selenocysteine were linked to the proline through sulfur or selenium atoms [39]. Radhakrishna and coworkers tested the antibacterial activity of 3-

[(phenylcarbonyl)selenyl] propanoic acid and its ester derivatives and 2-[(phenylcarbonyl)selenyl] acetic acid against *S. aureus*, *Salmonella typhimurium*, *E. coli* and *B. subtilis* showing moderate antimicrobial activity [40]. More recently 3,5-[dinitrobenzoylamino-selenomethyl] amino acids were synthesized, characterized and the antimicrobial activity has been tested by using agar diffusion. The new compounds showed moderate antibacterial activity against *S. aureus* and *E. coli* [38].

10.2.10. Dideoxynucleoside Analogues

Selenium deficiency of the host appears to enhance the virulence of some viruses and to increase the progression time of some viral infections. One of the hypotheses to explain this fact is that Se could increase the oxidative stress which could induce mutations or changes in the expression of some viral genes. When selenium-deficient mice are inoculated with a relatively harmless strain of coxsackievirus, mutations occurring in the viral genome result in a more virulent form of the virus, which causes an inflammation of the heart muscle known as myocarditis [41]. Moreover, coxsackievirus has been isolated from the blood of some sufferers of Keshan disease, suggesting that it may be a cofactor in the development of the cardiomyopathy associated with selenium deficiency in humans [42].

Low levels of plasma selenium have also been associated with a significantly increased risk of death from HIV. Declining selenium levels in HIV-infected individuals are sensitive markers of disease progression and severity, even before malnutrition becomes a factor. In HIV infection, the oxidative stress favors viral replication, possibly by activating specific transcription pathways. As an integral component of glutathione peroxidase and thioredoxin reductase, selenium plays an important role in decreasing oxidative stress in HIV-infected cells and possibly suppressing the rate of HIV replication [43].

AZT and other dideoxynucleoside analogues (ddC, ddI, d4T, 3TC and Abacavir) are efficient inhibitors of the reverse transcriptase of HIV. However these nucleoside analogues are cause of different adverse effects such as pancreatitis, peripheral neuropathy and appearance of drug-resistant viral strains.

On the basis of potent anti-HIV activity of 2',3'-dideoxynucleosides (ddNs), their bioisosteric analogues, 2',3'-dideoxy-4'-selenonucleosides, have been considered as possible inhibitors of reverse transcriptase of HIV, but no anti-HIV activity has been observed [44]. This could be explained by the fact that the cellular thymidine kinase is able to phosphorylate the 4'-seleno-ddNs in 4'-seleno-ddNTPs but the latter are not recognized by the reverse transcriptase of HIV.

Oxathiolane and dioxolane nucleosides, in which the 3'-CH_2 group is replaced by a sulfur or oxygen atom, have been tested for their anti-viral activity showing good results against cancer and virus [45-51]. Isosteric substitution of 3'-CH_2 with Se atom has been reported [52]. Various pyrimidine and purine oxaselenolane nucleosides were synthesized, the cytosine and 5-fluorocytosine analogues showed potent anti-HIV and anti-HBV activities in peripheral blood mononuclear cells and 2.2.15 cells respectively.

10.2.11. Imidazolium Ionic Liquids Containing Se

Ionic liquids (ILs) have been used as solvents in organic transformations and recently ILs have been studied for their antimicrobial activity. Imidazolium based ionic liquids functionalized with Se in the side chain were prepared and the antimicrobial activity was tested *in vitro*. These compounds resulted active against different species of bacteria and especially against the algae *Prototheca zopfii* [53].

10.3. APPLICATION OF ORGANOSELENIUM COMPOUNDS

10.3.1. Food Preservatives

In the last ten years organoselenium compounds have been studied for their potential use as food preservatives. Microbial activity and oxidation reactions are the cause of food degradation. A great deal of attention has been given to researching new alternative chemical additives able to preserve the microbiological quality while maintaining the organoleptic properties of food products [54]. Because of the residual effects of chemical food additives on the health of consumers, the research is focusing on natural antimicrobial substances such as plant essential oils [55].

Citronellal and citronellol are monoterpenes employed in the synthesis of several terpenoids (1-menthol, α -tocopherol and irones). Victoria and coworkers have used a green synthesis protocol for new organoselenium compounds derived from (R)-citronellal: α-phenylseleno citronellal and α-phenylseleno citronellol [56]. The presence of a phenylselenium group in the citronella molecule increases the antimicrobial activity compared with the natural compound against foodborne pathogenic bacteria such as the Gram-positive bacteria *Listeria monocytogenes*, *S. aureus* and the Gram-negative *S. typhimurium.* 2-phenylseleno citronellal showed very interesting antibacterial activity against *Salmonella,* one of the most important causes of gastroenteritis in the world [56].

10.3.2. Pesticides

Zakrewski and coworkers have described the synthesis and the pesticidal properties of thio- and selenoureas bearing 2,2,6,6-tetramethylpiperidine-1-oxyl(nitroxyl) moiety analogs of *N,N*-dimethyl-*N*'-aryl urea herbicides [31] suggesting a new interesting application of organoselenium compound.

10.3.3. Dietary Supplement

Se as a dietary supplement is already advised in all countries with a soil poor in Se. Recently a new formula for supplementary Se has been suggested, Se-enriched probiotics. Probiotics have attracted a lot of attention for their beneficial effect on human health. Oral administration of probiotics plays an important role in the balance of intestinal micro-biota, and in the prevention and treatment of certain diarrheal diseases as it inhibits the growth of pathogenic bacteria [57]. Recently, Se-enriched probiotics against *E. coli* have been tested. Probiotics enriched with Se were obtained growing probiotics in the presence of Se in appropriate growth media to allow the conversion of inorganic Se into selenomethionine and selenocysteine. Studies *in vitro* and *in vivo* have demonstrated that metabolites of probiotics reduce the number of pathogen *E. coli* bacteria. In conclusion Se-enriched probiotics combine the virtues of probiotics with the virtues of organic Se [58].

10.3.4. Wound Dressing

Another application of organoselenium is in wound dressings. The Gram-negative *P. aeruginosa* and the Gram-positive *S. aureus* often colonize wounds as a biofilm. At the moment many wound dressings are embedded with silver or antibiotics. Unfortunately, some of these agents can spread from the dressing to the injured tissue destroying cells. The advantage of using Se is in the fact that Se can be covalently linked to the dressing thereby exerting its effect over a long time period. Cellulose disks coated with organoselenium compounds have been used to inhibit the biofilm formation of *P. aeruginosa* and *S. aureus*, showing efficient activity. An organoselenium coating on cellulose inhibits bacterial attachment and biofilm formation [59].

10.4. CONCLUSION AND OUTLOOKS

The wide-spectrum antimicrobial activity of organoselenium is of great interest for the scientific community and many possible applications have been suggested. One of the limits of almost all studies carried out on organoselenium compounds is the determination of the *in vitro* cytotoxicity that most of the time was not evaluated. Cytotoxicity value is necessary to calculate the selectivity index to estimate the therapeutic window. Anyway, the possibility of using organoselenium compounds in the treatment of drug resistant microorganisms or bacteria difficult to fight such as *Mycobacterium tuberculosis* is an attractive prospect. Another attractive prospect is the possibility to use organoselenium compounds as pesticides. Hardly any of the new Se-compounds' mechanisms of action have been investigated. Knowledge of the mechanism of antimicrobial activity could inspire scientists to improve the structure of the molecules. Moreover, the possibility to use Se-containing compounds in dressings for healing wounds could be of great advantage. Last but not least, the potential use of synthetic organic compounds containing Se to enhance the safety and the shelf life of food products and their use as pesticide is a new interesting challenge for researchers, who must however exclude any residual toxicity of the compounds in food.

In conclusion organoselenium compounds are molecules which merit further studies due to their promising potential applications in different fields, from medical science to agronomy and the food industry.

ACKNOWLEDGEMENT

Declared none.

CONFLICT OF INTEREST

The authors confirm that this chapter contents have no conflict of interest.

REFERENCES

[1] Rayman, M. P. The importance of selenium to human health. *Lancet,* **2000**, *356,* 233-241.

[2] Birringer, M.; Pilawa, S.; Flohe, L. Trends in selenium biochemistry. *Nat Prod Rep*, **2002**,*19*, 693-718.

[3] Takahashi, H.; Nishina, A.; Fukumoto, R. H.; Kimura, H.; Koketsu, M.; Ishihara, H. Selenocarbamates are effective superoxide anion scavengers *in vitro*. *Eur J Pharm Sci,* **2005**, *24,* 291-295.

[4] Kiremidjian-Schumacher, L.; Roy, M.; Wishe, H. I.; Cohen, M. W.; Stotzky, G. Supplementation with selenium and human immune cell functions. II. Effect on cytotoxic lymphocytes and natural killer cells. *Biol Trace Elem Res,* **1994,** *41,* 115-127.

[5] Roy, M.; Kiremidjian-Schumacher, L.; Wishe, H. I.; Cohen, M. W.; Stotzky, G. Supplementation with selenium and human immune cell functions. I. Effect on lymphocyte proliferation and interleukin 2 receptor expression. *Biol Trace Elem Res,* **1994**, *41,* 103-114.

[6] Whanger, P. D. Selenium and its relationship to cancer: an update. *Br J Nutr*, **2004**, *91*, 11-28.

[7] Navarro-Alarcon, M.; Lopez-Martinez, M. C. Essentiality of selenium in the human body: relationship with different diseases. *Sci Total Environ,* **2000**, *249*, 347-371.

[8] Kaushal, N.; Bansal, M. P. Dietary selenium variation-induced oxidative stress modulates CDC2/cyclin B1 expression and apoptosis of germ cells in mice testis. *J Nutr Biochem,* **2007**, *18*, 553-564.

[9] Kaushal, N. Bansal, M. P. Inhibition of CDC2/Cyclin B1 in response to selenium-induced oxidative stress during spermatogenesis: potential role of Cdc25c and p21. *Mol Cell Biochem*, **2007**, *298*, 139-150.

[10] Brown, K. M.; Pickard, K.; Nicol, F.; Beckett, G. J.; Duthie, G. G.; Arthur, J. R. Effects of organic and inorganic selenium supplementation on selenoenzyme activity in blood lymphocytes, granulocytes, platelets and erythrocytes. *Clin Sci (Lond)*, **2000**, *98*, 593-599.

[11] Kim, H. Y.; Gladyshev, V. N. Different catalytic mechanisms in mammalian selenocysteine- and cysteine-containing methionine-R-sulfoxide reductases. *PLoS Biol,* **2005**, *3*: e375.

[12] Moro, A. V.; Nogueira, C. W.; Barbosa, N. B.; Menezes, P. H.; da Rocha, J. B.; Zeni, G. Highly stereoselective one-pot procedure to prepare bis- and tris-chalcogenide alkenes *via* addition of disulfides and diselenides to terminal alkynes. *J Org Chem*, **2005**, *70,* 5257-5268.

[13] Nogueira, C. W.; Zeni, G.; Rocha, J. B. Organoselenium and organotellurium compounds: toxicology and pharmacology. *Chem Rev*, **2004**, *104*, 6255-6285.

[14] Brown, E. D.; Wright, G. D. New targets and screening approaches in antimicrobial drug discovery. *Chem Rev*, **2005**, *105*, 759-774.

[15] Nozawa, R.; Yokota, T.; Fujimoto, T. Susceptibility of methicillin-resistant *Staphylococcus aureus* to the selenium-containing compound 2-phenyl-1,2-benzoisoselenazol-3(2H)-one (PZ51). *Antimicrob Agents Chemother*, **1989**, *33,* 1388-1390.

[16] Billack, B.; Santoro, M.; Lau-Cam, C. Growth inhibitory action of ebselen on fluconazole-resistant *Candida albicans*: role of the plasma membrane H+-ATPase. *Microb Drug Resist,* **2009**, *15*, 77-83.

[17] Chan, G.; Hardej, D.; Santoro, M.; Lau-Cam, C.; Billack, B. Evaluation of the antimicrobial activity of ebselen: role of the yeast plasma membrane H+-ATPase. *J Biochem Mol Toxicol*, **2007**, *21*, 252-264.

[18] Wojtowicz, H.; Kloc, K.; Maliszewska, I.; Mlochowski, J.; Pietka, M.; Piasecki, E. Azaanalogues of ebselen as antimicrobial and antiviral agents: synthesis and properties. *Farmaco*, **2004**, *59*, 863-868.

[19] Bien, M.; Blaszczyk, B.; Kalinowska, K.; Mlochowski, J.; Inglot, A. D. Antifungal activity of 2-(4-chlorophenyl)-1,2-benzisoselenazol-3(2H)-one, the analog of Ebselen. *Arch Immunol Ther Exp (Warsz),* **1999**, *47*, 185-193.

[20] Pietka-Ottlik, M.; Wojtowicz-Mlochowska, H.; Kolodziejczyk, K.; Piasecki, E.; Mlochowski, J. New organoselenium compounds active against pathogenic bacteria, fungi and viruses. *Chem Pharm Bull (Tokyo),* **2008**, *56*, 1423-1427.

[21] Musiol, R.; Jampilek, J.; Buchta, V.; Silva, L.; Niedbala, H.; Podeszwa, B.; Palka, A.; Majerz-Maniecka, K.; Oleksyn, B.; Polanski, J. Antifungal properties of new series of quinoline derivatives. *Bioorg Med Chem,* **2006**, *14,* 3592-3598.

[22] Patel, R. V.; Kumari, P.; Chikhalia, K. H. New Quinolinyl-1,3,4-Oxadiazoles: Synthesis, *In Vitro* Antibacterial, Antifungal And Antituberculosis Studies. *Med Chem.,* **2012**, Aug 17. [Epub ahead of print].

[23] Dardari, Z.; Lemrani, M.; Bahloul, A.; Sebban, A.; Hassar, M.; Kitane, S.; Berrada, M.; Boudouma, M. Antileishmanial activity of a new 8-hydroxyquinoline derivative designed 7-[5'-(3'-phenylisoxazolino)methyl]-8-hydroxyquinoline: preliminary study. *Farmaco,* **2004**, *59*, 195-199.

[24] Abdel-Hafez, S. H.; Hussein, M. A. Selenium-containing heterocycles: synthesis and pharmacological activities of some new 4-methylquinoline-2(1H) selenone derivatives. *Arch Pharm (Weinheim)*, **2008**, *341*, 240-246.

[25] Naik, H. R.; Naik, H. S.; Naik, T. R.; Naika, H. R.; Gouthamchandra, K.; Mahmood, R.; Ahamed, B. M. Synthesis of novel benzo[h]quinolines: wound healing, antibacterial, DNA binding and *in vitro* antioxidant activity. *Eur J Med Chem*, **2009**, *44*, 981-989.

[26] [Xu, X. J.; Xue, Z.; Xiao, Q.; Hou, A. X.; Liu, Y. Antibacterial activities of novel diselenide-bridged bis(porphyrin)s on *Staphylococcus aureus* investigated by microcalorimetry. *Biol Trace Elem Res*, **2008**, *125*, 185-192.

[27] Plano, D.; Baquedano, Y.; Moreno-Mateos, D.; Font, M.; Jimenez-Ruiz, A.; Palop, J.A.; Sanmartín, C. Selenocyanates and diselenides: a new class of potent antileishmanial agents. *Eur J Med Chem*, **2011**, *46*, 3315-3323.

[28] Li, X.; Liu, Y.; Zhao, R.; Wu, J.; Shen, X.; Qu, S. Microcalorimetric study of *Escherichia coli* growth inhibited by the selenomorpholine complexes. *Biol Trace Elem Res,* **2000**, *75*, 167-175.

[29] Xi, L.; Yi, L.; Jun, W.; Songsheng, Q. The effect of the selenomorpholine derivatives on the growth of *Staphylococcus aureus* studied by microcalorimetry. *Thermochimica Acta*, **2001**, *375*, 109-113.

[30] Xi, L.; Yi, L.; Jun, W.; Songsheng, Q. Action of the Selenomorpholine Compounds on the bacterium growth by microcalorimetry. *Chinese Journal of Chemistry,* **2002**, *20*, 829-833.

[31] Zakrzewski, J.; Krawczyk, M. Reactions of nitroxides. Part X: antifungal activity of selected sulfur and selenium derivatives of 2,2,6,6-tetramethylpiperidine. *Bioorg Med Chem Lett,* **2011**, *21*, 514-516.

[32] Abdel-Hafez, Sh. H.; Selenium containing heterocycles: synthesis, anti-inflammatory, analgesic and anti-microbial activities of some new 4-cyanopyridazine-3(2H)selenone derivatives. *Eur J Med Chem*, **2008**, *43*, 1971-1977.

[33] Deidda, D.; Lampis, G.; Maullu, C.; Pompei, R.; Isaia, F.; Lippolis, V.; Verani, G. Antifungal, antibacterial, antiviral and cytotoxic activity of novel thio- and seleno-azoles. *Pharmacol Res*, **1997**, *36*, 193-197.

[34] Chitra, S.; Paul, N.; Muthusubramanian, S.; Manisankar, P.; Yogeeswari, P.; Sriram, D. A facile synthesis of carbocycle-fused mono and bis-1,2,3-selenadiazoles and their antimicrobial and antimycobacterial studies. *Eur J Med Chem*, **2011**, *46*, 5465-5472.

[35] Karnik, A. V.; Kulkarni, A. M.; Malviya, N. J.; Mourya, B. R.;, Jadhav, B. L. Synthesis and *in vitro* anti-bacterial evaluation of tetracyclic-ortho-fused 4H-naphtho[1',2'-5,6]pyrano[3,4-d](1,2,3)selenadiazole and its derivatives. *Eur J Med Chem,* **2008**, *43*, 2615-2617.

[36] Scovill, J. P.; Klayman, D. L.; Franchino, C. F. 2-Acetylpyridine thiosemicarbazones. 4. Complexes with transition metals as antimalarial and antileukemic agents. *J Med Chem*, **1982**, *25*, 1261-1264.

[37] Molter, A.; Rust, J.; Lehmann, C. W.; Deepa, G.; Chiba, P.; Mohr, F. Synthesis, structures and anti-malaria activity of some gold(I) phosphine complexes containing seleno- and thiosemicarbazonato ligands. *Dalton Trans*, **2011**, *40*, 9810-9820.

[38] Al-Mudhaffar, D. M. H.; Hasan, A. A. A. Synthesis, characterization, and biological activity of some new amino acid derivatives containing selenium. *International Journal of Applied Science and Technology*, **2010**, *2*, 83-89.

[39] Caputo, R.; Dellagreca, M.; de Paola, I.; Mastroianni, D.; Longobardo, L. Novel sulfur and selenium containing bis-alpha-amino acids from 4-hydroxyproline. *Amino Acids*, **2010**, *38*, 305-310.

[40] Radhakrishna, P. M.; Sharadamma, K. C.; Vagdevi, H. M.; Abhilekha, P. M.; Rubeena Mubeen, S.; Nischal, K. Synthesis and antibacterial activity of novel organoselenium compounds. *International Journal of Chemistry*, **2010**, *2*, 149-154.

[41] Jun, E. J.; Ye, J. S.; Hwang, I. S.; Kim, Y. K.; Lee, H. Selenium deficiency contributes to the chronic myocarditis in coxsackievirus-infected mice. *Acta Virol*, **2011**, *55,* 23-29.

[42] Levander, O. A.; Beck, M. A. Interacting nutritional and infectious etiologies of Keshan disease. Insights from coxsackie virus B-induced myocarditis in mice deficient in selenium or vitamin E. *Biol Trace Elem Res*, **1997**, *56*, 5-21.

[43] Allard, J. P.; Aghdassi, E.; Chau, J.; Tam, C.; Kovacs, C. M.; Salit, I. E.; Walmsley, S. L. Effects of vitamin E and C supplementation on oxidative stress and viral load in HIV-infected subjects. *AIDS*, **1998**, *12*, 1653-1659.

[44] Jeong, L. S.; Choi, Y. N.; Tosh, D. K.; Choi, W. J.; Kim, H. O.; Choi, J. Design and synthesis of novel 2',3'-dideoxy-4'-selenonucleosides as potential antiviral agents. *Bioorg Med Chem*, **2008**, *16*, 9891-9897.

[45] Dienstag, J. L.; Perrillo, R. P,; Schiff, E. R.; Bartholomew, M.; Vicary, C.;, Rubin, M. A preliminary trial of lamivudine for chronic hepatitis B infection. *N Engl J Med*, **1995**, *333,* 1657-1661.

[46] Schinazi, R. F.; Boudinot, F. D.; Ibrahim, S. S.; Manning, C.; McClure, H. M.; Liotta, D. C. Pharmacokinetics and metabolism of racemic 2',3'-dideoxy-5-fluoro-3'-thiacytidine in rhesus monkeys. *Antimicrob Agents Chemother*, **1992**, *36*, 2432-2438.

[47] Schinazi, R. F.; Mead, J. R.; Feorino, P. M. Insights into HIV chemotherapy. *AIDS Res Hum Retroviruses*, **1992**, *8,* 963-990.

[48] Furman, P. A.; Davis, M.; Liotta, D. C.; Paff, M.; Frick, L. W.; Nelson, D. J.; Dornsife, R. E.; Wurster, J. A.; Wilson, L. J.;, Fyfe, J. A. The anti-hepatitis B virus activities, cytotoxicities, and anabolic profiles of the (-) and (+) enantiomers of *cis*-5-fluoro-1-[2-(hydroxymethyl)-1,3-oxathiolan-5-yl]cytosine. *Antimicrob Agents Chemother*, **1992**, *36*, 2686-2692.

[49] Grove, K. L.; Guo, X.; Liu, S. H.; Gao, Z.; Chu, C. K.; Cheng, Y. C. Anticancer activity of beta-L-dioxolane-cytidine, a novel nucleoside analogue with the unnatural L configuration. *Cancer Res*, **1995**, *55*, 3008-3011.

[50] Schinazi, R. F.; Lloyd, R. M. Jr.; Ramanathan, C. S.; Taylor, E. W. Antiviral drug resistance mutations in human immunodeficiency virus type 1 reverse transcriptase occur in specific RNA structural regions. *Antimicrob Agents Chemother*, **1994**, *38*, 268-274.

[51] Mansour, T. S.; Jin, H.; Wang, W.; Hooker, E. U.; Ashman, C.; Cammack, N.; Salomon, H.; Belmonte, A. R.; Wainberg, M. A. Anti-human immunodeficiency virus and anti-hepatitis-B virus activities and toxicities of the enantiomers of 2'-deoxy-3'-oxa-4'-thiocytidine and their 5-fluoro analogues *in vitro*. *J Med Chem*, **1995**, *38*, 1-4.

[52] Chu, C. K.; Ma, L.; Olgen, S.; Pierra, C.; Du, J.; Gumina, G.; Gullen, E.; Cheng, Y. C.; Schinazi, R. F. Synthesis and antiviral activity of oxaselenolane nucleosides. *J Med Chem*, **2000**, *43*, 3906-3912.

[53] Alberto, E. E.; Rossato, L. L.; Alves, S. H.; Alves, D.; Braga, A. L. Imidazolium ionic liquids containing selenium: synthesis and antimicrobial activity. *Org Biomol Chem*, **2011**, *9*, 1001-1003.

[54] Rahaman, A.; Kang, S. C. *In vitro* control of food-borne and food spoilage bacteria by essential oil and ethanol extracts of *Lonicera japonica Thunb*. *Food Chemistry,* **2009**, *116*, 670-675.

[55] Ebrahimabadi, A. H.; Mazoochi, A.; Kashi, F. J.; Djafari-Bidgoli, Z.; Batooli, H. Essential oil composition and antioxidant and antimicrobial properties of the aerial parts of *Salvia eremophila Boiss*. from Iran. *Food Chem Toxicol*, **2010**, *48*, 1371-1376.

[56] Victoria, F. N.; Radatz, C. S.; Sachini, M.; Jacob, R. G.; Alves, D.; Savegnago, L.; Perin, G.; Motta, A. S.; Silva, W. P. ; Lenardão, E. J. Further analysis of the antimicrobial activity of alpha-phenylseleno citronellal and alpha-phenylseleno citronellol. *Food Control*, **2012**, *23*, 95-99.

[57] Reid, G.; Burton, J. Use of Lactobacillus to prevent infection by pathogenic bacteria. *Microbes Infect*, **2002**, *4*, 319-324.

[58] Yang, J.; Huang, K.; Qin, S.;, Wu, X.; Zhao, Z.; Chen, F. Antibacterial action of selenium-enriched probiotics against pathogenic *Escherichia coli. Dig Dis Sci,* **2009**, *54,* 246-254.

[59] Tran, P. L.; Hammond, A. A.; Mosley, T.; Cortez, J.; Gray, T.; Colmer-Hamood, J. A.; Shashtri, M.; Spallholz, J. E.; Hamood, A. N.; Reid, T. W. Organoselenium coating on cellulose inhibits the formation of biofilms by *Pseudomonas aeruginosa* and *Staphylococcus aureus*. *Appl Environ Microbiol*, **2009**, *75,* 3586-3592.

Send Orders for Reprints to reprints@benthamscience.net
Organoselenium Chemistry Between Synthesis and Biochemistry, 2014, 345-360

CHAPTER 11

Selenium and "Bio-Logic" Catalysis: New Bioinspired Catalytic Reactions

Caterina Tidei and Claudio Santi*

Department of Pharmaceutical Sciences, University of Perugia, Via del Liceo 1-06123 Perugia, Italy

Abstract: The discovery in the early sixties of the fundamental role that selenium plays in many living systems has led to investigate the chemical mechanisms of action of the most important selenoenzymes. These mechanisms together with the possibility to develop biomimetic molecules to be used as useful catalysts in green synthetic protocols will be explored in this chapter.

Keywords: ROS, Antioxidant, Selenoenzymes, Oxidative stress, Glutathione Peroxidase, Prooxidants, Oxidation, Selenolate, Selenyl radical, EPR spectroscopy.

11.1. INTRODUCTION

The expression "biomimetic chemistry" was introduced for the first time by Breslow in 1972 to indicate the design of organic molecules able to imitate the reactivity of biological systems [1]. The purpose of this field of chemistry is on the one hand, to better understand and elucidate the physiological mechanisms, on the other, to apply these latter to the organic synthesis. Nature gives to the chemists appealing hints in order to realize high selective, fast and clean reactions. These results have been successfully achieved through actual biocatalytic approaches, however the use of enzymes is connected with a number of limitations such as limited stability and high costs of the biological matrices often associated to a reduced versatility of the method due to the high substrate specifities. Therefore, the design of artificial enzymes has represented in the last few years an interesting alternative, paving the way to the field of "biomimetic

***Address correspondence to Claudio Santi:** Department of Pharmaceutical Sciences, University of Perugia, *Via* del Liceo 1-06123 Perugia, Italy; Tel: +39 075-5855102; Fax: +39 075-5855116; E-mails:santi@unipg.it; tideicaterina@gmail.com.

catalysis". Artificial enzymes have been developed with the aim of reproducing the fundamental enzymatic features such as the high enzyme-substrate binding affinity and the high catalytic turnover, which converge into the significant rate acceleration compared to the uncatalyzed process. The first mimics were mainly constituted by large-size molecules, generally organic macrocycles such as cyclodextrins [2,3], cucurbiturils [4] or polymers like polyethyleneimine [5,6], which are able to generate a supramolecular environment favourable for the catalysis. However, this topic has been well discussed in a number of excellent reviews and books [2,7,8] and will be not reported in details on this chapter. More recently, the use of small organic molecules as bioinspired catalysts has captured the attention of several research groups, creating a bridge between biomimetic catalysis and organocatalysis.

The last decade developments in the fields of selenium chemistry have been achieved both in the context of supramolecular chemistry and design of artificial enzymes, and respect to the synthesis of small organoselenium compounds which are able to reproduce the activity of the most important selenoproteins, such as thioredoxin reductase (TrxR), iodothyronine deiodinase (ID) and glutathione peroxidase (GPx). The recent works on Gpx mimics will be treated in the last part of this chapter and will represent the connection with the new concept of "bio-logic" catalysis.

11.2. ARTIFICIAL SELENOENZYMES

The term "artificial enzyme" can be referred to a wide number of chemical identities, synthetic and semi-synthetic, but all designed with the aim of combining the substrate binding ability and the catalytic properties of the natural enzyme. These fundamental features can be acquired by reproducing the active site, which is generally small and constituted by few amino acids or in some cases by a single functional group, and the surrounding proteic backbone. In order to mimic the GPx-like activity, the artificial enzyme needs to contain a seleno or telluro core, able to imitate the residue of selenocysteine (Sec) that is responsible for the reduction of peroxides in Gpx [9]; the scaffold can be variable, but usually generates a hydrophobic microenvironment analogous to that in which enzymatic reactions take place. The first models of artificial selenoenzymes were the semisynthetic ones, in which scaffolds were based on naturally occurring proteins and obtained by chemical or genetic modifications. The pioneering work was the design of selenosubtilisin by Wu and Hilvert [10], followed by the preparation of selenotrypsin and tellurosubtilisin by Liu and coworkers [11,12]. In these cases, the scarce affinity for glutathione (GSH) was detected as the major problem and it was overcome few years later by Luo and coworkers, with the conversion of glutathione transferase (GST) into the seleno counterpart [13]. This latter was also obtained by the same research group through a genetic engineering approach, the auxotrophic expression system that was already used by Bock and coworkers to insert Sec into thioredoxin [14]. Also antibodies were employed as proteic scaffolds, incorporating Sec into monoclonal antibodies generated using hydrophobically modified GSH as haptens [15].

Another efficient strategy to simulate the activity of GPx is represented by the design of non proteic scaffolds, such as macrocyclic compounds. Between them, cyclodextrins (CDs) have been widely used because of their favourable features: they are readily available; they can bind a number of different guests through their hydrophobic cavity and possess two rims of hydroxyl groups, responsible for their water solubility and that can be variously modified using well-known procedures. Liu and coworkers were the first to attach a diselenide group to the primary [16] and then to the secondary face [17] of a β-CD. This resulted in a significant increase of the antioxidant activity respect to that of ebselen, the most successful small-size Gpx mimic. On this basis, the next step was combining the cyclodextrin system with a series of analogous of ebselen, all bearing a cyclic selenamidic core, bonded with β-CD through different linkers, and owning a good water solubility [18]. Also some examples of telluro-cyclodextrin models have been reported [19].

High catalytic activities have been achieved also with other kinds of synthetic host systems, such as dendrimers and hyperbranched polymers [20,21].

The influence of the rapid and considerable developments of recent years in the fields of nanotechnology and supramolecular chemistry has resulted in the preparation firstly of elemental micellar and nanosized vesicles with a seleno or telluro catalytic center [22,23], then of more complex and smarter systems such as temperature responsive microgel, whose GPx activity can be easily modulated [24], and nanotubes formed through self-assembly of CD-based supramolecular host-guest complexes in aqueous solution [25]; this latter example has also required the application of molecular imprinting strategies for the functionalization of the surface with selenium/tellurium catalytic groups, representing one of the most recent and interesting works in the design of artificial selenoenzymes.

11.3. BIOINSPIRED SELENOCATALYSTS

The characteristics of complexity, stability and water solubility of artificial selenoenzymes allow the achievement of high catalytic activities and the possibility of clinical applications. However, these features are not at all required in the optics of their use in organic synthesis. In this case, simpler structures, which can reproduce the non-bonding interactions of the catalytic sites of enzymes, should be more convenient. On this basis, several research groups are currently involved in the synthesis of small organoselenium compounds able to mimic the action of the most important selenoenzymes. In this chapter, we will explore the possibility to mimic natural reactions through small organic molecules in order to develop new eco-friendly protocols suggesting this way as a convenient strategy for green chemistry.

11.3.1. ID Mimics

Iodothyronine deiodinases (IDs) represent one of the most important class of selenoenzymes and are involved in the activation-inactivation pathway of thyroid hormones. In particular, 3 types of this enzyme (ID-1, ID-2, ID-3) mediate a series of outer- and inner-ring deiodinations of prohormone thyroxine (T4): ID-1 and ID-2 catalyze the outer-ring deiodination of T4 converting it into the active form 3,5,3'-triiodothyronine (T3); this latter then can be inactivated by ID-3 *via* inner ring deiodination to produce 3-3'-diiodothyronine (T2). ID-3 also promotes the inner ring deiodination of T4, producing 3,3',5'-triiodothyronine (reverse T3 or rT3), that can be converted by ID-1 and ID-2, through a secondary outer-ring deiodination, into T2 (Scheme **11.1**) [26].

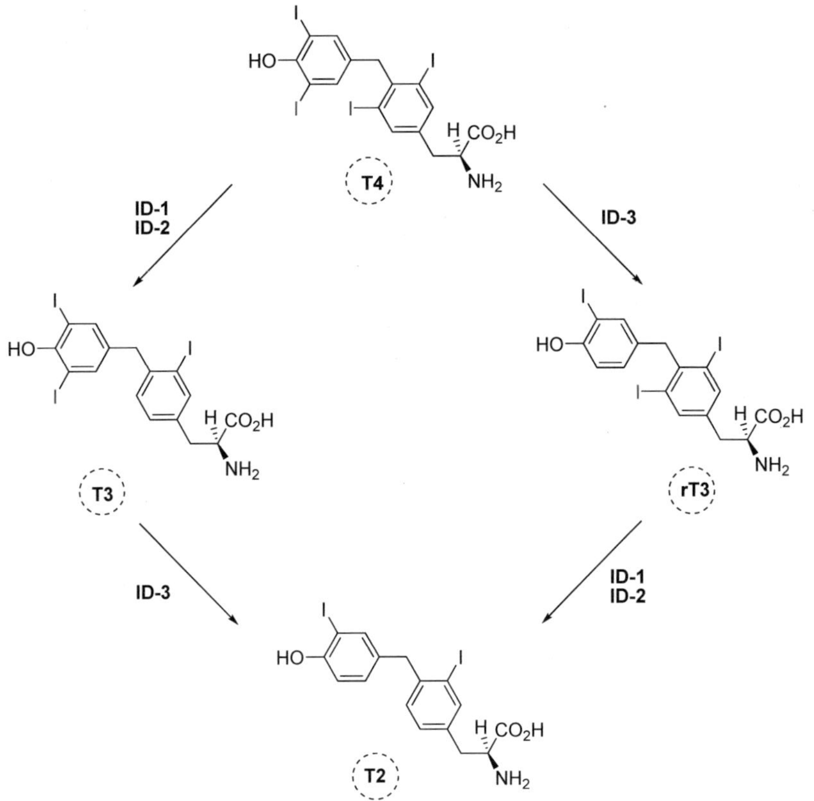

Scheme 11.1

In this complex framework, several attempts to realize simple organoselenium mimics have been made. Different kinds of selenols, starting from the simpler PhSeH, have shown to be efficient in the removal of iodine in thyroxine derivatives [27]. In the first chemical model for the deiodination of T4, Mugesh and Manna [28] demonstrated that the naphthyl-based selenol **1**, bearing also a thiol group closer to selenium atom, is able to catalyze selective inner-ring deiodinations, converting T4 in rT3 and T3 in T2, which is the main physiological inactivation pathway. The replacement of selenol group of **1** with a thiol (**2**) produces a significant decrease of the ID-3 like activity, in accordance with the critical role of Sec in the catalytic site of the enzyme [29], while the substitution with a phenyl selenenyl group causes the total loss of the activity. Moreover, the presence of two vicinal selenol group (**3**) determines a high enhancement [30], indicating the importance of selenol moiety.

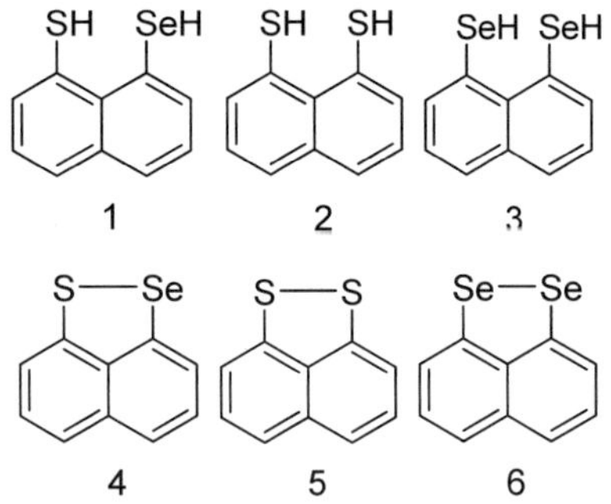

Figure 11.1: Naphthyl derivatives with ID-like activity.

This latter has resulted to be activated as selenolate in the native enzyme by one or more histidine residues [31]. However, the same authors demonstrated that the presence of a basic functionality not always determines an increase of deiodinase activity. The introduction of *N*-methylamino group in *ortho* to selenol in compound **3** and to thiol in **2** produces an increase of activity because it strengths the chalcogen bond and, as consequence, also the halogen bond, leading to the cleavage of the C-I bonds in T4 and T3 (Fig. **11.2**). On the contrary, in the presence of the thiol-selenol pair **1**, the secondary amino group may determine a steric hindrance which disfavours the interaction of selenol with the substrate; in this case, the thiol moiety should interact with the iodine causing the reduction of deiodinase activity (Fig. **11.1**). The mechanism proposed was also confirmed by the recovery of dichalcogenides **4**, **5** and **6** when the reaction of T4 deiodination was performed in the absence of dithithreitol (DTT) as cofactor [32].

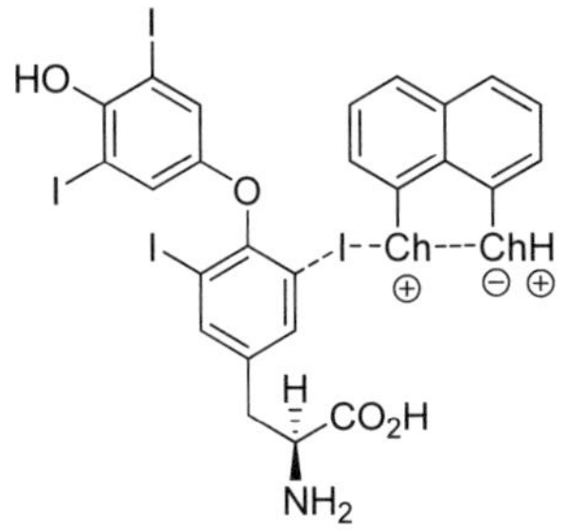

Figure 11.2: Mechanism of naphthyl derivatives deiodination.

In the last few years, the mechanisms of selenium-mediated deiodination have been largely studied, however the number of ID mimics is still limited and they have still not found application in synthetic chemistry as well as in therapy.

11.3.2. GPx Mimics and "Bio-Logic" Approach

Glutathione Peroxidases are probably the most studied selenoproteines from a structural as well as mechanistical point of view. Their physiological role is to protect the living systems against the oxidative stress produced by the peroxides. They exploit their functions by the reduction of peroxides into the more safe molecules of water or alcohols and this reduction is associated to a simultaneous oxidation of a thiol, generally a molecule of glutathione (Scheme **11.2**).

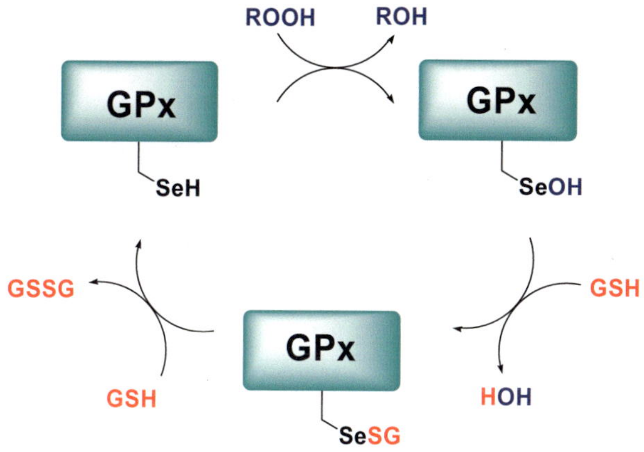

Scheme 11.2

From a chemical point of view the catalytic cycle depicted in the Scheme **11.2** represent an oxidation reaction in which the selenium is the catalyst, the hydrogen peroxide the stoichiometric oxidant and the glutathione the substrate. Based on this consideration, the oxygen transfer reaction played by the enzyme was attempted using small organoselenium Gpx mimics. Some of these molecules were designed in order to reproduce the condition of the selenium in the catalytic triade of the enzyme [33]. Starting from Ebselen (**7**), that has been discussed in detail in Chapter 12 as well as in a number of interesting reviews [34], arriving to PhSeZnCl (**8**) recently synthesized by us [35]. This latter represents probably the molecule in which the selenium an oxidation state closer to that present on the active site of the enzyme (Fig. **11.3**).

The catalytic activity of the zinc selenate **8** in the oxidation of endogenous and non-endogenous thiols has been demonstrated using as stoichiometric oxidant oxygen (Scheme **11.3**) or hydrogen peroxide (Scheme **11.4**).

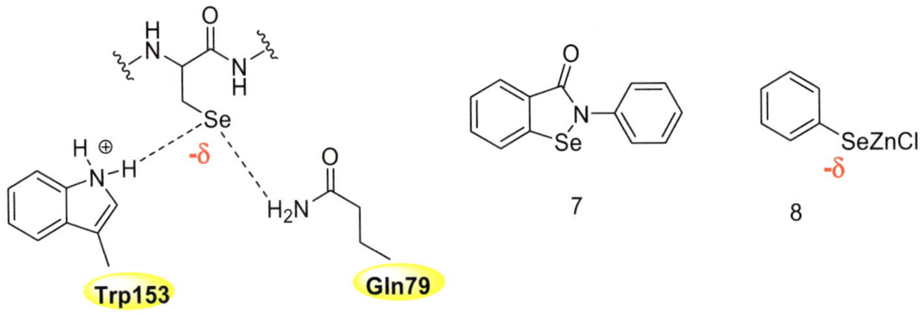

Figure 11.3: The catalytic triade of Gpx; Ebselen; PhSeZnCl.

In particular, it was proven that in 24 hours glutathione (GSH) is quantitatively oxidized by the air in the presence of a stoichiometric amount of **8** and that the reaction is considerably faster when hydrogen peroxide is the oxidant (Scheme **11.4**).

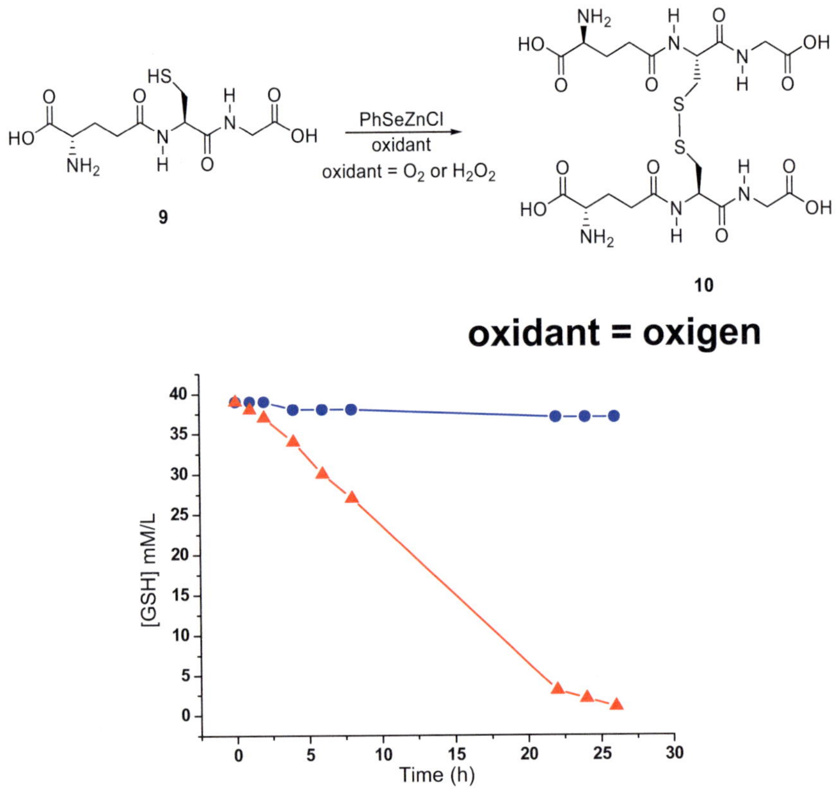

without catalyst (blue line) with PhSeZnCl 100 mol% (red line)

Scheme 11.3

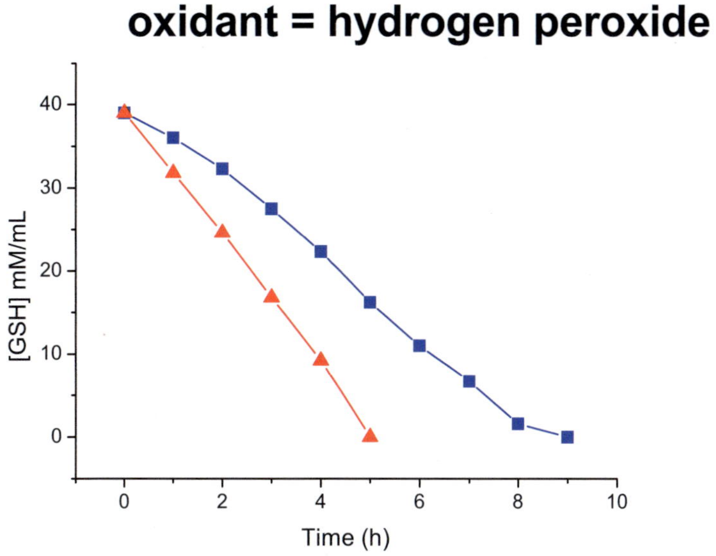

without catalyst (blue line) with PhSeZnCl 1mol% (red line)

Scheme 11.4

In NMR experiments affected using dithiotreitol as co-factor PhSeZnCl showed a Gpx-like activity higher than that of diphenyl diselenide that was known to be a good Gpx mimic (Scheme **11.5**).

This was explained by the fact that **8** acts not simply as precursor of the selenenic acid intermediate but the zinc chloride moiety is a good Lewis acid and can activate the catalytic cycle interacting with the selenium-sulfur intermediate as depicted in Scheme **11.5** [36].

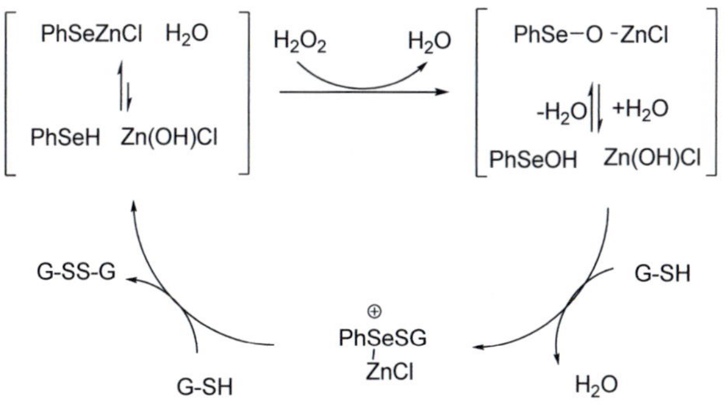

Scheme 11.5

According to Mugesh [37] removing the thiol from a Gpx biomimetic system the selenium atom, in the presence of an excess of hydrogen peroxide, undergoes overoxidation affording a perselenenic derivative and it was reported that perselenenic acid could be efficiently used for the epoxidation of carbon carbon double bonds [38]. On the base of this consideration selenols or diselenides (as precatalysts) can be used as precursors of the actual catalyst, a perselenenic acid providing a potential useful tool for the realization of a number of oxidative processes connected to the higher oxidant power of the perselenenic acid respect the selenenic and the seleninic one. In addition, this process should be affected in water producing as side products only molecules of water increasing the greenest of the oxidation reaction (Scheme **11.6**).

Because this approach is basically inspired by the logic of the mechanism of the glutathione peroxidase and it is not limited to reproduce this mechanism, in this case we don't refer to a biomimetic approach but to a "Bio-Logic" one.

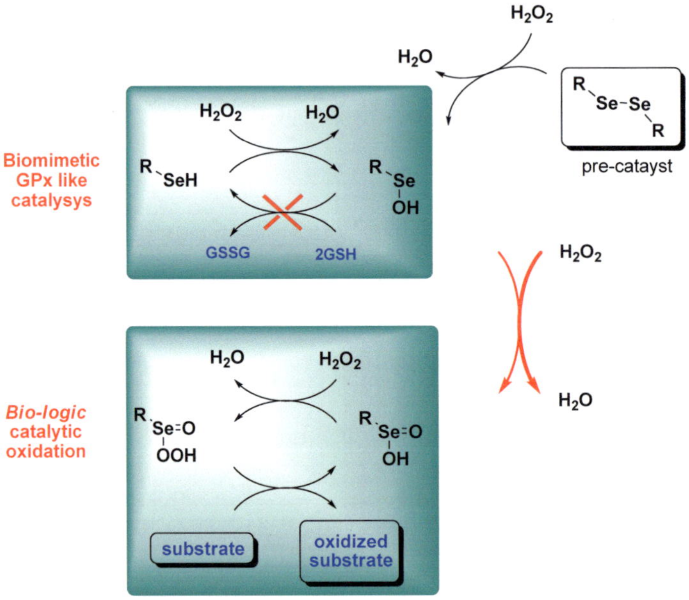

Scheme 11.6

This catalytic system has been applied for the first time to the oxidation of carbon-carbon double bonds using selenocystine (L-Sec)$_2$ as catalyst, H_2O_2 as oxidant in water at room temperature. In these conditions it was not possible to isolate the epoxide intermediate **12** [38] and the reaction afforded directly the stereospecific formation of *anti*-diols **13a-d** [39].

Using selenocysteine to catalyse the dihydroxilation of substrate containing endocyclic double bonds, like methyl cyclohexene **11b** and methyl cyclopentene **11c**, the corresponding diols were obtained in good enantiomeric excess. Unfortunately, for linear olefins as well as when the double bond have a phenyl group as substituent the selectivity strongly decreased (Table **11.1**).

Table 11.1: Selenocystine-promoted dihydroxylation of olefins

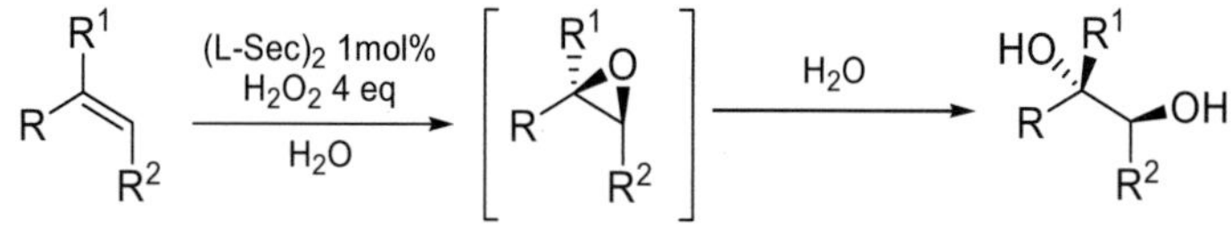

11 **12** **13**

Olefin	Product	Time (h)	Yield %	ee%
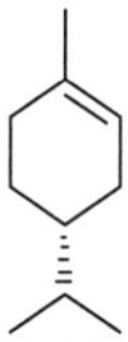	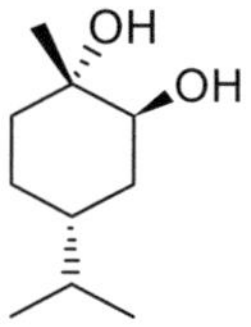	168	80	>99
11a	(+) *anti*-**13a**			
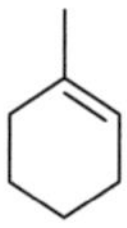	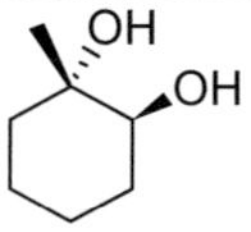	48	80	87
11b	(+) *anti*-**13b**			
	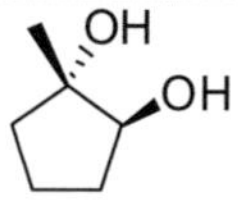	48	70	62
11c	(+) *anti*-**13c**			
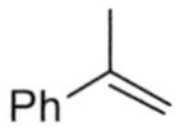	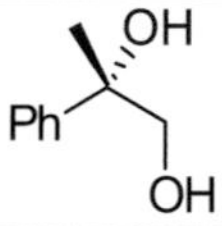	240	68	20
11d	**(+) 13d**			

Interestingly the catalyst is water soluble; after the reaction and the extraction of the organics with ethyl acetate the aqueous medium could be reused for at least five cycles without loss of yield and stereoselectivity.

This aspect strongly enhances the eco-sustainability of the process combining the use of a friendly reaction solvent with a reduced production of wastes.

With a similar procedure α-methoxy alcohols could be obtained in good yields replacing the water with methanol (Table **11.2**). The stereoselectivity are in this case significantly reduced respect to the dihydroxylation process and this was explained by the occurring in methanol of a competitive non-catalyzed side reaction which affords the racemic epoxide. This latter, only in the presence of the catalyst, is subjected to a regio- and stereospecific alcoholysis according to the Furst-Plattner rule (Scheme **11.7**).

Table 11.2: Selenocystine-promoted hydroxymethoxylation of olefins

Olefin	Product	Time (h)	Yield %	ee%
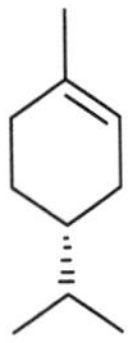	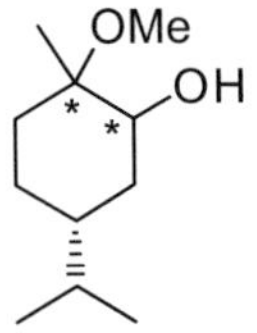	168	70	-
11a	*(1S,2S,4R)*-**14a** *(1R,2R,4R)*-**15a**		(60/40)	
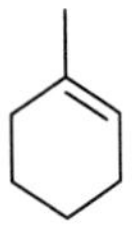	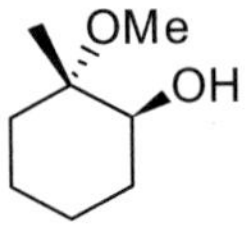	48	93	37
11b	(+)-*anti*-**14b**			
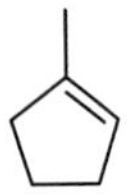	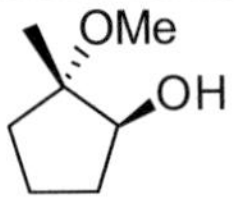	48	80	38
11c	(-)-*anti*-**14c**			
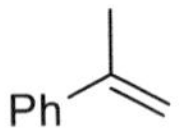	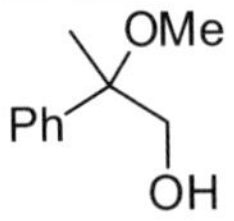	168	80	0
11d	**14e**			

11.4. CONCLUSIONS AND OUTLOOKS

In conclusion, we have provided a general view about the advantages and applicabilities of biomimetics in synthesis. In particular, we have focused on the possibility to overcome the problems connected with the classical methodologies and to achieve a broader versatility. This can be obtain with the "bio-logic" approach, applied for the first time in the context of selenium chemistry, in particular regarding the field of Gpx-mimics. We hope that it can represent a stimulating example for the realization of green processes using Nature as the first source of inspiration.

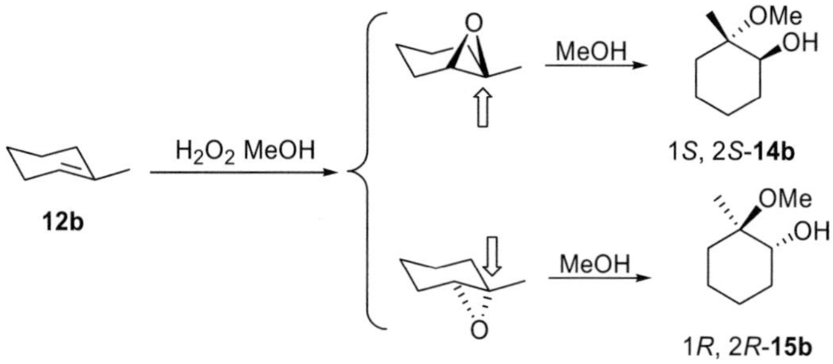

Scheme 11.7

ACKNOWLEDGEMENT

Declared none.

CONFLICT OF INTEREST

The authors confirm that this chapter contents have no conflict of interest.

REFERENCES

[1] Breslow, R. Biomimetic chemistry. *Chem. Soc. Rev.* **1972**, *1*, 553-580.
[2] Breslow, R.; Dong, S. D. Biomimetic reactions catalyzed by cyclodextrins and their derivatives. *Chem. Rev.* **1998**, *98*, 1997-2011.
[3] Bresolw, R. Biomimetic chemistry and artificial enzymes: catalysis by design. *Acc. Chem. Res.* **1995**, *28*, 146-153.
[4] Klock, C.; Dsouza, R. N.; Nau, W. M. Cucurbituril-mediated supramolecular acid catalysis. *Org. Lett.* **2009**, *11*, 2595-2598.
[5] Liu, L.; Rozenman, M.; Breslow, R. Hydrophobic effects on rates and substrate selectivities in polymeric transaminase mimics. *J. Am. Chem. Soc.* **2002**, *124*, 12660-12661.
[6] Klotz, I. M.; Sush, J. Evolution of synthetic polymers with enzyme-like catalytic activities. In: *Artificial enzymes*; Breslow, R. Ed.; Wiley-VCH: Weinheim, Germany, **2005**; pp. 63-88.
[7] Breslow, R. *Artificial enzymes*; Wiley-VCH: Weinheim, Germany, **2005.**

[8] Marchetti, L.; Levine, M. Biomimetic catalysis. *ACS Catal.* **2011**, *1*, 1090-1118.

[9] Flohé, L.; Günzler, W. A.; Schock, H. H. Glutathione peroxidase: a selenoenzyme. *FEBS Lett.,* **1973**, *32*, 132-134.

[10] Wu, Z.; Hilvert, D. Selenosubtilisin as a glutathione peroxidase mimic. *J. Am. Chem. Soc.*, **1990**, *112*, 5647-5648.

[11] Liu, J. Q.; Jiang, M. S,.; Luo, G. M.; Yan, G. L.; Shen, J. C. Conversion of trypsin into a seleniumcontaining enzyme by using chimical mutation. *Biotechnol. Lett.*, **1998**, *20*,693-696.

[12] Mao, S. Z.; Dong, Z. Y.; Liu, J. Q.; Li, X. Q.; Liu, X. M.; Luo, G. M.; Shen, J. C. Semisynthetic tellurosubtilisin with glutathione peroxidase activity. *J. Am. Chem. Soc.,* **2005**, *127*, 11588-11589.

[13] [13] Yu, H. J.; Liu, J. Q.; Böck, A.; Li, J.; Luo, G. M.; Shen, J. C. Engineering glutathione transferase to a novel glutathione peroxidase mimic with high catalytic efficiency. *J. Biol. Chem.,* **2005**, *280*, 11930-11935.

[14] Muller, S.; Senn, H.; Gsell, B.; Vetter, W.; Baron, C.; Böck, A. The formation of diselenide bridges in proteins by incorporation of selenocysteine residues: biosynthesis and characterization of $(Se)_2$-thioredoxin. *Biochemistry*, **1994**, *33*, 3404-3412.

[15] Luo, G. M.; Zhou, Z. Q.; Ding, L.; Gao, G.; Sun, Q. A.; Liu, Z.; Yang, T. S.; Shen, J. C. Generation of selenium-containing abzyme by using chemical mutation. *Biochem. Biophys. Res. Commun.* **1994**, *198*, 1240-1247.

[16] Liu, J. Q.; Gao, S. J.; Luo, G. M.; Yan, G. L.; Shen, J. C. Artificial imitation of glutathione peroxidase with 6-selenium-bridged β-cyclodextrin. *Biochem. Biophys. Res. Commun.* **1998**, *247*, 397-400.

[17] Liu, J. Q.; Luo, G. M.; Ren, X. J.; Mu, Y.; Bai, Y.; Shen, J. C. A bis-cyclodextrin diselenide with glutathione peroxidase-like activity. *Biochim. Biophys. Acta.*, **2000**, *1481*, 222-228.

[18] Liu, Y.; Li, B.; Li, L.; Zhang, H.-Y. Synthesis of organoselenium-modified β-cyclodextrins possessing a 1,2-benzisoselenazol-3(2H)-one moiety and their enzyme-mimic study. *Helv. Chim. Acta.,* **2002**, *85*, 9-18.

[19] Ren, X. J.; Yang, J. Q.; Liu, J. Q.; Su, D.; You, D. L.; Liu, C.; Zhang, K.; Luo, G. M.; Mu, Y.; Yan, G. L. A novel glutathione peroxidase mimic with antioxidant activity. *Arch. Biochem. Biophys.,* **2001**, *387*, 250-256.

[20] Zhang, X.; Xu, H. P.; Dong, Z. Y.; Wang, Y. P.; Liu, J. Q.; Shen, J. C. Highly efficient dendrimer-based mimic of glutathione peroxidase. *J. Am. Chem. Soc.*, **2004**, *126*, 10556-10557.

[21] Xu, H. P.; Gao, J.; Wang, Y. P.; Wang, Z. Q.; Smet, M.; Dehaen, W.; Zhang, X. Hyperbranched polyselenides as glutathione peroxidase mimics. *Chem. Commun.,* **2006**, 796-798.

[22] Huang, X.; Dong, Z. Y.; Liu, J. Q.; Mao, S. Z.; Xu, J. Y.; Luo, G. M.; Shen, J. C. Selenium-mediated micellar catalyst: an efficient enzyme model for glutathione peroxidase-like catalysis. *Langmuir*, **2007**, *23*, 1518-1522.

[23] Yin, Y. Z.; Huang, X.; Xu, J. Y.; Liu, J. Q.; Shen, J. C. Construction of an artificial glutathione peroxidase active site on copolymer vesicles. *Macromol. Biosci.,* **2010**, *10*, 1505-1516.

[24] Huang, X.; Yin, Y. Z.; Tang, Y.; Bai, X. L.; Zhang, Z. M.; Xu, J. Y.; Liu, J. Q.; Shen, J. C. Smart microgel catalyst with modulatory glutathione peroxidase activity. *Soft Matter*, **2009**, *5*, 1905-1911.

[25] Tang, Y.; Zhou, L. P.; Li, J. X.; Luo, Q.; Huang, X.; Wu, P., Wang, Y. G.; Xu, J. Y.; Liu, J. Q.; Shen, J. C. Giant nanotubes loaded with artificial peroxidase centers: self-assembly of supramolecular amphiphiles as a tool to functionalize nanotubes. *Angew. Chem. Int. Ed.*, **2010**, *49*, 3920-3924.

[26] a) Behne, D., Kyriakopoulos, A.; Meinhold, H.; Köhrle, J. Identification of type i iodothyronine 5′-deiodinase as a selenoenzyme. *Biochem. Biophys. Res. Commun.* **1990**, *173*, 1143-1149. b) Berry, M. J.; Banu, L.; Larsen, P. R. Type I iodothyronine deiodinase is a selenocysteine-containing enzyme. *Nature*, **1991**, *349*, 438-440. c) Larsen, P.R.; Berry, M.J. Nutritional and hormonal regulation of thyroid hormone deiodinases. *Annu. Rev. Nutr.,* **1995**, *15*, 323–352. d) Salvatore, D.; Bartha, T.; Harney, J. W.; Larsen, P. R. Molecular biological and biochemical characterization of the human type 2 selenodeiodinase. *Endocrinology*, **1996**, *137*, 3308–3315. e) Salvatore, D.; Low, S. C.; Berry, M. J.; Maia, A. L.; Harney, J. W.; Croteau, W.; St Germain, D. L.; Larsen, P. R. Type 3 iodothyronine deiodinase: cloning, *in vitro* expression and functional analysis of the placental selenoenzyme. *J Clin Invest*, **1995**, *95*, 2421–2430. e) Bianco, A. C.; Salvatore, D.; Gereben, B.; Berry, M. J.; Larsen, P. R. Biochemistry, cellular and molecular biology, and physiological roles of the iodothyronine selenodeiodinases. *Endocr. Rev.* **2002**, *23*, 38-89.

[27] Beck, C.; Jensen, S. B.; Reglinski, J. The selenium mediated de-iodination of iodophenols: a model for the mechanism of 5′ thyronine de-iodinase. *Bioorg. Med. Chem. Lett.*, **1994**, *4*, 1353-1356.

[28] a) Goto, K.; Sonoda, D.; Shimada, K.; Sase, S.; Kawashima, T. Modeling of the 5'-deiodination of thyroxine by iodothyronine deiodinase: chemical corroboration of a selenenyl iodide intermediate. *Angew. Chem.* **2010**, *122*, 555-557. b) Manna, D.; Mugesh, G. A chemical model for the inner-ring deiodination of thyroxine by iodothyronine deiodinase. *Angew. Chem.*, **2010**, *122*, 9432-9435.

[29] Kuiper, G. G. J. M.; Klootwijk, W.; Visser, T. J. Substitution of cysteine for selenocysteine in the catalytic center of type III iodothyronine deiodinase reduces catalytic efficiency and alters substrate preference. *Endocrinology*, **2003**, *144*, 2505-2513.

[30] Manna, D.; Mugesh, G. Deiodination of thyroid hormones by iodothyronine deiodinase mimics. Does an increase in the reactivity alter the regioselectivity? *J. Am. Chem. Soc.,* **2011**, *133*, 9980-9983.

[31] Köhrle, J.; Hesch, R. D. Biochemical characteristics of iodothyronine monodeiodination by rat liver microsomes: the interaction between iodothyronine substrate analogs and the ligand binding site of the iodothyronine deiodinase resembles that of the TBPA-iodothyronine ligand binding**.** *Horm. Metab. Res., Suppl. Ser.* **1984**, *14*, 42-55.

[32] Manna, D.; Mugesh, G. Regioselective deiodination of thyroxine by iodothyronine deiodinase mimics: an unusual mechanistic pathway involving cooperative chalcogen and halogen bonding. *J. Am. Chem. Soc.,* **2012**, *134*, 4269-4279.

[33] Epp, O.; Ladenstein, R.; Wendel, A. The refined structure of the selenoenzyme glutathione peroxidase at 0.2-nm resolution. *Eur. J. Biochem.*, **1983**, *133*, 51-69.

[34] a) Mugesh, G.; du Mont, W.-W.; Sies, H. Chemistry of biologically important synthetic organoselenium compounds. *Chem. Rev.*, 2011, *101*, 2125-2179. b) Mugesh, G.; Singh, H. B. Synthetic organoselenium compounds as antioxidants: glutathione peroxidase activity. *Chem. Soc. Rev.*, 2000, *29*, 347-357.

[35] Santi, C.; Santoro, S.; Battistelli, B.; Testaferri, L.; Tiecco, M. Preparation of the first bench-stable phenyl selenolate: an interesting "on water" nucleophilic reagent. *Eur. J. Org. Chem.* **2008**, 5387–5390.

[36] Tidei, C.; Piroddi, M.; Galli, F.; Santi, C. Thiols oxidation promoted by PhSeZnCl. *Tetrahedron Lett.*, **2012**, *53*, 232-234.

[37] Bhabak, K. P.; Mugesh, G. A simple and efficient strategy to enhance the antioxidant activities of amino-substituted glutathione peroxidase mimics. *Chem. Eur. J.*, **2008**, *14*, 8640-8651.

[38] Grieco, P. A.; Yokoyama, Y.; Gilman, S.; Nishizawa, M. Organoselenium chemistry. Epoxidation of olefins with benzeneseleninic acid and hydrogen peroxide ("benzeneperoxyseleninic acid"). *J. Org. Chem.*, **1977**, *42*, 2034-2036.

[39] Santi, C.; Di Lorenzo, R.; Tidei, C.; Bagnoli, L.; Wirth, T. Stereoselective selenium catalyzed dihydroxylation and hydroxymethoxylation of aliene. *Tetrahedron,* **2012**, *68*, 10530-10535.

Organoselenium Chemistry Between Synthesis and Biochemistry, 2014, 361-378

CHAPTER 12

Antioxidant Organoselenium Molecules

Michio Iwaoka*

Department of Chemistry, School of Science, Tokai University, Kitakaname, Hiratsuka-shi, Kanagawa 259-1292, Japan

Abstract: A number of selenoenzyme mimics, mostly of glutathione peroxidase (GPx), have been developed in decades so as to probe the molecular mechanism of the catalytic cycle as well as to utilize them as selenium antioxidants (SeAO), expecting that those organoselenium compounds would reduce reactive oxygen species (ROS) in living cells. Meanwhile, the design of antioxidant selenoenzyme mimics has encountered several problems relating to the biologically multimodal functions of SeAO. In this chapter, recent progress in the development of SeAO as ROS scavengers is overviewed. After brief explanation for the regulation of oxidative stress by SeAO, biological effects of SeAO and the proposed action mechanisms are described. Possible prooxidant (PO) effects of selenium compounds are also surveyed. Finally, the SeAO are classified in two types, *i.e.*, type A which is easily reducible to selenolate species (RSe^-) and type B which is not easily reducible, on the basis of their AO and PO functions in order to summarize the current status in the SeAO design and give some guidelines for the future study.

Keywords: Biomimetic, Bio-logic, Artificial enzymes, GPx mimics, Selenocystine, Selenomethione, Selenyl radicals, 2-Center-3-electron bond, Cytotoxicity, Time-resolved pulse radiolysis.

12.1. INTRODUCTION

Selenium is an essential micronutrient for most species of life on the earth. Twenty-five selenoproteins, including glutathione peroxidase (GPx), iodothyronine deiodinase (ID) and thioredoxin reductase (TrxR), have been identified in the human genome [1]. Many of these selenoenzymes are involved in biologically important redox processes and are assigned to antioxidant enzymes, although some of them are still not well understood for their biological functions. GPx, which reduces hydrogen peroxide (H_2O_2) or lipid hydroperoxides (LOOH)

***Address correspondence to Michio Iwaoka:** Department of Chemistry, School of Science, Tokai University, Kitakaname, Hiratsuka-shi, Kanagawa 259-1292, Japan; Tel: +81-463-58-1211; Fax: +81-463-50-2094; E-mail: miwaoka@tokai.ac.jp

Claudio Santi (Ed)

to non-toxic H_2O or LOH, respectively, by using glutathione (GSH) as a reducing cofactor (eq. 12.1), is the representative antioxidant selenoenzyme, whose structure and function have been elaborately studied [2].

$$H_2O_2 + 2GSH \xrightarrow{GPx} 2H_2O + GSSG \qquad (12\text{-}1)$$

Selenium is incorporated into selenoproteins as selenocysteine (SeCys), which is the selenium analogue of amino acid cysteine. Surprisingly, it was discovered that SeCys shares the RNA codon table with twenty proteinogenic amino acids [3]. UGA, the opal stop codon, indeed invokes SeCys tRNA to the ribosome when the selenocysteine insertion sequence (SECIS) is present in the mRNA chain. Thus, SeCys is recognized as the twenty-first amino acid [4]. The latest recombinant technology enables artificial expression of various selenoproteins by cloning the SECIS [5].

In the meantime, a number of selenoenzyme mimics, mostly of GPx, have been developed in decades so as to probe the molecular mechanism of the catalytic cycle as well as to utilize them as selenium antioxidants (SeAO), expecting that those organoselenium compounds would reduce reactive oxygen species (ROS) in living cells [6]. Such compounds, most of which are aromatic diselenides (ArSeSeAr) or their derivatives, have already been listed up in several reviews [6-8]. Ebselen (**1**) [9,10] and diphenyl diselenide (**2**) [11,12] are representative GPx mimics widely used as the standard molecules. Recently, aliphatic selenium compounds, such as SeCys derivatives, selenoglutathione (GSeH) and selenopeptides [13], as well as macromolecular systems containing selenium, *i.e.*, artificial selenoenzymes [14], have also been applied as GPx mimics because of their resemblance to the active site structure. Meanwhile, the design of antioxidant selenoenzyme mimics has encountered several problems (P1-3) relating to the biologically multimodal functions of SeAO:

P1. SeAO reduce not only H_2O_2 and LOOH but also radical species. This means that SeAO can work as both one-electron (1e) and two-electron (2e) reductants against ROS [15].

P2. In the 1e oxidation of SeAO, various transient intermediates are generated [16]. However, their molecular identities as well as their cascade reactions are not yet fully understood.

P3. SeAO sometimes work as a prooxidant (PO) too by the reaction with oxygen (O_2), producing superoxide ($\cdot O_2^-$) [17]. This reaction competes with the antioxidative function of the SeAO. The bimodal actions as AO and PO have been pointed out in recent literature [18].

According to these complications, the practical influence of SeAO on biological systems is not as straightforward as usually described by using the GPx-like 2e reaction (*i.e.*, eq. 12.1). Therefore, to design practical SeAO it is necessary to elucidate the mechanisms of the interaction between SeAO and ROS not only through 2e but also through 1e transfer. In this chapter, recent progress in the development of SeAO as ROS scavengers is overviewed. After brief explanation for the regulation of oxidative stress by SeAO, biological effects of SeAO and the proposed action mechanisms are described. Possible PO effects of selenium compounds are also surveyed. Finally, the SeAO are classified in two types on the basis of their AO and PO functions in order to summarize the current status in the SeAO design and give some guidelines for the future study.

12.2. REGULATION OF OXIDATIVE STRESS USING SeAO

ROS (*i.e.*, $\cdot O_2^-$, $\cdot OH$, H_2O_2, $LOO\cdot$, LOOH, *etc.*) are always generated from oxygen (O_2) in cells, but their concentrations are maintained low thanks to the existence of the antioxidative processes that convert ROS to fully reduced oxygen species, such as water (eq. 12.2). The high concentrations of ROS cause serious damages on DNA, proteins and membranes, leading the cells to the eventual apoptotic death [19]. Therefore, the antioxidative processes are important for keeping the redox balance in cells.

$$O_2 \xrightarrow{k_1} \text{ROS} \xrightarrow{k_2} H_2O \qquad (12\text{-}2)$$

The k_1 process corresponds to ROS formation due to respiration, normal cell activities and exposure to PO, such as radical initiators, metal ions and toxic chemicals. PO are the substances that generate ROS. Superoxide $\cdot O_2^-$, the first ROS generated from O_2 through 1e reduction, is subsequently converted to H_2O_2 by the second 1e reduction or the function of superoxide dismutase (SOD). In the presence of Fe^{2+}, H_2O_2 is further converted to hydroxyl radical ($\cdot OH$) through

Fenton reaction. The k_2 process corresponds to ROS elimination by the function of antioxidant enzymes, such as catalases and GPx, or small molecule AO, such as vitamins, polyphenols, sulfur compounds, *etc.*, which reduce ROS to non-toxic oxygen species.

The k_1 and k_2 processes are well balanced in normal cells ($k_1 = k_2$), hence the concentration of ROS is maintained low. However, due to several reasons, the imbalance sometimes happens. Illness, oncogenesis, inflammation and exposure to chemicals (*i.e.*, PO) accelerate the k_1 process, while the lack of AO, including antioxidant selenoenzymes, decelerates the k_2 process. Such conditions ($k_1 > k_2$) are called oxidative stress [20]. Obviously, oxidative stress is not preferable for normal cells.

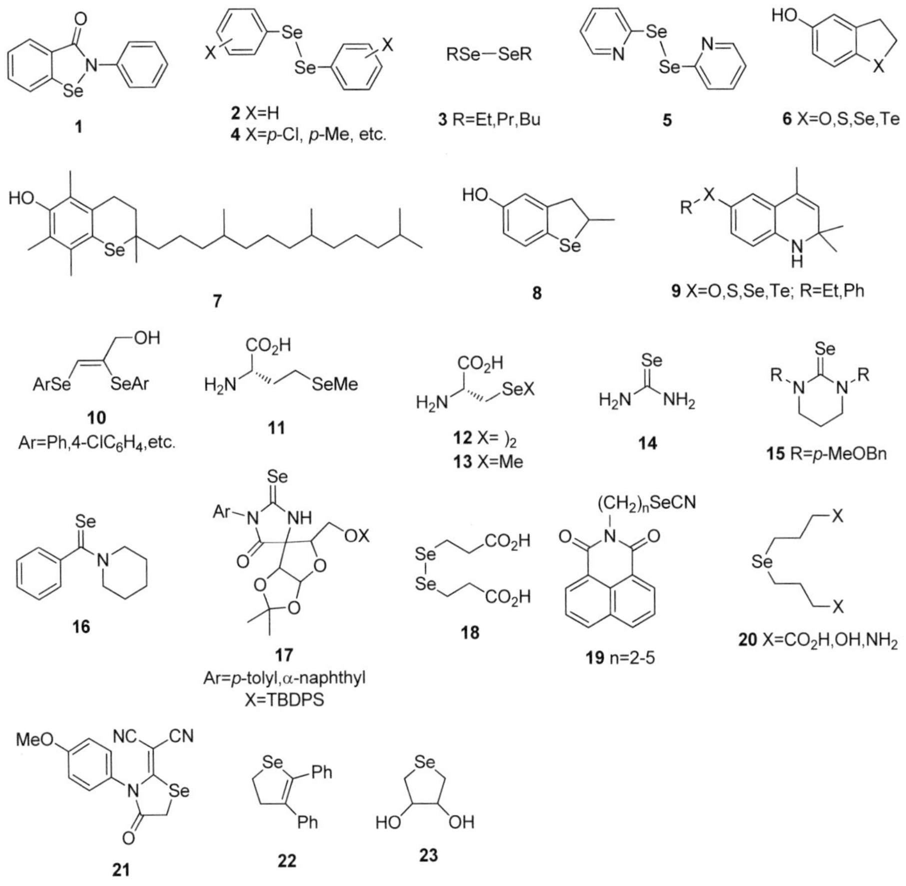

Figure 12.1: Selenium antioxidants (SeAO).

Various organoselenium molecules, such as ebselen (**1**) [9,10] and diphenyl diselenide (**2**) [11,12], have already been tested as a possible AO to secure the cells from oxidative stress. Fig. **12.1** shows the structures of organoselenium compounds, which are discussed in the following sections.

They were expected to behave as SeAO to decrease the concentration of ROS by accelerating the k_2 process in cells. However, several reports now indicate that the same compounds can also work as PO to accelerate the k_1 process [19]. Thus, the practical outcome from the treatment of the cells with selenium compounds would depend on the relative activities as AO and PO. If the AO activity is dominant, it brings benefits to normal or slightly stressed cells. On the contrary, if the PO activity is dominant, ROS are accumulated in cells, leading to the cell death. Such SePO can be applied for chemoprevention of cancer because the increase of the ROS concentration would result in serious damages on the cancer cells suffering from elevated oxidative stress but not on the normal cells with the capacity to endure the oxidative stress [21].

12.3. SELENIUM AO

As aforementioned, SeAO, which were developed as GPx mimics (*i.e.*, 2e SeAO; see eq. 12.1), are already summarized in several reviews [6-8]. Herein, from a practical point of view, SeAO, which have the ability to work as a 1e reductant (or a radical scavenger) against ROS *in vitro* and *in vivo*, are collected from the recent literature. Their interactions with ROS and the reactive transients produced by the reactions are discussed in the subsequent section.

Antioxidant functions of **1** were demonstrated by the decrease of focal cerebral ischemia of rats [9] and the antioxidant ability against Cd-induced oxidative stress in mice [10]. Similar antioxidant functions were also reported for **2**. Nogueira and coworkers examined the antioxidant effect of **2** against $CdCl_2$-induced oxidative damage [11] and the minimization of menopause related symptoms for female rates with ovariectomization [12]. On the other hand, diselenide **2** had no or low toxic effect on the body. The same group also investigated the antioxidant ability of various diselenides using mice and rats. Aliphatic diselenides **3** and aromatic diselenides **4** exhibited the antioxidant activity, comparable to **1**, against lipid peroxidation as well as the SH peroxidase activity [22]. However, at the low concentrations the PO activity of **3** was also observed [22]. 2,2'-Dipyridyl diselenide (**5**) showed proton-radical-scavenging and enhanced antioxidant activities in various assays using thiobarbituric acid reactive substances (TBARS), protein carbonyl levels and 2,2'-diphenyl-1-picrylhydrazyl (DPPH) and 2,2'-azino-*bis*(3-ethylbenzthiazoline-6-sulphonic acid) (ABTS) radicals as a probe of the AO activity [23].

Aromatic monoselenides were also investigated for their antioxidant functions. Engman and coworkers synthesized a series of chalcogen compounds **6** and compared their antioxidant capacity with each other *in vitro* [24]. It was found that the antioxidant capacity decreases in the order, Te > Se=S > O, and that Te and Se derivatives exhibit the capacity in a catalytic manner. Similar trends were observed in the antioxidant assay using liver microsome. In addition, the telluride had GPx-like 2e reduction activity, but the others showed essentially no GPx-like activity [24]. Engman's group then synthesized a variety of aromatic chalcogenides. *All-rac*-α-selenotocopherol (**7**), which does not work as a regenerable catalyst, exhibited less antioxidant activity than α-tocopherol (vitamin E) due to the stronger OH bond of **7** [25]. On the other hand, selenophenol derivative **8**, an intrinsically poor GPx mimic and a less active AO than α-tocopherol (vitamin E), was found to be a regenerable antioxidant catalyst and showed a high score of antioxidant activity in the two-phase (*i.e.*, water-chlorobenzene) assay system [26]. For dihydroquinoline ethoxyquin derivatives **9**, the best antioxidant capacity was obtained when X=O, while when X=Te, it exhibited both antioxidant (1e) and GPx-like (2e) activities [27]. On the other hand, Nogueira found that conjugated vinyl polyaryl selenides, such as **10**, show an antioxidant effect against lipid peroxidation and also in the SH peroxidase assay [28].

Aliphatic selenium compounds, such as selenomethionine (SeMet) (**11**), selenocystine ($SeCys_2$) (**12**), *Se*-methyl selenocysteine (MeSeCys) (**13**) and selenourea (SeU) (**14**) are also representative SeAO that have been applied in various antioxidant assays. The GPx-like activity (2e), anti-hemolytic ability (1e) and peroxyl radical scavenging ability (1e) of **11-13** were in the order, **12** > **13** > **11** [29]. Recently, these compounds were found to suppress Cu(I) or $Fe(II)/H_2O_2$ induced DNA damage through the coordination of Se to the metal center [30]. On the other hand, antioxidant properties of SeU derivatives, such as **15**, were shown by using the 7,12-dimethylbenz[a]anthracene (DMBA)-induced oxidative stress on mice [31]. Selenoamides, such as **16**, showed good scavenging effects against $\cdot O_2^-$ generated in polymorphonuclear leukocytes (PMNs), while they exhibited low cytotoxicity to human cell lines [32]. Selenohydantoin (**17**) is also a candidate of a good free radical scavenger [33].

Aliphatic diselenide **18** was found to be a good SeAO against lipid peroxidation with low cytotoxicity, although it has low GPx-like activity, suggesting that **18** is a free radical scavenger [34]. Correlation was found between the antioxidant activity and the 1e redox potential among the corresponding dihydroxy and diamino analogs of **18** [35].

Some aliphatic monoselenides have also been tested as SeAO. Selenocyanates **19** exhibited the maximum antioxidant properties against DMBA-PMA-induced oxidative stress on mice when n=5 [36]. DMBA and PMA (phorbol 12-myristate 13-acetate) are the chemicals that strongly promote cancer in cells. Symmetric aliphatic selenides **20** showed DNA protecting ability, peroxyl radical scavenging and GPx mimicking ability in water [37]. The activities decreased in the order, X=COOH > OH > NH_2, which was explained by the calculated HOMO energy level [37].

Cyclic selenium compounds were synthesized and tested for the antioxidant activity. Five-membered ring SeTCNQ derivatives, such as **21**, work as antioxidant or neurotrophic-like molecules to scavenge $\cdot O_2^-$ and suppress serum deprivation-induced apoptosis in rat pheochromocytoma cell line [38]. Dihydroselenophene **22** showed antioxidant activities against 2,3-dichloro-5,6-dicyano-1,4-benzoquinone (DDQ) and ABTS [39]. On the other hand, a water-soluble five-membered ring selenide (**23**) accelerated healing of indomethacin-induced stomach ulceration in mice [40].

12.4. INTERACTION OF SELENIUM AO WITH ROS

The molecular mechanisms for the antioxidative functions of SeAO, which are described above, have not been well understood yet. The species, which would be generated by the reaction of SeAO and ROS through 1e transfer, are usually highly reactive, hence rapidly converted to other stable species through the cascade reactions or the decomposition pathways. In general, the fate of the initial transients, usually selenyl radical cations, largely depends on the substituents. Aliphatic selenyl radicals are unstable and search a partner atom having a lone pair to gain the stability, while aromatic selenyl radicals are more stable due to the conjugation with the aromatic rings.

The time-resolved pulse radiolysis method combined with an absorption detector was applied by Brede to detect the transient formed in the reaction of benzeneselenol with 1,2-dichloroethane radical cation (eq. 12.3) [41]. The reaction afforded 25% of benzeneselenol radical cation, which showed absorption at 775 nm, and 75% of phenylselenyl radical, which showed absorption at 480 nm.

$$\mathrm{PhSeH} \xrightarrow{\mathrm{ClCH_2CH_2Cl}^{\bullet +}} \underset{775\ \mathrm{nm}}{\mathrm{PhSeH}^{\bullet +}} + \underset{480\ \mathrm{nm}}{\mathrm{PhSe}^{\bullet}} \qquad (12\text{-}3)$$

Koppenol carried out a similar pulse radiolysis study on $SeCys_2$ (**12**) to identify some selenium transients, such as CysSe· (460 nm) and CysSe.˙.SeCys$^-$ (455 nm) [42]. It was proposed that CysSe· is stabilized by formation of the two-center-three-electron (2c-3e) bond with another selenolate (CysSe:$^-$). $SeCys_2$ (**12**) was indeed found to catalyze the reduction of benzyl viologen (BV^{2+}) with dithiothreitol (DTT) to BV^+ [42]. Koppenol also observed the fast electron transfer (ET) from the selenolate anion, generated from SeCys or GSeH, to a tyrosine radical (TyrO·) (eq. 12.4) [43].

$$\mathrm{CysSe{:}^{-}} + \mathrm{TyrO}^{\bullet} \longrightarrow \mathrm{CysSe}^{\bullet} + \mathrm{TyrO{:}^{-}} \qquad (12\text{-}4)$$

Priyadarsini applied the pulse radiolysis technique to the 1e oxidation of SeMet (**11**), $SeCys_2$ (**12**), MeSeCys (**13**), SeU (**14**) with ·OH (or $\cdot Cl_2^-$, $\cdot Br_2^-$) [16]. When **11** was oxidized, selenium radical cations, which are stabilized by intermolecular coordination to Se atom or by intramolecular coordination to N or O atom, were produced, depending on the pH condition (eq. 12.5). At pH 1, the species stabilized by intermolecular 2c-3e Se-Se bond was observed with absorption at ~480 nm, while at neutral or basic pHs, the species were intramolecularly stabilized, showing the absorption at ~380 nm. Similar reactions were observed for **12-14** (eqs. 12.6, 12.7, 12.8).

11 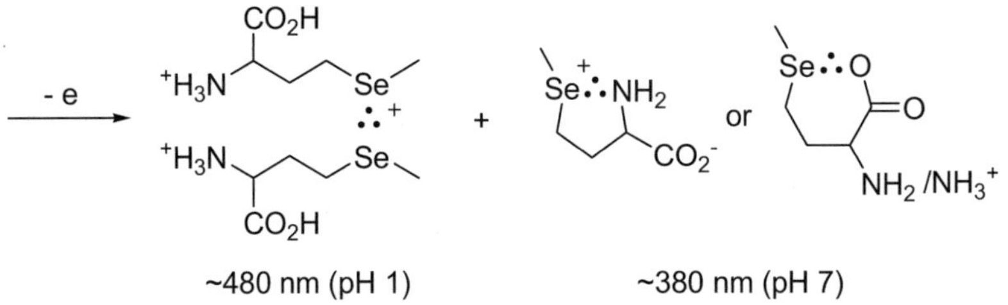

(12-5)

12 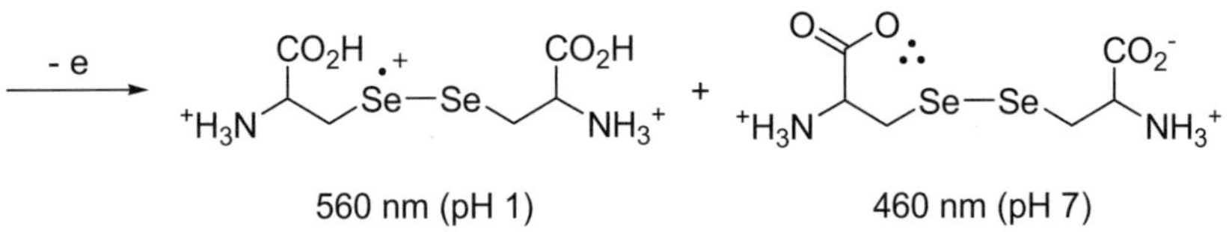

(12-6)

13 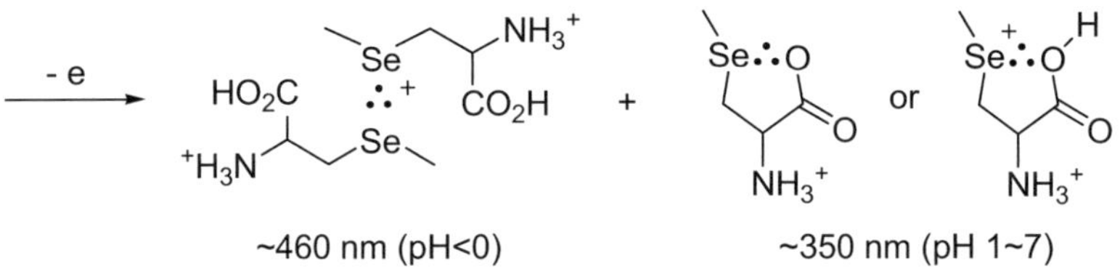

(12-7)

14 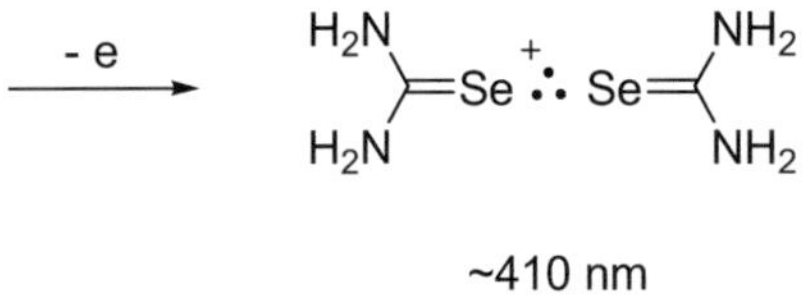

(12-8)

The behaviors of **11-14** in the 1e oxidation are similar to those observed for the sulfur analogs, but the transients produced from the selenium compounds are more stable with longer lifetimes [16]. The dimeric radical cation generated from *N,N*-dimethylselenourea shows the absorption at 430 nm and can be converted to selenium nano particles by the reaction with O_2 in poly vinyl alcohol (PVA) [44].

Priyadarsini's group also investigated the 1e oxidation process of diselenide **18** by pulse radiolysis using trichloromethyl peroxyl radical ($CCl_3OO\cdot$) as an oxidant to observe **24** as a transient species (eq. 12.9) [34]. Radical cation **24** had absorption at 560 nm, similarly to the radical cation generated from **12**.

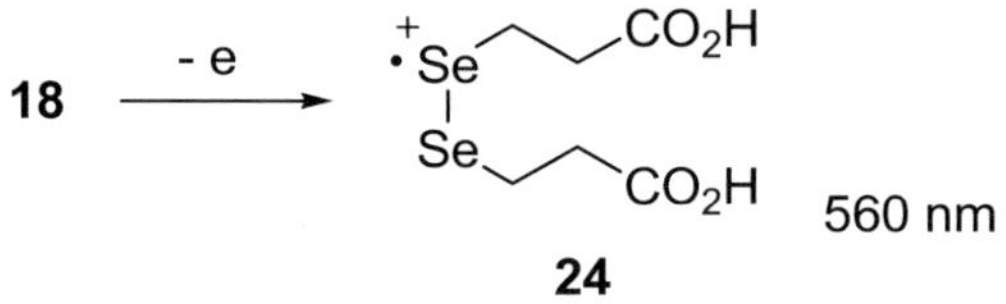

(12-9)

The transient species generated by 1e oxidation of diselenides, such as **24**, would decompose in the cascade reactions to produce selenyl radicals (RSe·) and other degradation products [45,46]. On the other hand, the transient species from monoselenides could be transformed to the selenoxides through the second 1e oxidation. For example, SeMet Se-oxide was likely to form in the reaction mixture of SeMet (**11**) and ·OH [47]. Cyclic selenide **23** was converted to the corresponding selenoxide through 1e oxidation with ·OH and the subsequent reaction with $\cdot O_2^-$ (eq. 12.10) [48].

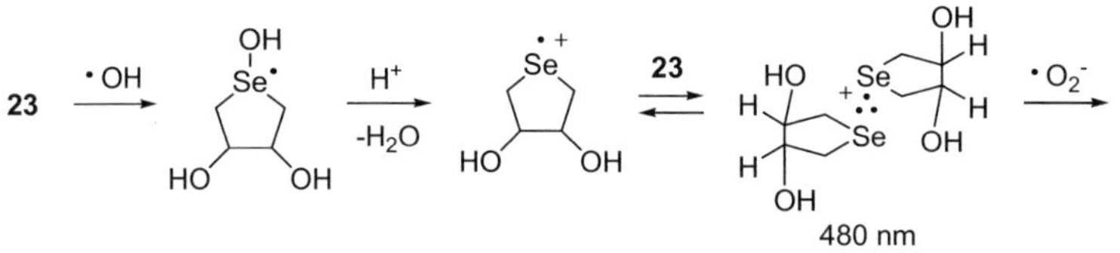

(12-10)

Similar reactions were reported for dimethyl selenide (Me_2Se). When Me_2Se reacted with ·OH, formation of the dimer radical cation with absorption at 470 nm and the subsequent disproportionation to the selenide and selenoxide were observed (eq. 12.11) [49]. On the other hand, when Me_2Se reacted with hydrogen radical (·H), the Se—C covalent bond was cleaved to produce methaneselenol (MeSeH) and methyl radical ($\cdot CH_3$) [49]. Similarly, the reaction of dimethyl chalcogenides with acetyl radical resulted in formation of chalcogenoacetate and $\cdot CH_3$ (eq. 12.12) [50]. These reactions suggest the possibility that an alkyl group eliminates from aliphatic monoselenides in the radical reactions.

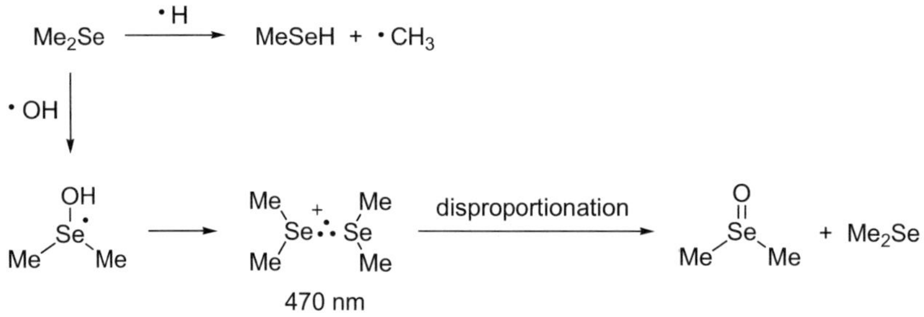

(12-11)

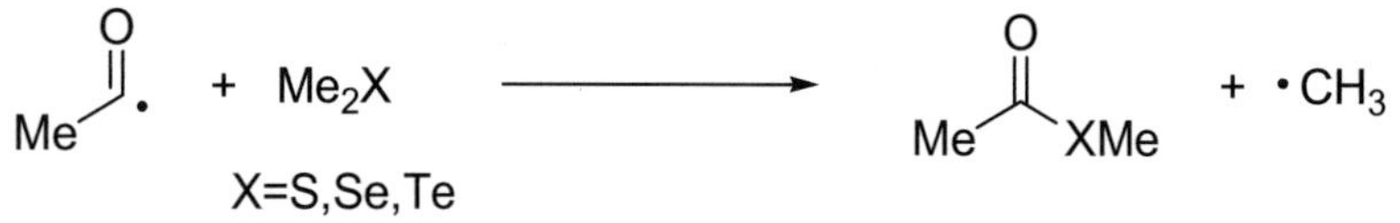

(12-12)

Aromatic SeAO, reacting with ROS, would produce rather stable radicals, which can be detected by electron paramagnetic resonance (EPR) spectroscopy. For example, the radicals generated from **6** (X = Se and Te) by the 1e oxidation and deprotonation, were detected by EPR (eq. 12.13) [24]. They could be converted back to **6** by the reaction with thiol. Thus, **6** will be useful as catalytic SeAO.

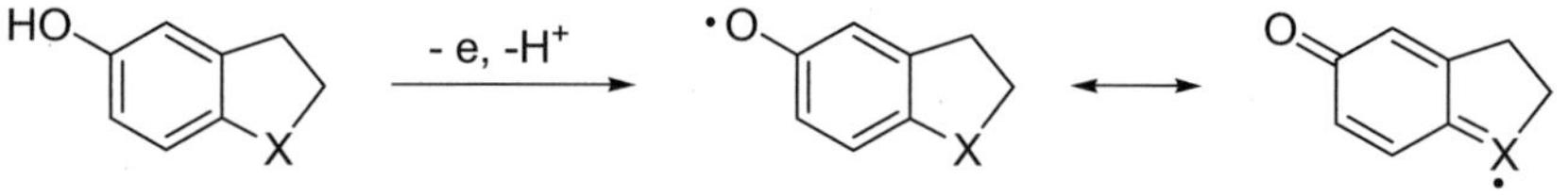

6 X=Se,Te

(12-13)

The catalytic cycle of aromatic SeAO was studied in detail by using selenide **8** [26]. In the cycle (eq. 12.14), the selenyl radical generated from **8** is further oxidized to the selenium cation, which is then converted to the selenoxide. The selenoxide can be easily reduced with thiol to reproduce **8**. This cycle is quite different from that of tocopherols, in which the radical directly reacts with thiol to reproduce the antioxidants.

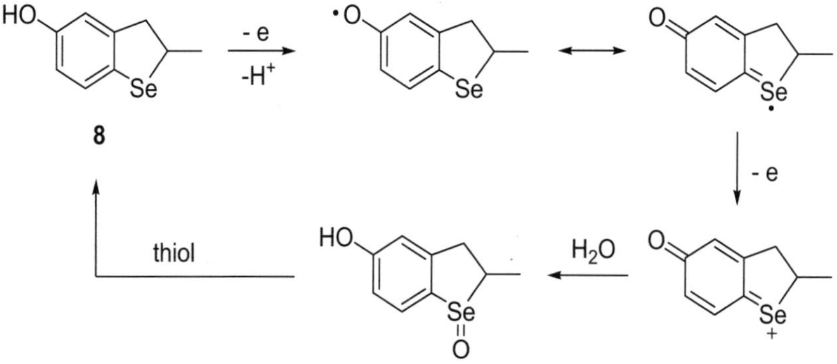

(12-14)

Dihydroselenophene **22** would establish the catalytic cycle as shown in eq. 12.15 [39]. The radical formed from **22** by the reaction with ABTS radical is reduced with GSH to reproduce **22**, whereas when reacted with DDQ, **22** would produce the selenophene through 2e oxidation.

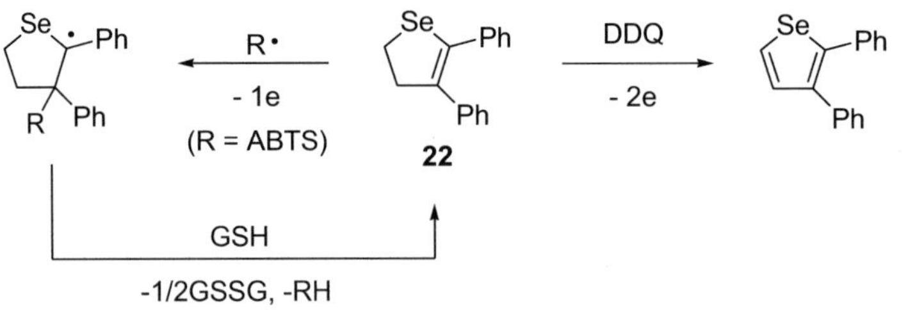

(12-15)

12.5. SELENIUM PO

SeAO, such as SeMet (**11**) and $SeCys_2$ (**12**), provide antioxidant benefits either directly (*i.e.*, as radical and/or hydroperoxide scavengers) or indirectly (*i.e.*, as a selenium source to synthesize antioxidant selenoenzymes). However, SeAO would have another function as PO to produce $\cdot O_2^-$ and ·OH [51]. There have been many evidences for the bimodal actions (*i.e.*, SeAO *vs.* SePO). $SeCys_2$ (**12**) produces ROS in cancer cells, causing DNA damage-mediated apoptosis [52]. *p*-Aminophenyl diselenide and some selenourea derivatives work as $\cdot O_2^-$ generators and cytotoxic agents against human prostate cancer cells (PC-3) and breast adenocarcinoma (MCF-7) [53].

Chaudiere previously reported that the selenolate (RSe^-), which can be formed by the reaction of diselenide (RSeSeR) with thiol, reacts with O_2 to afford $\cdot O_2^-$ (eq. 12.16) [17]. This type of ET reaction would largely explain the PO function of selenium compounds.

$$O_2 + RSe^- \longrightarrow \cdot O_2^- + RSe\cdot \qquad (12\text{-}16)$$

Methaneseleninic acid ($MeSeO_2H$) can be reduced in cell environments to methylselenolate ($MeSe^-$), which would produce $\cdot O_2^-$ through the reaction shown in eq. 12.16 and lead to the apoptosis of prostate cancer cells. Thus, $MeSeO_2H$ with an excess amount can be usable for chemoprevention against cancer [54]. SeMet (**11**) and MeSeCys (**13**) are not reducible, but by the function of methioninase or β-lyase, they are converted to MeSeH, which would react with O_2 to produce $\cdot O_2^-$ [54].

Another action mechanism of SePO was proposed for selenite (Na_2SeO_3) and diselenide **18** [55]. These selenium compounds act as a catalyst in the reaction between GSH and O_2 (eq. 12.17). $GSSe^-$ or RSe^- would participate in the catalytic cycle as a possible intermediate.

$$GSH + O_2 \xrightarrow{\text{SePO}} GSSG + \bullet O_2^- \qquad (12\text{-}17)$$

When selenium compounds are medicated as SePO, some portions should be metabolized to various selenium species, such as Me_2Se, SeCys, GSSeSG and hydrogen selenide (H_2Se) [56]. The metabolic pathways of H_2Se, which include production of selenoenzymes and selenosugars as well as MeSeH, Me_2Se and Me_3Se^+, have also been well elucidated [57]. Some of these species exhibit toxicity by increasing oxidative stress.

12.6. CONCLUSIONS AND OUTLOOKS

Since most of selenoproteins belong to antioxidant enzymes, antioxidant functions of selenium compounds have been elaborately studied to develop synthetic mimics of the selenoenzymes, especially of GPx [6-8]. However, the status of the SeAO design is not yet matured with complications, arising from multimodal functions of SeAO *in vitro* and *in vivo*, as mentioned in Introduction (*i.e.*, P1-3). To make the problems a little clearer, it would be meaningful to classify SeAO in two different types:

Type A: SeAO that are easily reducible to selenolate species (RSe^-). This type includes diselenides and their equivalents, such as **1-5**, $SeCys_2$ (**12**) and **18**.

Type B: SeAO that are not easily reducible to RSe^-. This type includes monoselenides, selenoureas and selenoamides, such as Me_2Se, **6-10**, SeMet (**11**), MeSeCys (**13**), SeU (**14**), **15-17** and **19-23**.

In the presence of reducing substrates, such as GSH and thioredoxin, diselenides and their derivatives (type A) can be converted to RSe^-, which would readily react with H_2O_2 (or LOOH) to produce H_2O (or LOH) through the 2e process like GPx enzyme (eq. 12.18). However, the RSe^- can also react with O_2 to produce $\cdot O_2^-$ through the 1e process as shown in eq. 12.16 [17]. This possible PO function of type A should compete to the AO function (eq. 12.18) *in vitro* and *in vivo*, making the net outcome from the application to biological systems not straightforward.

$$RSe^{-} + H^{+} + H_2O_2 \longrightarrow RSeOH + H_2O \qquad (12\text{-}18)$$

On the other hand, type B SeAO are not good 2e antioxidants. For example, the oxidation of monoselenides with H_2O_2 to selenoxides usually proceeds slowly. However, the type B would function as 1e donors to scavenge ROS. The afforded selenyl radical can be stabilized by the conjugation with the attached aromatic ring or by formation of an intermolecular or intramolecular 2c-3e bond with a lone pair of heteroatoms. Moreover, in some particular cases, it has been demonstrated that the selenyl radical can be reduced back to the SeAO in the presence of thiol substrates directly or indirectly through selenoxides [26,39,48]. Such catalytic behaviors of type B will be usable in developing new SeAO. On the other hand, the PO functions of type B SeAO would be limited, although in the presence of some enzymes, such as methioninase or β-lyase, the SeAO can release selenolate species as possible PO [54].

In conclusion, type A SeAO would function as both AO and PO, while type B SeAO can be potential 1e AO. Although the 1e oxidation process of SeAO with ROS and the cascade reactions are not fully understood at this moment, various transients generated in the reaction have already been characterized by the time-resolved pulse radiolysis and EPR spectroscopy as summarized here. These latest achievements will be helpful for the design of more effective and practical SeAO in future works.

ACKNOWLEDGEMENT

Declared none.

CONFLICT OF INTEREST

The authors confirm that this chapter contents have no conflict of interest.

REFERENCES

[1] Kryukov, G. V.; Castellano, S.; Novoselov, S. V.; Lobanov, A. V.; Zehtab, O.; Guigó, R.; Gladyshev, V. N., Characterization of mammalian selenoproteomes, *Science,* **2003**, *300*, 1439-1443.

[2] Toppo, S.; Flohé, L.; Ursini, F.; Vanin, S.; Maiorino, M., Catalytic mechanisms and specificities of glutathione peroxidases: variations of a basic scheme, *Biochim. Biophys. Acta*, **2009**, *1790*, 1486-1500.

[3] Böck, A.; Forchhammer, K.; Heider, J.; Baron, C., Selenoprotein synthesis: an expansion of the genetic code, *Trends Biochem. Sci.*, **1991**, *16*, 463-467.

[4] Böck, A.; Forchhammer, K.; Heider, J.; Leinfelder, W.; Sawers, G.; Veprek, B.; Zinoni, F., Selenocysteine: the 21st amino acid, *Mol. Microbiol.*, **1991**, *5*, 515-520.

[5] Zinoni, F.; Birkmann, A.; Stadtman, T. C.; Böck, A., Nucleotide sequence and expression of the selenocysteine-containing polypeptide of formate dehydrogenase (formate-hydrogen-lyase-linked) from *Escherichia coli*, *Proc. Natl. Acad. Sci. U.S.A.*, **1986**, *83*, 4650-4654.

[6] Mugesh, G.; Du Mont, W. W.; Sies, H., Chemistry of biologically important synthetic organoselenium compounds, *Chem. Rev.*, **2001**, *101*, 2125-2179.

[7] Mugesh, G.; Singh, H. B., Synthetic organoselenium compounds as antioxidants: glutathione peroxidase activity, *Chem. Soc. Rev.*, **2000**, *29*, 347-357.

[8] Bhuyan, B. J.; Mugesh, G. *Biological and biochemical aspects of selenium compounds* in *Organoselenium Chemistry*; Wirth, T. Ed.; Wiley-VCH Verlag GmbH & Co. KGaA: Weinheim, **2012**; chapter *9*, pp. 361-396.

[9] Imai, H.; Graham, D. I.; Masayasu, H.; Macrae, I. M., Antioxidant ebselen reduces oxidative damage in focal cerebral ischemia, *Free Rad. Biol. Med.*, **2003**, *34*, 56-63.

[10] Ardais, A. P.; Santos, F. W.; Nogueira, C. W., Ebselen attenuates cadmium-induced testicular damage in mice, *J. Appl. Toxicol.*, **2008**, *28*, 322-328.

[11] Santos, F. W.; Zeni, G.; Rocha, J. B. T.; Weis, S. N.; Fachinetto, J. M.; Favero, A. M.; Nogueira, C. W., Diphenyl diselenide reverses cadmium-induced oxidative damage on mice tissues, *Chem. Biol. Inter.*, **2005**, *151*, 159-165.

[12] Da Rocha, J. T.; Pinton, S.; Mazzanti, A.; Mazzanti, C. M.; Beckemann, D. V.; Nogueira, C. W.; Zeni, G., Effects of diphenyl diselenide on lipid profile and hepatic oxidative stress parameters in ovariectomized female rats, *J. Pharm. Pharmacol.*, **2011**, *63*, 663-669.

[13] Yoshida, S.; Kumakura, F.; Komatsu, I.; Arai, K.; Onuma, Y.; Hojo, H.; Singh, B. G.; Priyadarsini, K. I.; Iwaoka, M., Antioxidative glutathione peroxidase activity of selenoglutathione, *Angew. Chem. Int. Ed.*, **2011**, *50*, 2125-2128.

[14] Huang, X.; Liu, X.; Luo, Q.; Liu, J.; Shen, J., Artificial selenoenzymes: designed and redesigned, *Chem. Soc. Rev.*, **2011**, *40*, 1171-1184.

[15] Arnér, E. S. J., Selenoproteins - what unique properties can arise with selenocysteine in place of cysteine?, *Exp. Cell. Res.*, **2010**, *316*, 1296-1303.

[16] Mishra, B.; Priyadarsini, K. I.; Mohan, H., Effect of pH on one-electron oxidation chemistry of organoselenium compounds in aqueous solutions, *J. Phys. Chem. A*, **2006**, *110*, 1894-1900.

[17] Chaudiere, J.; Courtin, O.; Leclaire, J., Glutathione oxidase activity of selenocystamine: a mechanistic study, *Arch. Biochem. Biophys.*, **1992**, *296*, 328-336.

[18] Lee, K. H.; Jeong, D., Bimodal actions of selenium essential for antioxidant and toxic pro-oxidant activities: the selenium paradox (review), *Mol. Med. Rep.*, **2012**, *5*, 299-304.

[19] Stewart, M. S.; Spallholz, J. E.; Neldner, K. H.; Pence, B. C., Selenium compounds have disparate abilities to impose oxidative stress and induce apoptosis, *Free Rad. Biol. Med.*, **1999**, *26*, 42-48.

[20] Sies, H., Oxidative stress: oxidants and antioxidants., *Exper. Phys.*, **1997**, *82*, 291-295.

[21] Jamier, V.; Ba, L. A.; Jacob, C., Selenium- and tellurium-containing multifunctional redox agents as biochemical redox modulators with selective cytotoxicity, *Chem. Eur. J.*, **2010**, *16*, 10920-10928.

[22] Meotti, F. C.; Stangherlin, E. C.; Zeni, G.; Nogueira, C. W.; Rocha, J. B. T., Protective role of aryl and alkyl diselenides on lipid peroxidation, *Environ. Res.*, **2004**, *94*, 276-282.

[23] Luchese, C.; Brandão, R.; Acker, C. I.; Nogueira, C. W., 2,2′-dipyridyl diselenide is a better antioxidant than other disubstituted diaryl diselenides, *Mol. Cell. Biochem.*, **2012**, *367*, 153-163.

[24] Malmström, J.; Jonsson, M.; Cotgreave, I. A.; Hammarström, L.; Sjödin, M.; Engman, L., The antioxidant profile of 2,3-dihydrobenzo[*b*]furan-5-ol and its 1-thio, 1-seleno, and 1-telluro analogues, *J. Am. Chem. Soc.*, **2001**, *123*, 3434-3440.

[25] Shanks, D.; Amorati, R.; Fumo, M. G.; Pedulli, G. F.; Valgimigli, L.; Engman, L., Synthesis and antioxidant profile of all-*rac*-α-selenotocopherol, *J. Org. Chem.*, **2006**, *71*, 1033-1038.

[26] Kumar, S.; Johansson, H.; Engman, L.; Valgimigli, L.; Amorati, R.; Fumo, M. G.; Pedulli, G. F., Regenerable chain-breaking 2,3-dihydrobenzo[*b*]selenophene-5-ol antioxidants, *J. Org. Chem.*, **2007**, *72*, 2583-2595.

[27] Kumar, S.; Engman, L.; Valgimigli, L.; Amorati, R.; Fumo, M. G.; Pedulli, G. F., Antioxidant profile of ethoxyquin and some of its S, Se, and Te analogues, *J. Org. Chem.*, **2007**, *72*, 6046-6055.

[28] Savegnago, L.; Jesse, C. R.; Moro, A. V.; Borges, V. C.; Santos, F. W.; Rocha, J. B. T.; Nogueira, C. W., Bis selenide alkene derivatives: a class of potential antioxidant and antinociceptive agents, *Pharmacol. Biochem. Behav.*, **2006**, *83*, 221-229.

[29] Kumar, B. S.; Kunwar, A.; Singh, B. G.; Ahmad, A.; Priyadarsini, K. I., Anti-hemolytic and peroxyl radical scavenging activity of organoselenium compounds: an *in vitro* study, *Biol. Trace Elem. Res.*, **2011**, *140*, 127-138.

[30] Battin, E. E.; Zimmerman, M. T.; Ramoutar, R. R.; Quarles, C. E.; Brumaghim, J. L., Preventing metal-mediated oxidative DNA damage with selenium compounds, *Metallomics*, **2011**, *3*, 503-512.

[31] Talas, Z. S.; Ozdemir, I.; Yilmaz, I.; Gok, Y.; Orun, I., The investigation of the antioxidative properties of the novel synthetic organoselenium compounds in some rat tissues, *Exp. Biol. Med.*, **2008**, *233*, 575-579.

[32] Tsukagoshi, H.; Koketsu, M.; Kato, M.; Kurabayashi, M.; Nishina, A.; Kimura, H., Superoxide radical-scavenging effects from polymorphonuclear leukocytes and toxicity in human cell lines of newly synthesized organic selenium compounds, *FEBS Journal*, **2007**, *274*, 6046-6054.

[33] Merino-Montiel, P.; López, Ó.; Álvarez, E.; Fernández-Bolaños, J. G., Synthesis of conformationally-constrained thio(seleno)hydantoins and α-triazolyl lactones from D-arabinose as potential glycosidase inhibitors, *Tetrahedron*, **2012**, *68*, 4888-4898.

[34] Kunwar, A.; Mishra, B.; Barik, A.; Kumbhare, L. B.; Pandey, R.; Jain, V. K.; Priyadarsini, K. I., 3,3▯ -diselenodipropionic acid, an efficient peroxyl radical scavenger and a GPx mimic, protects erythrocytes (RBCs) from AAPH-induced hemolysis, *Chem. Res. Toxicol.*, **2007**, *20*, 1482-1487.

[35] Mishra, B.; Barik, A.; Kunwar, A.; Kumbhare, L. B.; Priyadarsini, K. I.; Jain, V. K., Correlating the GPx activity of selenocystine derivatives with one-electron redox reactions, *Phosph. Sulf. Silic. Relat. Elem.*, **2008**, *183*, 1018-1025.

[36] Hossain, S. U.; Sengupta, S.; Bhattacharya, S., Synthesis and evaluation of antioxidative properties of a series of organoselenium compounds, *Bioorg. Med. Chem.*, **2005**, *13*, 5750-5758.

[37] Prabhu, P.; Bag, P. P.; Singh, B. G.; Hodage, A.; Jain, V. K.; Iwaoka, M.; Priyadarsini, K. I., Effect of functional groups on antioxidant properties of substituted selenoethers, *Free Radic. Res.*, **2011**, *45*, 461-468.

[38] Nishina, A.; Kimura, H.; Kozawa, K.; Sommen, G.; Nakamura, T.; Heimgartner, H.; Koketsu, M.; Furukawa, S., A superoxide anion-scavenger, 1,3-selenazolidin-4-one suppresses serum deprivation-induced apoptosis in PC12 cells by activating MAP kinase, *Toxicol. Appl. Pharmacol.*, **2011**, *257*, 388-395.

[39] Schumacher, R. F.; Rosário, A. R.; Souza, A. C. G.; Acker, C. I.; Nogueira, C. W.; Zeni, G., The potential antioxidant activity of 2,3-dihydroselenophene, a prototype drug of 4-aryl-2,3-dihydroselenophenes, *Bioorg. Med. Chem.*, **2011**, *19*, 1418-1425.

[40] Chakraborty, S.; Yadav, S. K.; Subramanian, M.; Priyadarsini, K. I.; Iwaoka, M.; Chattopadhyay, S., DL-*trans*-3,4-dihydroxy-1-selenolane (DHS_{red}) accelerates healing of indomethacin-induced stomach ulceration in mice, *Free Radic. Res.*, **2012**, *46*, 1378-1386.

[41] Brede, O.; Hermann, R.; Naumov, S.; Mahal, H. S., Discrete ionization of two different short-living conformers of selenophenol by rapid free electron transfer to solvent parent radical cations, *Chem. Phys. Lett.*, **2001**, *350*, 165-172.

[42] Nauser, T.; Dockheer, S.; Kissner, R.; Koppenol, W. H., Catalysis of electron transfer by selenocysteine, *Biochemistry*, **2006**, *45*, 6038-6043.

[43] Steinmann, D.; Nauser, T.; Beld, J.; Tanner, M.; Günther, D.; Bounds, P. L.; Koppenol, W. H., Kinetics of tyrosyl radical reduction by selenocysteine, *Biochemistry*, **2008**, *47*, 9602-9607.

[44] Mishra, B.; Kumar, B. S.; Priyadarsini, K. I., Pulse and γ-radiolysis studies on aqueous solution of 1,1′-dimethyl-2-selenourea, *Radiat. Phys. Chem.*, **2008**, *77*, 125-130.

[45] Mishra, B.; Kumbhare, L. B.; Jain, V. K.; Priyadarsini, K. I., Pulse radiolysis studies on reactions of hydroxyl radicals with selenocystine derivatives, *J. Phys. Chem. B*, **2008**, *112*, 4441-4446.

[46] Priyadarsini, K. I.; Mishra, B., Radical cations of some water-soluble organoselenium compounds: Insights from pulse radiolysis studies, *Radiat. Phys. Chem.*, **2008**, *77*, 1294-1299.

[47] Mishra, B.; Sharma, A.; Naumov, S.; Priyadarsini, K. I., Novel reactions of one-electron oxidized radicals of selenomethionine in comparison with methionine, *J. Phys. Chem. B*, **2009**, *113*, 7709-7715.

[48] Singh, B. G.; Thomas, E.; Kumakura, F.; Dedachi, K.; Iwaoka, M.; Priyadarsini, K. I., One-electron redox processes in a cyclic selenide and a selenoxide: A pulse radiolysis study, *J. Phys. Chem. A*, **2010**, *114*, 8271-8277.

[49] Tobien, T.; Bonifačić, M.; Naumov, S.; Asmus, K.-D., Time-resolved study on the reactions of organic selenides with hydroxyl and oxide radicals, hydrated electrons, and H-atoms in aqueous solution, and DFT calculations of transients in comparison with sulfur analogues, *Phys. Chem. Chem. Phys.*, **2010**, *12*, 6750-6758.

[50] Horvat, S. M.; Schiesser, C. H., An ab initio and DFT study of homolytic substitution reactions of acyl radicals at sulfur, selenium, and tellurium, *New J. Chem.*, **2010**, *34*, 1692-1699.

[51] Rahmanto, A. S.; Davies, M. J., Selenium-containing amino acids as direct and indirect antioxidants, *IUBMB Life*, **2012**, *64*, 863-871.

[52] Chen, T.; Wong, Y.-S., Selenocystine induces reactive oxygen species-mediated apoptosis in human cancer cells, *Biomed. Pharmacother.*, **2009**, *63*, 105-113.

[53] Plano, D.; Baquedano, Y.; Ibánez, E.; Jiménez, I.; Palop, J. A.; Spallholz, J. E.; Sanmartín, C., Antioxidant-prooxidant properties of a new organoselenium compound library, *Molecules*, **2010**, *15*, 7292-7312.

[54] Drake, E. N., Cancer chemoprevention: selenium as a prooxidant, not an antioxidant, *Medical Hypotheses*, **2006**, *67*, 318-322.

[55] Chen, J.-J.; Boylan, L. M.; Wu, C.-K.; Spallholz, J. E., Oxidation of glutathione and superoxide generation by inorganic and organic selenium compounds, *Biofactors*, **2007**, *31*, 55-66.

[56] Mézes, M.; Balogh, K., Prooxidant mechanisms of selenium toxicity - a review, *Acta Biol. Szeged.*, **2009**, *53*, 15-18.

[57] Suzuki, K. T.; Tsuji, Y.; Ohta, Y.; Suzuki, N., Preferential organ distribution of methylselenol source Se-methylselenocysteine relative to methylseleninic acid, *Toxicol. Appl. Pharmacol.*, **2008**, *227*, 76-83.

Organoselenium Chemistry Between Synthesis and Biochemistry, 2014, 379-417

CHAPTER 13

^{77}Se NMR: Theoretical Aspects and Practical Applications

Waro Nakanishi* **and Satoko Hayashi**

Department of Material Science and Chemistry, Faculty of Systems Engineering, Wakayama University, 930 Sakaedani, Wakayama 640-8510, Japan

Abstract: ^{77}Se NMR spectroscopy has been established as a powerful tool in the study of the selenium chemistry. ^{77}Se NMR chemical shifts (δ(Se)) are sharply sensitive to the structural changes in selenium compounds. Therefore, they are widely applied to analyze the chemical bonds around Se atoms and to determine the structures. On the other hand, the calculated absolute magnetic shielding tensors of Se (σ(Se)) become much reliable. It is now possible to assign the spectra based not only on the empirical rules but also on the theoretical supports. Interpretation of the shift values separately by the contributions from the occupied orbitals and the orbital-orbital transitions is the typical example of the theoretical treatment. The theoretical background for the chemical shifts is explained, together with the coupling constants (*J*(Se, X)), although briefly. ^{77}Se NMR spectra are measured on a daily basis and the structures of selenium compounds are determined through the assignment of the spectra. Applications of ^{77}Se NMR are mainly discussed so as to give good hints when the spectra are assigned, for the better understanding of the selenium chemistry. Orientational effect on σ(Se) and δ(Se) by aryl groups, mechanisms for *J*(Se, X), NICS and relativistic effect on σ(Se) are also discussed, together with the usual applications.

Keywords: Relativistic effect, *Ab initio* calculations, Ligand effects, NMR spectroscopy, Selenium, Chemical shift, Spin-spin coupling, Isotopic effect, Natural bond orbital analysis, Structures.

13.1. INTRODUCTION

NMR spectroscopy has been established as an extremely powerful tool to study physical, chemical and biological sciences. ^{77}Se NMR spectroscopy is one of typical cases [1-5]. ^{77}Se NMR chemical shifts (δ(Se)) are sharply sensitive to the structural changes in selenium compounds. Therefore, they are widely applied to analyze the chemical bonds around Se atoms and/or to determine the structures [6-21].

***Address correspondence to Waro Nakanishi:** Department of Material Science and Chemistry, Faculty of Systems Engineering, Wakayama University, 930 Sakaedani, Wakayama 640-8510, Japan; Fax: +81 73 457 8253; Tel: +81 73 457 8252; E-mail: nakanisi@sys.wakayama-u.ac.jp

[6-21]. Indirect nuclear spin-spin coupling constants containing Se atoms (*J*(Se,X)) also provide highly important information around coupled nuclei, containing strongly bonded and weakly interacting states, since the values depend on the electron distribution between the nuclei [7,22-26]. Such NMR parameters are widely determined experimentally. McFarlane and Wood measured ^{77}Se NMR chemical shifts (δ(Se)) of various organic selenium compounds systematically using the INDOR method [6]. The reported values in the early stage were summarized by Lardon [1], McFarlane [27] and then Iwamura [7]. The measurements of δ(Se) increased rapidly as the development of NMR instrument employing the pulse FT method [28]. The reported values have been summarized on a case-by-case basis [1,8,10,29]. The *J*(Se,X) values have been determined experimentally and summarized similarly [10,27,29,30].

The observed values must be absolutely true in the experimental errors. However, they will not tell the reason why they appear at the values, although some possibilities could be erased experimentally. Therefore, it becomes very important not only for the better understanding of the experimental results but also for the further development of the investigations. The reason is often practically necessary to assign the spectra and to determine the structures through the assignments. Quantum chemical (QC) calculations are necessary for the analysis and for the interpretation of δ(Se) and *J*(Se,X) with physical meanings. It is now possible to predict the δ(Se) and *J*(Se,X) values reliably, although it may be better to compare the relative values from a standard, in some cases.

This chapter explains the theoretical background and practical applications of ^{77}Se NMR. The δ(Se) and *J*(Se,X) values in solutions are mainly discussed under the nonrelativistic conditions from the viewpoint of experimental chemists. Relaxation times are not contained due to limitations of space, although they provide important informations from organic selenium compounds concerning structures, interactions, molecular dynamics and so on. The parameters in the solid states are not discussed due to limitations of space. Please confer more specialized books and reviews [31-39], if necessary.

13.2. THEORETICAL BACKGROUND OF NMR SPECTROSCOPY

13.2.1. Hamiltonian for Diamagnetic Molecules in the External Magnetic Field

In the high-resolution NMR measurements of diamagnetic molecules, the Hamiltonian can be expressed by Equation 13.1, where σ_N, $\boldsymbol{B}$, $\boldsymbol{I}_N$, $\boldsymbol{I}_L$ and $\mathbf{J}_{NL}$ are magnetic shielding tensors of nucleus N, the external magnetic field, the nuclear spin momentum operators of nuclei N and L and the indirect nuclear spin-spin coupling tensor between the nuclei N and L, respectively [38]. Equation 13.1 tells us that the nuclear spin I_N must be non-zero for the appearance of NMR signals and the nuclear spins of both I_N and I_L are non-zero for the observation of nuclear spin-spin couplings. NMR chemical shifts (σ_N) and coupling constants ($\mathbf{J}_{NL}$) will be discussed for $N = {}^{77}$Se ($I = 1/2$), which satisfies the conditions.

$$H = -\Sigma_N(h/2\pi)\gamma_N(1 - \sigma_N)\boldsymbol{B}\bullet\boldsymbol{I}_N + \Sigma_{N>L}h\mathbf{J}_{NL}(\boldsymbol{I}_N\bullet\boldsymbol{I}_L) \qquad (13.1)$$

$$\boldsymbol{B} = \nabla\times\boldsymbol{A} \qquad (13.2)$$

$$\boldsymbol{A'} = \boldsymbol{A} + \nabla f(\boldsymbol{r}) = 1/2\boldsymbol{B}\times(\boldsymbol{r} + \boldsymbol{r_0})\ (\boldsymbol{A} = 1/2\boldsymbol{B}\times\boldsymbol{r} \text{ and } f(\boldsymbol{r}) = 1/2\boldsymbol{B}\times\boldsymbol{r_0}\bullet\boldsymbol{r}) \qquad (13.3)$$

However, we must be careful when NMR parameters are calculated under the external magnetic field $\boldsymbol{B}$ according to Equation 13.1 [24,36,39a]. $\boldsymbol{B}$ is assumed to be derived from the vector potential $\boldsymbol{A}$, according to Equation 13.2. In this case, $\boldsymbol{A}$ is not determined uniquely by $\boldsymbol{B}$, since $\boldsymbol{A'}$ also satisfies Equation 13.2, where $\boldsymbol{A'}$ is given by the addition of the gradient of any scalar function $f(\boldsymbol{r})$ to $\boldsymbol{A}$ ($\boldsymbol{A'} = \boldsymbol{A} + \nabla f(r)$: see Equation 13.3), since $\boldsymbol{B}^` = \boldsymbol{B} + \nabla \times \nabla f(\boldsymbol{r}) = \boldsymbol{B}$ where $\nabla\times\nabla f(\boldsymbol{r}) = 0$. This is the reason to arise the gauge problem in the calculations of the NMR parameters. For homogeneous magnetic field, $\boldsymbol{A} = 1/2\boldsymbol{B}\times\boldsymbol{r}$ is the simplest choice for the vector potential $\boldsymbol{A}$. $\boldsymbol{A'}$ is given by $1/2\boldsymbol{B}\times(\boldsymbol{r} + \boldsymbol{r_0})$ as shown by Equation 13.3, if the scalar function is chosen as $f(\boldsymbol{r}) = 1/2\boldsymbol{B}\times\boldsymbol{r_0}\bullet\boldsymbol{r}$. Equation 13.3 is equivalent to shifting the origin of the coordinate system to the new gauge origin $-\boldsymbol{r_0}$, or alternatively translating the molecule by $\boldsymbol{r_0}$ in the Cartesian frame. This treatment is the outline of the gauge–including/independent atomic orbital (GIAO) [39a] method. GIAO is most popular in the calculations of the NMR parameters. The individual gauge for localized orbitals (IGLO) [39b] method and the localized orbital–local origin (LORG) [39c] method were also proposed, together with others.

The relativistic effect is an important issue [40] when NMR parameters are discussed especially when the system contains heavier atoms [41]. It should be essentially important to clarify the contributions of the relativistic effect also on the ^{77}Se NMR parameters. The effect will be discussed in the last section of this chapter.

13.2.2. Theoretical Treatments of Magnetic Shielding Tensors σ

The absolute magnetic shielding tensors σ are calculated theoretically to understand or explain the observed chemical shifts [42]. The diamagnetic and paramagnetic shielding tensors (σ^d and σ^p, respectively) are exactly expressed by the Ramsey's Equation [43] and they are approximately calculated in the framework of the Hartree–Fock (HF) or DFT theory. Total absolute magnetic shielding tensors (σ^t) can be decomposed into σ^d and σ^p, in the HF and DFT calculations, as shown in Equation 13.4 [44,45]. σ^p is practically given as the averaged value of the three components, as shown in Equation 13.5, so are σ^d and σ^t. As shown by Equation 13.6, σ^d is expressed simply as the sum of contributions over the occupied orbitals. Parameters σ^p can be decomposed into the contribution from the occupied orbitals or the orbital–orbital transitions through the treatment of the coupled Hartree–Fock (CPHF) method, as shown by Equation 13.7. Therefore, σ^p will be discussed separately from each MO (ψ_i) and $\psi_i \rightarrow \psi_a$ transition, where ψ_i and ψ_a are occupied and unoccupied MO's, respectively. Indeed, σ^t is also evaluated at the MP2 level, but σ^t cannot be decomposed into σ^d and σ^p at the level.

$$\sigma^t = \sigma^d + \sigma^p \tag{13.4}$$

$$\sigma^p = (\sigma^p{}_{xx} + \sigma^p{}_{yy} + \sigma^p{}_{zz})/3 \tag{13.5}$$

$$\sigma^d = \Sigma_i^{occ} \sigma^d{}_i \tag{13.6}$$

$$\sigma^p = \Sigma_i^{occ} \Sigma_a^{unocc} \sigma^p{}_{i\rightarrow a} = \Sigma_i^{occ} \sigma^p{}_i \tag{13.7}$$

$$\sigma^p{}_{zz}(N) = -(\mu_o e^2/2m_e^2)\Sigma_i^{occ}\Sigma_a^{unocc}(\varepsilon_a - \varepsilon_i)^{-1} \times$$

$$\{<\psi_i|\hat{L}_z|\psi_a><\psi_a|\hat{L}_{z,N} r_N^{-3}|\psi_i> + <\psi_i|\hat{L}_{z,N}/_N^{-3}|\psi_a><\psi_a|\hat{L}_z|\psi_i>\} \tag{13.8}$$

$$\sigma^{p}_{zz}(N) = -(\mu_0 e^2/2m_e^2) <r_N^{-3}> \Sigma_i^{occ}\Sigma_a^{unocc}(\varepsilon_a - \varepsilon_i)^{-1}\times$$

$$\{<\psi_i|\hat{L}_z|\psi_a><\psi_a|\hat{L}_{z,N}|\psi_i> + <\psi_i|\hat{L}_{z,N}|\psi_a><\psi_a|\hat{L}_z|\psi_i>\} \quad (13.9)$$

Based on the second-order perturbation theory at the level of the HF and single-excitation CI (configuration interaction) approximation [46], $\sigma^{p}_{i\rightarrow a}$ on a resonance nucleus N is shown to be proportional to reciprocal orbital energy gaps $(\varepsilon_a - \varepsilon_i)^{-1}$. σ^p can be evaluated accurately by the CPHF method and given by Equation 13.7. However, σ^p will be discussed with the approximated image derived from Equation 13.8, if suitable, where ψ_k is the kth orbital function, $\hat{L}_{z,N}$ is orbital angular momentum around the resonance nucleus N and r_N is the distance from the nucleus N [47]. Since $\sigma^{p}_{zz,N}$ contains the $\hat{L}_{z,N}$ operator, $\sigma^{p}_{zz,N}$ arises from admixtures between atomic p_x and p_y orbitals of N in various molecular orbitals. When a magnetic field is applied on a selenium compound, mixing of unoccupied molecular orbitals (ψ_a) into occupied molecular orbitals (ψ_i) will occur. Such admixtures generate $\sigma^{p}_{zz,N}$, if ψ_i and ψ_a contain p_x and p_y of N, for example. The parameters $\sigma^{p}_{xx,N}$ and $\sigma^{p}_{yy,N}$ are similarly understood. Equation 13.8 is approximately rewritten to Equation 13.9, if $<r_N^{-3}>$ can be treated as a constant for a molecule. Equation 13.9 tells us that the main factors to control σ^p are the electron population terms around the resonance nucleus N ($<r_N^{-3}>$), reciprocal orbital energy gaps $(\varepsilon_a - \varepsilon_i)^{-1}$ and the orbital overlaps controlled by the angular momentum operators ($\{<\psi_i|\hat{L}_z|\psi_a><\psi_a|\hat{L}_{z,N}|\psi_i> + <\psi_i|\hat{L}_{z,N}|\psi_a><\psi_a|\hat{L}_z|\psi_i>\}$.

13.2.3. Theoretical Treatments of Indirect Nuclear Spin-Spin Coupling Tensor *J*

According to the non-relativistic theory, there are distinct several mechanisms contributing to the spin-spin coupling constants. The total value ($^nJ_{TL}$) is composed of contributions from the diamagnetic spin-orbit (DSO) term, the paramagnetic spin-orbit (PSO) term, the spin-dipolar (SD) term and the Fermi contact (FC) term, which is expressed in Equation 13.10 [7,22-26,47].

$$^nJ_{TL} = {}^nJ_{DSO} + {}^nJ_{PSO} + {}^nJ_{SD} + {}^nJ_{FC} \quad (13.10)$$

Scheme **13.1** summarizes the mechanism of the indirect nuclear spin–spin couplings and the origin of the $^nJ_{DSO}$, $^nJ_{PSO}$, $^nJ_{SD}$ and $^nJ_{FC}$ terms (see Equation

13.10) [47]. The singlet state (S_0) of a molecule (**M**) is the ground state if the nuclei (N) in **M** have no magnetic moments ($\mu_N = 0$), but the ground state cannot be the absolute S_0 if $\mu_N \neq 0$. The DSO term arises from the reorganization of S_0 in the μ_N-perturbed ground state, so it is usually very small. The PSO term arises from the mixing of upper singlet states (S_1, S_2, S_3, *etc.*) and the FC and SD terms do from admixtures of upper triplet states (T_1, T_2, T_3, *etc.*), in which only s-type atomic orbitals contribute to FC. The contributions from ${}^nJ_{PSO}$ and ${}^nJ_{FC}$ will also be discussed separately from each MO (ψ_i) and $\psi_i \rightarrow \psi_a$ transition, where ψ_i and ψ_a are occupied and unoccupied MO's, respectively.

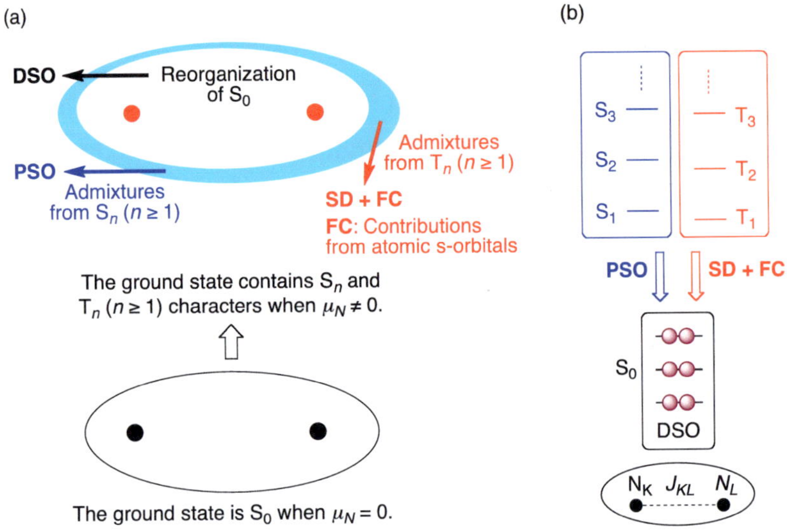

Scheme 13.1: Mechanisms for the ${}^nJ_{DSO}$, ${}^nJ_{PSO}$, ${}^nJ_{SD}$ and ${}^nJ_{FC}$ terms, contributing to ${}^nJ_{TL}$.

13.3. DEVELOPMENTS IN CALCULATIONS OF ^{77}Se NMR CHEMICAL SHIFTS

13.3.1. Early Development to Calculate σ(Se) with Gaussian-Type Atomic Orbitals

After essential establishment of the reliability in the calculations of σ(N) for N of first and/or second row nuclei, the theoretical investigations were extended to ^{77}Se nuclei around 1990s. Nakatsuji and coworkers calculated σ(Se) of nine selenium compounds [45], SeRR' (R, R' = H, Me, Et, Ph, SiH_3, GeH_3) by ab initio Hartree-Fock/finite perturbation method with the gauge origin being placed at the position of the Se atom, for example. The employed basis sets was the (13sl0p4d)/[6s5p1d] plus two polarization d-functions (ζ = 0.144 and 0.489) for Se [48]. The calculated values agreed well with the experimental values. The Se chemical shift was dominated by the Se valence 4p AO contribution to the paramagnetic term and shows a parallelism with net charge of the central Se atom. The Se chemical shift moves downfield as an increase in the electron-withdrawing ability of the ligands attached to the selenium atom (*cf*: the 4p-hole population).

The simple CPHF method [45] is pointed out to be not sufficient since very large basis sets would be required to overcome the "gauge problem" and obtain reliable results [49,50]. Ellis, Odom and coworkers proved the "gauge problem" through experimental and theoretical studies [51]. The theoretical results were carefully compared with δ_{obsd}(Se) for CH_3SeCH_3, CH_3SeH and SeH_2 measured in gas phase. These pioneering experiments undoubtedly represented one strong motivation for other theoretical investigations [45,52-55].

13.3.2. Application of σ(Se) to Orientational Effect in Aryl Selenides

Nakanishi and Hayashi reported σ(Se) calculated with the GIAO-DFT (B3LYP) method to examine the applicability of the basis set system to their aim of investigations [21], employing Gaussian 94 program package [56]. The method with the 6-311++G(3df,2pd) basis set at the B3LYP level gave excellent results [57] (r = 0.998), so did the 6-311++G(d,p) basis set. The method is applied to the planar (**pl**) and perpendicular (**pd**) structures of *p*-YC_6H_4SeH with various Y, together with the torsional angle (C_oC_iSeH) of every 15°. The orientational effect on the various *p*-YC_6H_4SeR and *p*-YC_6H_4SeAr, by skillful combination of the calculated σ(Se) values with the experimental values. More detailed applications will be explained later again.

The results of calculations facilitated the further methodological improvement, although the limited number of selenium compounds examined must also be responsible for such excellent results.

13.3.3. Examination of Basis Set System to Obtain the Reliable σ(Se)

Bayse examined the basis sets and the levels suitable to predict the δ(Se) values employing large number of selenium compounds using GIAO-MP2 and GIAO-

DFT methods [58]. Reliable chemical shifts are calculated for many selenium compounds. However, the reliability of GIAO-DFT is limited in many of the same cases as MP2 but outperforms for $[SeO_3]^{2-}$, $[SeO_4]^{2-}$, $SeCF_2$, 1,2,3-selenadiazole, cyclic dication Se_4^{2+} and related ions but underperforms in several notable cases (*e.g.*, Me_2SeO, SeO_3, $SeCl_2$ and MeSeSeMe).

However, it is difficult to suggest a generally reliable choice due to significant variance in performance with method and functional, but a basis set of at least triple-ζ quality would be recommended. Systems involving 3c–4e bonding should be augmented with diffuse functions and theoretical chemical shifts of terminal selenium anions should be treated with skepticism due to the absence of solvation effects. The limited RECP (relativistic effective core potentials) basis set [59] gives surprisingly good results, a promising development for the reliable calculation of theoretical chemical shifts for the larger molecules of interest to synthetic, biological and medicinal chemists.

13.3.4. Development to Calculate σ(Se) with Slater-Type Atomic Orbitals

Ziegler and coworkers summarized the early stage in the calculations of σ(Se):[60]. The ^{77}Se chemical shifts became to calculate employing the HF based IGLO-HF [52], GIAO-HF [51,52,54], and the correlated second order perturbation theory approaches (GIAO-MP2) [52,54], together with the GIAO based coupled cluster singles and doubles (GIAO-CCSD) method [53,61]. It is concluded that correlation effects must be included for a quantitative description of ^{77}Se chemical shifts, in particular, for a proper description of the selenium lone pairs [52-54]. The inclusion of correlation effects at the MP2 level is shown to be sufficient for many selenium systems, although higher levels such as GIAO-CCSD are required to describe the electron correlation properly of cyclic dication Se_4^{2+}, for example [52]. On the other hand, density functional theory (DFT) accounts for correlation effects implicitly and should thus be able to predict the ^{77}Se chemical shifts [52,62-64].

They tested the GIAO-DFT method [51,52,54] against other theoretical approaches and reported calculations of ^{77}Se chemical shifts and shieldings for a number of selenium containing compounds. The calculated shifts span a range of about 2800 ppm and therefore cover almost the complete range of known ^{77}Se chemical shifts. Compared to other methods, DFT is seen to give generally results of better quality than the GIAO-HF scheme. DFT is even superior to both GIAO-HF and GIAO-MP2 for highly correlated molecules, CSe_2 and Se_4^{2-}, which are the most prominent examples. This can be explained by the overestimation of correlation at the MP2 level of theory [52]. The highly correlated and expensive GIAO-CCSD method should be able to give results of comparable or better quality for such systems. Further improvement of the method containing the SOS-DFPT method (sum-over-states density-functional-perturbation-theoretical method) was also discussed. Calculations with STO can be performed in ADF program packages [65].

13.4. BASIC INFORMATION DERIVED FROM CALCULATED σ(Se)

13.4.1. Origin of δ(Se): pre-α, α, β, γ and δ Effects

The origin of δ(Se) can be clarified by analyzing σ^p(Se) with physical meanings. The α and β effects [28,66] are well known as the downfield shifts in the processes from H_2Se to MeSeH then to EtSeH, respectively, for example. An idea called the "pre-α effect" is proposed to clarify the origin of δ(Se) in a unified form [67]. The pre-α effect is defined as the downfield shift by a proton added to a lone pair orbital of Se^{2-}. Scheme **13.2** explains the process. The δ(Se) value of H_2Se (δ(Se: H_2Se)), relative to Se^{2-}, corresponds to twice as large as the pre-α effect, which would be most practical. The γ effect of the upfield shifts is also discussed, together with the mechanism, which corresponds to the process from EtSeH to *n*-PrSeH, for example [28,66].

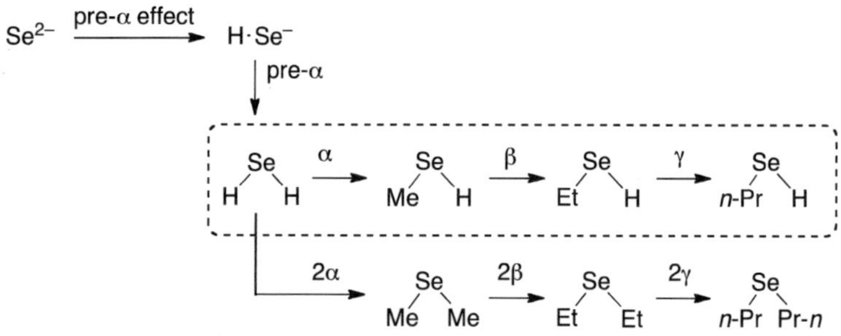

Scheme 13.2: Pre-α, α, β and γ effects.

Table **13.1** collects σ^d(Se), σ^p(Se) and σ^t(Se) of various selenides calculated by the GIAO–DFT (B3LYP) method, together with σ^t(Se) by the GIAO–MP2 method [68]. Table **13.2** shows the contributions from each ψ_i of the valence orbital to σ^p(Se) and the components ($\sigma^p(Se)_{xx}$, $\sigma^p(Se)_{yy}$ and $\sigma^p(Se)_{zz}$) in H_2Se, together with the energies

(ε_i) and the characters of ψ_i. The pre-α effect with two protons in H_2Se is explained by the generation of double σ(Se–H) and σ*(Se–H) through the protonation to the spherical Se^{2-}. Such orbitals lead effective transitions for the pre-α effect. Fig. **13.1** explains the pre-α effect.

Table 13.1: Calculated pre-α, α and β effects of various selenides, together with the observed values[a,b]

Method	**GIAO-DFT**				**GIAO-MP2**		**Observed**		
Compd	**σ^d(Se)**	**σ^p(Se)**	**σ^t(Se)**	**Effect[c,d]**	**σ^t(Se)**	**Effect[c,d]**	**δ(Se)**	**Effect[d,e]**	**Solvent**
Se^{2-} (O_h)	3005.7	0.0	3005.7		3005.1				
H_2Se (C_{2v})	2998.0	–931.3	2066.7	–469.5: pα	2252.7	–376.2: pα	–225.5[f]	222: pα[g]	nedt
MeSeH (C_s)	2998.2	–1155.0	1843.2	–223.5: α	2072.7	–180.0: α	–115[f]	111: α	$CDCl_3$
EtSeH (C_s)	3000.1	–1235.0	1765.1	–78.1: β	1995.5	–77.2: β	36[f]	151: β	$CDCl_3$
i-PrSeH (C_s)	3004.5	–1469.7	1534.8	–154.2: β	1776.2	–148.3: β	161[f]	138: β	$CDCl_3$
t-BuSeH (C_s)	3009.3	–1553.5	1455.8	–129.1: β	1696.2	–125.3: β	289[f]	135: β	$CDCl_3$
Me_2Se (C_{2v})	2999.1	–1349.0	1650.1	–208.3: α	1907.4	–172.7: α	0.0[h]	113: α	$CDCl_3$
Et_2Se (C_{2v})	3006.2	–1516.6	1489.6	–80.3: β	1747.6	–79.9: β	230[f]	115: β	$CDCl_3$
i-Pr_2Se (C_2)	3015.3	–1777.1	1238.2	–103.0: β	1476.1	–107.8: β	429[f]	107: β	$CDCl_3$
t-Bu_2Se (C_2)	3027.1	–1970.7	1056.4	–99.0: β	[i]	[i]	614[f]	102: β	$CDCl_3$

[a] The 6-311+G(3df) basis sets being employed for Se and the 6-311+G(3d,2p) basis sets for other nuclei of the Gaussian 03 program [69]. [b] In ppm. [c] In σ(Se) scale. [d] Values and the corresponding effects are shown. [e] In δ(Se) scale. [f] Ref. [70a]. [g] From δ(Se:HSeNa) = -447 in DMF (Ref. [10]). [h] δ(Se) = 13.1 in gas phase (Ref. [70b]). [i] Not calculated due to large memory requirements.

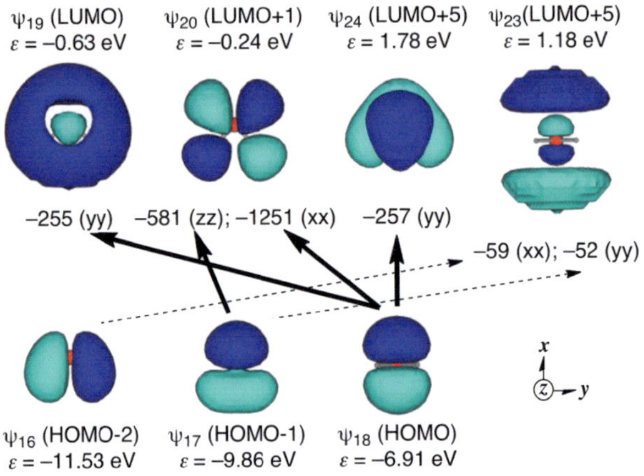

Figure 13.1: Pre-α effect in H_2Se clarified through the contribution from each $\psi_i \rightarrow \psi_a$ transition. While ψ_{17} and ψ_{22} correspond to double σ(Se–H) and σ*(Se–H) orbitals in H_2Se, although ψ_{17} and ψ_{18} are called s-type and p-type lone pair orbitals, respectively. ψ_{18} is drawn from another direction.

It is also demonstrated that the α effect is the downfield shift caused by the replacement of Se–H by Se–Me. The extension of HOMO-2 ($4p_y$(Se)), HOMO-1 ($4p_x$(Se)) and HOMO ($4p_z$(Se)) over the whole molecule of Me_2Se is mainly responsible for the α effect. The β effect originates not from the occupied-to-unoccupied ($\psi_i \rightarrow \psi_a$) transitions but from the occupied-to-occupied ($\psi_i \rightarrow \psi_j$) transitions, although not shown in Table **13.2**. The γ effect of the upfield shifts is also analyzed, although complex. The δ effect in *n*-BuSeR relative to *n*-PrSeR is negligibly small.

Table 13.2: Contributions from ψ_i to σ^p(Se) and the components in H_2Se, together with the energies and characters of ψ_i[a–d]

i in ψ_i	ε/eV	$\sigma^p(Se)_{xx}$	$\sigma^p(Se)_{yy}$	$\sigma^p(Se)_{zz}$	σ^p(Se)	Sym	Character
$\psi_i \rightarrow \psi_j$[e]		317.9	314.8	318.7	317.1		
1–14		–15.9	–5.8	–7.0	–9.6		Inner orbitals
15	–19.58	–1.4	–17.0	–29.4	–15.9	A1	σ(H_2Se: 4s(Se))
16	–11.53	–247.4	–9.6	–482.3	–246.4	B2	σ(H_2Se b_2: $4p_y$(Se))
17	–9.86	–4.5	–215.4	–831.5	–350.5	A1	σ(H_2Se a_1: $4p_x$(Se))
18	–6.91	–1312.4	–565.0	–0.3	–625.9	B1	$4p_z$(Se)
1–18	–1581.7	–812.9	–1350.6	–1248.4			
Total	–1263.8	–498.1	–1031.9	–931.3			

[a] Optimized with the 6-311+G(3df) basis sets for Se and 6-311+G(3d,2p) basis sets for other nuclei of the Gaussian 03 program [69]. [b] A utility program of Gaussian 03 (NMRANAL-NH03G) [67] is applied to separate the contributions from each molecular orbital. [c] The contribution from each molecular orbital contains only that from the occupied to unoccupied ($\psi_i \rightarrow \psi_a$) transitions. [d] σ(Se) and the components are given in ppm. [e] Corresponding to transitions from occupied to occupied orbitals.

13.4.2. Contributions from Atomic p(Se), d(Se) and f(Se) Orbitals to σ^p(Se)

Contributions from atomic p(Se), d(Se) and f(Se) orbitals to σ^p(Se) are evaluated for various selenium containing compounds, where σ^p(Se) can be imaged according to Equation 13.4 [71]. The 6-311+G(3df,3pd) basis set is employed with the GIAO-DFT (B3LYP) method for the evaluation separately by the types of atomic orbitals (s(Se), p(Se), d(Se) and f(Se)) employing the orthogonality of the AOs. The contributions from s(Se) will be zero due to the spherical distribution of electrons. The contributions are evaluated for neutral and charged Se^*H_n (* = null, +, or –) and some oxides to build the image of the contributions. The effect of methyl and halogen substitutions is also examined employing $R_rSe^*X_xO_o$ (* = null, +, or –) where R = H or Me; X = F, Cl, or Br.

Table **13.3** shows the results for Se*H$_n$ (* = null, +, or –). The p(Se) contributions are larger than 96% for SeH^- ($C_{\infty v}$), SeH_2 (C_{2v}), SeH_3^+ (C_{3v}), SeH_3^+ (D_{3h}) and SeH_4 (T_d). Therefore, σ^p(Se) of these compounds can be analyzed based on p(Se). The p(Se) contributions are 79–75% for SeH_4 (TBP), SeH_5^+ (TBP), SeH_5^+ (SP) and SeH_5^- (SP). Methyl and halogen substitutions increase the contributions by 1–2% (per Me) and 4–7% (per X), respectively. The contributions are 92–79% for H_2SeO (C_s), H_2SeO_2 (C_{2v}) and H_4SeO (C_{2v}). The values are similarly increased by the substitutions. Consequently, σ^p(Se) of these compounds can be analyzed based on p(Se) with some corrections by d(Se). The p(Se) contribution of SeH_6 (O_h) is 52%: σ^p(Se: SeH_6 (O_h)) must be analyzed based on both p(Se) and d(Se). The contributions for the Me and X derivatives of SeH_6 amount to 86–77%. Therefore, σ^p(Se) of the derivatives can also be analyzed mainly based on p(Se) with some corrections by d(Se) in this case. Contributions from f(Se) are negligible.

Table 13.3: Contributions from each MO and from atomic p(Se), d(Se) and f(Se) orbitals, together with χ(H), to σ^p(Se) in Se*H$_n$ (* = null, +, or –)[a–c]

Species	σ^p(Se)	From p(Se)	From d(Se)	From f(Se)	From χ(H)[d]	% p(Se)
SeH^- ($C_{\infty v}$)	–501.5	–496.0	–11.0	0.2	5.3	98.9
SeH_2 (C_{2v})	–928.8	–906.1	–27.5	–0.4	5.2	97.6
SeH_3^+ (C_{3v})	–1079.9	–1039.0	–48.1	0.4	6.8	96.2
SeH_4 (T_d)	–1666.8	–1629.5	–40.4	0.5	2.6	97.8
SeH_4 (C_{2v})[e]	–966.3	–765.8	–203.7	–0.1	3.3	79.3
SeH_5^+ (D_{3h})[e]	–1207.4	–911.4	–294.8	–0.1	–1.1	75.5
SeH_5^+ (C_{4v})[f]	–1159.9	–867.9	–291.3	0.0	–0.8	74.8
SeH_5^- (C_{4v})[f]	–905.9	–674.5	–235.6	–0.2	4.3	74.5
SeH_6 (O_h)	–983.6	–514.1	–467.4	0.2	–2.3	52.3

[a] Calculated with the DFT-GIAO method employing the 6-311+G(3df,3pd) basis sets. [b] A utility program (NMRANAL-NH03G) [67] being employed. [c] In ppm. [d] Contribution from AOs at H. [e] Trigonal bipyramidal (TBP). [f] Square pyramidal (SP).

13.4.3. Charge Effect on σ(Se): Evaluation of Electron Population Terms $<r_N^{-3}>$

The electron population term $<r_N^{-3}>$ is one of most important factors to control σ^p(Se), as expressed in Equations 13.8 and 13.9. Nakanishi and coworkers proposed the method to evaluate $<r_N^{-3}>$ in **M**, which employs pure p atomic orbitals [72]. They recognized that Equation 13.11 will fold for the pure p atomic orbitals for an example. Therefore, the $<r_N^{-3}>_{ij}$ values averaged by ψ_i and ψ_j, which contain pure p atomic orbitals, can be calculated according to Equation 13.12, although $<r_N^{-3}>_{\mathrm{HOMO}}$ ($\psi_i = \psi_j$ = HOMO) and $<r_N^{-3}>_{\mathrm{LUMO}}$ ($\psi_i = \psi_j$ = LUMO) will be used in place of $<r_N^{-3}>_{ij}$, if they are suitable.

$$<4p_x|\hat{L}_{z,N}\, r_N^{-3}|4p_y> = <4p_x|r_N^{-3}|4p_x><4p_x|\hat{L}_z|4p_y> \quad (13.11)$$

$$<r_N^{-3}>_{ij} = <\psi_i|r_N^{-3}|\psi_j> \quad (13.12)$$

$$y = 2.713\exp(0.0602x) - 1.713\ (R^2 = 0.999) \quad (13.13)$$

The $<r_{Se}^{-3}>_{4p}$ values are calculated for HOMO and LUMO constructed by the pure atomic 4p(Se) orbitals in Se^{4+}, Se^{2+}, Se^0 and Se^{2-} of the O_h symmetry, together with HSe^+, H_2Se and HSe^- of the $C_{\infty v}$ or C_{2v} symmetry, which satisfy the conditions to calculate $<r_{Se}^{-3}>_{4p}$, shown in Equations 13.11 and 13.12. The $<r_N^{-3}>$ values evaluate how molecular orbitals (MOs) shrink or expand. Therefore, it must be crucial to clarify whether $<r_N^{-3}>$ or the relative values ($<r_N^{-3}>_{\mathrm{rel}}$) correlate with $Q(N)$. Fig. **13.2** shows the plot of $<r_{Se}^{-3}>_{\mathrm{4p;HOMO;rel}}$ and $<r_{Se}^{-3}>_{\mathrm{4p;LUMO;rel}}$ *versus* $Q(Se)$ (natural charge evaluated with NPA analysis), although the values are not given in Table. The regression curve is given by Equation 13.13, which connects $<r_{Se}^{-3}>_{4p}$ to $Q(Se)$ through an exponential function.

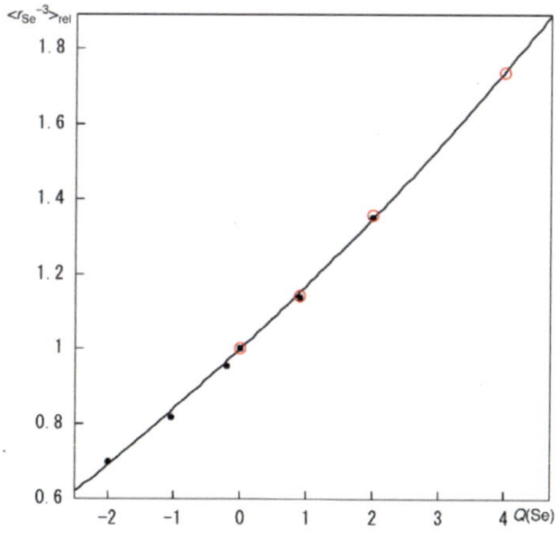

Figure 13.2: Plot of $<r_{Se}^{-3}>_{\mathrm{4p;\ HOMO;\ rel}}$ (●) and $<r_{Se}^{-3}>_{\mathrm{4p;\ LUMO;\ rel}}$ (○) *versus* Q(Se) for Se^{4+}, Se^{2+}, Se^0 and Se^{2-}, together with HSe^+, H_2Se and HSe^- evaluated with the B3LYP/6-311+G(3d) method.

13.5. APPLICATIONS OF ^{77}Se NMR

13.5.1. Determination of Structures and Detection of Intermediates in Solutions

^{77}Se NMR chemical shifts were calculated for chalcogen cation ring systems $S_xSe_{4-x}^{2+}$ (x = 0–3) with CAS, BPW91 and B3PW91 methods using the GIAO formalism [73]. The pure density functional BPW91 and CAS perform substantially better and are able to predict of these computationally with superior accuracy. BPW91 and CAS methods not only show the best reported performance to date but also unquestionably confirm the tentative assignment of the experimental NMR spectrum of $S_xSe_{4-x}^{2+}$ cations. Table **13.4** shows the calculated ^{77}Se NMR chemical shifts and the observed values. The electronic structures and molecular properties of square-planar 6π-electron ring molecules and ions E_2N_2 and E_4^{2+} (E = S, Se, Te) were also studied using various ab initio methods and density functionals.

Table 13.4: ^{77}Se NMR chemical shifts calculated at different levels of theory using cc-pVTZ basis set[a]

Species	BPW91	B3PW91	CAS	exp[b]
Se_4^{2+}	1941	2120	1893	1936
SSe_3^{2+} (<u>Se</u>Se<u>Se</u>)	1924	2087	1892	1939
SSe_3^{2+} (Se<u>Se</u>Se)	2049	2229	1965	2032
cis-$S_2Se_2^{2+}$	2042	2198	1967	2023
trans-$S_2Se_2^{2+}$	1873	2013	1858	1890
S_3Se^{2+}	2001	2135	1941	1954

[a] Values are from MeSeMe. [b] Ref. [74].

The ^{77}Se chemical shifts are also calculated for individual selenium containing compounds with various methods to predict the structures. The conformer was successfully assigned to the *anti,anti*-isomer of Se[N(mes*)]$_2$ (**1**) and *syn*- and *anti*-isomers of Se[NH(mes)]$_2$ (**2**) through the observed ^{77}Se chemical shifts supported by the calculated values (Fig. **13.3**) [75]. In this investigations, the PBE0/TZVP calculations were carried out on 28 molecules. The computed ^{77}Se shielding tensors exhibit a good linear relationship with the observed chemical shifts. The inclusion of solvent effects during geometry optimization improved the accuracy of the calculations.

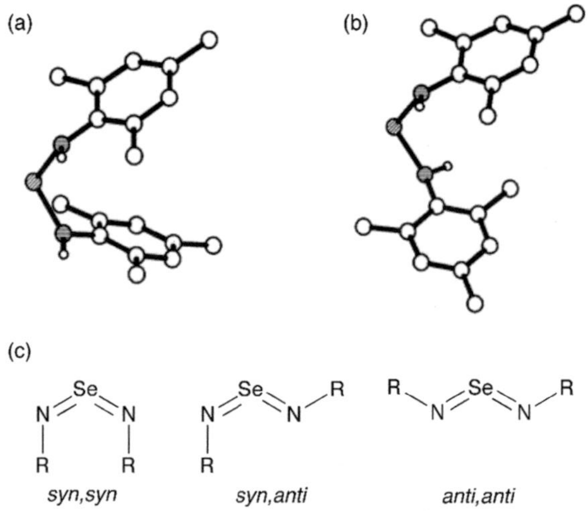

Figure 13.3: Crystal structures of *syn*- and *anti*-isomers of Se{NH(mes)}$_2$ (**2**) (a and b, respectively) with three possible conformations of Se{N(mes*)}$_2$ (**1**) (c).

The oxidation products of selenomethionine (SeMet) have been studied *via* experimental ^{77}Se NMR and theoretical ^{77}Se chemical shifts [76]. Four signals are observed: a diastereomeric pair of selenoxides at 840 ppm and two unidentified resonances at 703 and 716 ppm. Theoretical ΔG and chemical shifts suggest the 703 and 716 ppm resonances correspond to hypervalent selenuranes, formed by reaction of the selenoxide with the amine or acid group of the amino acid. The N-formyl SeMet formed only the selenoxide pair at 840 ppm. The oxidized SeMet methyl ester produced signals at 703 and 716 ppm, which are assigned as the Se–N selenurane.

13.5.2. Structural Determination of Arylselenides in Solutions: Orientational Effect on δ(Se)

Many series of δ(Se) have been reported for *para*-substituted phenyl selenides (*p*-YC_6H_4SeR; ArSeR). Odom tried to explain δ(Se) of *p*-YC_6H_4SeR based on those of the related compounds such as *p*-YC_6H_4SeMe [14]. However, the trial seemed not succeeded for *p*-$YC_6H_4SeCOPh$. The plot of δ(Se: *p*-$YC_6H_4SeCOPh$) *versus* δ(Se: *p*-YC_6H_4SeMe) did not give a good linear correlation. Nakanishi and Hayashi considered that the electronic effect on δ(Se) in *p*-$YC_6H_4SeCOPh$ must be very different from that of *p*-YC_6H_4SeMe due to the different orientation between the two. They tried to explain the structures of *p*-YC_6H_4SeR in a unified form based on the orientational effect on δ(Se) by the *p*-YC_6H_4Se group [16,20,21,77]. After elucidation of the characteristic orientational effect of arylselanyl groups on the basis of the theoretical background [77], a method was proposed and established to determine the structures of various aryl selenides in solutions [78].

It is inevitable to fix the conformation of all *p*-YC_6H_4Se groups in *p*-YC_6H_4SeR examined to clarify the relationship between δ(Se) and the structures (or conformers). Typical conformers in relation to the orientational effect of ArSeR are planar (**pl**) and perpendicular (**pd**) conformers, where the Se–C_R bond in ArSeR is on the Ar plane in **pl** and it is perpendicular to the plane in **pd**, together with the non-planer and non-perpendicular ones (**np**). 9-(Arylselanyl)anthracenes (**3**: *p*-YC_6H_4SeAtc) and 1-(arylselanyl)anthraquinones (**4**: *p*-YC_6H_4SeAtq) are chosen as the candidates for **pl** and **pd**, respectively: Y in **3** and **4** are H (**a**), NMe_2 (**b**), OMe (**c**), Me (**d**), F (**e**), Cl (**f**), Br (**g**), COOEt (**h**), CN (**i**) and NO_2 (**j**) (Chart **13.1**) [77]. 9-(arylselanyl)triptycenes (**5**: *p*-YC_6H_4SeTpc) are also employed as another standard for $\delta(Se)_{SCS}$ of **pl** [78].

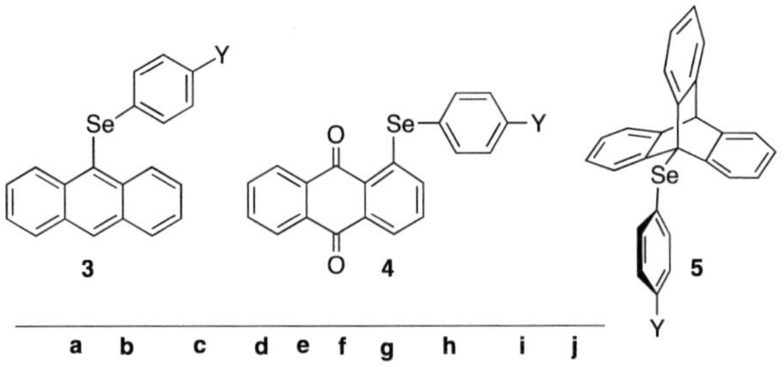

3 4 5

	a	b	c	d	e	f	g	h	i	j
Y =	H	NMe_2	OMe	Me	F	Cl	Br	COOEt	CN	NO_2

Chart 13.1: Compounds **3**–**5**.

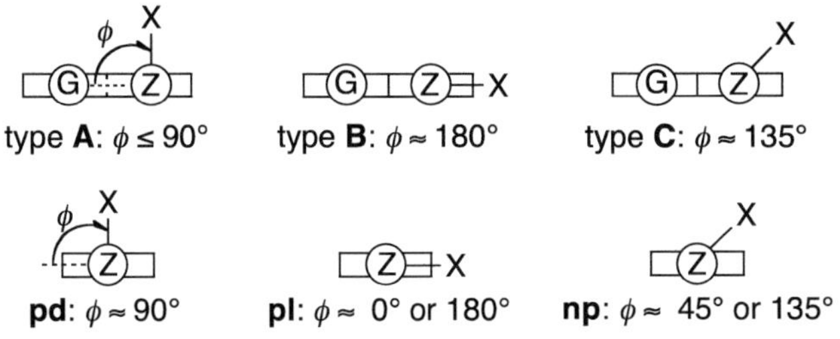

Scheme 13.3: The **pd, pl** and **np** notation in the benzene system and types **A**–**C** in the naphthalene system.

The notation of type **A** (**A**), type **B** (**B**) and type **C** (**C**) is applied on the conformers of the 9-anthracenyl (9-Atc) and 1-anthraquinoyl (1-Atq) groups in **3** and **4**, which is proposed for 1-(arylselanyl)naphthalenes (*p*-YC_6H_4SeNap) [15–21,79,80]. Scheme **13.3** explains the **A**, **B** and **C** notation with **pl**, **pd** and **np**. The structure of **3** is type **A** for 9-Atc and **pl** for Ar, which is denoted by **3** (**A**: **pl**). That of **4** is type **B** for the 1-Atq and **pd** for Ar (**4** (**B**: **pd**)). The series of δ(Se) in **3** and **4** must be typical for **pl** and **pd**, respectively, and demonstrated to serve as the standards for δ(Se) of **pl** and **pd**, respectively. The notation is also applied to the triptycyl group in **5**, although the angles between the two benzene rings are not 180º but 120º.

The driving force for (**A**: **pl**) in **3** and **5** must be the steric factor, which prevents to stabilize (**A**: **pd**) due to the repulsive force between the C_6H_4Y group and the adjacent benzene rings in the anthryl and triptycyl groups. On the other hand, (**B**: **pd**) is stabilized by the n_p(O)---σ*(Se–C) 3c–4e type interaction in **4**, which is further stabilized by the p–π conjugation of the p(Se)–π(Atq) type. The p–π conjugation of the p(Se)–π(Ar) type operates also in *p*-YC_6H_4SeR such as 1-(arylselanyl)naphthalenes (*p*-YC_6H_4SeNap), which controls the fine detail of the structures of *p*-YC_6H_4SeR especially in solutions. Scheme **13.4** illustrates the p–π conjugation in **3** and **4**, together with the steric effect in **3** [77].

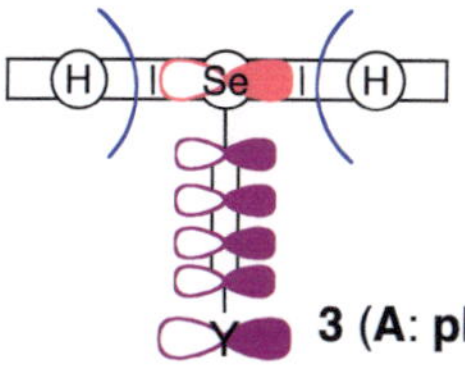

3 (A: pl)

Stabilized by steric repulsion

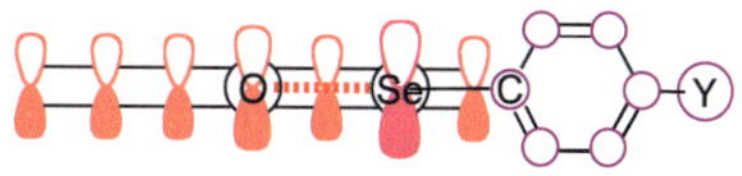

4 (B: pd)

Stabilized through n_p(O)---σ*(Se–C) interaction

Scheme 13.4: Factors to stabilize **3 (A: pl)** and **4 (B: pd)** conformations.

Table **13.5** collects the δ(Se)scs values of **3-5**, which contain various Y. The plots of δ(Se: **3**) *versus* calculated σ^t(Se: *p*-YC_6H_4SeH (**pl**))$_{rel}$ ($y = 0.932x + 0.18$: $R^2 = 0.990$) and δ(Se: **4**) *versus* calculated σ^t(Se: *p*-YC_6H_4SeH (**pd**))$_{rel}$ ($y = 0.610x - 1.73$: $R^2 = 0.976$) gave very good linear correlations, although not shown in Figures. The results confirm the (**A**: **pl**) and (**B**: **pd**) structures for **3** and **4**, respectively with all Y examined, theoretically. Fig. **13.4** shows the plots of δ(Se: **4**)$_{SCS}$ *versus* δ(Se: **3**)$_{SCS}$ and Fig. **13.5** does the plots of δ(Se: **5**)$_{SCS}$ *versus* δ(Se: **3**)$_{SCS}$. An excellent correlation is observed in the plot of Fig. **13.4**, which demonstrates that the structures of **5** are all (**A**: **pl**) for all Y examined, similarly to the case of **3**. Little difference is detected between δ(Se: **5**)$_{SCS}$ and δ(Se: **3**)$_{SCS}$, although the C(sp^2)–Se–C(sp^2) bond constructs **3**, whereas the C(sp^2)–Se–C(sp^3) bond does **5**.

Y = NMe_2 (**b**) << OMe (**c**) << Me (**d**) < F (**e**) < H (**a**)

< Cl (**f**) ≈ Br (**g**) << COOEt (**h**) << CN (**i**) << NO_2 (**j**) (for **pl**) (13.14)

Y = NMe_2 (**b**) << OMe (**c**) << F (**e**) < Me (**d**) < Cl (**f**) ≈ Br (**g**)

<< H (**a**) ≈ COOEt (**h**) < NO_2 (**j**) << CN (**i**) (for **pd**) (13.15)

Table 13.5: The δ(Se)$_{SCS}$ values of **3**–**5**[a]

Compd	NMe_2	OMe	Me	H	F	Cl	Br	CO_2Et	CN	NO_2
3	–22.7	–12.7	–6.3	0.0 (245.3)	–3.3	1.9	2.4	17.4	27.7	32.7
4	–20.6	–15.5	–9.2	0.0 (511.4)	–10.5	–7.1	–6.4	0.1	8.5	2.7
5	–21.6	–11.6	–5.9	0.0 (257.1)	–0.7	3.7	4.2	18.7	29.9	36.6

[a] δ(Se)$_{SCS}$ are given for **3**–**5**, together with δ(Se) for **3a**–**5a** in parenthesis, measured in chloroform-*d* (0.050 M) at 213K.

After establishment of the standard δ(Se)$_{SCS}$ for (**A**: **pl**) in **3** and **5** and for (**B**: **pd**) in **4**, the structural determinations of arylselenides (**6**–**21**) in solutions are examined (Chart **13.2**). Some of them are discussed next. An excellent correlation is also observed in the plot of δ(Se)$_{SCS}$ of **6** *versus* δ(Se: **3**)$_{SCS}$, which shows that the structure of **6** is all (**A**: **pl**) for all Y examined. The correlation in the plot of δ(Se: **13** with R = Me)$_{SCS}$ *versus* δ(Se: **3**)$_{SCS}$ is also very good, which demonstrates that **13** exists in (**A**: **pl**) for all Y examined in the solutions. In the case of δ(Se: **17** with R = Bz)$_{SCS}$, the plot *versus* δ(Se: **4**)$_{SCS}$ gave a very good correlation as shown in Fig. **13.6**, although the plot *versus* δ(Se: **3**)$_{SCS}$ did not give a good correlation. The results demonstrate that the structure of **17** is in (**B**: **pd**) in solutions for all Y examined. This is the reason why the plot of δ(Se: **17**)$_{SCS}$ *versus* δ(Se: **13**)$_{SCS}$ did not give good correlation in the plot [77b, 78].

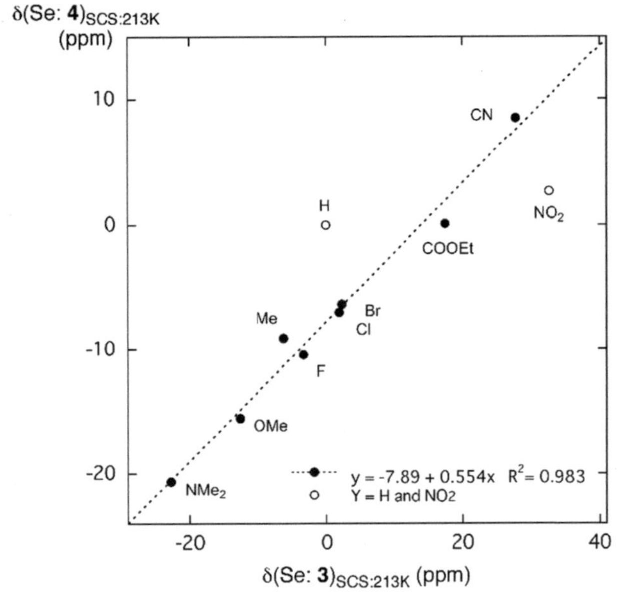

Figure 13.4: Plot of $\delta(\text{Se}:\mathbf{4})_{\text{SCS:213K}}$ *versus* $\delta(\text{Se}:\mathbf{3})_{\text{SCS:213K}}$.

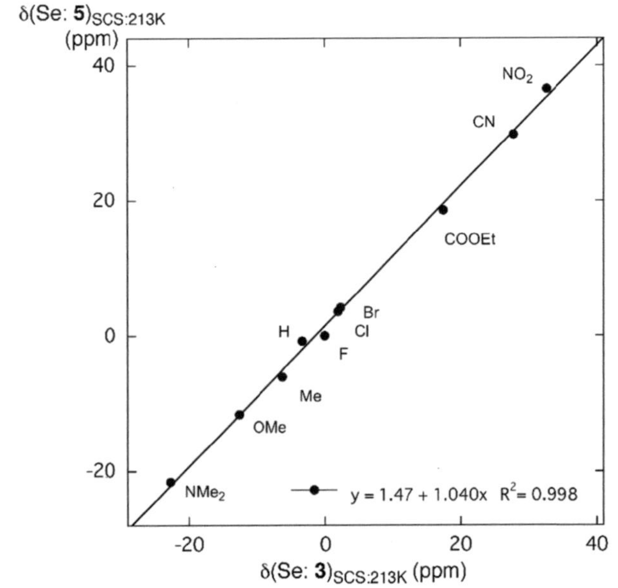

Figure 13.5: Plot of $\delta(\text{Se})_{\text{SCS:213}}$ of **5** *versus* $\delta(\text{Se})_{\text{SCS:213}}$ of **3**.

Compounds **8** and **9** contain halogens at the 8-position of 1-arylselanylnaphthlenes (**7**). It is expected that **8** and **9** exist in (**B**: **pd**) in solutions if the electron accepting ability of halogens is strong enough to keep the structures (**B**: **pd**) as in **4**. Fig. **13.7** shows the plot of δ(Se: **9**)$_{SCS}$ *versus* δ(Se: **4**)$_{SCS}$. Data for Y = OMe, Me, Cl, Br and H gave an excellent correlation, showing the (**B**: **pd**) structure for the derivatives. However, data for Y = COOEt deviate slightly from the correlation, which must be the reflection of the slight equilibrium with (**A**: **pl**). The deviations in δ(Se: **9**)$_{SCS}$ seem substantial for Y = CN and NO_2. The δ(Se: **9**)$_{SCS}$ values are almost equal with each other for Y = CN and NO_2. The results must arise from the substantial equilibrium with (**A**: **pl**). δ(Se: **8**)$_{SCS}$ showed similar trend to the case of δ(Se: **9**)$_{SCS}$. The results demonstrate that the halogens seem not so effective to keep the structure (**B**: **pd**) for all Y examined in solutions, relative to the case of **4**.

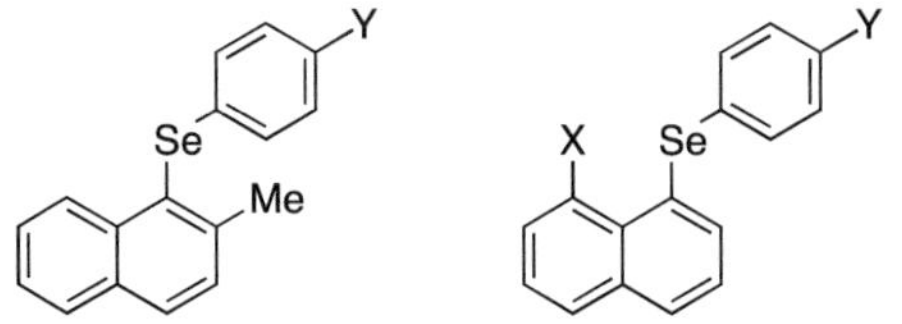

6 **7** (X = H), **8** (X = Cl)
9 (X = Br), **10** (X = SeMe)

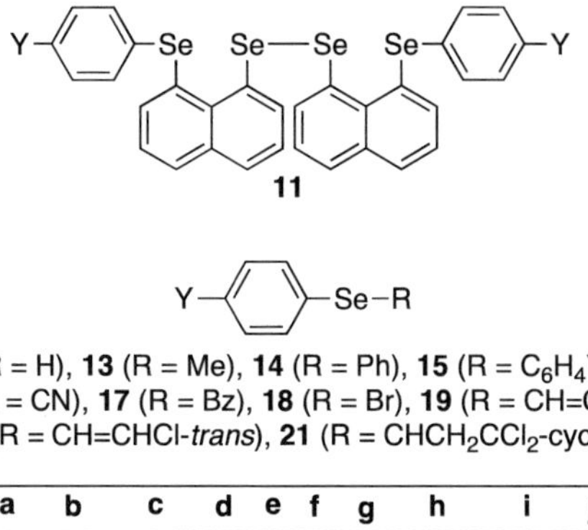

11

12 (R = H), **13** (R = Me), **14** (R = Ph), **15** (R = C_6H_4Y-*p*)
16 (R = CN), **17** (R = Bz), **18** (R = Br), **19** (R = CH=CH_2)
20 (R = CH=CHCl-*trans*), **21** (R = $CHCH_2CCl_2$-cyclo)

	a	**b**	**c**	**d**	**e**	**f**	**g**	**h**	**i**	**j**
Y =	H	NMe_2	OMe	Me	F	Cl	Br	COOEt	CN	NO_2

Chart 13.2: Compounds **6–21**.

Structures of **7** in solutions were also examined similarly. The structures are concluded to be (**B**: **pd**) for Y = NMe_2, OMe and F but (**A**: **pl**) for Cl, Br, COOEt, CN and NMe_2 with in equilibrium between (**B**: **pd**) and (**A**: **pl**) for Y = Me and H, although there are some possibility of the equilibrium for Y = F. Scheme **13.5** shows the p–π conjugation in **7**. The relative contribution controls the predominant structure of **7** in solutions.

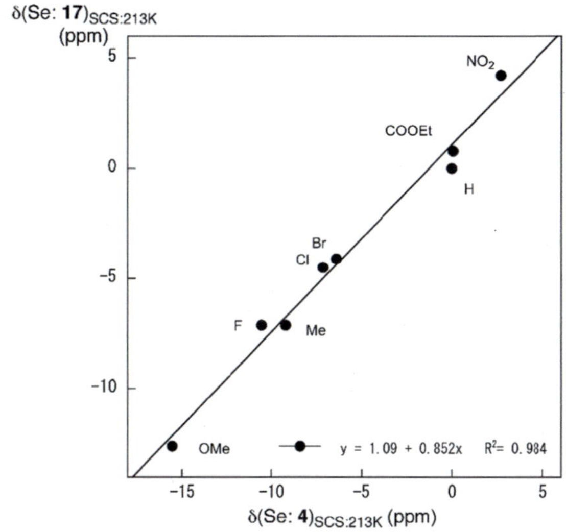

Figure 13.6: The plot of δ(Se: **17**)$_{SCS:213K}$ *versus* δ(Se: **4**)$_{SCS:213K}$.

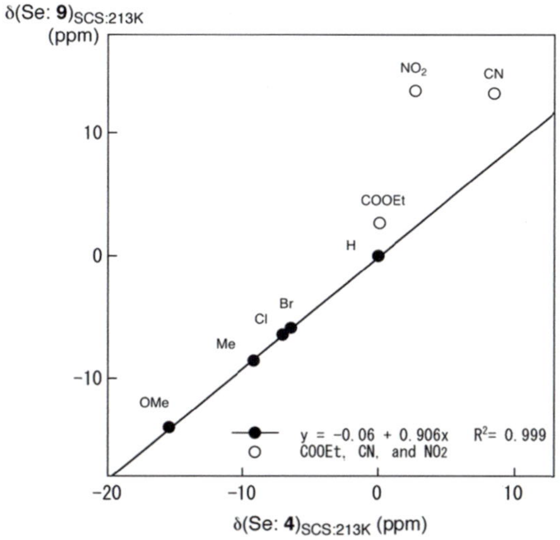

Figure 13.7: Plot of δ(Se: **9**)$_{SCS:213K}$ *versus* δ(Se: **4**)$_{SCS:213K}$.

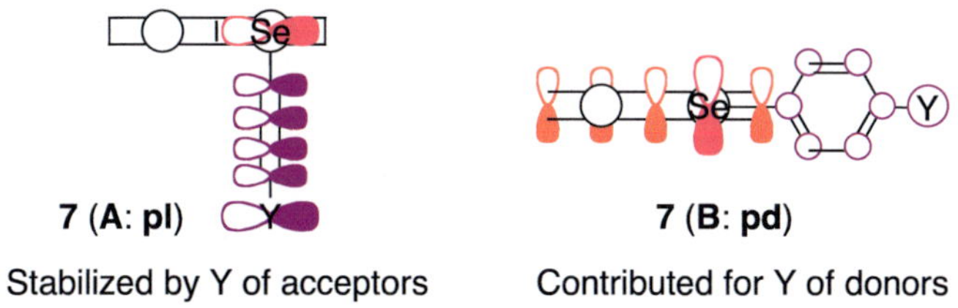

Scheme 13.5: Factors to stabilize (**A**: **pl**) and (**B**: **pd**) conformations.

Structures of other selenides are similarly examined employing the δ(Se)$_{SCS}$ values.

13.6. CALCULATIONS OF NUCLEAR SPIN-SPIN COUPLING CONSTANTS

13.6.1. ^1J(Se, X: X = C and F)

The ^1J(Se, C) values were calculated by DFT methods at the B3LYP/6-311+G(d,p) level of theory for 1-cyclohepta-2,4,6-trienyl-selanes (Se(C_7H_7)$_2$) and RSeC_7H_7 (R = *n*-Bu, *t*-Bu, Ph, *p*-FC_6H_4), together with those for various organic selenium compounds [81]. The calculated values were confirmed to be negative due to the large negative contributions from the Fermi contact (FC) term with smaller positive magnitude contributions from the spin–dipole (SD) and the paramagnetic spin-orbital (PSO) terms. The calculated ^1J(Se, C) values well reproduced the observed values. Following correlation was obtained for ca. 20 compounds: $^1J(\mathrm{Se, C})_{\mathrm{calcd}} = 0.98\,^1J(\mathrm{Se, C})_{\mathrm{obsd}} + 10.3$ ($R^2 = 0.997$).

The temperature dependence of the nuclear spin coupling between Se and F ($J_{\mathrm{Se\cdots F}}$) was investigated by Tomoda and coworkers for 2-(fluoromethyl)phenylselenenyl cyanate and bis[2-(fluoromethyl)phenyl] diselenide in CD_2Cl_2 and CD_3CN, to evaluate weak intramolecular nonbonded Se···F interactions [82-84]. Interactions of the n(G)---σ*(Se–F) type were discussed.

13.6.2. ^1J(Se, Se)

The observed ^1J(Se, Se) values ($^1J_{\mathrm{obsd}}$(Se, Se)) are reported to be rather small (≤ 64 Hz) for noncyclic RSeSeR' (**22**). On the other hand, large $^1J_{\mathrm{obsd}}$(Se, Se) values (330–380 Hz) are observed for 4-substituted naphto[1,8-*c*,*d*]-1,2-diselenoles (**23**: 379.4 Hz for **23c**, 375.9 Hz for **23d** and 330.8 Hz for **23g**), which correspond to symperiplanar diselenides (Chart **13.3**). ^1J(Se, Se: **23**) becomes larger as the electron density on Se increases [85].

The ^1J(Se, Se) values are analyzed on the basis of the molecular orbital (MO) theory with the triple ζ basis sets of the Slater type at the DFT level. ^1J(Se, Se) are calculated modeled by MeSeSeMe (**22a**), which shows the typical torsional angular dependence on ϕ(C_{Me}SeSeC_{Me}) [85]. Equation 13.10 is applied to analyze the total value ($^nJ_{TL}$). The paramagnetic spin-orbit terms contribute predominantly. Fig. **13.8** draws the plots of $^1J_{\mathrm{DSO}}$, $^1J_{\mathrm{PSO}}$, $^1J_{\mathrm{SD}}$, $^1J_{\mathrm{FC}}$, $^1J_{\mathrm{SD+FC}}$ and $^1J_{\mathrm{TL}}$ *versus* ϕ in **22a**. It is well demonstrated that $^1J_{\mathrm{TL}}$ changes depending on ϕ, similarly to the case of ^3J(H, H) [22]. The results exhibit that $^1J_{\mathrm{PSO}}$(Se, Se: **22a**) and $^1J_{\mathrm{SD+FC}}$(Se, Se: **22a**) contribute 65% and 35% to $^1J_{\mathrm{TL}}$(Se, Se: **22a**), respectively, irrespective of the ϕ(CSeSeC) values. Whereas $^1J_{\mathrm{TL}}$(Se, Se: **22a**) is predicted to be less than 44 Hz for $\phi = 90° \pm 15°$, it is very large at $\phi = 0°$ (684 Hz) and 180° (628 Hz). Therefore, $^1J_{\mathrm{obsd}}$(Se, Se: **22**) and $^1J_{\mathrm{obsd}}$(Se, Se: **23**) are explained substantially modeled by **22a** with $\phi \approx 90°$ and 0°, respectively.

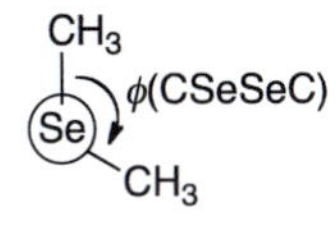

MeSeSeMe (**22a**)
RSeSeR' (**22**)

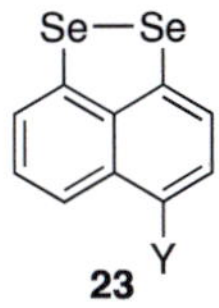

23

	a	b	c	d	e	f	g
Y =	H	OMe	Me	Cl	COOMe	CN	NO_2

Chart 13.3: Compounds **22** and **23**.

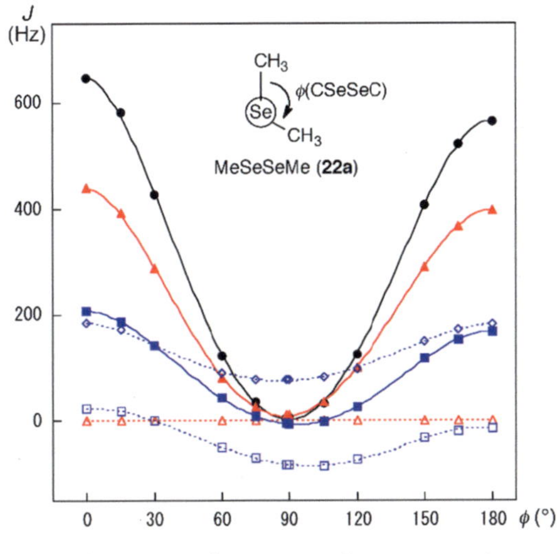

Figure 13.8: Plots of $^1J_{\mathrm{TL}}$ (●), $^1J_{\mathrm{DSO}}$ (△), $^1J_{\mathrm{PSO}}$ (▲), $^1J_{\mathrm{SD+FC}}$ (■), $^1J_{\mathrm{SD}}$ (□) and $^1J_{\mathrm{FC}}$ (◇) *versus* ϕ(CSeSeC) in ^1J(Se, Se: **22a**).

The contributions are evaluated separately from each MO (ψ_i) and each $\psi_i \rightarrow \psi_a$ transition, where ψ_i and ψ_a are occupied and unoccupied MO's, respectively. The separate evaluation of ^1J(Se, Se) enables us to recognize and visualize the origin and the mechanism of the couplings. The mechanism of ^1J(Se, Se: **22a**) is investigated employing $^1J_{PSO}$(Se, Se: **22a**), since J_{PSO}(Se, Se) contributes about 65% to $^1J_{TL}$(Se, Se: **22a**). Fig. **13.9** shows the $\psi_{42}\rightarrow\psi_{44}$ and $\psi_{43}\rightarrow\psi_{44}$ transitions at both ϕ = 0°, 90° and 180°. Characters of ψ_{42}(HOMO-1), ψ_{43}(HOMO) and ψ_{44}(LUMO) are π(Se–Se), π^*(Se–Se) and σ^*(Se–Se), respectively, at ϕ = 0° and 180°. The $\psi_{42}\rightarrow\psi_{44}$ and $\psi_{43}\rightarrow\psi_{44}$ transitions contribute almost exclusively to $^1J_{PSO}$(Se, Se: **22a**) at ϕ = 0° and 180°, whereas the transitions are almost cancelled as a whole at $\phi \approx$ 90°. This is why $^1J_{PSO}$(Se, Se: **22a**) is small at $\phi \approx$ 90°. The origin of $^1J_{PSO}$(Se, Se: **23**) is well analyzed through similar treatment. The Y dependence of ^1J (Se, Se: **23**) is also examined theoretically, which is controlled by $Qn(^1\text{Se})$.

^nJ(Se, Se) (n = 2–4) were analyzed similarly.

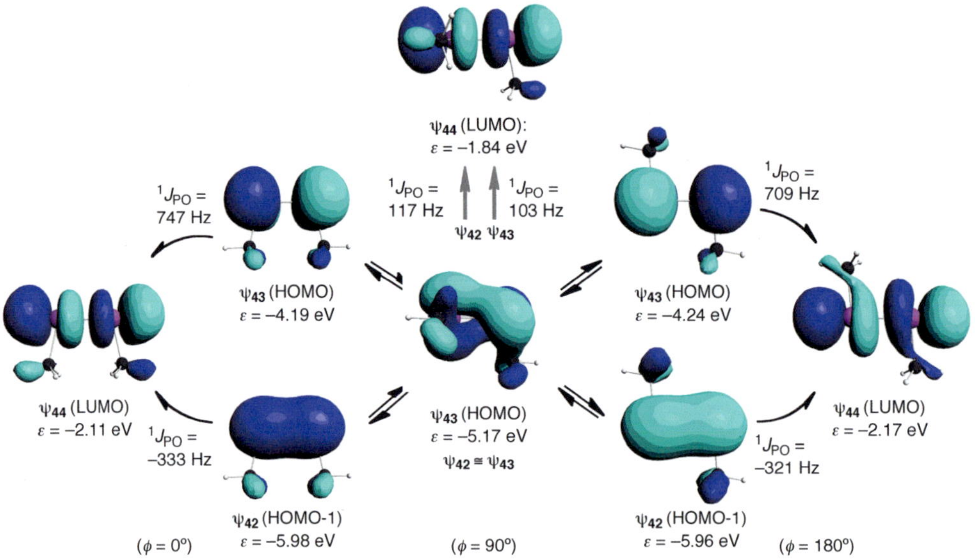

Figure 13.9: Contributions from the $\psi_{42}\rightarrow\psi_{44}$ and $\psi_{43}\rightarrow\psi_{44}$ transitions to $^1J_{PSO}$(Se, Se: **22a**) at ϕ = 0, 90 and 180°. ψ_{42} and ψ_{43} interconvert at $\phi \approx$ 90°.

13.7. INDICES FOR AROMATICITY: NUCLEUS INDEPENDENT CHEMICAL SHIFTS, NICS

Schleyer and coworkers introduced a new concept of nucleus-independent chemical shifts (NICS) in 1996 [86,87]. NICS means the magnetic shielding constants at the positions other than those of nucleus. They can be calculated but cannot usually be detected. NICS indices are the negative values of the magnetic shielding calculated at the chosen points in the vicinity of molecules. They are typically computed at ring centers, at point above and even on the grids in and around the molecules. Significantly, negative NICS values in interior positions of rings or cages indicate the presence of induced diatropic ring current, which is interpreted as aromaticity, whereas positive values denote paratropic ring currents and anti-aromaticity.

NICS values do not depend purely on the delocalized electron system such as the π system, but also on other magnetic shielding contributions due to local circulations of electron in bonds, lone pairs and atom cores. NICS(1) calculated at points 1 Å above the ring center is recommended as being better measures of π effect than NICS(0) calculated at the ring center. The shape of the magnetic shielding function provides information about the electron delocalization and molecular aromaticity. The direction and the size of the anisotropic effect of double, triple, or aromatic rings can be obtained by plotting iso-chemical-shielding surface (ICSS).

The aromatic ring-current shielding (ARCS) was introduced by Jusélius and Sundfolm in 1999 [88]. The long-range part of the shielding function (3–20 Å from the ring, for example) is used to provide information about the ring current. The isotropic magnetic shielding function $\sigma(z)$ is given by Equation 13.16 where z is the perpendicular distance from the center of the current loop, μ_0 is the vacuum permeability, $\partial I_{\text{ring}}/\partial B_{\text{ext}}$ is the current susceptibility, B_{ext} is the external magnetic field and R is the radius of the current loop. The strength of the ring-current susceptibility can be defined as the ARCS aromaticity index, which will be obtained by fitting NICS to Equation 13.16.

$$\sigma(z) = -(\mu_0/2)(\partial I_{\text{ring}}/\partial B_{\text{ext}})[R^2/(z^2 + R^2)^{3/2}] \qquad (13.16)$$

The molecular aromaticity can be studied in more detail by employing orbital decomposition of the shieldings inherent in the IGLO (individual gauge for localized orbitals) method [89,90]. The total shielding can be dissected into individual contributions from each localized molecular orbitals (LMO) associated with bonds, lone-pairs and core electrons. The procedure proposed by Pipek and Mezey for LMO-NICS separates s and p systems by means of orbital symmetry [91] when applied to planar molecules. NICS(1)$_\pi$ are very similar to NICS(1) in many cases, therefore, NICS(1) can also be used as the index of aromaticity. NICS can also be discussed separately by the contribution from each orbital (MO-NICS) [92]. The GIAO (gauge-including atomic-orbitals) method is applied in this case. NICS$_\pi$ indices can also be calculated with this method.

13.8. RELATIVISTIC EFFECT

13.8.1. Survey of Relativistic Effect

The relativistic effect will arise when the speed of an electron in an atom cannot be neglected relative to that of light. The evaluation of the relativistic effect must be very complex and difficult. However, the effect became to be evaluated successfully and dramatically under the zeroth-order regular approximation (ZORA) level [93]. The ZORA treatment is explained briefly, here.

$$E = (m_0{}^2c^4 + p^2c^2)^{1/2} - m_0c^2 + V \qquad (13.17)$$

$$= (p^2/2m_0)[1 + (E - V)/2m_0c^2]^{-1} + V \qquad (13.18)$$

$$E = p^2c^2[(2m_0c^2 - V)\{1 + E\,(2m_0c^2 - V)\}]^{-1} + V \qquad (13.19)$$

$$E_0 = p^2c^2(2m_0c^2 - V)^{-1} + V;\ E_1 = -E_0p^2c^2(2m_0c^2 - V)^{-2} \qquad (13.20)$$

Equation 13.17 shows the relativistic energy of a single electron in a scalar potential V. Equation 13.18 can be derived through dividing both numerator and denominator by c^2 after rationalizing the numerator. Equation 13.19 is derived from Equation 13.18. Equation 13.20 shows the zeroth-order and first-order terms (E_0 and E_1, respectively) of Equation 13.19, which are derived through the expansion by $E(2m_0c^2 - V)$. One realizes that E_0 in Equation 13.20 is different from the non-relativistic energy ($E = p^2/2m_0 + V$). Namely E_0 contains the relativistic effect in it. E_0 in Equation 13.20 is called zeroth-order regular approximation, which could be successfully applied to evaluate the relativistic effect.

13.8.2. Relativistic Effect on σ(Se)

The contribution of relativistic terms has been pointed out for heavier atoms [42, 94]. The effect is discussed under the zeroth-order regular approximation (ZORA). The outline of the ZORA Hamiltonian to evaluate $\sigma(N)$ with all electrons [95,96] is explained briefly in Equation 13.21, [97] where $V(\boldsymbol{r})$ is the effective Kohn-Sham potential given by the sum of the nuclear, Hartree and exchange-correlation potentials in external magnetic field $\boldsymbol{B}$ within the DFT theory, $\boldsymbol{\sigma}$ are the Pauli matrices and c is the speed of light. π and $K(\boldsymbol{r})$ are given by Equations 13.22 and 13.23, respectively. The first two terms in Equation 13.21 form the basis of the scalar relativistic approximation and the third term represents spin-orbit coupling. The last term is for the coupling between spin and magnetic field.

$\sigma^{t}(N)$ can be decomposed into $\sigma^{d}(N)$ and $\sigma^{p}(N)$ in the HF and DFT calculations under scalar relativistic conditions, [95-97] similar to the nonrelativistic case. On the other hand, $\sigma^{t}(N)$ are expressed as the sum of $\sigma^{d}(N)$, $\sigma^{p}(N)$ and $\sigma^{so}(N)$ (the contributions from the spin-orbit interaction terms) as shown in Equation 13.24, if evaluated at the spin-orbit ZORA relativistic level [95-97]. Equation 13.24 can be understood as based on Equation 13.21 by neglecting the fourth term. $\sigma^{d}(N)$ + $\sigma^{p}(N)$ (= $\sigma^{d+p}(N)$) correspond to the first two terms in Equation 13.21 and are affected by the scalar relativistic effect, whereas $\sigma^{so}(N)$ originate inherently as the spin-orbit effect.

$$H_{ZORA} = V(\boldsymbol{r}) + \pi(K(\boldsymbol{r})/2)\pi + (K^2(\boldsymbol{r})/4c^2)\boldsymbol{\sigma}\cdot[\nabla V(\boldsymbol{r})\times \boldsymbol{p}] - (K(\boldsymbol{r})/c)\boldsymbol{\sigma}\cdot\boldsymbol{p} \qquad (13.21)$$

$$\pi = \boldsymbol{p} + (1/c)\,\boldsymbol{A}(\boldsymbol{r}),\ \boldsymbol{B} = \nabla\times\boldsymbol{A}(\boldsymbol{r}) \qquad (13.22)$$

$$K(\boldsymbol{r}) = \{1 - (V(\boldsymbol{r})/2c^2)\}^{-1} \qquad (13.23)$$

$$\sigma^{t}(N) = \sigma^{d}(N) + \sigma^{p}(N) + \sigma^{so}(N) = \sigma^{d+p}(N) + \sigma^{so}(N) \qquad (13.24)$$

The relativistic effects are variously calculated and reported. A typical example for the relativistic effects on absolute magnetic shielding tensors (σ(Se)) is explained here, which are explicitly evaluated for various selenium species with the DFT(BLYP)-GIAO method [98]. Calculations are performed under relativistic and nonrelativistic conditions with the Slater-type basis sets in ADF 2010 [99] in the framework of ZORA, employing the optimized structures under nonrelativistic conditions at B3LYP of Gaussian 03 [69]. Quadruple zeta all electron with four polarization functions (QZ4Pae) are mainly applied to evaluate σ(Se). Fig. **13.10** shows the results. Ranges of the effect on diamagnetic (σ^d(Se)), paramagnetic shielding tensors (σ^p(Se)) and σ^{d+p}(Se) (= σ^d(Se) + σ^p(Se)) are –24 to –20 ppm, –115 to –3 ppm and –136 to –26 ppm, respectively. The spin-orbit terms (σ^{so}(Se)) are evaluated to be 92 – 225 ppm with QZ4Pae, which clarifies the effect on total shielding tensors (σ^t(Se) = σ^{d+p}(Se) + σ^{so}(Se)) to be –8 to 152 ppm, at the spin-orbit ZORA level. The calculated σ^t(Se) values reproduced well the observed values.

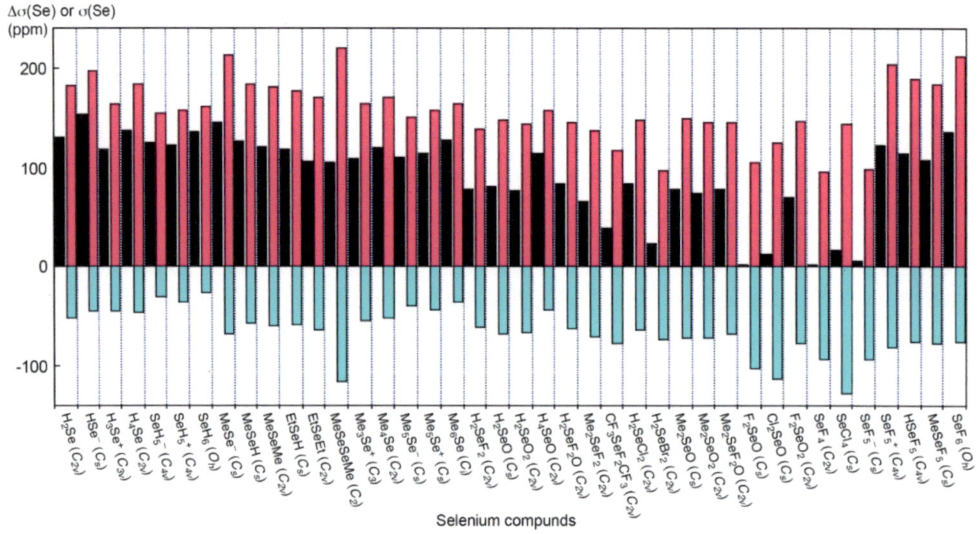

Figure 13.10: Relativistic effect on σ(Se) for various selenium compounds evaluated with the QZ4Pae basis sets: ■ stands for the total term ($\Delta\sigma^t(\mathrm{Se})_{\mathrm{Rlt\text{-}so}} = \Delta\sigma^{d+p}(\mathrm{Se})_{\mathrm{Rlt\text{-}so}} + \sigma^{so}(\mathrm{Se})_{\mathrm{Rlt\text{-}so}}$), ■ for the scalar term ($\Delta\sigma^{d+p}(\mathrm{Se})_{\mathrm{Rlt\text{-}so}}$) and ■ the spin-orbit term ($\sigma^{so}(\mathrm{Se})_{\mathrm{Rlt\text{-}so}}$).

13.8.3. Relativistic Effect on *J*(Se, X)

The relativistic ZORA-DFT method was employed to investigate indirect spin–spin coupling tensors of ^1J(Se, F) in SeF_6, together with ^1J(S, F) in SF_6 [100] and ^1J(Te, F) in TeF_6 [101]. The results demonstrate that the ZORA-DFT method reproduces the magnitude of $^1J(\mathrm{X, F})_{\mathrm{iso}}$. The calculated $^1J(\mathrm{X, F})_{\mathrm{iso}}$ for SF_6 and SeF_6 are in agreement with experimental values, $^1J(\mathrm{S, F})_{\mathrm{iso}}$ = (–)251.6 ± 0.2 Hz (liquid at 298 K) and $^1J(\mathrm{Se, F})_{\mathrm{iso}}$ = (–)1420.9 ± 0.5 Hz (liquid), deviating by –60 and –104 Hz, respectively.

13.9. CONCLUSION AND OUTLOOK

The theoretical background for ^{77}Se NMR parameters, chemical shifts (σ(Se) and δ(Se)) and coupling constants (*J*(Se, X)), is explained, containing historical developments, although briefly. Applications of ^{77}Se NMR closely relayed to the theoretical background are mainly discussed, together with usual applications, since the discussion is expected to give good hints when the spectra are assigned. It is another purpose to show the brief introduction of the theoretical treatment of ^{77}Se NMR to experimental chemists, who are interested in the theoretical background and willing to understand ^{77}Se NMR based on the theoretical support. It seems now not so difficult to measure ^{77}Se NMR spectra by the development of NMR machine and equipment. In addition, the assignment of spectra will be much easier by the support of the calculated values, through the rapid progress of computers and computational methods. It is our desire that the better understanding of selenium compounds through ^{77}Se NMR creates next idea to investigate the selenium chemistry.

ACKNOWLEDGEMENT

Declared none.

CONFLICT OF INTEREST

The authors confirm that this chapter contents have no conflict of interest.

REFERENCES

[1] *Organic Selenium Compounds: Their Chemistry and Biology*, Klayman, D.L.; Günther, W.H.H. Eds.; Wiley: New York, **1973**.

[2] *The Chemistry of Organic Selenium and Tellurium Compounds*, Patai, S.; Rappoport, Z. Eds.; Wiley: New York, **1986**; Vols. *1* and *2*.

[3] *Organic Selenium Chemistry,* Liotta, D. Ed.; Wiley-Interscience: New York, **1987**.

[4] *Organoselenium Chemistry, A practical Approach,* Back, T.G. Ed.; Oxford University Press: Oxford, **1999**.

[5] *Organoselenium Chemistry: Modern Developments in Organic Synthesis*, *Top. Curr. Chem.*, Wirth, T. Ed.; Springer, **2000**; Vol. *208*.

[6] MacFarlane, W.; Wood, R.J. Nuclear magnetic double-resonance studies of organo-selenium compounds. *J. Chem. Soc. Dalton Trans.,* **1972**, 1397-1401.

[7] Iwamura, H.; Nakanishi, W. Recent application of ^{77}Se NMR spectroscopy in organoselenium chemistry. *Yuki Gosei Kagaku Kyokaishi* (*J. Synth. Org. Chem. Jpn.*) **1981**, *39*, 795-804.

[8] *The Chemistry of Organic Selenium and Tellurium Compounds*, Patai, S.; Rappoport, Z. Eds.; Wiley: New York, **1986**; Vol. *1*, Chap. 6.

[9] Duddeck, H. Selenium-77 nuclear magnetic resonance spectroscopy. *Prog. Nucl. Magn. Reson. Spectrosc.*, **1995**, *27*, 1-323.

[10] Klapötke, T.M.; Broschag, M. *Compilation of Reported ^{77}Se NMR Chemical Shifts*, Wiley: New York, **1996**.

[11] Gronowitz, S.; Konar, A.; Hçrnfeldt, A.-B. ^{77}Se n.m.r. studies of organoselenium compounds: III-substituent effects in 4,4′-disubstituted diphenyl selenides and 4,4′-disubstituted diphenylmethanes studied by ^{1}H, ^{13}C and ^{77}Se spectroscopy. *Org. Magn. Reson.*, **1977**, *9*, 213-217.

[12] Kalabin, G.A.; Kushnarev, D.F.; Bzesovsky, V.M.; Tschmutova, G.A. ^{1}H, ^{13}C and ^{77}Se NMR spectra of substituted selenoanisoles. *Org. Magn. Reson.*, **1979**, *12*, 598-604;

[13] Kalabin, G.A.; Kushnarev, D.F.; Mannafov, T.G. NMR spectroscopy of selenium and tellurium organic compounds. III. parameters of proton, carbon-13, and selenium-77 spectra and structure of gem-dichlorocytopropylaryl selenides. *Zh. Org. Khim.*, **1980**, *16*, 505-512.

[14] Mullen, G.P.; Luthra, N,P.; Dunlap, R.B.; Odom, J.D. Synthesis and multinuclear magnetic resonance study of para-substituted phenyl selenobenzoates. *J. Org. Chem.*, **1985**, *50*, 811-816.

[15] Nakanishi, W.; Hayashi, S.; Uehara, T. Structure of 1-(arylselanyl)naphthalenes - Y dependence in 1-(p-$YC_6H_4Se)C_{10}H_7$. *Eur. J. Org. Chem.*, **2001**, 3933-3943.

[16] Hayashi, S.; Nakanishi, W. Novel substituent effect on ^{77}Se NMR chemical shifts caused by 4c-6e *versus* 2c-4e and 3c-4e in naphthalene peri positions: spectroscopic and theoretical study. *J. Org. Chem.*, **1999**, *64*, 6688-6696.

[17] Nakanishi, W.; Hayashi, S.; Yamaguchi, H. Inverse substituent effect on ^{77}Se NMR chemical shifts in naphthalene systems with linear 4c-6e Se_4 Bond: 1-[8-(p-$YC_6H_4Se)C_{10}H_6$] $SeSe[C_{10}H_6(SeC_6H_4Y$-p)-8]-1 *vs.*1-(MeSe)-8-(p-$YC_6H_4Se)C_{10}H_6$. *Chem. Lett.*, **1996**, 947-948.

[18] Nakanishi, W.; Hayashi, S.; Sakaue, A.; Ono, G.; Kawada, Y. Attractive interaction caused by the linear F···Se-C alignment in naphthalene peri positions. *J. Am. Chem. Soc.*, **1998**, *120*, 3635-3640.

[19] Nakanishi, W.; Hayashi, S. Structure of 1-(arylselanyl)naphthalenes. 2. G dependence in 8-G-1-(p-$YC_6H_4Se)C_{10}H_6$. *J. Org. Chem.*, **2002**, *67*, 38-48.

[20] Nakanishi, W.; Hayashi, S. On the factors to determine ^{77}Se NMR chemical shifts of organic selenium compounds: application of GIAO magnetic schielding tensor to the ^{77}Se NMR spectroscopy. *Chem. Lett.*, **1998**, 523-524.

[21] Nakanishi, W.; Hayashi, S. Structural study of aryl selenides in solution based on ^{77}Se NMR chemical shifts: application of the GIAO magnetic shielding tensor of the ^{77}Se nucleus. *J. Phys. Chem. A*, **1999**, *103*, 6074-6081.

[22] Karplus, M. Contact electron-spin coupling of nuclear magnetic moments. *J. Chem. Phys.,* **1959**, *30*, 11-15; Karplus, M. Vicinal proton coupling in nuclear magnetic resonance. *J. Am. Chem. Soc.,* **1963**, *85*, 2870-2871.

[23] a) Barfield, M.; Smith, W.B. Internal H-C-C angle dependence of vicinal ^{1}H-^{1}H coupling constants. *J. Am. Chem. Soc.*, **1992**, *114*, 1574-1581; b) Mallory, F.B.; Mallory, C.W.; Butler, K.E.; Lewis, M.B.; Xia, A.Q.; Luzik, Jr., E.D.; Fredenburgh, L.E.; Ramanjulu, M.M.; Van, Q.N.; Francl, M.M.; Freed, D.A.; Wray, C.C.; Hann, C.; Nerz-Stormes, M.; Carroll, P.J.; Chirlian, L.E. Nuclear spinspin coupling *via* nonbonded interactions. 8. The distance dependence of through-space fluorine-fluorine coupling. *J. Am. Chem. Soc.*; **2000**, *122*, 4108-4116.

[24] a) Helgaker, T.; Jaszunski, M.; Ruud, K. Ab initio methods for the calculation of NMR shielding and indirect spin-spin coupling constants. *Chem. Rev.*, **1999**, *99*, 293-352; b) Fukui, H. Theory and calculation of nuclear spin–spin coupling constants. *Prog. Nucl. Magn. Reson. Spectrosc.*, **1999**, *35*, 267-294; c) Bienati, M.; Adamo, C.; Barone, V. Performance of a new hybrid Hartree-Fock/Kohn-Sham model (B98) in predicting vibrational frequencies, polarisabilities and NMR chemical shifts. *Chem. Phys. Lett.*, **1999**, *311*, 69-76; d) Contreras, R.H.; Peralta, J.E.; Giribet, C.G.; Ruiz de Azua, M.C.; Facelli, J.C. Advances in theoretical and physical aspects of spin-spin coupling constants. In *Annual Reports on NMR Spectroscopy*, Webb, G.A. Ed.; Academic Press: New York, **2000**; Vol. *41*, pp 55-184.

[25] Mallory and coworkers demonstrated that the value of *J*(F,F) becomes exponentially smaller as the nonbonded F···F distance increases. The values are controlled by spin–orbit interactions. See also Ref. [23b].

[26] Helgaker, T.; Pecul, M. In *Calculations of NMR and EPR Parameters: Theory and Applications*, Kaupp, M.; BThl, M.; Malkin, V.G. Eds.; Wiley-VCH: Weinheim, **2004**, Chap. 7.

[27] McFarlane, H.C.E.; McFarlane, W. *Selenium-77 and Tellurium-125*, In *NMR of Newly Accessible Nuclei*, Laszlo, P. Ed.; Academic Press: New York, **1983**; Vol. *2*, pp 275-299.

[28] Derome, A.E. *Modern NMR Techniques for Chemistry Research*, Pergamon Press: Oxford, **1987**.

[29] Duddeck, H. *Chemical Shifts and Coupling Constants for Selenium-77, Landolt-Bornstein*: *Nuclear Magnetic Resonance (Nmr) Data, Group III*, Gupta, R.R.; Lecher, M.D. Eds.; Springer-Verlag, **2004**; Vol. *35*, Subvolume G.

[30] Duddeck, H. *Sulfur, Selenium, & Tellurium NMR* In *Encyclopedia of Nuclear Magnetic Resonance,* Grant, D.M.; Harris, R.K. Eds.; Wiley, New York, **1996**; Vol. *7*, pp 4623-4636.

[31] Scuseria, G.E.; Staroverov, V.N. In *Theory and Application of Computational Chemistry: The First 40 Years*; Dykstra, C.E.; Frenking, G.; Kim, K.S.; Scuseria, G.E. Eds.; Elsevier: Amsterdam, **2005**, Chap. 24, pp 669-724.

[32] Hehre, W.J.; Radom, L.; Schleyer, P.v.R.; Pople, J.A. *Ab Initio Molecular Orbital Theory*, John Willey & Sons: New York, **1986**.

[33] Eyring, H.; Walter, J.; Kimball, G.E. *Quantum Chemistry*, John Wiley & Sons. Inc.: New York, London, **1944**.

[34] Atkins, P.W.; Friedman, R.S. *Molecular Quantum Mechanics*, 4th ed. Oxford University Press: Oxford, New York, **2005**.

[35] Simons, J.; Nichols, J. *Quantum Mechanics in Chemistry*, Oxford University Press: New York, **1997**.

[36] Harada, Y. *Quantum Chemistry*, Shokabo, Tokyo, **2007**; Vols. *1* and *2*.

[37] Foresman, J.B.; Frisch, Æ. In *Exploring Chemistry with Electronic Structure Methods*, 2nd ed.; Gaussian, Inc.: Pittsburgh, PA, **1996**.

[38] For more detail, see Chap. 4 of Ref. [42].

[39] a) Pulay, P.; Hinton, J.F. *Shielding Theory: GIAO Method* In *Encyclopedia of Nuclear Magnetic Resonance,* Grant, D.M.; Harris, R.K. Eds.; Wiley: New York, **1996**; Vol. *7*, pp 4334-4339; b) Kutzelnigg, W.; Fleischer, U.; van Wüllen, C. *Shielding Calculations: IGLO Method* In *Encyclopedia of Nuclear Magnetic Resonance,* Grant, D.M.; Harris, R.K. Eds.; Wiley: New York, **1996**; Vol. *7*, pp 4284-4291; c) Hansen, A.E.; Bilde, M. *Shielding Calculations: LOGO & SOLO Approaches* In *Encyclopedia of Nuclear Magnetic Resonance,* Grant, D.M.; Harris, R.K. Eds.; Wiley: New York, **1996**; Vol. *7*, pp 4292-4299.

[40] a) *Introduction to Relativistic Quantum Chemistry*, Dyall, K.G.; Fægli, K. Eds.; Oxford University Press: Oxford, NY, **2007**; b) Reiher, M.; Wolf, A. *Relativistic Quantum Chemistry: The Fundamental Theory of Molecular Science*, Wiley-VCH: Weinheim, **2009**.

[41] a) Fukuda, R.; Hada, M.; Nakatsuji, H. Quasirelativistic theory for the magnetic shielding constant. I. Formulation of Douglas-Kroll-Hess transformation for the magnetic field and its application to atomic systems. *J. Chem. Phys.*, **2003**, *118*, 1015-1026; b) Fukuda, R.; Hada, M.; Nakatsuji, H. Quasirelativistic theory for magnetic shielding constants. II. Gauge-including atomic orbitals and applications to molecules. *J. Chem. Phys.*, **2003**, *118*, 1027-1035; c) Tanaka, S.; Sugimoto, M.; Takashima, H.; Hada, M.; Nakatsuji, H. Theoretical study on metal NMR chemical shifts. Electronic mechanism of the Xe chemical shift. *Bull. Chem. Soc. Jpn.*, **1996**, *69*, 953-959; d) Ballard, C.C.; Hada, M.; Kaneko, H.; Nakatsuji, H. Relativistic study of nuclear magnetic shielding constants: hydrogen halides. *Chem. Phys. Lett.*, **1996**, *254*, 170-178; e) Nakatsuji, H.; Hada, M.; Kaneko, H.; Ballard, C.C. Relativistic study of nuclear magnetic shielding constants: mercury dihalides. *Chem. Phys. Lett.*, **1996**, *255*, 195-202; f) Hada, M.; Kaneko, H.; Nakatsuji, H. Relativistic study of nuclear magnetic shielding constants: tungsten hexahalides and tetraoxide. *Chem. Phys. Lett.*, **1996**, *261*, 7-12.

[42] *Calculations of NMR and EPR Parameters: Theory and Applications*, Kaupp, M.; Bühl, M.; Malkin, V.G. Eds.; Wiley-VCH: Weinheim, Germany, **2004**.

[43] a) Ramsey, N.F. The internal diamagnetic field correction in measurements of the proton magnetic moment. *Phys. Rev.*, **1950**, *77*, 567; b) Ramsey, N.F. Magnetic shielding of nuclei in molecules. *Phys. Rev.*, **1950**, *78*, 699-703; c) Ramsey, N.F. Dependence of magnetic shielding of nuclei upon molecular orientation. *Phys. Rev.*, **1951**, *83*, 540-541; d) Ramsey, N.F. Chemical effects in nuclear magnetic resonance and in diamagnetic susceptibility. *Phys. Rev.*, **1952**, *86*, 243-246; e) Ramsey, N.F. Interactions between nuclear spins in molecules. *Phys. Rev.*, **1952**, *85*, 143-144; f) Ramsey, N.F. Electron coupled interactions between nuclear spins in molecules. *Phys. Rev.*, **1953**, *91*, 303-307.

[44] This decomposition includes small arbitrariness due to the coordinate origin dependence, though it does not damage usual chemical analyses and insights into ^{77}Se NMR spectroscopy.

[45] Nakatsuji, H.; Higashioji, T.; Sugimoto, M. Theoretical study of metal NMR chemical shifts. Selenium compounds. *Bull. Chem. Soc. Jpn.,* **1993**, *66*, 3235-3240.

[46] Foresman, J. B.; Head-Gordon, M.; Pople, M. J.; Frisch, J. A. Toward a systematic molecular orbital theory for excited states. *J. Phys. Chem.*, **1992**, *96*, 135-149. For CI: Foresman, J.B.; Frisch, Æ. *Exploring Chemistry with Electronic Structure Methods*, 2nd ed.; Gaussian, Inc.: Pittsburgh, PA, **1996**, pp 278-283.

[47] See also Refs. [67], [72], [77] and [85].

[48] *Gaussian Basis Sets for Molecular Calculations, Physical Sciences Data 16*, Fuzinaga, S. Ed.; Elsevier Science Publishing Co., **1984**.

[49] Fukui, H. Methods of calculating NMR chemical shifts. *Magn. Res. Rev.*, **1987**, *11*, 205-274.

[50] Chesnut, D.B. In *Annual Reports on NMR Spectroscopy*, Webb, G.A. Ed.; Academic: New York, **1989**; Vol. *21*, pp 51-97.

[51] Ellis, P.D. Odom, J.D. Lipton, A.S. Chen, Q. Gulick, J.M. *Gas Phase Measurement and ab initio Calculations* In *Nuclear Magnetic Shieldings and Molecular Structure*, Tossell, J.A. Ed.; Kluwer Academic Publishers: Dordrecht, **1993**; pp 539-555.

[52] Bühl, M.; Thiel, W.; Fleischer, U.; Kutzelnigg, W. Ab Initio computation of ^{77}Se NMR chemical shifts with the IGLO-SCF, the GIAO-SCF, and the GIAO-MP2 methods. *J. Phys. Chem.*, **1995**, *99*, 4000-4007.

[53] Bühl, M.; Gauss, J.; Stanton, J.F. Accurate computations of ^{77}Se NMR chemical shifts with the GIAO-CCSD method. *Chem. Phys. Lett.*, **1995**, *241*, 248-252.

[54] Magyarfalvi, G.; Pulay, P. Basis set and correlation effects in the calculation of selenium NMR shieldings. *Chem. Phys. Lett.*, **1994**, *225*, 280-284.

[55] Malkin, V.G.; Malkina, O.L.; Casida, M.E.; Salahub, D.R. Nuclear magnetic resonance shielding tensors calculated with a sum-over-states density functional perturbation theory. *J. Am. Chem. Soc.*, **1994**, *116*, 5898-5908.

[56] *Gaussian 94, Revision D.4*; Frisch, M.J.; Trucks, G.W.; Schlegel, H.B.; Gill, P.M.W.; Johnson, B.G.; Robb, M.A.; Cheeseman, J.R.; Keith, T.; Petersson, G.A.; Montgomery, J.A.; Raghavachari, K.; Al-Laham, M.A.; Zakrzewski, V.G.; Ortiz, J.V.; Foresman, J.B.; Cioslowski, J.; Stefanov, B.B.; Nanayakkara, A.; Challacombe, M.; Peng, C.Y.; Ayala, P.Y.; Chen, W.; Wong, M.W.; Andres, J.L.; Replogle, E.S.; Gomperts, R.; Martin, R.L.; Fox, D.J.; Binkley, J.S.; Defrees, D.J.; Baker, J.; Stewart, J.P.; Head-Gordon, M.; Gonzalez, C.; Pople, J.A. Gaussian, Inc.: Pittsburgh, PA, **1995**.

[57] a) Pople, J.A.; Krishnan, R.; Schlegel, H.B.; Binkley, J.S. Electron correlation theories and their application to the study of simple reaction potential surfaces. *Int. J. Quantum Chem.*, **1978**, *14*, 545-560; b) Bartlett, R.J.; Purvis III, G.D. Many-body perturbation theory, coupled-pair many-electron theory, and the importance of quadruple excitations for the correlation problem. *Int. J. Quantum Chem.*, **1978**, *14*, 561-581.

[58] Bayse, C.A. Considerations for reliable calculation of ^{77}Se chemical shifts. *J. Chem. Theory Comput,* **2005**, *1*, 1119-1128.

[59] Schwerdfeger, P., Ed.; Relativistic Electronic Structure Theory (Part 1 and 2), Elsevier and references therein.

[60] Schreckenbach, G.; Ruiz-Morales, Y.; Ziegler, T. The calculation of ^{77}Se chemical shifts using gauge including atomic orbitals and density functional theory. *J. Chem. Phys.*, **1996**, *104*, 8605-8612.

[61] Gauss, J.; Stanton, J.F. Gauge-invariant calculation of nuclear magnetic shielding constants at the coupled-cluster singles and doubles level. *J. Chem. Phys.*, **1995**, *102*, 251-253.

[62] Malkin, V.G.; Malkina, O.L.; Salahub, D.R. Calculations of NMR shielding constants by uncoupled density functional theory. *Chem. Phys. Lett.*, **1993**, *204*, 80-86.

[63] Malkin, V.G.; Malkina, O.L.; Salahub, D.R. Calculations of NMR shielding constants beyond uncoupled density functional theory. IGLO approach. *Chem. Phys. Lett.*, **1993**, *204*, 87-95.

[64] Malkin, V.G.; Malkina, O.L.; Erikson, L.A.; Salahub, D.R. In *Modern Density Functional Theory: A Tool for Chemistry*, In *Theoretical and Computational Chemistry*, Politzer, P.; Seminario, J.M. Eds.; Elsevier: Amsterdam, The Netherlands, **1995**; Vol. *2*, pp 273-347.

[65] ADF SCM, *Theoretical Chemistry*. Baerends, E. J.; Autschbach, J.; Bêrces, A.; Berger, J. A.; Bickelhaupt, F. M.; Bo, C.; de Boeij, P. L.; Boerrigter, P. M.; Cavallo, L.; Chong, D. P.; Deng, L.; Dickson, R. M.; Ellis, D. E.; van Faassen, M.; Fan, L.; Fischer, T. H.; Fonseca Guerra, C.; van Gisbergen, S. J. A.; Groeneveld, J. A.; Gritsenko, O. V.; Grüning, M.; Harris, F. E.; van den Hoek, P.; Jacob, C. R.; Jacobsen, H.; Jensen, L.; Kadantsev, E. S.; van Kessel, G.; Klooster, R.; Kootstra, F.; van Lenthe, E.; McCormack, D. A.; Michalak, A.; Neugebauer, J.; Nicu, V. P.; Osinga, V. P.; Patchkovskii, S.; Philipsen, P. H. T.; Post, D.; Pye, C. C.; Ravenek, W.; Romaniello, P.; Ros, P.; Schipper, P. R. T.; Schreckenbach, G.; Snijders, J. G.; Solâ, M.; Swart, M.; Swerhone, D.; te Velde, G.; Vernooijs, P.; Versluis, L.; Visscher, L.; Visser, O.; Wang, F.; Wesolowski, T. A.; van Wezenbeek, E. M.; Wiesenekker, G.; Wolff, S. K.; Woo, T. K.; Yakovlev, A. L.; Ziegler, T. Vrije Universiteit, Amsterdam (Netherlands). See also Guerra, C.F.; Visser, O.; Snijders, J.G.; te Velde, G.; Baerends E.J. In *Methods and Techniques for Computational Chemistry*, Clementi, E.; Corongiu, C. Eds.; STEF: Cagliari, **1995**; pp 303-395.

[66] Wrackmeyer, B.; Klimkina, E.V.; Milius, W. Molecular structures of a 1,5-diselena-2,4-disila-3-oxa-cycloheptane derivative and a diselane. Behavior of the mixture containing 1,2-dilithio- and 1,2-bis(lithioseleno)-1,2-dicarba-closo-dodecaborane(12). *Z. Anorg. Allg. Chem.*, **2011**, 1895-1902.

[67] Nakanishi, W.; Hayashi, S.; Hada, M. How ^{77}Se NMR chemical shifts originate from pre-α, α, β, and γ effects: Interpretation based on molecular orbital theory. *Chem. Eur. J.*, **2007**, *13*, 5282-5293.

[68] For calculations of σ(Se: H_2Se), Wilson, P.J. Accurate non-relativistic density functional theory predictions for ^{77}Se NMR shielding parameters. *Mol. Phys.*, **2001**, *99*, 363-367. See also Refs. [39b], [51] and [67].

[69] *Gaussian 03, Revision B.05*; Frisch, M.J.; Trucks, G.W.; Schlegel, H.B.; Scuseria, G.E.; Robb, M.A.; Cheeseman, J.R.; Montgomery, Jr, J.A.; Vreven, T.; Kudin, K.N.; Burant, J.C.; Millam, J.M.; Iyengar, S.S.; Tomasi, J.; Barone, V.; Mennucci, B.; Cossi, M.; Scalmani, G.; Rega, N.; Petersson, G.A.; Nakatsuji, H.; Hada, M.; Ehara, M.; Toyota, K.; Fukuda, R.; Hasegawa, J.; Ishida, M.; Nakajima, T.; Honda, Y.; Kitao, O.; Nakai, H.; Klene, M.; Li, X.; Knox, J.E.; Hratchian, H.P.; Cross, J.B.; Bakken, V.; Adamo, C.; Jaramillo, J.; Gomperts, R.; Stratmann, R.E.; Yazyev, O.; Austin, A.J.; Cammi, R.; Pomelli, C.; Ochterski, J.W.; Ayala, P.Y.; Morokuma, K.; Voth, G.A.; Salvador, P.; Dannenberg, J.J.; Zakrzewski, V.G.; Dapprich, S.; Daniels, A.D.; Strain, M.C.; Farkas, O.; Malick, D.K.; Rabuck, A.D.; Raghavachari, K.; Foresman, J.B.; Ortiz, J.V.; Cui, Q.; Baboul, A.G.; Clifford, S.; Cioslowski, J.; Stefanov, B.B.; Liu, G.; Liashenko, A.; Piskorz, P.; Komaromi, I.; Martin, R.L.; Fox, D.J.; Keith, T.; Al-Laham, M.A.; Peng, C.Y.; Nanayakkara, A.; Challacombe, M.; Gill, P.M.W.; Johnson, B.; Chen, W.; Wong, M.W.; Gonzalez, C.; Pople, J.A. Gaussian, Inc.: Pittsburgh, PA, **2003**.

[70] a) Luthra, N.P.; Boccanfuso, A.M.; Dunlap, R.B.; Odom, J.D. Studies in selenium-77 and tellurium-125 nuclear magnetic resonance. Substituent effects and polarizability. *J. Organomet. Chem.*, **1988**, *354*, 51-62; b) Luthra, N.P.; Dunlap, R.B.; Odom, J.D. The use of dimethyl selenide as a chemical shift reference in ^{77}Se NMR spectroscopy. *J. Magn. Reson.*, **1983**, *52*, 318-322.

[71] Nakanishi, W.; Hayashi, S.; Narahara, K.; Hada, M. Contributions from atomic p(Se), d(Se) orbitals to absolute paramagnetic shielding tensors in neutral and charged SeH_n and some oxides including the effect of methyl and halogen substitutions on σ^{P}(Se). *Chem. Eur. J.*, **2008,** *14*, 9647-9655.

[72] Nakanishi, W.; Hayashi, S.; Narahara, K.; Yamaki, D.; Hada, M. Evaluation of electron population terms for $\langle r_{Se}^{-3}\rangle_{4p}$, $\langle r_{S}^{-3}\rangle_{3p}$, and $\langle r_{O}^{-3}\rangle_{2p}$: How do HOMO and LUMO shrink or expand depending on nuclear charges? *Chem. Eur. J.*, **2008**, *14*, 7278-7284.

[73] Tuononen, H.M.; Suontamo, R.; Valkonen, J.; Laitinen, R.S. Electronic structures and spectroscopic properties of 6π-electron ring molecules and ions E_2N_2 and E_4^{2+} (E = S, Se, Te). *J. Phys. Chem. A*, **2004**, *108*, 5670-5677.

[74] Collins, M.J.; Gillespie, R.J.; Sawyer, J.F.; Schrobilgen, G.J. Selenium-77 NMR and Raman study of $S_xSe_{4-x}^{2+}$ cations and crystal structure of $(S_{3.0}Se_{1.0})_2(Sb_4F_{17})(SbF_6)_3$ containing a disordered mixture of $S_xSe_{4-x}^{2+}$ cations. *Inorg. Chem.*, **1986**, *25*, 2053-2057.

[75] Maaninen, T.; Tuononen, H.M.; Kosunen, L.; Oilunkaniemi, R.; Hiitola, J.; Laitinen, R.; Chivers, T. Formation, structural characterization, and calculated NMR chemical shifts of selenium-nitrogen compounds from $SeCl_4$ and ArNHLi (Ar = supermesityl, mesityl). *Z. Anorg. Allg. Chem.*, **2004**, *630*, 1947-1954.

[76] Ritchey, J.A.; Davis, B.M.; Pleban, P.A.; Bayse, C.A. Experimental and theoretical evidence for cyclic selenurane formation during selenomethionine oxidation. *Org. Biomol. Chem.*, **2005**, *3*, 4337-4342.

[77] a) Nakanishi, W.; Hayashi, S.; Shimizu, D.; Hada, M. Orientational effect of aryl groups on ^{77}Se NMR chemical shifts: Experimental and theoretical investigations. *Chem. Eur. J.*, **2006**, *12*, 3829-3846; b) Hayashi, S.; Nakanishi, W. Proposal for sets of ^{77}Se NMR

chemical shifts in planar and perpendicular orientations of aryl group and the applications. *Bioinorg. Chem. Appl.*, **2006**, DOI: 10.1155/BCA/2006/79327.

[78] Nakamoto, T.; Hayashi, S.; Nakanishi, W. ^{77}Se NMR chemical shifts of 9-(arylselanyl)triptycenes: New standard for planar structures of ArSeR and applications to determine the structures in solutions. *J. Org. Chem.*, **2008**, *73*, 9259-9269.

[79] Hayashi, S.; Wada, H.; Ueno, T.; Nakanishi, W. Structures of 1-(arylseleninyl)naphthalenes: O, G, and Y dependences in 8-G-1-[*p*-$YC_6H_4Se(O)]C_{10}H_6$. *J. Org. Chem.*, **2006**, *71*, 5574-5585.

[80] Nakanishi, W.; Hayashi, S.; Uehara, T. Successive change in conformation caused by *p*-Y groups in 1-(MeSe)-8-(*p*-$YC_6H_4Se)C_{10}H_6$: Role of linear Se···Se-C three-center-four-electron *versus* n(Se)···n(Se) two-center-four-electron nonbonded interactions. *J. Phys. Chem. A*, **1999**, *103*, 9906-9912.

[81] Wrackmeyer, B.; Hernandez, Z.G.; Herberhold, M. 1-Cyclohepta-2,4,6-trienyl-selanes-a ^{77}Se NMR study: Indirect nuclear ^{77}Se-^{13}C spin-spin coupling constants and application of density functional theory (DFT) calculations. *Magn. Reson. Chem.*, **2007**, *45*, 198-204.

[82] Iwaoka, M.; Komatsu, H.; Katsuda, T.; Tomoda, S. Nature of nonbonded Se···O interactions characterized by ^{17}O NMR spectroscopy and NBO and AIM analyses. *J. Am. Chem. Soc.*, **2004**, *126*, 5309-5317.

[83] a) Iwaoka, M.; Tomoda, S. Direct observation of intramolecular interaction between a divalent selenium and a tertiary amine by means of single crystal X-ray analysis and NMR spectroscopy. *Phosphorus, Sulfur Silicon Relat. Elem.* **1992**, *67*, 125-130; b) Iwaoka, M.; Tomoda, S. Nature of the intramolecular Se···N nonbonded interaction of 2-selenobenzylamine derivatives. An experimental evaluation by ^{1}H, ^{77}Se, and ^{15}N NMR spectroscopy. *J. Am. Chem. Soc.,* **1996**, *118*, 8077-8084; c) Iwaoka, M.; Komatsu, H.; Tomoda, S. Characterization of an intramolecular nonbonded interaction between selenium and fluorine in solution. *Chem. Lett.*, **1998**, 969-970; d) Komatsu, H.; Iwaoka, M.; Tomoda, S. Intramolecular non-bonded interaction between selenium and oxygen as revealed by ^{17}O and ^{77}Se NMR spectroscopy and natural bond orbital analysis. *Chem. Commun.*, **1999**, 205-206; e) Iwaoka, M.; Katsuda, T.; Komatsu, H.; Tomoda, S. Experimental and theoretical studies on the nature of weak nonbonded interactions between divalent selenium and halogen atoms. *J. Org. Chem.*, **2005**, *70*, 321-327.

[84] Iwaoka, M.; Komatsu, H.; Katsuda, T.; Tomoda, S. Quantitative evaluation of weak nonbonded Se···F interactions and their remarkable nature as orbital interactions. *J. Am. Chem. Soc.*, 2002, **124**, 1902-1909. See also, Ref. [83c].

[85] a) Nakanishi, W.; Hayashi, S. Torsional angular dependence of ^{1}J(Se,Se) and Fermi contact control of ^{4}J(Se,Se): Analysis of ^{n}J(Se,Se) (n =1-4) based on molecular orbital theory. *Chem. Eur. J.*, **2008**, *14*, 5645-5658; b) Tanioku, A.; Hayashi, S.; Nakanishi, W. Analysis of one-bond Se-Se nuclear couplings in diselenides and 1,2-diselenoles on the basis of molecular orbital theory: Torsional angular dependence, electron density influence, and origin in ^{1}J(Se, Se). *Bioinorg. Chem. Appl.*, **2009**, DOI: 10.1155/2009/381925.

[86] Schleyer, P.v.R.; Maerker, C.; Dransfeld, A.; Jiao, H.; Hommes, N.J.; Hommes, R.v.E. Nucleus-independent chemical shifts: A simple and efficient aromaticity probe. *J. Am. Chem. Soc.*, **1996**, *118*, 6317-6318.

[87] Subramanian, G.; Schleyer, P.v.R.; Jiao, H. Are the most stable fused heterobicycles the most aromatic? *Angew. Chem., Int. Ed. Engl.*, **1996**, *35*, 2638-2641.

[88] a) Jusélius, J.; Sundholm, D. Ab initio determination of the induced ring current in aromatic molecules. *Phys. Chem. Chem. Phys.*, **1999**, *1*, 3429-3435; b) Jusélius, J.; Sundholm, D. The aromatic pathways of porphins, chlorins and bacteriochlorins. *Phys. Chem. Chem. Phys.*, **2000**, *2*, 2145-2151; c) Jusélius, J.; Sundholm, D. The aromatic character of magnesium porphyrins. *J. Org. Chem.*, **2000**, *65*, 5233-5237; d) Jusélius, J.; Sundholm, D. The aromaticity and antiaromaticity of dehydroannulenes. *Phys. Chem. Chem. Phys.*, **2001**, *3*, 2433-2437; e) Jusélius, J.; Straka, M.; Sundholm, D. Magnetic-Shielding Calculations on Al_4^{2-} and Analogues. A New Family of Aromatic Molecules? *J. Phys. Chem. A,* **2001**, *105*, 9939-9944; f) Berger, R.J.F.; Schmidt, M.A.; Jusélius, J.; Sundholm, D.; Schmidbaur, H. Tetraberyllium-η^4-oxo-hexa(arylcarboxylates). *Z. Naturforsch.,* **2001**, *56b*, 979-989; g) Jusélius, J.; Patzschke, M.; Sundholm, D. Calculation of ring-current susceptibilities for potentially homoaromatic hydrocarbons. *J. Mol. Struct.: THEOCHEM,* **2003**, *633*, 123-136.

[89] a) Kutzelnigg, W. Theory of magnetic-susceptibilities and NMR chemical-shifts in terms of localized quantities. *Isr. J. Chem.*, **1980**, *19*, 193-200; b) Schindler, M.; Kutzelnigg, W. Theory of magnetic susceptibilities and NMR chemical shifts in terms of localized quantities. II. Application to some simple molecules. *J. Chem. Phys.*, **1982**, *76*, 1919-1933; c) Schindler, M.; Kutzelnigg, W. Theory of magnetic susceptibilities and NMR chemical shifts in terms of localized quantities. 3. Application to hydrocarbons and other organic molecules. *J. Am. Chem. Soc.*, **1983**, *105*, 1360-1370; d) Schindler, M.; Kutzelnigg, W. Theory of magnetic susceptibilities and N.M.R. chemical shifts in terms of localized quantities IV. Some small molecules with multiple bonds (N_2, HCN, CO, C_2H_2, CO_2, N_2O, O_3, FNO). *Mol. Phys.*, **1983**, *48*, 781-798.

[90] For reviews, see: a) Kutzelnigg, W.; Fleischer, U.; Schindler, M. In *NMR Basic Principles and Progress*, Diehl, P.; Fluck, E.; Kosfeld, R. Eds.; Springer: Berlin, **1990**; Vol. *23*, pp 165-262; b) Kutzelnigg, W.; Wüllen, Ch.van; Fleischer, U.; Franke, R.; Mourik, T.v. In *NMR Shieldings and Molecular Structure*, Tossell, J. A. Ed.; Kluwer Academic Publishers: Norwell, MA, **1993**.

[91] Pipek, J.; Mezey, P.G. A fast intrinsic localization procedure applicable for ab initio and semiempirical linear combination of atomic orbital wave functions. *J. Chem. Phys.*, **1998**, *90*, 4916-4926.

[92] a) Heine, T.; Schleyer, P.v.R.; Corminboeuf, C.; Seifert, G.; Reviakine, R.; Weber, J. Analysis of aromatic delocalization: Individual molecular orbital contributions to nucleus-independent chemical shifts. *J. Phys. Chem. A*, **2003**, *107*, 6470-6475; b) Bohmann, J.A;. Weinhold, F.; Farrar, T.C. Natural chemical shielding analysis of nuclear magnetic resonance shielding tensors from gauge-including atomic orbital calculations. *J. Chem. Phys.*, **1997**, *107*, 1173-1184; c) Corminboeuf, C.; Heine, T.; Seifert, G.; Schleyer, P.v.R.; Weber, J. Induced magnetic fields in aromatic [n]-annulenes-interpretation of NICS tensor components. *Phys. Chem. Chem. Phys.*, **2004**, *6*, 273-276.

[93] a) van Lenthe, E.; Baerends, E.J.; Snijders, J.G. Relativistic regular two-component Hamiltonians. *J. Chem. Phys.*, **1993**, *99*, 4597-4610; b) van Lenthe, E.; Baerends, E.J.;

Snijders, J.G. Relativistic total energy using regular approximations. *J. Chem. Phys.*, **1994**, *101*, 9783-9792; c) van Lenthe, E.; van Leeuwen, R.; Baerends, E.J.; Snijders, J.G. Relativistic regular two-component Hamiltonians. *Int. J. Quantum Chem.*, **1996**, *57*, 281-293; d) van Lenthe, E.; Snijders, J.G.; Baerends, E.J. The zero-order regular approximation for relativistic effects: The effect of spin–orbit coupling in closed shell molecules. *J. Chem. Phys.*, **1996**, *105*, 6505-6516; e) van Lenthe, E.; Ehlers, A.E.; Baerends, E.J. Geometry optimizations in the zero order regular approximation for relativistic effects. *J. Chem. Phys.*, **1999**, *110*, 8943-8953.

[94] Boese, A.D.; Martin, J.M.L. Development of density functionals for thermochemical kinetics. *J. Chem. Phys.*, **2004**, *121*, 3405-3416.

[95] a) Schreckenbach, G.; Ziegler, T. Calculation of NMR shielding tensors using gauge-including atomic orbitals and modern density functional theory. *J. Phys. Chem.*, **1995**, *99*, 606-611; b) Schreckenbach, G.; Ziegler, T. The calculation of NMR shielding tensors based on density functional theory and the frozen-core approximation. *Int. J. Quantum Chem.*, **1996**, *60*, 753-766; c) Schreckenbach, G.; Ziegler, T. Calculation of NMR shielding tensors based on density functional theory and a scalar relativistic Pauli-type Hamiltonian. The application to transition metal complexes. *Int. J. Quantum Chem.*, **1997**, *61*, 899-918.

[96] a) Wolff, S.K.; Ziegler, T. Calculation of DFT-GIAO NMR shifts with the inclusion of spin-orbit coupling. *J. Chem. Phys.*, **1998**, *109*, 895-905; b) Wolff, S.K.; Ziegler, T.; van Lenthe, E.; Baerends, E.J. Density functional calculations of nuclear magnetic shieldings using the zeroth-order regular approximation (ZORA) for relativistic effects: ZORA nuclear magnetic resonance. *J. Chem. Phys.*, **1999**, *110*, 7689-7695.

[97] Yates, J.R.; Pickard, C.J.; Payne, M.C.; Mauri, F. Relativistic nuclear magnetic resonance chemical shifts of heavy nuclei with pseudopotentials and the zeroth-order regular approximation. *J. Chem. Phys.,* **2003, 118**, 5746-5753.

[98] Nakanishi, W.; Hayashi, S.; Katsura, Y.; Hada, M. Relativistic Effect on ^{77}Se NMR Chemical Shifts of Various Selenium Species in the Framework of Zeroth-Order Regular Approximation. *J. Phys. Chem. A*, **2011**, *115*, 8721-8730.

[99] Baerends, E.J.; Ziegler, T.; Autschbach, J.; Bashford, D.; Berces, A.; Bickelhaupt, F.M.; Bo, C.; Boerrigter, P.M.; Cavallo, L.; Chong, D.P.; Deng, L.; Dickson, R.M.; Ellis, D.E.; van Faassen, M.; Fan, L.; Fischer, T.H.; Fonseca Guerra, C.; Ghysels, A.; Giammona, A.; van Gisbergen, S.J.A.; Götz, A.W.; Groeneveld, J.A.; Gritsenko, O.V.; Grüning, M.; Gusarov, S.; Harris, F.E.; van den Hoek, P.; Jacob, C.R.; Jacobsen, H.; Jensen, L.; Kaminski, J.W.; van Kessel, G.; Kootstra, F.; Kovalenko, A.; Krykunov, M.V.; van Lenthe, E.; McCormack, D.A.; Michalak, A.; Mitoraj, M.; Neugebauer, J.; Nicu, V.P.; Noodleman, L.; Osinga, V.P.; Patchkovskii, S.; Philipsen, P.H.T.; Post, D.; Pye, C.C.; Ravenek, W.; Rodríguez, J.I.; Ros, P.; Schipper, P.R.T.; Schreckenbach, G.; Seldenthuis, J.S.; Seth, M.; Snijders, J.G.; Sola, M.; Swart, M.; Swerhone, D.; te Velde, G.; Vernooijs, P.; Versluis, L.; Visscher, L.; Visser, O.; Wang, F.; Wesolowski, T.A.; van Wezenbeek, E.M.; Wiesenekker, G.; Wolff, S.K.; Woo, T.K.; Yakovlev, A.L. ADF 2010.01 SCM Theoretical Chemistry; Vrije Universiteit: Amsterdam, Netherlands, **2010**.

[100] Jackowski, K.; Makulski, W.; Kozminski, W. An improved ^{33}S nuclear magnetic shielding scale from the gas-phase study of COS. *Magn. Reson. Chem.*, **2002**, *40*, 563-565.

[101] Feindel, K.W.; Wasylishen, R.E. A relativistic DFT study of one-bond fluorine-X indirect spin–spin coupling tensors. *Magn. Reson. Chem.*, **2004**, *42*, S158-S167.

Index

A

Claudio Santi (Ed)

B

C

M

N

O

P

R

S

T

U

V

W

Z

Made in the USA
Las Vegas, NV
03 November 2026

57424407R00264